Lecture Notes in Mathematics

E[...]n

Springer-Verlag
Berlin Heidelberg New York London Paris Tokyo Hong Kong

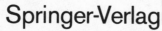

Editor

Herbert Heyer
Mathematisches Institut, Universität Tübingen
Auf der Morgenstelle 10, 7400 Tübingen, Federal Republic of Germany

Mathematics Subject Classification (1980):
Primary: 60B15, 60J15, 60J30, 60G60, 43A05, 43A33, 43A85
Secondary: 31C25, 33A45, 33A75, 46L50, 62H10

ISBN 3-540-51401-5 Springer-Verlag Berlin Heidelberg New York
ISBN 0-387-51401-5 Springer-Verlag New York Berlin Heidelberg

Printing and binding: Druckhaus Beltz, Hemsbach/Bergstr.
2146/3140-543210 – Printed on acid-free paper

P R E F A C E

The 1988 Conference "Probability Measures on Groups" was designed, in deviation from the previous meetings on the subject, as a forum on which the interplay between structural probability theory and seemingly quite distant topics of pure and applied mathematics were to be discussed and documented. The various interrelations were made explicit in the expository talks presented on special invitation. The organizers are very grateful to their colleagues for having prepared these special talks and for having submitted the underlying manuscripts in elaborated form to the Proceedings. Here is the list of the survey speakers and the titles of their talks

N.H. BINGHAM, The Royal Holloway and Bedford New College, Egham, England

Tauberian methods in probability theory

T. HIDA, Nagoya University, Nagoya, Japan

Infinite dimensional rotation group and unitary group

M.M. RAO, University of California at Riverside, Riverside, USA

Bimeasures and harmonizable processes

A. TERRAS, University of California at San Diego, La Jolla, USA

The central limit theorem for symmetric spaces of GL(3)

G.S. WATSON, Princeton University, Princeton, USA

Statistics of rotations.

In the following we shall briefly comment on the research contributions contained in this volume.

There has been a wide range of topics with some emphasis on convolution semigroups of measures, their potential theory and harmonic analysis. While at the preceding conferences on probability measures on groups the basic structure was mainly a group or a semigroup, the recent development emphasizes more general group-like structures.

These generalized translation or hypergroup structures, probabilistic in nature, appear in duals of non-Abelian groups, orbit and double-coset spaces, but also in the spaces $\mathbb{Z}_+$, $\mathbb{R}_+$, the unit interval or the unit disk where they differ significantly from the usual semigroup structures.

The research articles of this volume can be classified in four sections.

Probabilities and potentials on hypergroups. In this new branch of structural probability theory the concentration is on the study of random walks and more general space-homogeneous Markov processes. W.R. BLOOM and H. HEYER characterize the potentials of transient convolution semigroups in terms of Deny type fundamental families. L. GALLARDO supplements his previous paper on the Poisson representation of infinitely divisible probability measures. A. KUMAR and A.I. SINGH present a dichotomy theorem for random walks. In R. LASSER's paper stationary processes over polynomial hypergroups are studied. M. VOIT contributes to the problem of Schoenberg's duality for positive and negative definite functions. An extension of the theory of L^p-improving measures to hypergroups is given by R.C. VREM. Hm. ZEUNER continues in his article his previous work on one-dimensional hypergroups.

Probability measures on semigroups and groups. In this section the authors present new results pertaining to well-established open problems. M.S. BINGHAM proves the existence of a conditional probability distribution on an Abelian group. S.G. DANI and M. McCRUDDEN deal with a special aspect of the embedding problem for infinitely divisible probability measures on Lie groups. The paper of W. HAZOD and S. NOBEL concerns the convergence-of-types theorem which has implications to the theory of stable and semistable measures, a topic which is covered in E. SIEBERT's article. A. MUKHERJEA's results on the convergence of convolution products of probability measures shed new light on an important problem. To study roots of Haar measure on a compact group is the aim of a paper by G. TURNWALD.

Structure of special distributions and processes. This section documents in a particularly impressive way a variety of problems where general structural insight helps analyzing classical problems. J.L. DUNAU and H. SENATEUR discuss some generalizations of the Cauchy distribution. Y. HE applies an extended Delphic theory to point processes. In G. HÖGNÄS' paper the semigroup of analytic mappings is studied from some probabilistic point of view. J. KISYŃSKI presents in his contribution a detailed treatment of Feller generators and their processes. G. LETAC deals with the classification problem for exponential families. Divisibility problems immanent to the Hungarian contributions to structural probability theory are the topic of A. ZEMPLENI's article.

Harmonic analysis in probability theory on semigroups, groups and algebras. A. DERIGHETTI's paper is on the induction of convolution operators. Ph. FEINSILVER and R. SCHOTT treat stochastic processes with values in a Lie group by replacing their trajectories by smooth vector-valued curves. N. OBATA studies the Lévy Laplacian and its potential theory. Mixtures of characters of semigroups constitute the contents of P. RESSEL's article. In M. SCHÜRMANN's contribution infinitely divisible states and their impact to quantum stochastics are studied. K. YLINEN presents further results in the theory of random fields over a locally compact group.

From this short survey we are justified to conclude that the theory of probability on algebraic-topological structures not only develops by deepening and by extending the fundamental problems towards important generalizations and applications, but also by broadening its scope into other fields of mathematics, physics and statistics. Here we mention classical analysis (Tauberian theory), number theory (Helgason transfrom), stochastic analysis (Hida calculus), quantum stochastics (infinitely divisible states and instruments), data analysis (stationary processes and fields) and invariant decision theory (statistics on homogeneous spaces).

It goes without special mention that the 52 participants of the Conference who travelled to Oberwolfach from 13 different countries (including Germany) enjoyed the outstanding hospitality of the Forschungsinstitut and appreciated its inspiring atmosphere. The well-known recognition of Oberwolfach as a research center of international rank made it possible to attract specialists from all over the world to cooperate and give shape to the meeting.

The organizers of the Conference and the editor of these Proceedings extend their heartfelt thanks to all participants and contributors.

Tübingen, October 1988

Herbert Heyer

C O N T E N T S

The authors Y. He and A. Zempléni were kind enough to submit their manuscripts although they were prevented from participating at the conference.

A Fourier-analytic proof that conditional probability distributions exist on a group

Michael S. Bingham
Department of Statistics
University of Hull, England, HU6 7RX

Let X be a random variable defined on a probability space $(\Omega, \mathcal{F}, P)$ and taking values in a measurable space $(G, \mathcal{B})$. If $\mathcal{A}$ is a sub-σ-field of $\mathcal{F}$, it is well known that the conditional probability distribution of X given $\mathcal{A}$ does not necessarily exist unless suitable extra conditions are assumed. The purpose of this paper is to present a new proof based on Fourier analysis of the following existence result.

__Theorem__ Let G be a locally compact second countable abelian group with Borel σ-field $\mathcal{B}$ and let X be a G-valued, $\mathcal{B}$-measurable random variable defined on a probability space $(\Omega, \mathcal{F}, P)$. Then for any sub-σ-field $\mathcal{A}$ of $\mathcal{F}$ there exists on $(G, \mathcal{B})$ a conditional probability distribution of X given $\mathcal{A}$.

From the above theorem we can deduce the following well known more general result.

__Corollary__ Let X be a random variable defined on the probability space $(\Omega, \mathcal{F}, P)$ and taking values in the standard Borel space $(E, \mathcal{E})$. Then for any sub-σ-field $\mathcal{A}$ of $\mathcal{F}$ there exists on $(E, \mathcal{E})$ a conditional probability distribution of X given $\mathcal{A}$.

Conditional probability distributions on standard Borel spaces are discussed in Parthasarathy (1967). See also Breiman (1968).

Throughout this article we shall suppose that G is a locally compact second countable abelian group and that $\hat{G}$ is the dual group of G. Thus $\hat{G}$ consists of all the (continuous) characters of G, i.e. the continuous homomorphisms of G into the unit circle group. $\hat{G}$ is endowed with its natural group structure and the compact open topology, which together make $\hat{G}$ into a locally compact second countable abelian group. Complex finite linear combinations of elements of $\hat{G}$ are called trigonometric polynomials on G. References for information on locally compact abelian groups and their duals include Hewitt and Ross (1963, 1970) and Rudin (1962).

__Proof of the Theorem.__ Choose a countable dense subset of C of $\hat{G}$ and assume without loss of generality that the identity e of $\hat{G}$ is in C. Choose also an increasing sequence (L_m) of compact subsets of G with union G and define

$$M_{m+1} := L_{m+1} \setminus L_m \quad (m \geq 1), \; M_1 := L_1.$$

For each $y \in C$ and $m = 1,2,\ldots$ choose a version $\phi(y,m)$ of the conditional expectation $E[y(X)1(X \in M_m)|\mathcal{A}]$, where $1(F)$ indicates the indicator function of the event F. Then there exists $A \in \mathcal{A}$ with $P(A) = 1$ such that

$$\sum_{m=1}^{\infty} \phi\,(e,m) = 1 \tag{1}$$

holds for all $\omega \in A$. For each $\omega \notin A$ redefine $\phi(e,m)$ to be $P(X \in M_m)$ for all m. Then (1) holds for all $\omega \in \Omega$ and $\phi(e,m)$ is a version of $E[1(X \in M_m)|\mathcal{A}]$ for all m.

Now choose a sequence (U_k) of subsets of $\hat{G}$ which form a basis for the topology of $\hat{G}$ at e. Then for all positive integers m, ℓ we have

$$\sup_{x \in M_m} |y_1(x) - y_2(x)| \leq \sup_{x \in L_m} |(y_1 - y_2)(x) - 1| < \frac{1}{\ell}$$

whenever $y_1, y_2 \in C$, $y_1 - y_2 \in U_k$ and k is sufficiently large. Then

$$|\phi(y_1,m) - \phi(y_2,m)| < 1/\ell \quad \text{a.s.P}$$

whenever $y_1, y_2 \in C$, $y_1 - y_2 \in U_k$ and k is sufficiently large. As C is countable, it follows that there exists $A_m \in \mathcal{A}$ with $P(A_m) = 1$ such that

$$\phi(y_1,m) - \phi(y_2,m) \to 0 \text{ for all } \omega \in A_m$$

as $y_1 - y_2 \to e$ with $y_1, y_2 \in C$. Now take $A' := \bigcap_{m=1}^{\infty} A_m \in \mathcal{A}$.

Then $P(A') = 1$ and, for each $\omega \in A'$, each of the functions $y \to \phi(y,m)$ is uniformly continuous on C. For $\omega \notin A'$ redefine $\phi(y,m)$ to be $P(X \in M_m)$.

Next,

$$|y(X)1(X \in M_m)| = 1(X \in M_m)$$

so

$$|\phi(y,m)| \leq \phi(e,m) \tag{2}$$

holds a.s.P for each $y \in C$ and each m. As C is countable there exists $A'' \in \mathcal{A}$ with $P(A'') = 1$ such that (2) holds for all $\omega \in A''$, all $y \in C$ and all m. Define

$$\phi(y) := \begin{cases} \sum_{m=1}^{\infty} \phi(y,m) & \text{if } \omega \in A'' \\ 1 & \text{if } \omega \notin A'' \end{cases}$$

Then $\phi(y)$ is always finite and, by the dominated convergence theorem, ϕ is a uniformly continuous function of y on C for each $\omega \in \Omega$. Also $\phi(e) = 1$ for all $\omega \in \Omega$ and $\phi(y)$ is a version of $E[y(X)|\mathcal{A}]$ for all $y \in C$.

For each positive integer n, complex numbers $\alpha_1, \ldots, \alpha_n$ and characters $y_1,\ldots,y_n$ in C we have

$$\sum_{r=1}^{n}\sum_{s=1}^{n} \alpha_r \overline{\alpha_s}\, \phi(y_r - y_s) = E[|\sum_{r=1}^{n} \alpha_r y_r(X)|^2 | \mathcal{A}] \geq 0 \text{ a.s.P.}$$

Therefore there exists $A''' \varepsilon \mathcal{A}$ with $P(A''') = 1$ such that

$$\sum_{r=1}^{n}\sum_{s=1}^{n} \alpha_r\, \overline{\alpha_s}\, \phi(y_r - y_s) \geq 0 \tag{3}$$

holds for all positive integers n, all complex numbers $\alpha_1,\ldots\alpha_n$ (with rational real and imaginary parts) and all $y_1,\ldots y_n$ εC whenever $\omega \varepsilon A'''$. Redefine $\phi(y) = 1$ for all $y \varepsilon C$ whenever $\omega \notin A'''$.

For each $\omega \varepsilon \Omega$ the function $y \to \phi(y)$ is uniformly continuous on C and so has a unique extension to a continuous function on $\hat{G}$, which we again denote by ϕ. Then, for each fixed $y \varepsilon \hat{G}$, $\phi(y)$ is a version of $E[y(X)|\mathcal{A}]$ and for each fixed $\omega \varepsilon \Omega$ we have

(i) $\phi(e) = 1$

(ii) ϕ is continuous

(iii) (3) holds for all choices of positive integer n, complex numbers $\alpha_1,\ldots\alpha_n$ and $y_1\ldots,y_n \varepsilon \hat{G}$.

By Bochner's theorem (see for example Parthasarathy (1967), Heyer (1978), Rudin (1962) or Hewitt and Ross (1970)) there is, for each fixed $\omega \varepsilon \Omega$, a probability measure

$$B \to \mu(B,\omega) \quad , \quad B \varepsilon \mathcal{B}$$

on $(G,\mathcal{B})$ such that

$$\int y(x)\mu(dx,\omega) = \phi(y)(\omega)$$

for all $y \varepsilon \hat{G}$ and all $\omega \varepsilon \Omega$.

To complete the proof we show that $\mu : \mathcal{B} \times \Omega \to [o,1]$ is a conditional probability distribution of X given $\mathcal{A}$. In order to do so it remains to be proved that, for each fixed $B \varepsilon \mathcal{B}$,

$$\omega \to \mu(B,\omega) \quad \text{is} \quad \mathcal{A}\text{-measurable} \tag{4}$$

and

$$\int_A \mu(B,\omega)\, dP(\omega) = P([X \varepsilon B] \cap A) \text{ for all } A \varepsilon \mathcal{A} \tag{5}$$

or, in other words, that

$$\omega \to \int f(x)\mu(dx,\omega) \text{ is } \mathcal{A}\text{-measurable} \tag{6}$$

and

$$\int_A (\int f(x)\mu(dx,\omega))dP(\omega) = \int_A f(X)\, dP \text{ for all } A \varepsilon \mathcal{A} \tag{7}$$

whenever f is the indicator function of a set $B \varepsilon \mathcal{B}$.

By our choices of μ and ϕ we know that (6) and (7) hold whenever $f \varepsilon \hat{G}$. Therefore they also hold whenever f is a trigonometric polynomial on G.

Let B be a closed subset of G and let f be the indicator function of B. For each positive integer n define

$$g_n(x) = \frac{1}{nd(x,B) + 1}, \quad x \in G$$

where d is a metric for the topology of G. Each g_n is a real bounded continuous function on G and by the proof of (31.4) in Hewitt and Ross (1970) there exists a trigonometric polynomial f_n on G such that

$$\sup_{x \in L_n} |g_n(x) - f_n(x)| < 1/n$$

and

$$\sup_{x \in G} |f_n(x)| \leq \sup_{x \in G} |g_n(x)| + \frac{1}{n} \leq 2.$$

Thus, f is the pointwise limit of a uniformly bounded sequence of trigonometric polynomials (f_n). The dominated convergence theorem enables us to deduce the validity of (6) and (7) for f from their validity for each f_n.

The class of sets B in $\mathcal{B}$ for which (4) and (5) hold is obviously a sub-σ-field of $\mathcal{B}$ and the last paragraph shows that it contains all the closed subsets of G. Therefore (4) and (5) hold for all $B \in \mathcal{B}$ and the proof is complete.

To deduce the corollary, first note that any standard Borel space is Borel isomorphic to a Borel subspace of $(M, \mathcal{B}_M)$ where M denotes the space of all infinite sequences of 0's and 1's (with the product topology) and $\mathcal{B}_M$ is the Borel σ-field of M (See Theorem 2.3, page 8 of Parthasarathy (1967)). In turn $(M, \mathcal{B}_M)$ can obviously be identified with a Borel subspace of $(T^\infty, \mathcal{B}_{T^\infty})$, the compact second countable abelian group obtained by taking the product of an infinite sequence of copies of the unit circle group T. Thus, it is enough to establish the corollary for the case when E is a Borel subset of T^∞ and $\mathcal{E} = E \cap \mathcal{B}_{T^\infty}$, where $\mathcal{B}_{T^\infty}$ is the Borel σ-field of T^∞.

Let μ be the conditional distribution of X given by the theorem when X is regarded as T^∞-valued. Since $P(X \in E) = 1$ there is a set $A \in \mathcal{A}$ with $P(A) = 1$ such that

$$\mu(E, \omega) = 1 \quad \text{for all } \omega \in A.$$

Choose an arbitrary probability measure ν on $(E, \mathcal{E})$ and for each $\omega \notin A$ redefine $\mu(.,\omega) = \nu(.)$. Then μ restricted to $(E, \mathcal{E})$ is a conditional probability distribution for X.

Alternatively the corollary can be deduced as in Breiman (1968) from the special case of the theorem in which G is the real line.

References

Breiman L, (1968): 'Probability', Addison-Wesley Publishing Co., Reading, Massachusetts.

Hewitt E. and Ross, K. A., (1963, 1970): 'Abstract Harmonic Analysis' Vols. 1 and 2, Springer Verlag, Berlin, Heidelberg, New York.

Heyer H., (1977): 'Probability Measures on Locally Compact Groups', Springer Verlag, Berlin, Heidelberg, New York.

Parthasarathy K.R., (1967): 'Probability Measures on Metric Spaces', Academic Press, New York.

Rudin W., (1962): 'Fourier Analysis on Groups', Interscience (Wiley), New York.

TAUBERIAN THEOREMS IN PROBABILITY THEORY

N. H. Bingham
Department of Mathematics
Royal Holloway and Bedford New College
Egham Hill, Egham, Surrey TW20 OEX

§1. WIENER THEORY

We begin with the classical Wiener Tauberian theory for the real line, in various formulations. The basic result is

Wiener's Approximation Theorem for $L^1(\mathbb{R})$. For $f \in L^1$, the following are equivalent:

(i) The Fourier transform $\hat{f}(t) := \int_{\mathbb{R}} e^{ixt} f(x)dx$ has no zeros $t \in \mathbb{R}$,

(ii) Linear combinations of translates of f are dense in L^1,

(iii) For $g \in L^\infty(\mathbb{R})$, the integral equation

$$(f*g)(x) := \int_{\mathbb{R}} f(x-y)g(y)dy \equiv 0$$

has only the trivial solution $g = 0$ a.e.

For proof, see Wiener (1932), (1933), Reiter (1968), I. As a corollary, one obtains

Wiener's First Tauberian Theorem. If $f \in L^1$ and $\hat{f}$ has no real zeros, $g \in L^\infty$, then

$$(f*g)(x) \to c \int f \qquad (x \to \infty)$$

implies

$$(h*g)(x) \to c \int h \qquad (x \to \infty)$$

for all $h \in L^1$.

A useful variant is obtained by demanding more of f and less of g. Write M for the subclass of L^1 consisting of continuous functions f with

$$\Sigma_{n=-\infty}^{\infty} \sup_{[n,n+1)} |f(.)| < \infty, \tag{1}$$

M* for the class of signed measures $\mu = \mu_+ - \mu_-$ whose 'modulus' $|\mu| := \mu_+ + \mu_-$ satisfies

$$\sup_x |d\mu|([x,x+1)) < \infty \tag{2}$$

(thus if $\mu = \int g$ with $g \in L^{\infty}$, then $\mu \in M^*$). One has (Wiener (1932), Reiter (1968))

<u>Wiener's Second Tauberian Theorem.</u> If $f \in M$ with $\hat{f}$ non-vanishing, $\mu \in M^*$, then

$$(f*\mu)(x) := \int f(x-y)d\mu(y) \to c \int f \qquad (x \to \infty)$$

implies

$$(h*\mu)(x) \to c \int h \qquad (x \to \infty)$$

for all $h \in M$.

The Wiener theory above extends from $L^1(\mathbb{R})$ to $L^1(G)$ with G a locally compact abelian group; see e.g. Rudin (1963), Reiter (1968). For generalisations to locally compact groups, we refer to Reiter (1974), Leptin & Poguntke (1979), and the survey Leptin (1984), §5. The appropriate context for generalisations of the second form of the Tauberian theorem is that of <u>Segal algebras</u>; see Reiter (1968), (1971), Liu et al. (1974), Feichtinger (1981). For the duality between M and M^*, see Edwards (1958), Goldberg (1967). The condition $\mu \in M^*$ is of <u>amalgam norm</u> type; for background see Fournier & Stewart (1985) and the references cited there.

§2. BENEŠ' THEOREM

We now examine Wiener's second Tauberian theorem in the case of positive measures, important for applications in probability theory and elsewhere. The result below is due to Beneš (1961); we sketch a streamlined approach to its proof.

<u>Beneš' Tauberian Theorem</u>. If μ is a positive measure satisfying

$$\sup_x \int_x^{x+1} d\mu < \infty, \qquad (2')$$

and f satisfies (1) with $\hat{f}$ non-vanishing, then

$$\int f(x-y)d\mu(y) \to c \int f$$

implies

$$\int g(x-y)d\mu(y) \to c \int g$$

for all g satisfying (1) and continuous almost everywhere.

<u>Proof</u>.
Step 1. Replace f by its convolution with the Gaussian kernel $e^{-\frac{1}{2}x^2}$. This preserves our hypothesis and non-vanishing of transforms, and induces continuity. We may thus suppose without loss that f is con-

tinuous.

Step 2. For g continuous, we can use Wiener's second theorem above. For g an indicator of an interval [a,b], approximate g above and below by continuous piecewise linear functions g_i, $g_1 \le g \le g_2$:

$$g_1 := 0 \text{ outside } [a,b], \quad 1 \text{ in } [a+\varepsilon, b-\varepsilon],$$
$$g_2 := 0 \text{ outside } [a-\varepsilon, b+\varepsilon], \quad 1 \text{ in } [a,b].$$

Then as μ is positive,

$$\int g_1(x-y)d\mu(y) \le \int g(x-y)\, d\mu(y) \le \int g_2(x-y)\, d\mu(y).$$

By continuity of g_i, $\int g_i(x-y)d\mu(y) \to c \int g_i$, and $\int g_i$ may be made arbitrarily close to $\int g$ by choice of ε. The result follows for indicators g.

Step 3. The result extends to simple functions (linear combinations of indicators) by linearity.

Step 4. If g has compact support and is continuous almost everywhere, g is Riemann integrable, and so is approximable above and below by simple functions g_i. Extend from g_i to g as above.

Step 5. If g is continuous a.e. and satisfies (1), we may for each $\varepsilon > 0$ use (2') to choose n so large that contributions from outside $[-n,n]$ are less than ε. We approximate to g by $g_n := g\, I_{[-n,n]}$ and use the step above.

Step 6. Finally, a counter-example due to Smith (see Beneš (1961)) shows that the result may fail if the discontinuity-set of g has positive measure. //

Note that, writing λ for Lebesgue measure, it suffices to have positivity of $\mu+b\lambda$ for some b rather than of μ (as $\int f(x-y)d\lambda(y) = \int f$). A variant on this yields the following classical one-sided form of Wiener's theorem, due to Pitt (see e.g. Widder (1941), V.13).

<u>Corollary</u>. If f is non-negative, continuous, satisfies (1) and has $\hat{f}$ non-vanishing, p(.) is bounded below, and $\int f(x-y)p(y)dy$ is bounded, then

$$\int f(x-y)\, p(y)\, dy \to c \int f$$

implies

$$\int g(x-y)\, p(y)\, dy \to c \int g$$

for all g satisfying (1) and continuous a.e.

<u>Proof</u>. Since f is continuous and non-negative with $\hat{f}$ non-vanishing, f is positive on some interval: $f(.) \ge \varepsilon > 0$ on $[a-\delta, a]$ say. If $p(.) \ge -b$, replacing p(.) by p(.)+b we may suppose $p(.) \ge 0$. Then for some M and all x,

$$M \ge \int f(x+a-y)p(y)dy = \int f(y)p(x+a-y)dy \ge \int_{a-\delta}^{a} \ge \varepsilon \int_0^{\delta} p(x+t)dt,$$

so $\int_0^\delta p(x+t)dt$ is bounded above by M/ε. If $n := [1/\delta]$, $\int_0^1 p(x+t)dt$ is bounded above by $(n+1)M/\varepsilon$, which is (2') for the measure μ with density p. Now use Beneš' theorem. //

§3. RENEWAL THEORY

We illustrate the usefulness of these ideas by using them to derive an effortless proof of a major result in probability theory.

Blackwell's Renewal Theorem. If F is a probability law on R with renewal function $U := \Sigma_0^\infty F^{n*}$, and F is non-lattice with mean $m > 0$, then for all $b > 0$,

$$U(x+b) - U(x) \to \begin{cases} b/m & (x \to +\infty), \\ 0 & (x \to -\infty). \end{cases}$$

Proof. That the (positive) measure U satisfies (2) is easy to show (see e.g. Feller (1971), XI.1 (1.7)).

Define $f := \delta_0 - F$, $\mu := U$; the renewal equation $f*\mu \equiv \delta_0$ holds.

Write $\hat{F}$ for the Fourier-Stieltjes transform (characteristic function) of F; then integrating by parts,

$$\hat{f}(t) = (\hat{F}(t) - 1)/(it).$$

As the mean m of F exists, f satisfies (1), and $\int f = m > 0$. Also $\hat{f}$ is non-vanishing: at the origin (as $\hat{f}(0) = \hat{F}'(0)/i = m$), and elsewhere (as F is non-lattice). Now use Beneš' theorem with $g := I_{[0,b]}$: as $f*\mu \to m^{-1} \int f$ at $+\infty$ and 0 at $-\infty$,

$$(g*\mu)(x) = U(x+b) - U(x) \to m^{-1} \int g = b/m \text{ at } +\infty,\ 0 \text{ at } -\infty. \quad //$$

The lattice case is similar but simpler. One may use the Erdös-Feller-Pollard theorem; for details see Feller (1957), XIII.4.

This justly famous result was proved by Blackwell (1953). The relevance of Wiener methods was immediately realised; cf. Smith (1954), (1955), Karlin (1954). The argument above is due (implicitly) to Beneš (1961); it was discovered independently by T.-L. Lai (unpublished), from whom I learned it in 1976.

Of course, alternative proofs of Blackwell's theorem exist; among the recent ones we single out the approach via coupling due to Lindvall (1977), (1986), Athreya et al. (1978), Thorisson (1987). Such proofs may well be preferred by probabilists because they are probabilistic rather than analytic. However, I know of no proof which compares in brevity with that above. Note that, while the Tauberian content of the proof above is Beneš' theorem, the Tauberian content of the coupling proof is nil.

<u>More general groups</u>. The renewal theorem for $\mathbb{R}^2$ was proved by Chung (1952); here the limit is 0 for all approaches to ∞. The case of $\mathbb{R}^d$ was considered by Doney (1966), Stam (1969), (1971), Nagaev (1979).

The renewal theorem for locally compact abelian groups G is due to Port & Stone (1969) (cf. Revuz (1975)). Call G of <u>type II</u> if it is capable of supporting a random walk - with law μ, say - whose renewal measure $\nu := \Sigma_0^\infty \mu^{n*}$ does not tend to 0 at ∞, <u>type I</u> otherwise (thus $\mathbb{R}^d$ and $\mathbb{Z}^d$ are type II for $d = 1$, type I for $d \geq 2$). Then G is of type II iff G is isomorphic to $\mathbb{R} \oplus K$ or $\mathbb{Z} \oplus K$ with K compact.

This dichotomy has been studied for non-abelian groups (see e.g. Brunel & Revuz (1975), Crepel & Lacroix (1976)). The non-amenable case - always type I - is due to Derriennic & Guivarc'h (1973). The nil-potent case was studied by Guivarc'h et al. (1977), the unimodular case (type II groups are of the form KxE with K compact and E isomorphic to $\mathbb{R}$ or $\mathbb{Z}$) by Sunyach (1981). For a definitive treatment of the amenable, non-unimodular case, see Elie (1982) (in particular p.260 and §§1.6, 3.14, 3.22, 5.1).

<u>Variants and applications</u>. Approaches to renewal theory based on the renewal equation (Choquet-Deny theorem, direct Riemann integrability) are in Feller (1971), XI.1,9 and the references cited there. Smith (1955) has a 'key renewal theorem', of the form $(g*U)(x) \to m^{-1} \int g$ at $+\infty$, 0 at $-\infty$. Classical applications include work of Lotka and others on demography. Renewal theory in settings such as products of random matrices and Markov chains is developed in Kesten (1973), (1974). Berbee (1979) considers the dependent case.

§4. BEURLING'S TAUBERIAN THEOREM.

The Wiener theory is a powerful tool for studying the asymptotics of convolution integrals. One often needs extensions which will handle more complicated integrals which approximate to convolutions in some sense.

Call $\phi: \mathbb{R} \to \mathbb{R}_+$ <u>self-neglecting</u> ($\phi \in SN$) if it is continuous, $o(x)$ at ∞, and

$$\phi(x+t\phi(x))/\phi(x) \to 1 \qquad (x \to \infty) \qquad \forall t \in \mathbb{R}$$

(for background, see e.g. Bingham et al. (1987), §2.11).

Beurling's Tauberian Theorem. If $f \in L^1$, $\hat{f}$ is non-vanishing, $g \in L^\infty$, and $\phi \in SN$,

$$\int g(y)\, f((x-y)/\phi(x))\, dy/\phi(x) \to c \int f \quad (x \to \infty)$$

implies

$$\int g(y)\, h((x-y)/\phi(x))\, dy/\phi(x) \to c \int h$$

for all $h \in L^1$.

For proof, see Peterson (1972), Moh (1972).

Application to summability methods. Recall (e.g. Hardy (1949)) the classical summability methods of Borel:

$$s_n \to s \quad (B) \quad \text{means} \quad \Sigma_0^\infty s_k . e^{-x} x^k/k! \to s \qquad (x \to \infty),$$

Euler: for $0 < p < 1$,

$$s_n \to s \quad (E_p) \quad \text{means} \quad \Sigma_0^n s_k . \binom{n}{k} p^k (1-p)^{n-k} \to s \quad (n \in \mathbb{Z},\ n \to \infty),$$

Valiron: for $a > 0$,

$$s_n \to s \quad (V_a) \quad \text{means} \quad (a/\pi x)^{\frac{1}{2}} \Sigma_0^\infty s_k . \exp\{-a(x-k)^2/x\} \to s \quad (x \to \infty),$$

and consider also the 'moving average' method:

$$s_n \to s \quad (M(\sqrt{n})) \text{ means } \frac{1}{\varepsilon\sqrt{n}} \sum_{n \le k < n+\varepsilon\sqrt{n}} s_k \to s \quad (n \to \infty)\ \forall \varepsilon > 0.$$

The basic Tauberian result linking these methods is the following

Theorem. For bounded sequences (s_n), the following are equivalent:

(i) $s_n \to s$ (E_p) for some (all) p,

(ii) $s_n \to s$ (B),

(iii) $s_n \to s$ (V_a) for some (all) a,

(iv) $s_n \to s$ $(M(\sqrt{n}))$.

The equivalence of (i), (ii) and (iii) is due to Meyer-König (1949) (he considers too other 'circle methods' or Kreisverfahren); that of (i), (ii) and (iv) was proved by Diaconis & Stein (1978). For a short proof of a more general result - using Beurling's theorem and local forms of the central limit theorem - see Bingham (1981).

Riesz means. If $\lambda = (\lambda_n) \uparrow \infty$, the Riesz (typical) mean of order 1, $R(\lambda,1)$ (which we abbreviate to $R(\lambda)$ as we shall only consider order 1) is defined by writing (for $s_n = \Sigma_0^n a_k$)

$$s_n \to s \qquad (R(\lambda))$$

for

$$\frac{1}{x} \int_0^x \{ \sum_{n:\lambda_n \le y} a_n \}\, dy \to s \qquad (x \to \infty).$$

The moving-average method $M(\sqrt{n})$ above may be written in more classical language as $R(e^{\sqrt{n}})$. For proof see Bingham (1981); the result is substantially due to Karamata (1937), (1938). It turns out that the moving-average method $M(\phi)$ based on any $\phi \in SN$ is the Riesz mean $R(\exp\{\int_1^x dt/\phi(t)\})$; for proof and applications, see Bingham & Goldie (1988).

One-sided Tauberian condition. The Tauberian condition $s_n = O(1)$ in the Theorem is too restrictive for some purposes, and may be weakened to boundedness below ('$s_n = O_L(1)$') and beyond. For details and applications, see Tenenbaum (1980), Bingham & Goldie (1983), Bingham (1984a), (1984b), Bingham & Tenenbaum (1986).

Laws of large numbers. The Theorem above may be compared with the following result of Chow (1973): for $X, X_1, ..$ independent and identically distributed, the following are equivalent:

$$E(|X|^2) < \infty \ \&\ EX = \mu,$$
$$X_n \to \mu \quad \text{a.s.} \quad (B),$$
$$X_n \to \mu \quad \text{a.s.} \quad (E_p),$$
$$X_n \to \mu \quad \text{a.s.} \quad (M(\sqrt{n})).$$

Thus the condition 'i.i.d. with finite variance' plays the role of a probabilistic Tauberian condition, allowing one to pass between the Euler and Borel methods. One may compare this with the known best-possible Tauberian condition, seek to weaken independence, etc.; for details see Bingham (1985). In the L^p case ($p > 1$), there are similar results involving Riesz and Valiron means (Bingham & Tenenbaum (1986)), and Cesàro means (Déniel & Derriennic (1988+)).

Weighted versions. For interesting extensions of Beurling's theorem, see Feichtinger & Schmeisser (1986).

Random-walk methods. One may complement the work above by applying probabilistic ideas in pure summability (or Tauberian) theory. For a case in point, see Bingham (1984c).

Combinatorial optimisation. The Borel method occurs in the technique of 'Poissonisation'. For applications of this in conjunction with Tauberian arguments, see Steele et al. (1987), Steele (1988+).

§5. OCCUPATION TIMES

We illustrate by considering various limit theorems for occupation times of random walks. We consider $X_1, X_2, ..$ independent and identically distributed with mean μ and variance σ^2, and write $S_n := \Sigma_1^n X_k$ for the generated random walk.

I. Large sets, positive drift: $\mu > 0$.

a. Central Limit Theorems. For simplicity, consider the $\mathbb{Z}$-valued case. For $A \subset \mathbb{N}$,

$$P(S_n \in A) = \Sigma_0^\infty I_A(k) \cdot P(S_n = k).$$

Use the Theorem (extended as in Bingham (1981)) with $s_n := I_A(n)$:

$$P(S_n \in A) \to c$$

iff

$$I_A(n) \to c \qquad (B, \text{ or } E_p, \text{ or } V_a, \text{ or } R(e^{\sqrt{n}}))$$

iff (Bingham (1981))

$$\exists\, \varepsilon_n \to 0 \text{ with } \frac{1}{n} \Sigma_1^n (I_A(k) + \varepsilon_k) = c + o(1/\sqrt{n}).$$

These equivalences express a sense (stronger than the usual one) in which $A \subset \mathbb{N}$ 'has density c'; in view of the circle methods of summability involved, we say that A has circle density c. This density concept has been used in various contexts in analytic number theory; see Diaconis & Stein (1978), Tenenbaum (1980), Knopfmacher & Schwarz (1981a,b).

In view of the Theorem: the Tauberian content here is Beurling's theorem.

b. Law of Large Numbers. In the same context as above, compare the weaker 'Cesàro forms' of the above equivalences:

$$\frac{1}{n} \Sigma_1^n I_A(S_k) \to c \quad \text{a.s.}$$

iff

$$\frac{1}{n} \Sigma_1^n P(S_k \in A) \to c$$

iff

$$\frac{1}{n}|A \cap \{1,2,..,n\}| \to c.$$

This law of large numbers is due to Stam (1968), Meilijson (1973). For extensions, see Bingham & Goldie (1982), Högnäs & Mukherjea (1984), Berbee (1987). The Tauberian content (perhaps surprisingly, in view of the above) is nil.

II. Large sets, zero drift: $\mu = 0$.

We now have oscillation rather than drift (recurrence rather than transience). The corresponding equivalences are:

$$\frac{1}{n}|A \cap \{1,..,n\}| \to c$$

iff

$$P(S_n \in A) \to \tfrac{1}{2}c$$

iff

$$\frac{1}{n} \Sigma_0^n I_A(S_k) \text{ converges in distribution,}$$

and then the limit law is that of cX where X has the arc-sine law. This result is due to Davydov & Ibragimov (1971), Davydov (1973), (1974); its Tauberian content is Wiener's theorem.

We note the special case $A = \mathbb{N}$ (so $c = 1$): this is Spitzer's arc-sine law (Spitzer (1956)). Spitzer's full result is

$$\frac{1}{n} \Sigma_0^n P(S_k > 0) \to \rho \in [0,1]$$

iff

$$\frac{1}{n}\Sigma_0^n I_{[0,\infty)}(S_k) \text{ converges in distribution,}$$

and then the limit law is F_ρ, the generalised arc-sine law with mean ρ (cf. Bingham et al. (1987), §8.11). In the above we have $\rho = \frac{1}{2}$ (since $\mu = 0$ and $\sigma^2 < \infty$, S_n is attracted without centring to normality). When S_n is attracted without centring to a stable process X_t, one has

$$P(S_n > 0) \to \rho := P(X_1 > 0).$$

The interesting question of the equivalence of ordinary and Cesàro convergence of $P(S_n > 0)$ remains open.

III. Small sets, zero drift: $\mu = 0$.

Suppose now that A is compact (for a walk on $\mathbb{R}$) or finite (for one on $\mathbb{Z}$). It was shown by Darling & Kac (1957) that

$$\Sigma_0^n I_A(S_k), \text{ suitably normed, converges in distribution}$$

iff

$$\Sigma_0^n P(S_k \in A) \in R_\rho \qquad (0 < \rho \leq \tfrac{1}{2})$$

(the class of functions varying regularly with index ρ), and that the limit law is then the Mittag-Leffler law with index ρ.

The local limit theorem readily shows that such behaviour occurs if S_n is attracted to a stable law with index $\alpha \in (1,2]$ (and then $\rho = 1 - 1/\alpha$). The converse question, which is difficult and Tauberian in character, was considered by Kesten (1968). He showed that for symmetric random walk the converse holds, and conjectured that symmetry is unnecessary. This is so, in that the converse also holds with complete asymmetry (Bingham & Hawkes (1983)), but the general question remains obscure. The Tauberian content may be reduced to Karamata's theorem, but note that the proof of the Darling-Kac theorem (obtaining a Mellin convolution, taking Mellin transforms, dividing by a non-zero Mellin transform) is itself of Wiener character.

§6. TAUBERIAN THEOREMS OF EXPONENTIAL TYPE

If F is a probability law on $[0, \infty)$ whose tail is small enough (exponentially small), its characteristic function is entire. One may consider the moment-generating function

$$\hat{F}(s) := \int_0^\infty e^{sx} dF(x),$$

and link its rate of growth to the rate of decay of the tail. The result, due to Kasahara (1978), is (writing $f^{\leftarrow}$ for the inverse function of f)

Kasahara's Theorem. If $0 < \alpha < 1$, $\phi \in R_\alpha$, $\psi(x) := x/\phi(x) \in R_{1-\alpha}$,

$$-\log(1-F(x)) \sim B\,\phi^{\leftarrow}(x) \in R_\alpha \qquad (x \to \infty)$$

iff

$$\log \hat{F}(s) \sim (1-\alpha)(\alpha/B)^{\alpha/(1-\alpha)} \psi^{\leftarrow}(s) \in R_{1/(1-\alpha)} \quad (s \to \infty).$$

The passage from F to $\hat{F}$ is Abelian, that from $\hat{F}$ to F Tauberian, the Tauberian condition being positivity of the measure F. Limit-of-oscillation results are also possible, when the relevant ratios do not converge (see e.g. Bingham et al. (1987), §4.12).

We note an application to the Mittag-Leffler law F_ρ of §5.III. Here

$$\hat{F}_\rho(s) = \Sigma_0^\infty s^n/\Gamma(1+n\rho) \sim \exp(s^{1/\rho})/\rho \qquad (s \to \infty),$$

whence

$$-\log(1-F_\rho(x)) \sim (1-\rho)\rho^{\rho/(1-\rho)} x^{1/(1-\rho)} \qquad (x \to \infty).$$

Another application arises in studying supercritical branching processes Z_n. Write $\mu \in (1,\infty)$ for the mean, W for the a.s. limit of Z_n, suitably normed. When the probability generating function f(s) of Z_1 is a polynomial of degree d, the moment generating function $\hat{W}(s)$ of W is an entire function of order γ, where $\mu^\gamma = d$. The limit-of-oscillation form of Kasahara's theorem may be used to translate information on $\log \hat{W}(s)$ into information on $-\log P(W > x)$ $(x \to \infty)$; see Bingham (1988), §4, following Harris (1948).

A similar result enables us to compare the asymptotics of F(x) as $x \downarrow 0$ (for a probability law F on $[0,\infty)$) with that of its Laplace-Stieltjes transform $\hat{F}(s)$ as $s \to \infty$.

<u>De Bruijn's Theorem</u>. If $\alpha < 0$,

$$-\log F(x) \sim B/\phi^{\leftarrow}(1/x) \in R_{1/\alpha}(0+) \qquad (x \downarrow 0)$$

iff

$$-\log \hat{F}(s) \sim (1-\alpha)(B/-\alpha)^{\alpha/(\alpha-1)}/\psi^{\leftarrow}(s) \in R_{\alpha-1} \qquad (s \to \infty).$$

For proof, see de Bruijn (1959), or Bingham et al. (1987), §4.12. Again, there is a limit-of-oscillation form (Bingham (1988), Appendix).

We give two examples. First, consider the stable subordinator X_t of index $\rho \in (0,1)$:

$$E \exp(-sX_t) = \exp(-ts^\rho), \qquad s \geq 0.$$

Then (cf. Bingham & Teugels (1975), §5)

$$-\log P(X_1 \leq x) \sim (1-\rho)\rho^{\rho/(1-\rho)}/x^{\rho/(1-\rho)} \qquad (x \downarrow 0)$$

(there is even a stronger form, for the density). One may actually translate this into the above result on the Mittag-Leffler tail (cf. Feller (1971), XIII.8 Ex. b).

A second application concerns the behaviour of the left tail $P(0 < W \leq x)$ as $x \downarrow 0$ of the branching-process limit W above, in the case where the minimum family size m is at least two (the "Böttcher case"). For details, see Bingham (1988).

§7. SPHERICAL SYMMETRY

We turn now to random walks in $\mathbb{R}^d$ with spherical symmetry, which have been studied extensively by Kingman (1963). In $\mathbb{R}^d$, let $\underline{X}_n$ be independent vectors whose laws are spherically symmetric, and consider the random walk $\underline{S}_n = \Sigma_1^n \underline{X}_k$ they generate. If we look only at norms, $X_n := \|\underline{X}\|_n$ etc., the appropriate analogue of the characteristic function is the Hankel transform

$$\phi_X(t) := \int_0^\infty \Lambda_{\frac{1}{2}d}(xt)\, dP(X \le x),$$

where (J_ν being the usual Bessel function)

$$\Lambda_\nu(x) := \Gamma(\nu+1)\, J_\nu(x)/(\tfrac{1}{2}x)^\nu:$$

$$\phi_{S_n}(t) = \Pi_1^n\, \phi_{X_k}(t).$$

Here ϕ_X is the radial part of the Fourier-Stieltjes transform in $\mathbb{R}^d$; see e.g. Bochner & Chandrasekharan (1949). With an appropriate definition of <u>stability</u>, the <u>domain of attraction</u> condition is

$$1 - \phi_X(t) \sim t^\alpha L(1/t) \in R_\alpha(0+) \qquad (t \downarrow 0) \qquad (0 < \alpha \le 2)$$

(where L is slowly varying). Then

$$\int_0^\infty [1-\Lambda_\nu(t)]\, dt/t^{1+s} = \frac{\Gamma(1+\nu)\Gamma(1-\frac{1}{2}s)}{s.2^s\Gamma(1+\nu+\frac{1}{2}s)} \neq 0 \qquad (0 < \mathrm{Re}\, s < 2)$$

enables one to translate this into

$$1-F(x) \sim \frac{L(x)}{x^\alpha}.\frac{2^\alpha\Gamma(1+\nu+\frac{1}{2}\alpha)}{\Gamma(1+\nu)\Gamma(1-\frac{1}{2}\alpha)} \in R_{-\alpha} \qquad (x \to \infty)$$

by Wiener theory (Bingham (1984d)). The original proof, Bingham (1972a), is much harder.

<u>Generalised convolutions.</u> This example motivated Urbanik (1964) to study generalised convolutions. Here one studies an associative map

$$\circ : (F_1, F_2) \to F_1 \circ F_2,$$

a binary operation on measures on $[0,\infty)$, and a characteristic function ϕ such that

$$\phi_{F_1 \circ F_2}(t) = \phi_{F_1}(t)\, \phi_{F_2}(t);$$

this is an integral transform

$$\phi_F(t) = \int_0^\infty \Omega(tx)\, dF(x)$$

generalising the Hankel transform above. For further background, see e.g. Bingham (1984d), Urbanik (1986).

<u>Symmetric spaces of Euclidean type</u>. Write G_d for the group of Euclidean motions of $\mathbb{R}^d$. Then $(G_d, SO(d))$ is a symmetric space (G,K) of Euclidean type (Helgason (1962)), and $\Lambda_{\frac{1}{2}d}(t\|.\|)$ are the zonal spherical functions. One may compare this situation with that of the symmetric space $(G,K) := (SO(d+1), SO(d))$ of compact type, with zonal spherical functions the Gegenbauer polynomials. There is a theory of random

walks on hyperspheres analogous to the Kingman theory above, and even containing it in the limit as the radius of the hypersphere tends to infinity; for details see Bingham (1972b). The analysis involved goes back to Bochner (1954), (1955), and has recently aroused renewed interest from the point of view of hypergroups (cf. Heyer (1984)). As it happens, this was my introduction both to Tauberian theory and to probability theory on groups.

REFERENCES

K. B. ATHREYA, D. McDONALD & P. E. NEY (1978): Coupling and the renewal theorem. Amer. Math. Monthly 85, 809-814.

V. E. BENEŠ (1961): Extensions of Wiener's Tauberian theorem for positive measures. J. Math. Anal. Appl. 2, 1-20.

H. C. P. BERBEE (1979): Random walks with stationary increments and renewal theory. Math. Centrum, Amsterdam.

____________ (1987): Convergence rates in the strong law for bounded mixing sequences. Prob. Th. Rel. Fields 74, 255-270.

N. H. BINGHAM (1972a): Tauberian theorems for integral transforms of Hankel type. J. London Math. Soc. (2) 5, 493-503.

____________ (1972b): Random walk on spheres. Z. Wahrschein. 22, 169-192.

____________ (1981): Tauberian theorems and the central limit theorem. Ann. Probab. 9, 221-231.

____________ (1984a): On Euler and Borel summability. J. London Math. Soc. (2) 29, 141-146.

____________ (1984b): On Valiron and circle convergence. Math. Z. 186, 273-286.

____________ (1984c): Tauberian theorems for summability methods of random-walk type. J. London Math. Soc. (2) 30, 281-287.

____________ (1984d): On a theorem of Kłosowska about generalised convolutions. Colloq. Math. 48, 117-125.

____________ (1985): On Tauberian theorems in probability theory. Nieuw Arch. Wiskunde (4) 3, 157-166.

____________ (1988): On the limit of a supercritical branching process. J. Appl. Probab. 25A.

____________ & C. M. GOLDIE (1982): Probabilistic and deterministic averaging. Trans. Amer. Math. Soc. 269, 453-480.

________________________ (1983): On one-sided Tauberian conditions. Analysis 3, 159-188.

________________________ (1988): Riesz means and self-neglecting functions. Math. Z.

________________________ & J. L. TEUGELS (1987): Regular variation. Cambridge Univ. Press.

____________ & J. HAWKES (1983): Some limit theorems for occupation times. London Math. Soc. Lecture Notes 79, 46-62.

____________ & G. TENENBAUM (1986): Riesz and Valiron means and fractional moments. Math. Proc. Cambridge Phil. Soc. 99, 143-149.

____________ & J. L. TEUGELS (1975): Duality for regularly varying functions. Quart. J. Math. (3) 26, 333-353.

D. BLACKWELL (1953): Extension of a renewal theorem. Pacific J. Math. 3, 315-320.

S. BOCHNER (1954): Positive zonal functions on spheres. Proc. Nat. Acad. Sci. 40, 1141-1147.

__________ (1955): Sturm-Liouville and heat equations whose eigenvalues are ultraspherical polynomials and associated Bessel functions. Proc. Conf. Differential Equations 23-48, Univ. Maryland Press.

__________ & K. CHANDRASEKHARAN (1949): Fourier transforms. Princeton Univ. Press.

N. G. de BRUIJN (1959): Pairs of slowly oscillating functions occurring in asymptotic problems concerning the Laplace transform. Nieuw Arch. Wiskunde (3) 7, 20-26.

A. BRUNEL & D. REVUZ (1975): Sur la théorie de renouvellement pour les groupes non abéliens. Israel J. Math. 20, 46-56.

Y.-S. CHOW (1973): Delayed sums and Borel summability of independent identically distributed random variables. Bull. Inst. Math. Acad. Sinica 1, 207-220.

K. L. CHUNG (1952): On the renewal theorem in higher dimensions. Skand. Akt. 35, 188-194.

P. CREPEL & J. LACROIX (1976): Théorème de renouvellement pour les marches aléatoires sur les groupes localement compacts. Lecture Notes in Math. 563, 27-42, Springer.

D. A. DARLING & M. KAC (1957): On occupation times of Markov processes. Trans. Amer. Math. Soc. 84, 444-458.

Yu. A. DAVYDOV (1973): Limit theorems for functionals of processes with independent increments. Th. Probab. Appl. 18, 431-441.

____________ (1974): Sur une classe des fonctionelles des processus stables et des marches aléatoires. Ann. Inst. H. Poincaré (B) 10, 1-29.

____________ & I. A. IBRAGIMOV (1971): On asymptotic behaviour of some functionals of processes with independent increments. Th. Probab. Appl. 16, 162-167.

Y. DÉNIEL & Y. DERRIENNIC (1988+): Sur la convergence presque sure, au sens de Cesàro d'ordre α, $0 < \alpha < 1$, des variables indépendantes et identiquement distribueés. Ann. Probab.

Y. DERRIENNIC & Y. GUIVARC'H (1973): Théorème de renouvellement pour les groupes non moyennables. Comptes Rendus 277A, 613-615.

P. DIACONIS & C. STEIN (1978): Some Tauberian theorems related to coin tossing. Ann. Probab. 6, 483-490.

R. A. DONEY (1966): An analogue of the renewal theorem in higher dimensions. Proc. London Math. Soc. (3) 16, 669-684.

R. E. EDWARDS (1958): Comments on Wiener's Tauberian theorems. J. London Math. Soc. 33, 462-468.

L. ELIE (1982): Comportement asymptotique du noyau potentiel sur les groupes de Lie. Ann. Sci. École Norm. Sup. (4) 15, 257-364.

H. G. FEICHTINGER (1981): On a new Segal algebra. Monatsh. Math. 92, 269-289.

_______________ & H. J. SCHMEISSER (1986): Weighted versions of Beurling's Tauberian theorem. Math. Ann. 275, 353-363.

W. FELLER (1957): An introduction to probability theory and its applications. Vol. 1, 2nd ed., Wiley.

________ (1971): An introduction to probability theory and its applications. Vol. 2, 2nd ed., Wiley.

J. J. F. FOURNIER & J. STEWART (1985): Amalgams of L^p and ℓ^q. Bull. Amer. Math. Soc. 13, 1-21.

R. R. GOLDBERG (1967): On a space of functions of Wiener. Duke Math. J. 34, 683-691.

Y. GUIVARC'H, M. KEANE & B. ROYNETTE (1977): Marches aléatoires sur les groupes de Lie. Lecture Notes in Math. 624, Springer.

G. H. HARDY (1949): Divergent series. Oxford Univ. Press.

T. E. HARRIS (1948): Branching processes. Ann. Math. Statist. 41, 474-494.

S. HELGASON (1962): Differential geometry and symmetric spaces. Academic Press.

H. HEYER (1984): Probability theory on hypergroups: a survey. Lecture Notes in Math. 1064, 481-550, Springer (Prob. Groups VII).

G. HÖGNÄS & A. MUKHERJEA (1984): On the limit of the average of values of a function at random points. Lecture Notes in Math. 1064, 204-218, Springer (Prob. Groups VII).

J. KARAMATA (1937): Sur les théorèmes inverses des procedes de sommabilité. Act. Sci. Indust. 450, Hermann, Paris.

__________ (1938): Allgemeine Umkehrsatze der Limitierungsverfahren. Abh. Math. Sem. Hansischen Univ. 12, 48-63.

S. KARLIN (1955): On the renewal equation. Pacific J. Math. 5, 229-257.

Y. KASAHARA (1978): Tauberian theorems of exponential type. J. Math. Kyoto Univ. 18, 209-219.

H. KESTEN (1968): A Tauberian theorem for random walk. Israel J. Math. 6, 278-294.

________ (1973): Random difference equations and renewal theory for products of random matrices. Acta Math. 131, 207-248.

________ (1974): Renewal theory for functionals of a Markov chain with general state space. Ann. Probab. 2, 355-386.

J. F. C. KINGMAN (1963): Random walks with spherical symmetry. Acta Math. 109, 11-53.

J. KNOPFMACHER & W. SCHWARZ (1981a,b): Binomial expected values of arithmetic functions. I: J. reine ang. Math. 328, 84-98. II: Coll. Math. Soc. J. Bolyai 34, 863-905 (Topics in classical number theory).

H. LEPTIN (1984): The structure of $L^1(G)$ for locally compact groups. Monogr. Stud. Math. 18, 48-61, Pitman.

________ & D. POGUNTKE (1979): Symmetry and non-symmetry for locally compact groups. J. Functional Analysis 33, 119-134.

T. LINDVALL (1977): A probabilistic proof of Blackwell's renewal theorem. Ann. Probab. 5, 482-485.

__________ (1986): On coupling of renewal processes with use of failure rates. Stoch. Proc. Appl. 22, 1-15.

T. S. LIU, A. van ROOIJ & J.-K. WANG (1974): On some group algebra modules related to Wiener's algebra M_1. Pacific J. Math. 55, 507-520.

I. MEILIJSON (1973): The average of the values of a function at random points. Israel J. Math. 15, 193-203.

W. MEYER-KÖNIG (1949): Untersuchungen über einige verwandte Limitierungsverfahren. Math. Z. 52, 257-304.

T. T. MOH (1972): On a general Tauberian theorem. Proc. Amer. Math. Soc. 36, 167-172.

A. V. NAGAEV (1979): Renewal theorems in $\mathbb{R}^d$. Th. Probab. Appl. 24, 572-581.

G. E. PETERSON (1972): Tauberian theorems for integrals II. J. London Math. Soc. (2) 5, 182-190.

S. C. PORT & C. J. STONE (1969): Potential theory of random walks on abelian groups. Acta Math. 122, 19-114.

H. REITER (1968): Classical harmonic analysis and locally compact groups. Oxford Univ. Press.

________ (1971): L^1-algebras and Segal algebras. Lecture Notes in Math. 231, Springer.

________ (1974): Über den Satz von Wiener auf lokalkompakte Gruppen. Comm. Math. Helvet. 49, 333-364.

D. REVUZ (1975): Markov chains. North Holland (2nd ed. 1984).

W. RUDIN (1963): Fourier analysis on groups. Wiley.

W. L. SMITH (1954): Asymptotic renewal theorems. Proc. Roy. Soc. Edinburgh A 64, 9-48.

________ (1955): Renewal theory and its ramifications. J. Roy. Statist. Soc. B 20, 243-302.

F. SPITZER (1956): A combinatorial lemma and its application to probability theory. Trans. Amer. Math. Soc. 82, 323-339.

A. J. STAM (1968): Laws of large numbers for functions of random walks with positive drift. Comp. Math. 19, 299-333.

________ (1969): Renewal theory in r dimensions, I. Comp. Math. 21, 383-399.

________ (1971): Renewal theory in r dimensions, II. Comp. Math. 23, 1-13.

J. M. STEELE (1988+): Probabilistic and worst-case analysis of classical problems of combinatorial optimisation in Euclidean space. Math. Oper. Res.

________, L. A. SHEPP & W. F. EDDY (1987): On the number of leaves of a Euclidean minimal spanning tree. J. Appl. Probab. 24, 809-826.

C. SUNYACH (1981): Capacité et théorie du renouvellement. Bull. Soc. Math. France 109, 283-296.

G. TENENBAUM (1980): Sur le procédé de sommation de Borel et la répartition du nombre des facteurs premiers des entiers. Enseign. Math. 26, 225-245.

K. URBANIK (1964): Generalised convolution algebras. Studia Math. 23, 217-245.

________ (1986). Generalised convolution algebras IV. Studia Math. 83, 57-95.

D. V. WIDDER (1941): The Laplace transform. Princeton Univ. Press.

N. WIENER (1932): Tauberian theorems. Ann. Math. 33, 1-100.

________ (1933): The Fourier integral and certain of its applications. Cambridge Univ. Press.

CHARACTERISATION OF POTENTIAL KERNELS OF TRANSIENT CONVOLUTION SEMIGROUPS ON A COMMUTATIVE HYPERGROUP

Walter R Bloom
School of Mathematical
and Physical Sciences
Murdoch University
Perth WA 6150
AUSTRALIA

Herbert Heyer
Mathematisches Institut
der Universität Tübingen
Auf der Morgenstelle 10
7400 Tübingen 1
Federal Republic of Germany

1. Introduction

In his fundamental paper [5] J. Deny extended, within a functional-analytic framework, G.A. Hunt's potential theory of submarkovian kernels to continuous generally unbounded potential kernels that are not necessarily submarkovian. He also in an earlier paper [4] applied his general theory to convolution kernels on an abelian group. It was this specialisation that motivated C. Berg and G. Forst in their monograph [1] to develop Deny's theory exclusively for locally compact abelian groups

In the present note we study convolution kernels on a commutative hypergroup with the view to extending the characterisation of potential kernels of transient convolution semigroups as perfect kernels (cf Berg and Forst [1], Theorem 17.8) to these more general structures.

For our presentation we follow the scheme of Berg and Forst [1]. While most of the details carry over to commutative hypergroups with little change some care has to be taken with the basic notions of convolvability and uniform continuity, and this is the subject of the next section.

One attractive feature of our approach is the application of strongly negative definite functions on the dual $K^\wedge$ of the commutative hypergroup K, leading to a simpler proof of the characterisation Theorem 5.3 below. In Section 3 on transient convolution semigroups we give a shorter proof of Bloom and Heyer [3], Theorem 7.10 and at the same time take care of an omission, where we required the Fourier transform of a (possibly) unbounded measure for which we proved positive definiteness but neglected to show shift boundedness.

The analysis throughout will be carried out on a (locally compact) commutative hypergroup K . For a definition and properties refer to Jewett [7] whose notation we follow. We write $K^\wedge$ for the dual hypergroup of K and π for the Plancherel measure on $K^\wedge$ corresponding to the Haar measure ω_K on K , guaranteed by Spector [8], Theorem III.4 together with Levitan's theorem (Jewett [7], Theorem 7.3I).

For $\mu \in M(K)$ (the space of bounded Radon measures on K) and $\sigma \in M(K^\wedge)$ we have

$$\hat{\mu}(\gamma) = \int_K \bar{\gamma}(x) d\mu(x)$$

for all $\gamma \in K^\wedge$ and

$$\check{\sigma}(y) = \int_{K^\wedge} \chi(y) d\sigma(\chi)$$

for all $y \in K$. The functions $\hat{\mu}, \check{\sigma}$ define the Fourier transform and inverse Fourier transform respectively, and have obvious extensions to the respective subspaces of integrable functions.

The spaces of nonnegative contraction measures on K , and of regular Borel measures on K with values in $[0,\infty]$, will be denoted by $M^1(K)$, $M^\infty(K)$ respectively.

2. Convolvability and uniform continuity

For each $x,y \in K$ write

$$f(x*y) = \int_K f \, d\epsilon_x * \epsilon_y$$

and

$$\mu * f(x) = \mu^-(f_x) = \int_K f(x*z^-) d\mu(z)$$

where $f \in C(K)$, $\mu \in M(K)$. We denote by f_x , μ_x the x-translates of f,μ respectively, defined by $f_x(y) = f(x*y)$ and $\mu_x(f) = \mu(f_x)$.

2.1 Definition The convolution of $\mu,\nu \in M^\infty(K)$ exists if

$$\int_K \int_K f(x*y) d\mu(x) d\nu(y) < \infty$$

for all $f \in C_c^+(K)$ in which case the measure defined by the above integral will be denoted by $\mu * \nu$.

The above definition can be extended to (convolvable) complex measures μ, ν by requiring that $|\mu| * |\nu|$ exist. It is easy to see that if $\mu * \nu$ exists then so does $\nu * \mu$, and that

$$(\mu * \nu)(f) = \mu(\nu^- * f) = \nu(\mu^- * f)$$

for all $f \in C_c(K)$. It is also the case that convolution is associative in the sense that if $\mu, \nu, \tau \in M^\infty(K)$ and if $\mu * (\nu * \tau)$ and $\mu * \nu$ exist then $(\mu * \nu) * \tau$ exists and

$$\mu * (\nu * \tau) = (\mu * \nu) * \tau$$

This can be proved using Lemma 2.2 below.

For $\mu \in M^\infty(K)$ write $D^+(\mu)$ for the set of positive measures ν for which $\mu * \nu$ exists. The following lemma will be needed in the sequel.

2.2 Lemma Let $\mu \in M^\infty(K)$ be nonzero.

(a) For every compact $C \subset K$ there exists $f \in C_c^+(K)$ such that $\mu * f \geq 1$ on C .

(b) If $\mu * \nu = 0$ for $\nu \in D^+(\mu)$ then $\nu = 0$.

Proof (a) Choose $x \in C$ and let $z \in \mathrm{supp}(\mu^-)$. Consider $w \in \mathrm{supp}(\epsilon_x * \epsilon_z)$ and $f \in C_c^+(K)$ with $f(w) > 0$. Then $\epsilon_x * \epsilon_z(f) > 0$ (for otherwise $\epsilon_x * \epsilon_z$ would vanish on a neighbourhood of w) which is equivalent to $f_x(z) > 0$. Since $z \in \mathrm{supp}(\mu^-)$ we must have $\mu^-(f_x) > 0$ or, equivalently, $\mu * f(x) > 0$.

It follows from Corollary 2.8(ii) below that $\mu * f$ is continuous, so that $\mu * f > 0$ on a neighbourhood V_x of x . A standard compactness argument now gives the existence of $f \in C_c^+(K)$ for which $\mu * f > 0$ on C , and scaling gives the result.

(b) Let $\nu \in D^+(\mu)$ satisfy $\mu * \nu = 0$ and choose compact $C \subset K$. By (a) there exists $f \in C_c^+(K)$ such that $\mu^- * f \geq 1$ on C . Thus

$$\nu(C) \leq \nu(\mu^- * f) = \mu * \nu(f) = 0$$

and, since the compact set C was chosen arbitrarily, $\nu = 0$. //

2.3 Definition A measure $\mu \in M^{\infty}(K)$ is called (vaguely) shift bounded if the set $\{\mu_x : x \in K\}$ of translates of μ is vaguely bounded.

It is straightforward to show that $\mu \in M^{\infty}(K)$ is shift bounded if and only if $\mu * f \in C_b(K)$ for all $f \in C_c(K)$. An important consequence of this and the equality

$$\int_K \int_K f(x*y)\,d\mu(x)\,d\nu(y) = \int_K \mu^{-} * f\, d\nu \quad , \quad f \in C_c^{+}(K)$$

is that the convolution of a nonnegative shift bounded measure μ and a bounded measure ν exists.

2.4 Lemma Let μ be a shift bounded measure in $M^{\infty}(K)$. The linear map $f \to \mu * f$ of $C_c(K)$ into $C_b(K)$ is continuous.

The lemma can be proved exactly as in the locally compact abelian group case; see Berg and Forst [1], Proposition 1.12.

2.5 Definition We call f uniformly continuous if given $\epsilon > 0$ and $x_0 \in K$ there exists a neighbourhood U of x_0 such that $\|f_{x_0} - f_x\|_{\infty} < \epsilon$ for all $x \in U$.

Write $C_u(K)$ for the space of uniformly continuous functions on K. Jewett [7], Lemma 4.3B showed that continuous functions with compact support satisfy a weak version of uniform continuity. We can strengthen Jewett's result as follows.

2.6 Theorem Let $f \in C(K)$ and let C be a compact subset of K. Then for $x_0 \in K$ and $\epsilon > 0$ there exists a compact neighbourhood U of x_0 such that for all $x \in U$, $y \in C$

$$|f(y^{-}*x) - f(y^{-}*x_0)| < \epsilon$$

Proof First note that $(x,y) \to f(y^{-}*x)$ is continuous on $K \times K$. Given (x_0,y_0) there exist neighbourhoods V_{x_0}, V_{y_0} of x_0, y_0 respectively such that

$$(x,y) \in V_{x_0} \times V_{y_0} \Rightarrow |f(y^{-}*x) - f(y_0^{-}*x_0)| < \tfrac{1}{2}\epsilon$$

Choose such neighbourhoods $V_{x_0,y}, V_y$ for each $y \in C$. Then by compactness

$C \subset \bigcup_{i=1}^{n} V_{y_i}$; put $V = \bigcap_{i=1}^{n} V_{x_0,y_i}$.

Now for $x \in V$, $y \in C$ we have $y \in V_{y_i}$ for some i and $x \in V_{x_0,y_i}$. Thus by the choice of these neighbourhoods (note that $x_0 \in V_{x_0,y_i}$)

$$\begin{aligned}|f(y^{-}*x) - f(y^{-}*x_0)| &\le |f(y^{-}*x) - f(y_i^{-}*x_0)| + |f(y_i^{-}*x_0) - f(y^{-}*x_0)| \\ &< \tfrac{1}{2}\epsilon + \tfrac{1}{2}\epsilon = \epsilon \qquad //\end{aligned}$$

2.7 <u>Corollary</u> $C_c(K) \subset C_u(K)$

Theorem 2.6 leads to the following standard convolution results.

2.8 <u>Corollary</u>

(i) If $f \in C_b(K)$, $\mu \in M^+(K)$ then $\mu * f \in C(K)$

(ii) If $f \in C_c(K)$, $\mu \in M^\infty(K)$ then $\mu * f \in C(K)$

<u>Proof</u> (i) Let $\epsilon > 0$ be given and assume without loss of generality that $\mu, f \neq 0$. Choose a compact set $C \subset K$ such that $\mu(K\backslash C) < \epsilon/(4\|f\|_\infty)$ and consider $x_0 \in K$. By Theorem 2.6 there exists a compact neighbourhood V of x_0 such that for $x \in V$, $y \in C$

$$|\mu * f(x) - \mu * f(x_0)| \le \tfrac{1}{2}\epsilon + \int_C |f(y^{-}*x) - f(y^{-}*x_0)|\,d\mu(y) < \epsilon$$

(ii) Let $\epsilon > 0$ be given and assume without loss of generality that $\mu, f \neq 0$. Choose $x_0 \in K$, any compact neighbourhood W of x_0 and write $L = W*\mathrm{supp}(f)^{-}$. For $x \in W$

$$(\{y^{-}\}*\{x\}) \cap \mathrm{supp}(f) \neq \phi \iff y \in L$$

Thus

$$\mu * f(x) - \mu * f(x_0) = \int_L (f(y^{-}*x) - f(y^{-}*x_0))\,d\mu(y)$$

By Theorem 2.6 choose a compact neighbourhood V of x_0 such that for $x \in V$, $y \in L$

$$|f(y^- * x) - f(y^- * x_0)| < \epsilon/|\mu|(L)$$

Then for $x \in V \cap W$

$$|\mu * f(x) - \mu * f(x_0)| < \frac{\epsilon|\mu|(L)}{|\mu|(L)} = \epsilon \qquad //$$

We are now in a position to prove the following result.

2.9 Theorem For each $f \in C_c(K)$ the map $\mu \to \mu * f$ of $M^\infty(K)$ (with the vague topology) into $C(K)$ (with the topology of compact convergence) is continuous.

Proof Let $D \subset K$ be compact, so that $D*(\text{supp } f)^-$ is compact, and choose $g \in C_c^+(K)$ such that $g = 1$ on $D*(\text{supp } f)^-$. Then for $\mu_0 \in M^\infty(K)$ and $\epsilon > 0$

$$W = \{\mu \in M^\infty(K) : \left|\int g \, d\mu - \int g \, d\mu_0\right| < \epsilon\}$$

is a neighbourhood of μ_0 in $M^\infty(K)$

Now by Corollary 2.7, f is uniformly continuous. Thus for each $x \in D$ there exists a neighbourhood U_x of x such that for $y \in U_x$

$$\|f_x^- - f_y^-\|_\infty < \epsilon$$

As D is compact, $D \subset \bigcup_{i=1}^{n} U_{x_i}$ for some $x_1, x_2, \ldots, x_n \in D$. By our choice of g

$$\mu(D*(\text{supp } f)^-) \le \int g \, d\mu < \int g \, d\mu_0 + \epsilon$$

so that for $y \in U_{x_i} \cap D$, $\mu \in W$

$$\left|\int f_{x_i}^- \, d\mu - \int f_y^- \, d\mu\right| \le \|f_{x_i}^- - f_y^-\|_\infty \, \mu(D*(\text{supp } f)^-)$$
$$< \epsilon\left(\int g \, d\mu_0 + \epsilon\right)$$

(For the first inequality note that $\text{supp}(f_{x_i}^-) = \{x_i\}*(\text{supp } f)^- \subset D*(\text{supp } f)^-$ with a similar inclusion holding for $\text{supp}(f_y^-)$.)

Let V be the neighbourhood of μ_0 in $M^\infty(K)$ consisting of all μ for which

$$\left|\int f^-_{x_i}\, d\mu - \int f^-_{x_i}\, d\mu_0\right| < \epsilon \quad \text{for all} \quad i = 1,2,\ldots,n$$

Then for each $\mu \in V \cap W$ and all $y \in D$

$$\left|\int f^-_y\, d\mu - \int f^-_y\, d\mu_0\right| \le 2\epsilon\left(\int g\, d\mu_0 + \epsilon\right) + \epsilon$$

But the lefthand side of this inequality is just

$$|\mu * f(y) - \mu_0 * f(y)|$$

and this proves the theorem. //

We end this section with two results on the vague convergence of nets of convolutions. Their proofs are straightforward and will be omitted.

2.10 Lemma Let $(\mu_\alpha),(\nu_\alpha)$ be increasing nets in $M^\infty(K)$ converging vaguely (τ_V) to μ,ν respectively. Suppose that $\mu_\alpha * \nu_\alpha$ exists for all α and there exists $\lambda \in M^\infty(K)$ such that $\mu_\alpha * \nu_\alpha \le \lambda$ for all α. Then the convolution $\mu * \nu$ exists and

$$\mu * \nu = \tau_V - \lim_\alpha \mu_\alpha * \nu_\alpha$$

2.11 Lemma Let (μ_α) be a net in $M^\infty(K)$ and suppose $\kappa \in M^\infty(K)$ is a nonzero measure such that $\kappa * \mu_\alpha$ exists for all α and $(\kappa * \mu_\alpha)$ is τ_V-bounded. Then (μ_α) is τ_V-bounded.

3. Transient convolution semigroups

We consider a continuous convolution semigroup $(\mu_t)_{t\ge0}$ of nonnegative contraction measures such that $\tau_V - \lim_{t\to0} \mu_t = \epsilon_e$, which we refer to from now on as a *convolution semigroup*. As detailed in Bloom and Heyer [3] there are one-to-one correspondences between such convolution semigroups, strongly negative definite functions ψ on $K^\wedge$, and resolvent families $(\rho_\lambda)_{\lambda>0}$ in $M^+(K)$, given as follows:

3.1 Schoenberg correspondence

$$\hat{\mu}_t(\gamma) = \exp(-t\,\psi(\gamma)) \quad \text{for all} \quad \gamma \in K^\wedge \text{ and } t > 0$$

3.2 Resolvent correspondence

$$\rho_\lambda(f) = \int_0^\infty e^{-\lambda t}\mu_t(f)\,dt \qquad \text{for all} \quad f \in C_c(K)$$

with

$$\hat{\rho}_\lambda(\gamma) = (\lambda + \psi(\gamma))^{-1} \qquad \text{for all} \quad \gamma \in K^\wedge\ ,\ \lambda > 0$$

For every convolution semigroup $(\mu_t)_{t\geq 0}$ there is defined the extended real number

$$\lim_{\lambda\to 0} \rho_\lambda(f) = \int_0^\infty \mu_t(f)dt \quad , \quad f \in C_c(K)$$

If this limit is finite for every $f \in C_c(K)$ then

$$\kappa(f) = \lim_{\lambda\to 0} \rho_\lambda(f)$$

defines a measure $\kappa \in M^\infty(K)$. In this case $(\mu_t)_{t\geq 0}$ is said to be transient and κ is called the potential kernel of $(\mu_t)_{t\geq 0}$.

The following properties of potential kernels of transient convolution semigroups hold just as in the abelian group case.

3.3 (Berg and Forst [1], Section 13.3). For all $\lambda > 0$ the convolution $\kappa * \rho_\lambda$ exists and

$$\kappa - \rho_\lambda = \lambda\, \kappa * \rho_\lambda$$

3.4 (Bloom and Heyer [3], Proposition 7.2). For all $\lambda > 0$

$$\mathrm{supp}(\kappa) = \mathrm{supp}(\rho_\lambda) = \overline{(\bigcup_{t>0} \mathrm{supp}(\mu_t))}$$

and $[\mathrm{supp}(\kappa)]$, the smallest subhypergroup of K containing $\mathrm{supp}(\kappa)$, is σ-compact.

3.5 (Berg and Forst [1], Proposition 13.7). For all $\lambda > 0$, $\lambda\kappa + \epsilon_e$ is the potential kernel of the elementary kernel determined by $\lambda\rho_\lambda$.

Note that the elementary kernel defined by $\mu \in M^1(K)$ is the measure

$$\kappa = \sum_{n=0}^{\infty} \mu^n$$

the series being τ_v-convergent. Clearly any elementary kernel is the potential kernel of a transient Poisson semigroup $(\mu_t)_{t\geq 0}$ of the form

$$\mu_t = e^{-t}\exp(t\mu) \quad , \quad t \geq 0$$

where $\mu \in M^1(K)$ defines the elementary kernel. The transience of $(\mu_t)_{t\geq 0}$ is equivalent to the τ_v-convergence of $\sum_{n=0}^{\infty} \mu^n$ and hence to the existence of κ .

3.6 (Gallardo and Gebuhrer [6], Theorem 3.4). The potential kernel κ of a transient convolution semigroup is shift bounded.

Write

$$M_p^{\infty}(K) = \{\mu \in M^{\infty}(K) : \mu \text{ is positive definite and } \mu * f * f^{\sim} \text{ is bounded for all } f \in C_c(K)\}$$

where $f^{\sim}(x) = \overline{f(x^{-})}$. In Bloom and Heyer [3], Theorem 7.10 we presented part of the following result:

3.7 Theorem Let K be a strong hypergroup and let (μ_t) be a continuous convolution semigroup on K such that $\mu_0 = \epsilon_e$ with corresponding strongly negative definite function ψ on $K^{\wedge}$. Assume $\frac{1}{\psi}$ to be locally integrable. Then

(a) $\frac{1}{\psi}\,\omega_{K^{\wedge}} \in M_p^{\infty}(K^{\wedge})$

(b) (μ_t) is transient and the potential kernel κ of (μ_t) is given by

$$\kappa = (\frac{1}{\psi}\,\omega_{K^{\wedge}})^{\vee}$$

(c) If in addition (μ_t) is symmetric (which is equivalent to ψ being real valued) then $\kappa \in M_p^{\infty}(K)$, and both κ and $\frac{1}{\psi}\,\omega_{K^{\wedge}}$ are shift bounded.

Proof (a) For every $\lambda > 0$ the resolvent measure ρ_λ has Fourier transform

$$\hat{\rho}_\lambda = \frac{1}{\lambda + \psi}$$

which is bounded and, by Bloom and Heyer [3], Proposition 2.5, positive definite. We also see that $(\lambda + \psi)^{-1}\omega_{K^{\wedge}} \in M_p^{\infty}(K^{\wedge})$ since for all $g \in C_c(K^{\wedge})$ and $\chi \in K^{\wedge}$

$$\left|\left(\frac{1}{\lambda + \psi} \omega_{K^\wedge}\right) * g * g^{\sim}(\chi)\right| = \left|\frac{1}{\lambda + \psi} * g * g^{\sim}(\chi)\right| \leq \left\|\frac{1}{\lambda + \psi}\right\|_\infty \|g * g^{\sim}\|_1 < \infty$$

But the local integrability of $\frac{1}{\psi}$ guarantees that

$$\tau_v - \lim_{\lambda \to 0} \frac{1}{\lambda + \psi} \omega_{K^\wedge} = \frac{1}{\psi} \omega_{K^\wedge}$$

It is easily shown that $M_p^\infty(K)$ is τ_v-closed in $M^\infty(K)$ whence (a) follows.

(b) We first note from Bloom and Heyer [3], Proposition 3.4(d) that $\omega_{A(K,\Lambda)} = \epsilon_e$ so that $A(K,\Lambda) = \{e\}$ and, using [3], Propositions 3.4(b) and 3.1(b), $\Lambda = K^\wedge$. Taking $g \in C_c(K^\wedge)$ we have

$$\lim_{\lambda \to 0} \rho_\lambda(|\check{g}|^2) = \lim_{\lambda \to 0} \int_K |\check{g}|^2 \, d\rho_\lambda$$

$$= \lim_{\lambda \to 0} \int_{K^\wedge} g * g^{\sim} \frac{1}{\lambda + \overline{\psi}} \, d\omega_{K^\wedge}$$

$$= \int_{K^\wedge} g * g^{\sim} \frac{1}{\overline{\psi}} \, d\omega_{K^\wedge} < \infty \qquad (\#)$$

since $g * g^{\sim} \in C_c(K^\wedge)$ and $1/\overline{\psi}$ is assumed to be locally integrable, so that the last integral exists.

Now let $f \in C_c^+(K)$ and choose $g \in C_c(K^\wedge)$ such that $|\check{g}|^2 \geq f$. Then

$$\lim_{\lambda \to 0} \rho_\lambda(f) \leq \lim_{\lambda \to 0} \rho_\lambda(|\check{g}|^2) < \infty$$

shows that (μ_t) is transient with potential measure κ say. Appealing to (#) we have for $g \in C_c(K^\wedge)$

$$\int_K |\check{g}|^2 \, d\kappa = \int_{K^\wedge} g * g^{\sim} \frac{1}{\overline{\psi}} \, d\omega_{K^\wedge}$$

which just says that $\kappa = \left(\frac{1}{\psi} \omega_{K^\wedge}\right)^\vee$ as required.

(c) By assumption $\hat{\rho}_\lambda = (\lambda + \psi)^{-1} \geq 0$ for all $\lambda > 0$ whence it follows that ρ_λ is positive definite (Bloom and Heyer [3], Proposition 6.11) for every $\lambda > 0$. Since $\tau_v - \lim_{\lambda \to 0} \rho_\lambda = \kappa$ we have $\kappa \in M_p^\infty(K)$. Both κ and $\frac{1}{\psi} \omega_{K^\wedge}$ are nonnegative, and shift boundedness follows easily. //

4. Potential theory

The potential theoretic principles formulated in Deny [4] and [5] in terms of operators can be adapted to the hypergroup setting along the lines of Section 15 in Berg and Forst [1]. In the following it will be assumed that the measure κ is either an elementary kernel or the potential kernel of a transient convolution semigroup although the principles below can be formulated for arbitrary non-negative κ .

4.1 Principle of unicity of mass (PUM)

For $\sigma_1,\sigma_2 \in D(\kappa)$

$$\kappa * \sigma_1 = \kappa * \sigma_2 \Rightarrow \sigma_1 = \sigma_2$$

Every elementary kernel satisfies PUM (see Berg and Forst [1], Theorem 15.19).

4.2 Balayage principle (BP)

A measure $\mu' \in D(\kappa)$ is said to be a κ-balayaged measure of $\mu \in D(\kappa)$ on an open set V if

(a) $\operatorname{supp}(\mu') \subset \overline{V}$

(b) $\kappa * \mu' \le \kappa * \mu$ on K

(c) $\kappa * \mu' = \kappa * \mu$ on V

We say that κ satisfies the balayage principle if for every $\mu \in M_c^\infty(K)$ and for every open relatively compact set V there exists a κ-balayaged measure of μ on V .

Every potential kernel of a transient convolution semigroup satisfies BP (Berg and Forst [1], Theorem 16.14).

4.3 Equilibrium principle (EP)

A measure κ satisfies the equilibrium principle if for every relatively compact open subset V of K there exists $\mu_V \in D(\kappa)$ such that

(a) $\operatorname{supp}(\mu_V) \subset \overline{V}$

(b) $\kappa * \mu_V \leq \omega_K$ on K

(c) $\kappa * \mu_V = \omega_K$ on V

The potential kernel of a transient convolution semigroup satisfies EP (see Berg and Forst [1], Theorem 16.22). It follows that given a relatively compact open set V and a compact neighbourhood U of e there exists $g \in C_c^+(K)$ satisfying

(a) $\operatorname{supp}(g) \subset \overline{V} * U$

(b) $\kappa * g \leq 1$ on K

(c) $\kappa * g = 1$ on V

4.4 (Berg and Forst [1], Corollary 16.26) A κ-balayaged measure on the open subsets of K diminishes the total mass, that is for all $\sigma \in D^+(\kappa)$ and every κ-balayaged measure σ' of σ on V, $\sigma'(K) \leq \sigma(K)$.

The fundamental convolution theorems having been established, the proofs of all of the above results carry over from abelian groups to commutative hypergroups with little change.

5. Characterisation theorem

5.1 Definition Let $\kappa \in M^\infty(K)$. A fundamental family associated with κ is a net $(\sigma_V)_\Omega \subset M^1(K)$, where the V range over a base Ω of compact open neighbourhoods of e, satisfying the following properties. For all $V \in \Omega$

(a) $\sigma_V \in D^+(\kappa)$, $\sigma_V * \kappa \leq \kappa$ and $\sigma_V * \kappa \neq \kappa$

(b) $\sigma_V * \kappa = \kappa$ on V^c

(c) $\tau_v - \lim_{n\to\infty} \sigma_V^n * \kappa = 0$

We refer to $\kappa \in M^\infty(K)$ as a perfect kernel if there exists a fundamental family associated with it.

5.2 Example The constant net $(\mu_V)_\Omega$ with $\mu_V = \mu \in M^1(K)$ for all $V \in \Omega$ is a fundamental family associated with the elementary kernel

$$\kappa = \sum_{n=0}^{\infty} \mu^n$$

5.3 Theorem For any $\kappa \in M^\infty(K)$ the following statements are equivalent:

(a) κ is a perfect kernel;

(b) κ is the potential kernel of a uniquely determined transient convolution semigroup.

Proof We outline a proof of Theorem 5.3 following Berg and Forst [1], Theorems 17.4 and 17.8, which are given there for the locally compact abelian group case.

We first note that κ is the potential kernel of at most one transient convolution semigroup. Indeed if there were two such semigroups $(\mu_t),(\mu_t')$ with corresponding resolvent families $(\rho_\lambda),(\rho_\lambda')$ respectively then 3.3 applies to show

$$\kappa = (\lambda\kappa + \epsilon_e) * \rho_\lambda = (\lambda\kappa + \epsilon_e) * \rho_\lambda'$$

for all $\lambda > 0$. Since $\lambda\kappa + \epsilon_e$ is a potential kernel (3.5), it satisfies PUM (4.2) and hence $\rho_\lambda = \rho_\lambda'$ for all $\lambda > 0$. Thus $\mu_t = \mu_t'$ for all $t > 0$.

To show that the potential kernel κ for a transient convolution semigroup (μ_t) on K is a perfect kernel we choose Ω to be a base of compact neighbourhoods at e and, for each $V \in \Omega$, σ_V to be a κ-balayaged measure of ϵ_e on V^c (see 4.2). It is not hard to see that $(\sigma_V)_\Omega$ is a fundamental family associated with κ.

Finally we show that every perfect kernel κ is the potential kernel of a transient convolution semigroup. This is achieved by considering the fundamental family $(\sigma_V)_\Omega$ associated with κ and showing the existence of a uniquely determined function $\psi : \Gamma \to C$ such that

$$\hat{\eta}_V(\gamma)\, \psi(\gamma) = a_V(1 - \hat{\sigma}_V(\gamma))$$

where $\eta_V = a_V \kappa * (\epsilon_e - \sigma_V)$ with a_V chosen such that $\|\eta_V\| = 1$. It is easy to see that ψ is continuous and, appealing to [3], Proposition 2.4, $\hat{\eta}_V\psi$ is strongly negative definite. Clearly $\lim_\Omega \hat{\eta}_V = 1$ uniformly on compact sets, so that $\psi = \lim_\Omega \hat{\eta}_V\psi$ is also strongly negative definite (Bloom and Heyer [3], Theorem 6.5). The required convolution semigroup is then (μ_t) where $\hat{\mu}_t = \exp(-t\psi)$ is just the Schoenberg correspondence. Now use 3.6 to show that $\kappa_V = a_V^{-1} \sum_{n=0}^{\infty} \sigma_V^n$ is shift bounded, from which it follows that κ is shift bounded, and the proof of the locally compact abelian group case carries over to hypergroups to show that (μ_t) is transient. //

As an application of Theorem 5.3 we can show that every transient convolution semigroup on a commutative hypergroup is the weak limit of Poisson semigroups.

5.4 <u>Theorem</u> Let $(\mu_t)_{t\geq 0}$ be a transient convolution semigroup with potential kernel κ, associated fundamental family $(\sigma_V)_\Omega$ and corresponding norming sequence $(a_V)_\Omega$ chosen as in the proof of Theorem 5.3 so that $\eta_V = a_V(\kappa - \sigma_V * \kappa)$ has norm 1 for all $V \in \Omega$. Then

$$\mu_t = \tau_w - \lim_\Omega e^{-ta_V} \exp(ta_V\sigma_V)$$

for all $t \geq 0$.

<u>Proof</u> Writing $\mu_t^V = e^{-ta_V} \exp(ta_V\sigma_V)$ for each $V \in \Omega$, it is easily checked that (μ_t^V) is a continuous convolution semigroup on K and

$$\left(\mu_t^V\right)^\wedge = \exp[-ta_V(1 - \hat{\sigma}_V)] = \exp(-t\hat{\eta}_V\psi)$$

for all $t \geq 0$, where ψ denotes the negative definite function corresponding to $(\mu_t)_{t\geq 0}$. Now $\lim_\Omega \hat{\eta}_V = 1$ uniformly on compact sets, so that

$\lim_\Omega \left(\mu_t^V\right)^\wedge = \exp(-t\psi)$ uniformly on compact sets for all t. From the continuity theorem for nets (see the proof of Bloom and Heyer [2], Theorem 4.6) we have $\tau_w\text{-}\lim_\Omega \mu_t^V = \mu_t$ for all $t \geq 0$. //

REFERENCES

[1] Christian Berg and Gunnar Forst, Potential theory on locally compact abelian groups. Ergebnisse der Mathematik und ihrer Grenzgebiete, Band 87. Springer, Berlin, Heidelberg, New York, 1975.

[2] Walter R. Bloom and Herbert Heyer, The Fourier transform for probability measures on hypergroups. Rend. Mat. Ser.VII, **2** (1982), 315-334.

[3] Walter R. Bloom and Herbert Heyer, Convolution semigroups and resolvent families of measures on hypergroups. Math. Z. **188** (1985), 449-474.

[4] Jacques Deny, Familles fondamentales. Noyaux associés. Ann. Inst. Fourier (Grenoble) **3** (1951), 73-101.

[5] Jacques Deny, Noyaux de convolution de Hunt et noyaux associés à une famille fondamentale. Ann. Inst. Fourier (Grenoble) **12** (1962), 643-667.

[6] Leonard Gallardo and Olivier Gebuhrer, Analyse harmonique et marches aléatoires sur les hypergroupes. Prépublication IRMA de Strasbourg (1985).

[7] Robert I. Jewett, Spaces with an abstract convolution of measures. Adv. in Math. **18** (1975), 1-101.

[8] R. Spector, Mesures invariantes sur les hypergroupes. Trans. Amer. Math. Soc. **239** (1978) 147-165.

The work for this paper was partly carried out while the first author held an Alexander von Humboldt fellowship at the University of Tübingen.

EMBEDDING INFINITELY DIVISIBLE PROBABILITIES ON THE AFFINE GROUP.

S.G. Dani and M.McCrudden.

For any locally compact Hausdorff topological group G let $\mathcal{P}(G)$ denote the topological semigroup of all probability measures on G, where $\mathcal{P}(G)$ is given the weak topology and the multiplication is convolution of measures.

A measure $\mu \in \mathcal{P}(G)$ is called infinitely divisible on G if and only if, for each natural number n, there exists $\nu \in \mathcal{P}(G)$ such that $\nu^n = \mu$. A measure $\mu \in \mathcal{P}(G)$ is said to be continuously embedded on G if and only if there is a continuous homomorphism $t \longrightarrow \mu_t$ of $\mathbb{R}_+$ (the non-negative reals under addition) into $\mathcal{P}(G)$ such that $\mu = \mu_1$. The embedding problem is the problem of determining the class of groups G which have the property that every $\mu \in \mathcal{P}(G)$ which is infinitely divisible on G is also continuously embedded on G. Any group G with this property is said to have the embedding property.

A restricted version of the problem asks the same question when we restrict attention to connected Lie groups G. Even in this case the problem is still open, though many partial results are known. A unified approach to the problem in the connected Lie case appears in a recent paper by the present authors [1], from which most of the previously known partial results can be deduced.

Nevertheless, we are still a long way from a complete solution. One interesting test case is the case when G is the (classical) connected affine group, thought of as $\mathbb{R}^n \circledS GL^+(n,\mathbb{R})$, and μ is a measure which is infinitely divisible on G but is supported on the subgroup $\mathbb{R}^n$. If μ has an mth-root supported on $\mathbb{R}^n$, for all $m \geqslant 1$, then μ is certainly continuously embedded on $\mathbb{R}^n$, and so on G, because $\mathbb{R}^n$ is a root compact group ([3], Chapter III). However if the roots of μ lie only inside G and not necessarily inside the subgroup $\mathbb{R}^n$, the situation is rather more complicated. This test case was the basic motivating question for the present paper, in which we obtain results which do indeed imply that such a measure is continuously embedded on G.

To describe our results explicitly, we recall that a (real) almost algebraic group is a subgroup of some $GL(d,\mathbb{R})$ which has finite index in its algebraic closure in $GL(d,\mathbb{R})$.

<u>Theorem 1</u>. Let G be a connected almost algebraic group with compact centre, let H be a (topologically) closed subgroup of G, and let $\mu \in \mathbb{P}(H)$ be infinitely divisible on H. If $Z^\circ(\mu) \subseteq H$, where $Z^\circ(\mu)$ is the identity component of

$$Z(\mu) = \{x \in G : xy = yx, \text{ all } y \in \operatorname{supp}(\mu)\},$$

and if $Z^\circ(\mu)$ is abelian, then μ is continuously embedded on H.

<u>Theorem 2</u>. Let G be an almost algebraic group, let H be a (topologically) closed subgroup of G, and let $\mu \in \mathbb{P}(H)$ such that $\operatorname{supp}(\mu)$ is abelian i.e. for all $x, y \in \operatorname{supp}(\mu)$, $xy = yx$. Suppose that $\tilde{G}(\mu)$, the smallest almost algebraic subgroup of G containing $\operatorname{supp}(\mu)$, is connected and that $Z(\mu) \cap M(\mu) \subseteq H$, where $M(\mu)$ is the identity component of the normaliser of $\tilde{G}(\mu)$ in G.

If μ is infinitely divisible on H, there exists $y \in M(\mu) \cap Z(\mu)$ such that

(i) $y\mu$ is continuously embedded on H, and

(ii) y has finite order not exceeding the order of $Z(\mu) \cap M(\mu)/Z^\circ(\mu)$.

In particular, if $Z(\mu)$ is also connected, then $y = 1$, and μ is continuously embedded on H.

The main ingredients of the proof of theorem 1 are a generalisation of the methods of [5] and a recent theorem of the present authors ([2], Theorem 1.1). Theorem 2 is then obtained by applying theorem 1 to $(\operatorname{Aff}(\tilde{G}(\mu)))^\circ$, the connected affine group of $\tilde{G}(\mu)$, and "coming back" to G. The solution of the problem stated above for the classical connected affine group is then an easy consequence of theorem 2.

§1. <u>Construction of a root set sequence</u>

Let H be a Lie group, let $\mu \in \mathbb{P}(H)$, and let $Z^\circ(\mu)$ be the connected centraliser in H of $\operatorname{supp}(\mu)$. If λ is an mth-root of μ, then λ is supported on $N(\mu)$, the normaliser in H of $H(\mu)$, where $H(\mu)$ is the closed subgroup of H generated by $\operatorname{supp}(\mu)$ ([1], Proposition 1.1). Since $N(\mu)$ normalises $Z^\circ(\mu)$ we have a homomorphism $p : N(\mu) \longrightarrow \operatorname{Aut}(Z^\circ(\mu))$, given by

$$\forall\, x \in N(\mu), \quad \forall\, h \in Z^\circ(\mu), \quad p(x)(h) = xhx^{-1}.$$

Let $L(\mu)$ denote the Lie algebra of $Z^\circ(\mu)$, and let δ: $\mathrm{Aut}(Z^\circ(\mu)) \longrightarrow \mathrm{Aut}(L(\mu))$ be the natural homomorphism which associates with an automorphism its derivative. We write $\tilde{p}$ for $\delta p : N(\mu) \longrightarrow \mathrm{Aut}(L(\mu))$. We note that $\tilde{p}(\mu)$ is the identity of $\mathrm{Aut}(L(\mu))$, so for any m^{th}-root λ of μ, $\tilde{p}(\lambda)$ is an element of $\mathrm{Aut}(L(\mu))$ of finite order. It follows that $\tilde{p}(\lambda)$ is a semisimple linear map of $L(\mu)$, all of whose eigenvalues are root of unity. We write $\sigma(\lambda)$ for the spectrum of $\tilde{p}(\lambda)$, thought of as a set of pairs $\{(\lambda_i, t_i) : 1 \leqslant i \leqslant r\}$, where $\lambda_1,\ldots,\lambda_r$ are the distinct (possibly complex) eigenvalues of $\tilde{p}(\lambda)$ and t_i is the multiplicity of λ_i, $1 \leqslant i \leqslant r$.

<u>Proposition 1.</u> Let $(m_k)_{k\geqslant 1}$ be a sequence of positive integers and suppose $\mu \in \mathcal{P}(H)$ is infinitely divisible on H. Then there exists a sequence $(R_k^*(\mu))_{k\geqslant 0}$ of closed subsets of $\mathcal{P}(H)$ such that

(i) $R_0^*(\mu) = \{\mu\}$,

(ii) for all $k \geqslant 1$, $\lambda \in R_k^*(\mu) \Rightarrow \lambda^{m_k} \in R_{k-1}^*(\mu)$,

(iii) for all $k \geqslant 1$, $\lambda, \nu \in R_k^*(\mu) \Rightarrow \sigma(\lambda) = \sigma(\nu)$,

(iv) for all $k \geqslant 1$, if $\lambda \in R_k^*(\mu)$ and $\nu \in \mathcal{P}(H)$ such that

(a) $\nu^{n_k} = \mu$, where $n_k = \prod_{i=1}^{k} m_i$

and (b) $\sigma(\nu) = \sigma(\lambda)$

then $\nu \in R_k^*(\mu)$,

(v) for all $k \geqslant 1$, $R_k^*(\mu)$ contains an n-divisible element, for every $n \geqslant 1$.

<u>Proof</u>. We construct the sequence $R_k^*(\mu)$ inductively. Let $R_0^*(\mu) = \{\mu\}$. To define $R_1^*(\mu)$, we write

$$R_1(\mu) = \{\lambda \in \mathcal{P}(H) : \lambda^{m_1} = \mu\},$$

and we note that infinite divisibility of μ on H implies that $R_1(\mu) \neq \emptyset$, and for each $n \geqslant 1$, there exists $\nu \in \mathcal{P}(H)$ such that $\nu^n \in R_1(\mu)$. The set $\{\sigma(\lambda) : \lambda \in R_1(\mu)\}$ is clearly finite (its cardinality is not more than m_1^d, where d is the dimension of $Z^\circ(\mu)$), let its elements be $\sigma_1, \sigma_2,\ldots,\sigma_{r_1}$, and for $1 \leqslant j \leqslant r_1$, write

$$R_1(\mu,\sigma_j) = \{\lambda \in R_1(\mu) : \sigma(\lambda) = \sigma_j\}.$$

Then

$$R_1(\mu) = \bigcup_{j=1}^{r_1} R_1(\mu,\sigma_j) \qquad \text{disjointly.} \qquad \ldots.(*)$$

We claim that for some $1 \leqslant j \leqslant r_1$, $R_1(\mu,\sigma_j)$ contains an n-divisible element for every $n \geqslant 1$. For if not then for each $1 \leqslant j \leqslant r_1$ there exists N_j such that no element of $R_1(\mu,\sigma_j)$ is N_j-divisible, and if we set $N = \prod_{j=1}^{r_1} N_j$, then by (*) no element of $R_1(\mu)$ is N-divisible, a contradiction.

We select $1 \leqslant j_0 \leqslant r_1$ such that for all $n \geqslant 1$, $R_1(\mu,\sigma_j)$ contains an n-divisible element and we write $R_1^*(\mu)$ for $R_1(\mu,\sigma_{j_0})$. It is easy to check that $R_1^*(\mu)$ is closed in $\mathcal{P}(H)$ and that (ii) (iii) (iv) (v) of the statement of Proposition 1 are satisfied for $k = 1$.

To finish the proof we must show how to construct $R_{k+1}^*(\mu)$, given that R_0, R_1, $R_2,\ldots,R_k$ have already been defined.

Let $R_1(R_k^*(\mu)) = \{\lambda \in \mathcal{P}(H) : \lambda^{m_{k+1}} \in R_k^*(\mu)\}$ and note that by property (v) for k, $R_1(R_k^*(\mu))$ contains n-divisible elements for all $n \geqslant 1$. The set $\{\sigma(\lambda) : \lambda \in R_1(R_k^*(\mu))\}$ is finite (its cardinality is not more than m_{k+1}^d), let its elements be $\sigma_1,\sigma_2,\ldots,\sigma_{r_{k+1}}$. If we write

$$R_1(R_k^*(\mu),\ \sigma_j) = \{\lambda \in R_1(R_k^*(\mu)) : \sigma(\lambda) = \sigma_j\}$$

then

$$R_1(R_k^*(\mu)) = \bigcup_{j=1}^{r_{k+1}} R_1(R_k^*(\mu),\ \sigma_j),$$

where the right hand side is a disjoint union of closed subsets of $\mathcal{P}(H)$. An argument as before shows that there is some $1 \leqslant j_0 \leqslant r_{k+1}$ such that $R_1(R_k^*(\mu),\ \sigma_{j_0})$ contains an n-divisible element for all $n \geqslant 1$. Now set $R_{k+1}^*(\mu)$ equal to $R_1(R_k^*(\mu),\sigma_{j_0})$. Given that, by construction, the sets $R_0,R_1,..,R_k$ satisfy properties (i)..(v) of the proposition up to k, it is now easy to see that $R_0,R_1,\ldots,R_{k+1}$ satisfy the same properties up to k+1.

This description of the inductive step completes the construct of the sequence $(R_k^*(\mu))_{k\geqslant 0}$.

§2. Proof of Theorem 1.

Throughout this section G denotes an almost algebraic group which is connected and has compact centre, H is a closed subgroup of G, and $\mu \in$ (H) such that μ is infinitely divisible on H and $H \supseteq Z^\circ(\mu)$ the connected centalizer of supp(μ) in G.

We begin by choosing a sequence $(m_k)_{k\geq 1}$ of positive integers with the property that for each $n \geq 1$, there exists $k_n \geq 1$ such that $n \mid (\prod_{i=1}^{k_n} m_i)$. For example, if $(p_n)_{n\geq 1}$ is the set of all positive primes in increasing order of magnitude, we might take $m_k = p_1, p_2, \ldots, p_k$. The sequence $(m_k)_{k\geq 1}$ being chosen we can then use Proposition 1 to find a sequence $(R_k^*(\mu))_{k\geq 0}$ of closed subsets of $\mathcal{P}(H)$ satisfying properties (i)..(iv) of Proposition 1. (Note that (v) is automatically satisfied by our choice of the sequence $(m_k)_{k\geq 1}$).

For each $k \geq 0$ let d_k be the dimension of the 1-eigenspace of $\tilde{p}(\lambda)$, for any $\lambda \in R_k^*(\mu)$. We note that d_k is independent of the choice of $\lambda \in R_k^*(\mu)$, by property (iii) of $(R_k^*(\mu))_{k\geq 0}$. By property (ii), for all $k \geq 0$, $d_{k+1} \leq d_k$, so there exists some $K > 0$ such that for all $k \geq K$, $d_k = d_K$.

For each $k \geq 1$ we select $\lambda_k \in R_{k+K}^*(\mu)$, and then by property (ii), $\lambda_k^{r_k} \in R_K^*(\mu)$, where $r_k = \prod_{i=K+1}^{k+K} m_i$. Write $\nu_k = \lambda_k^{r_k}$, then by properties (i) and (ii), each ν_k is a root of μ. Since G is connected almost algebraic and has compact centre, it follows from Theorem 1.1 of [2] that $\{\nu_k : k \geq 1\}$ is relatively compact modulo $Z(\mu)$, and since $Z(\mu)$ has a finite group of components, being almost algebraic, we may further deduce that $\{\nu_k : k \geq 1\}$ is relatively compact modulo $Z^\circ(\mu)$. Hence there is a sequence $\{h_k : k \geq 1\}$ in $Z^\circ(\mu)$ such that $\{\nu_k h_k : k \geq 1\}$ is a relatively compact subset of $\mathcal{P}(H)$.

<u>Proposition 2.</u> Suppose λ is a root of μ, so that $p(\lambda) \in \mathrm{Aut}\,(Z^\circ(\mu))$. For each $r \geq 1$ and all $t \in Z^\circ(\mu)$,

$$(t\lambda)^r = \lambda^r(p(\lambda)^{-r}(t)\ p(\lambda)^{-r+1}(t)\ldots p(\lambda)^{-1}(t))$$

and
$$(\lambda t)^r = (p(\lambda)(t)p(\lambda)^2(t)\ldots p(\lambda)^r(t))\lambda^r.$$

<u>Proof</u> By Proposition 1.1 of [1] we may write $\lambda = \gamma g$, where $g \in N(\mu)$ and γ is supported on $G(\mu)$, the closed subgroup of G generated by $\mathrm{supp}(\mu)$. Then

$$\begin{aligned}\lambda^r = (\gamma g)^r &= g^r(g^{-r}\gamma g^r)\ldots(g^{-1}\gamma g)\\ &= g^r(g^{-r}t\gamma g^r)\ldots(g^{-1}t\gamma g)((g^{-r}tg^r)\ldots(g^{-1}tg))^{-1}\\ &= (t\gamma g)^r((g^{-r}tg^r)\ldots(g^{-1}tg))^{-1}\\ &= (t\lambda)^r(p(\lambda)^{-r}(t)\ldots p(\lambda)^{-1}(t))^{-1}\end{aligned}$$

which gives the first formula. The derivation of the second formula is similar.

<u>From this point till the end of this section we assume that $Z^\circ(\mu)$ is abelian.</u>

Returning to our earlier notation and writing $M_K = \prod_{i=1}^{K} m_i$, we note that $\nu_k^{M_k} = \mu$ for all $k \geq 1$, and so $\tilde{p}(\nu_k)^{M_K}$ is the identity on L, where L denotes the Lie algebra of $Z^{\circ}(\mu)$. For each $k \geq 1$ we denote by L_k the linear subspace of L which is the 1-eigenspace of $\tilde{p}(\nu_k)$, and we write L_k^* for the unique $\tilde{p}(\nu_k)$-invariant subspace of L such that $L = L_k \oplus L_k^*$. We may then write

$$h_k = \exp(u_k + v_k) = \exp(u_k)\exp(v_k),$$

where $u_k \in L_k$, $v_k \in L_k^*$, and $\exp : L \longrightarrow Z^{\circ}(\mu)$ is the exponential map. Since $(\tilde{p}(\nu_k) - I)$ is clearly invertible on L_k^*, and

$$(\tilde{p}(\nu_k) - I)\Big(\sum_{s=1}^{M_K} \tilde{p}(\nu_k)^s\Big) = (\tilde{p}(\nu_k)^{M_K} - I)\tilde{p}(\nu_k),$$

we see that for all $k \geq 1$,

$$\sum_{s=1}^{M_K} \tilde{p}(\nu_k)^s\,(v_k) = 0$$

while clearly

$$\sum_{s=1}^{M_K} \tilde{p}(\nu_k)^s\,(u_k) = M_K\, u_k.$$

By Proposition 2 it follows that

$$\begin{aligned}
(\nu_k\, h_k)^{M_K} &= \Big[\prod_{s=1}^{M_K} p(\nu_k)^s(h_k)\Big]\mu \\
&= \Big[\prod_{s=1}^{M_K} \exp\,(\tilde{p}(\nu_k)^s(u_k + v_k))\Big]\mu \\
&= \exp\Big[\sum_{s=1}^{M_K} \tilde{p}(\nu_k)^s(u_k + v_k)\Big]\mu \\
&= \exp(M_K\, u_k)\mu \\
&= (\exp u_k)^{M_K}\mu.
\end{aligned}$$

Since $\{(\nu_k\, h_k)^{M_K} : k \geq 1\}$ is a relatively compact subset of $\mathcal{P}(H)$, we conclude from Theorem 2.1, Chapter III of [6] that $\{(\exp u_k)^{M_K} : k \geq 1\}$ is a relatively compact subset of $Z^{\circ}(\mu)$. But this implies that $\{\exp\,(u_k) : k \geq 1\}$ is relatively compact, since $Z^{\circ}(\mu) \cong V \oplus T$, where V is a vector group and T is a torus. Writing $y_k = \exp v_k$ we conclude that $\{\nu_k\, y_k : k \geq 1\}$ is a relatively compact subset of $\mathcal{P}(H)$, and the calculation above also shows that for all $k \geq 1$, $(\nu_k\, y_k)^{M_K} = \mu$. So if we write

$$C = \overline{\{\nu_k\, y_k : k \geq 1\}}$$

we see from property (iv) of Proposition 1 that C is a compact subset of $R_K^*(\mu)$.

For each $k \geq 1$, let us write

$$S_k(C) = \{\alpha \in R_{k+K}^*(\mu) : \alpha^{r_k} \in C\}.$$

<u>Proposition 3</u> For each $k \geq 1$, $S_k(C)$ is non-empty and compact.

<u>Proof</u> (i) To prove $S_k(C) \neq \emptyset$, it suffices to show that for each $k \geq 1$ we can find $z_k \in Z^\circ(\mu)$ such that $(z_k\lambda_k)^{r_k} = \nu_k y_k$. For then $z_k\lambda_k \in R^*_{k+K}(\mu)$, by property (iv) of Proposition 1.

But by Proposition 2 and the fact that $Z^\circ(\mu)$ is abelian, we have that for any $w \in L$,

$$((\exp w)\lambda_k)^{r_k} = \nu_k\Big[\exp\Big(\sum_{s=1}^{r_k} \tilde{p}(\lambda_k)^{-s}(w)\Big)\Big]$$

so since $y_k = \exp v_k$ with $v_k \in L^*_k$, it suffices to prove that the image of the linear map $\sum_{s=1}^{r_k} \tilde{p}(\lambda_k)^{-s}$ contains L^*_k. But this is clear since

$$\sum_{s=1}^{r_k} \tilde{p}(\lambda_k)^{-s} = \Big[\sum_{s=0}^{r_k-1} \tilde{p}(\lambda_k)^s\Big]\,\tilde{p}(\lambda_k)^{-r_k}$$

$$\Big[\sum_{s=0}^{r_k-1} \tilde{p}(\lambda_k)^s\Big]\Big[I - \tilde{p}(\lambda_k)\Big] = I - \tilde{p}(\lambda_k)^{r_k} = I - \tilde{p}(\nu_k),$$

and $\tilde{p}(\lambda_k)$ is invertible on L while $I - \tilde{p}(\nu_k)$ is invertible on L^*_k.

(ii) Suppose $S_k(C)$ is not compact, then since it is clearly closed in $\mathcal{P}(H)$ we may assume that $S_k(C)$ contains a sequence $\{\alpha_n : n \geq 1\}$ which is not relatively compact. Since each α_n is a root of μ, we deduce from Theorem 1.1 of [2] that there is a sequence $(t_n)_{n\geq 1}$ in $Z^\circ(\mu)$ such that $\{\alpha_n t_n : n \geq 1\}$ is relatively compact, and hence $\{t_n : n \geq 1\}$ is not relatively compact. Write $t_n = \exp b_n$, where $b_n \in L$, then by Proposition 2 and a calculation as earlier, we have

$$(\alpha_n t_n)^{r_k} = \exp\Big[\sum_{s=1}^{r_k} \tilde{p}(\alpha_n)^s(b_n)\Big]\alpha_n^{r_k},$$

so since C is compact and $\alpha_n^{r_k} \in C$, for all $n \geq 1$, we deduce that

$$\Big\{\exp\Big[\sum_{s=1}^{r_k} \tilde{p}(\alpha_n)^s(b_n)\Big] : n \geq 1\Big\}$$

is a relatively compact subset of $Z^\circ(\mu)$.

Since $p(\alpha_n) = \tilde{p}(\alpha_n t_n)$, we see that $\{\tilde{p}(\alpha_n) : n \geq 1\}$ is a relatively compact subset of Aut(L), and so $\{\sum_{s=1}^{r_k} \tilde{p}(\alpha_n)^s : n \geq 1\}$ is a relatively compact subset of End(L). To show that this set is in fact a relatively compact subset of Aut(L), it suffices to prove that there is a non-zero constant c such that for all $n \geq 1$, $\det\Big[\sum_{s=1}^{r_k} \tilde{p}(\alpha_n)^s\Big] = c$.

Let L_n be the 1-eigenspace of $\tilde{p}(\alpha_n)$ and note that since $\{\alpha_n : n \geq 1\} \subseteq R^*_{k+K}(\mu)$, all $\tilde{p}(\alpha_n)$ have the same spectrum. Note also that by choice of K, the 1-eigenspace of $\tilde{p}(\alpha_n^{r_k})$ equals L_n. Let L^c be the complexification of L, then in the obvious notation, $L^c = L_n^c \oplus L_n^{*c}$, and thinking of $\tilde{p}(\alpha_n)$ as a linear map on L^c, we can pick a $\mathbb{C}$-basis for L^c of the form $\{x_1,\dots,x_\ell, y_1,\dots,y_{d-\ell}\}$ where $\{x_1,\dots,x_\ell\}$ is a basis for L_n^c and $\{y_1,\dots,y_{d-\ell}\}$ is a basis for L_n^{*c}, both consisting of eigenvectors of $\tilde{p}(\alpha_n)$. Let β_j be the eigenvalue corresponding to y_j, and note that $\beta_j \neq 1$, but β_j is a root of unity, since $\tilde{p}(\alpha_n)$ has finite order. Then

$$\sum_{s=1}^{r_k} \tilde{p}(\alpha_n)^s(x_i) = r_k\, x_i \qquad (1 \leqslant i \leqslant \ell)$$

and
$$\sum_{s=1}^{r_k} \tilde{p}(\alpha_n)^s(y_j) = \beta_j\, (1-\beta_j^{r_k})/(1-\beta_j) \qquad (1 \leqslant j \leqslant d-\ell)$$

Note that $1-\beta_j^{r_k} \neq 0$, since $\beta_j^{r_k}$ is an eigenvalue of $\tilde{p}(\alpha_n^{r_k})$ but cannot equal 1 since the 1-eigenspace of $\tilde{p}(\alpha_n^{r_k})$ has dimension ℓ. We conclude that for all $n \geqslant 1$,

$$\det\Big[\sum_{s=1}^{r_k} \tilde{p}(\alpha_n)^s\Big] = r_k^{\ell} \prod_{j=1}^{d-\ell} \left[\frac{(1-\beta_j^{r_k})\beta_j}{1-\beta_j}\right]$$

and since the right hand side is non-zero and independent of n, we conclude that $\left\{\sum_{s=1}^{r_k} \tilde{p}(\alpha_n)^s : n \geqslant 1\right\}$ is indeed a relatively compact subset of $\mathrm{Aut}(L)$.

If we now write $\gamma_n(x) = \prod_{s=1}^{r_k} p(\alpha_n)^s(x)$, for $x \in Z^\circ(\mu)$, then γ_n is an endomorphism of $Z^\circ(\mu)$, and $\delta(\gamma_n) = \sum_{s=1}^{r_k} \tilde{p}(\alpha_n)^s$, where $\delta : \mathrm{End}(Z^\circ(\mu)) \longrightarrow \mathrm{End}(L)$ is the map taking any endomorphism of $Z^\circ(\mu)$ to its derivative. Since δ is a topological isomorphism from $\mathrm{End}(Z^\circ(\mu))$ onto its image, and this image is closed in End (L), we deduce from above that $\{\gamma_n : n \geqslant 1\}$ is a relatively compact subset of $\mathrm{Aut}(Z^\circ(\mu))$. But

$$\begin{aligned}\gamma_n(t_n) &= \gamma_n(\exp b_n)\\ &= \exp\Big[\sum_{s=1}^{r_k} \tilde{p}(\alpha_n)^s(b_n)\Big]\end{aligned}$$

and we saw earlier that this last set is relatively compact, hence $\{t_n : n \geqslant 1\} = \{\gamma_n^{-1}(\gamma_n(t_n)) : n \geqslant 1\}$ is relatively compact set in $Z^\circ(\mu)$. This contradiction completes the proof that $S_k(C)$ is compact.

If we now write $D_k = \{\alpha^{r_k} : \alpha \in S_k(C)\}$, then each D_k is compact by Proposition 3, and clearly $\{D_k : k \geqslant 1\}$ has the finite intersection property, so $\bigcap_{k=1}^{\infty} D_k \neq \emptyset$. We select $\nu \in \bigcap_{k=1}^{\infty} D_k$.

<u>Proposition 4.</u> For each $k \geqslant 1$ and each $\lambda \in R_{k+K}^{*}(\mu)$ such that $\lambda^{r_k} = \nu$, λ is

supported on $Z(Z^\circ(\nu))$, the centraliser of the connected centraliser of supp(ν).

__Proof__. For any root α of μ, $Z^\circ(\alpha)$ is just the analytic subgroup of $Z^\circ(\mu)$ whose Lie algebra is the subalgebra of L fixed by $\tilde{p}(\alpha)$ i.e. the 1-eigenspace of $\tilde{p}(\alpha)$. By choice of K, $\dim Z^\circ(\nu) = \dim Z^\circ(\lambda)$, and since $\lambda^{r_k} = \nu$, $Z^\circ(\lambda) \subseteq Z^\circ(\nu)$, from which the proposition is immediate.

__Proposition 5.__ Any $\alpha \in \mathcal{P}(G)$ is root compact on $Z(Z^\circ(\alpha))$.

__Proof__ Let $U = Z(Z^\circ(\alpha))$, then $W = UZ^\circ(\alpha)$ is an almost algebraic subgroup of G, hence is closed. We have the commutative diagram

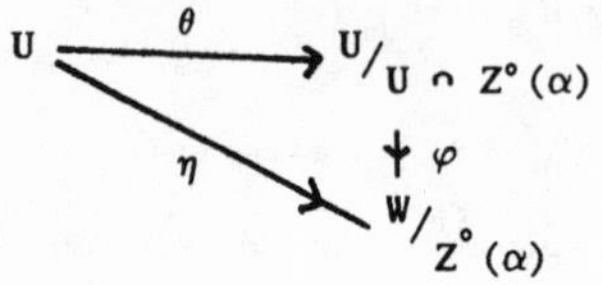

where for $x \in U$, $\eta(x) = xZ^\circ(\alpha)$, $\theta(x) = x(U \cap Z^\circ(\alpha))$ $\varphi(x(U \cap Z^\circ(\alpha))) = xZ^\circ(\alpha)$, and φ is a topological isomorphism. If R is the root set of α in U, then since $W/_{Z^\circ(\alpha)}$ is closed in $G/_{Z^\circ(\alpha)}$, we conclude from Theorem 1.1 of [2] that $\eta(R)$ is relatively compact in $\mathcal{P}(W/_{Z^\circ(\alpha)})$, so by the diagram $\theta(R)$ is relatively compact in $\mathcal{P}(U/_{U \cap Z^\circ(\alpha)})$. Since $U \cap Z^\circ(\alpha)$ is central in U this is sufficient sufficient to show that R is relatively compact in $\mathcal{P}(U)$, by Proposition 4.4 of [1].

We can now complete the proof of Theorem 1. Since ν is a root of μ, it is sufficient to show that ν is continuously embedded on $P = H \cap Z(Z^\circ(\nu))$. By construction of ν and Propositions 4 and 5, ν is root compact on P and is also r_k-divisible on P, for all $k \geq 1$. By choice of $(m_k)_{k\geq 1}$, given any $n \geq 1$, we can find $k_n \geq 1$ such that $n\left[\prod_{i=1}^{K} m_i\right]$ divides $\prod_{i=1}^{k_n} m_i$, and so n divides $\prod_{i=K+1}^{k_n} m_i$, which is $r_{(k_n-K)}$. It follows that ν is infinitely divisible on P, so since P is a Lie group we conclude that μ continuously embedded on P, by the reasoning of Theorem 3.5.8. of [3].

§3. __Proof of theorem 2__.

In this section G denotes an almost algebraic group, H is a closed subgroup of G, and $\mu \in \mathcal{P}(H)$ such that

$$\forall\, x, y \in \mathrm{supp}(\mu), \quad xy = yx.$$

We use $\tilde{G}(\mu)$ for the smallest almost algebraic subgroup of G containing supp(μ), and we note that $\tilde{G}(\mu)$ is abelian. We assume also that $\tilde{G}(\mu)$ is connected, and that $H \supseteq Z(\mu) \cap M(\mu)$, where $M(\mu)$ is the connected normaliser of $\tilde{G}(\mu)$ in G. We note that $\tilde{G}(\mu) \subseteq M(\mu) \cap Z(\mu)$. Finally we assume that μ is infintely divisible on H.

By Proposition 1.1 of [1], all roots of μ are supported on $\tilde{N}(\mu)$, the normaliser of $\tilde{G}(\mu)$ in G, and as $\tilde{N}(\mu)$ is almost algebraic, the index of $M(\mu)$ in $\tilde{N}(\mu)$ is finite. It follows that μ is infinitely divisible on $K = M(\mu) \cap H$, and that K is a closed subgroup of the almost algebraic group $M(\mu)$ with $Z_1(\mu) = M(\mu) \cap Z(\mu) \subseteq K$.

Let $G' = (\mathrm{Aff}(\tilde{G}(\mu)))^{\circ}$, thought of as the semidirect extension of $\tilde{G}(\mu)$ by $(\mathrm{Aut}(\tilde{G}(\mu)))^{\circ}$. Then G' is a connected almost algebraic group. Indeed if $\tilde{G}(\mu) \cong \mathbb{R}^m \times T^n$, then

$$(\mathrm{Aut}(\tilde{G}(\mu)))^{\circ} \cong \left\{\begin{bmatrix} I_n & B \\ 0 & C \end{bmatrix} : C \in GL^+(\mathbb{R}^m),\ B \text{ any } n\times m \text{ matrix}\right\}$$

It is elementary to check that the centre of $(\mathrm{Aut}(\tilde{G}(\mu)))^{\circ}$ is trivial. It follows that the centre of G' is all $x \in \tilde{G}(\mu)$ which are fixed by every $\alpha \in (\mathrm{Aut}(\tilde{G}(\mu)))^{\circ}$. Since this is just the maximal torus in $\tilde{G}(\mu)$, we conclude that G' has compact centre. Let $q : M(\mu) \longrightarrow (\mathrm{Aut}(\tilde{G}(\mu)))^{\circ}$ be the homomorphism taking $x \in M(\mu)$ to the automorphism of $\tilde{G}(\mu)$ which sends $t \in \tilde{G}(\mu)$ to xtx^{-1}.

For each $\nu \in \mathcal{P}(\tilde{G}(\mu))$ we write ν' for ν thought of a a measure on G'.

<u>Proposition 6.</u> μ' is infinitely divisible on $K' = q(K)\tilde{G}(\mu) \subseteq G'$.

<u>Proof</u> For $n \geq 1$ we pick $\lambda \in \mathcal{P}(K)$ such that $\lambda^n = \mu$. By Proposition 1.1 of [1] we can find $g \in K$ such that $\mathrm{supp}(\lambda) \subseteq g\tilde{G}(\mu)$, and then $g^n \in \tilde{G}(\mu)$.

Let A be the Zariski closure of $\{g\}$ in $M(\mu)$, then A is an abelian almost algebraic group of the form $A^{\circ} \times F$, where F is a finite cyclic subgroup. Since $g^n \in \tilde{G}(\mu)$ we see that $A \cap \tilde{G}(\mu)$ has the form $A^{\circ} \times F_1$, where F_1 is a subgroup of F whose index in F divides n. If $g = ax$ with $a \in A^{\circ}$ and $x \in F$ then x^n is an element of finite order in $\tilde{G}(\mu)$, so $x^n \in T$, the maximal torus of $\tilde{G}(\mu)$. Since $g \in M(\mu)$ and $M(\mu)$ is connected, g centralises T. We can pick $y \in T$ such that $y^n = x^n$, and then $(y^{-1}x)^n = 1$, and $y^{-1}x$ lies in the same $\tilde{G}(\mu)$-coset as g.

Let $h = y^{-1}x$, then we may write $\lambda = h\nu$, where ν is supported on $\tilde{G}(\mu)$. Let $\lambda_1 = q(h)\nu'$, then λ_1 is a measure on G' and

$$\lambda_1^n = (q(h)\nu')^n$$
$$= (q(h)\nu' q(h)^{-1}). \; . \; . \; .(q(h)^n \nu' q(h)^{-n}) q(h)^n$$
$$= (h\nu h^{-1})'. \; . \; . \; .(h^n \nu h^{-n})' = ((h\nu)^n h^{-n})' = \mu',$$

since $h^n = 1$. So μ' is indeed infinitely divisble on K'.

<u>Proposition 7</u>. This is a continuous map $t \to \lambda_t$ or $\mathbb{R}_+$ into $\mathcal{P}(\tilde{G}(\mu))$, and a continuous homomorphism $t \longrightarrow x_t$ of $\mathbb{R}$ into $q(K)$ such that $t \longrightarrow x_t\lambda_t'$ is a continuous homomorphism of $\mathbb{R}_+$ into $\mathcal{P}(K')$, with $x_1 = 1$ and $\lambda_1 = \mu$.

<u>Proof</u> In view of Lemma 3.1, and the remarks preceeding it of [1], it is easy to see that $q(M(\mu))$ is an almost algebraic subgroup of $(\text{Aut}(\tilde{G}(\mu)))^\circ$. Hence $K' = q(K)\tilde{G}(\mu)$ is a closed subgroup of G'. By Proposition 6, μ' is infinitely divisible on K', while the centraliser of $\text{supp}(\mu')$ in G' is just $\tilde{G}(\mu) \subseteq K'$. Furthermore G' is connected, almost algebraic and has compact centre, so by Theorem 1, μ' is continuously embedded on K' i.e. there exists a continuous homomorphism $t \longrightarrow \mu_t'$ or $\mathbb{R}_+$ into $\mathcal{P}(K')$ such that $\mu_1' = \mu'$.

Let $\varphi = G' \longrightarrow (\text{Aut}(\tilde{G}(\mu)))^\circ$ be the natural homomorphism, then $t \longrightarrow \varphi(\mu_t')$ is a continuous homomorphism from $\mathbb{R}_+$ into $q(K)$ such that $\varphi(\mu_1') = 1$. If we now let $x_t = \varphi(\mu_t')$ and $\lambda_t' = x_t^{-1}\mu_t'$, the result follows.

<u>Lemma 8</u>. Let R be a connected Lie group and let B be a closed connected normal subgroup of R such that R/B is compact and one-dimensional. Then there exists a circle subgroup U of R such that $R = BU$ and $B \cap U$ is trivial; that is R is the semi-direct product of B and U.

<u>Proof</u> Let $\tilde{R}$ be the universal covering group of R and let $\theta : \tilde{R} \longrightarrow R$ be the covering homomorphism. Let $D = \text{Ker}\,\theta$, $\tilde{B} = \theta^{-1}(B)$, and $B_1 = (\tilde{B})^\circ$, the connected component of the identity in $\tilde{B}$. Since B is connected we have $\theta(B_1) = B$ and so $\tilde{B} = DB_1$. The connectedness of B_1 implies that $\tilde{R}/B_1$ is simply connected ([4], remark after Theorem 1.2 of Chapter XII), and since this group is one-dimensional, we conclude that $\tilde{R}/B_1 \cong \mathbb{R}$. Furthermore,

$$[\tilde{R}/B_1] \Big/ [\tilde{B}/B_1] \cong \tilde{R}/\tilde{B} \cong R/B$$

and as R/B is compact, we conclude that $\tilde{B}/B_1$ is an infinite cycle group.

Pick $z \in D$ such that zB_1/B_1 generates $\tilde{B}/B_1$. Since D is central in $\tilde{R}$ we conclude from Theorem 1.2 of Chapter XVI of [4] that there exists a closed one-dimensional connected subgroup $\tilde{U}$ of $\tilde{R}$ such that $z \in \tilde{U}$. As $\tilde{U}B_1$ is then an analytic subgroup of $\tilde{R}$ containing B_1 but also containing z, while $z \notin B_1$, we conclude that $\tilde{R} = \tilde{U}B_1$. Hence

$$R \cong {}^{\tilde{U}B_1}/_{B_1} \cong {}^{\tilde{U}}/_{\tilde{U}\cap B_1}$$

and since $\tilde{U}$ is one-dimensional and connected, we conclude that $\tilde{U} \cap B_1$ is trivial.

If $g \in \tilde{B} \cap \tilde{U}$. then $g = z^k h$ for some $k \in \mathbb{Z}$ and $h \in B_1$, and so $z^{-k}g \in \tilde{U} \cap B_1$, hence by the last observation $z^k = g$. It follows that $\tilde{B} \cap \tilde{U} \subseteq \langle z\rangle \subseteq D$.

Set $U = \theta(\tilde{U})$, then as $z \in \operatorname{Ker}\theta$, U is a compact connected one-dimensional Lie group i.e. a circle group. We have $R = \theta(\tilde{R}) = \theta(\tilde{U})\theta(B_1) = UB$. Further $B \cap U = \theta(\tilde{B}) \cap \theta(\tilde{U}) = \theta(\tilde{B} \cap \tilde{U})$ since $\tilde{B}$ is 'saturated'. But $\tilde{B} \cap \tilde{U} \subseteq \operatorname{Ker}\theta$, hence $B \cap U$ is trivial.

<u>Proposition 9</u> There is a continuous homomorphism $t \longrightarrow y_t$ of $\mathbb{R}$ into K, and a continuous map $t \longrightarrow \lambda_t$ of $\mathbb{R}_+$ into $\mathcal{P}(\tilde{G}(\mu))$ such that

(i) the map $t \longrightarrow y_t\lambda_t$ is a continuous homomorphism of $\mathbb{R}_+$ into $\mathcal{P}(K)$

(ii) $\lambda_1 = \mu$

(iii) y_1 is an element of finite order not exceeding the order of $(Z(\mu) \cap M(\mu))/_{Z^\circ(\mu)}$.

<u>Proof</u> Let λ_t, x_t be as in Proposition 7, let $S = \{x_t : t \in \mathbb{R}\}$, and let $R = q^{-1}(S) \subseteq K$. Since S is compact and $\operatorname{Ker}q = Z(\mu) \cap M(\mu)$, which is almost algebraic, it follows that R is an almost algebraic subgroup of $M(\mu)$. If S is trivial the result follows by taking $y_t = 1$ for all $t \in \mathbb{R}$, so we may assume that $\dim S = 1$.

We note that $q(R^\circ)$ is an analytic subgroup of S of finite index, hence $q(R^\circ) = S$. By lemma 8 we can find a circle subgroup U of R° such that $R^\circ = (\operatorname{Ker}q)^\circ U$ (semidirect product). This is because the dimension of $R^\circ/_{(\operatorname{Ker}q)^\circ}$, is clearly one, and this group is a closed subgroup of $^R/_{(\operatorname{Ker}q)^\circ}$, which is compact, because it is an extension of the finite group $^{\operatorname{Ker}q}/_{(\operatorname{Ker}q)^\circ}$ by the compact group S.

If $f : U \longrightarrow S$ is the restriction of q to U, then f is surjective and

$\mathrm{Ker} f = U \cap \mathrm{Ker} q$, which is isomorphic to a subgroup of $\mathrm{Ker} q/(\mathrm{Ker} q)^\circ$, because $U \cap (\mathrm{Ker} q)^\circ$ is trivial. Now let $\{y_t : t \in \mathbb{R}\}$ be a parameterisation of U such that $f(y_t) = x_t$, for all $t \in \mathbb{R}$. As $x_1 = 1$, we have $y_1 \in \mathrm{Ker}\, f$, hence the order of y_1 is at most the order of $\mathrm{Ker} q/(\mathrm{Ker} q)^\circ$, which is itself a divisor of the order of $(Z(\mu) \cap M(\mu))/Z(\mu)^\circ$.

It remains only to verify that $t \longrightarrow y_t\lambda_t$ is a homomorphism. Let $\nu_t = y_t\lambda_t$, let $\mu_t' = x_t\lambda_t'$, then by Proposition 7, for all $s, t \geq 0$,

$$y_{s+t}^{-1}\,\nu_s\,\nu_t = y_{s+t}^{-1}\,y_s\,\lambda_s\,y_t\,\lambda_t$$
$$= (y_t^{-1}\,\lambda_s\,y_t)\lambda_t$$

and so $y_{s+t}^{-1}\,\nu_s\,\nu_t$ is a measure on $\tilde{G}(\mu)$. Further

$$(y_{s+t}^{-1}\,\nu_s\nu_t)' = (y_t^{-1}\lambda_s\,y_t)'\,\lambda_t'$$
$$= x_t^{-1}\,\lambda_s'\,x_t\,\lambda_t'$$
$$= x_{s+t}^{-1}x_s\,\lambda_s'\,x_t\,\lambda_t'$$
$$= x_{s+t}^{-1}\,\mu_s'\,\mu_t' = x_{s+t}^{-1}\mu_{s+t}' = \lambda_{s+t}'$$

We conclude that

$$y_{s+t}^{-1}\,\nu_s\,\nu_t = \lambda_{s+t}$$

giving

$$\nu_s\,\nu_t = y_{s+t}\,\lambda_{s+t} = \nu_{s+t}.$$

Theorem 2 is now immediate from Proposition 9.

§4. <u>Measures on the classical connected affine group</u>.

We return very briefly to the motivating problem mentioned in the introduction. So we let G be the connected (classical) affine group, thought of as $\mathbb{R}^n \circledS GL^+(n,\mathbb{R})$, where $\mathbb{R}^n$ is identified with the normal subgroup $\mathbb{R}^n \circledS I_n$. If $\mu \in \mathbb{P}(G)$ such that $\mathrm{supp}(\mu) \subseteq \mathbb{R}^n$, then $\tilde{G}(\mu) = \langle\mathrm{supp}(\mu)\rangle$, the linear subspace of $\mathbb{R}^n$ spanned by $\mathrm{supp}(\mu)$. If this subspace is m-dimensional then clearly $Z(\mu)$ is isomorphic to the group $\mathbb{R}^n \circledS H$, where

$$H = \left\{\left[\begin{array}{c|c} I_m & ? \\ \hline 0 & A \end{array}\right] : A \in GL^+(n-m, \mathbb{R})\right\} \subseteq GL^+(n,\mathbb{R}),$$

and so $Z(\mu)$ is connected. By theorem 2 we conclude that if μ is infinitely divisible on G then μ is continuously embedded on G.

<u>References</u>

1. Dani, S.G. and McCrudden, M. 'Factors, roots and embeddability of measures on Lie groups', to appear in Math. Zeit.

2. Dani, S.G. and McCrudden, M. 'On the factor sets of measures and local tightness of convolution semigroups over Lie groups', to appear in Journal of Theoretical Probability.

3. Heyer, H. 'Probability measures on locally compact groups', Springer Verlag, Berlin-Heidelberg, 1977.

4. Hochschild, G. 'The structure of Lie groups', Holden Day, San Francisco-London-Amsterdam, 1965.

5. McCrudden, M. 'Infinitely divisible probabilities on $SL(2,\mathbb{C})$ are continuously embedded', Math. Proc. Camb. Phil. Soc. 92, p.101-107, 1982.

6. Parthasarathy, K.R. 'Probability measures on metric spaces' Academic Press, 1967.

S. G. Dani,
School of Mathematics,
Tata Institute of Fundamental Research,
Homi Bhabha Road,
Bombay 400005,
INDIA.

M. McCrudden,
Dept. of Mathematics,
University of Manchester,
Oxford Road,
Manchester, M13 9PL,
ENGLAND.

A propos de l'induction des convoluteurs

Antoine Derighetti
Université de Lausanne, Institut de mathématiques
CH-1015 Lausanne-Dorigny (Suisse)

1. Introduction.

En généralisation d'un théorème célèbre de de Leeuw, nous avons défini dans [4] une isométrie i de $CV_p(H)$ dans $CV_p(G)$ où H est un sous-groupe fermé quelconque d'un groupe localement compact G.

Dans ce travail, nous montrons notamment les résultats suivants :

1) i est un morphisme d'algèbre (théorème 5),
2) i ne dépend que des choix des mesures de Haar de H et de G (proposition 7),
3) i commute à l'action de $A_p(G)$ sur $CV_p(G)$ (proposition 9).

Nous explicitons en outre les preuves de diverses assertions de [5] et [6] et *surtout de* [7] (cf le théorème 10 ci-dessous).

Dans le dernier paragraphe, nous illustrons ces résultats par un retour au cas des groupes abéliens.

2. Quelques précisions concernant la définition de i_q.

Toutes les mesures de Haar sont choisies invariantes à gauche. Soient dh une mesure de Haar de H, dx une mesure de Haar de G, q une fonction continue strictement positive sur G avec $q(xh) = q(x)\Delta_H(h)\Delta_G(h^{-1})$ pour tout $x \in G \quad h \in H$ et $d_q\dot{x}$ l'unique mesure sur G/H telle que $dhd_q\dot{x} = q(x)dx$ (pour tout $x \in G$ on pose $\dot{x} = xH = \omega(x)$). Les éléments de $\mathcal{L}^p(G) = \mathcal{L}^p(G, dx)$ sont des fonctions *finies complexes définies sur G tout entier* (cf [3] chap. IV paragraphe 3 no 4 Définition 2, p.129). On note $[f]$ l'élément correspondant de $L^p(G)$. Pour tout $f \in \mathcal{L}^p(G)$ et tout $g \in \mathcal{L}^{p'}(G)$, on pose $([f],[g]) = \int_G f(x)\overline{g(x)}dx$. On dit que T, opérateur borné de $L^p(G)$, appartient à $CV_p(G)$ si $T({}_a\phi) = {}_a(T\phi)$ pour tout $\phi \in L^p(G) \quad a \in G$. La norme d'opérateur est notée $|||T|||$. Si f

est une fonction finie définie sur G et $x \in G$, on note $f_{x,H}$ la fonction sur H définie par $f_{x,H}(h) = f(xh)$.

Proposition 1 et définition 1. Soit $T \in CV_p(H)$. Les assertions suivantes sont vérifiées:

a) Pour tout $\phi, \psi \in C_{oo}(G)$ $\quad x \in G \quad \dot{x} \longmapsto \left(T\left[\left(\frac{\phi}{q^{1/p}}\right)_{x,H}\right], \left[\left(\frac{\psi}{q^{1/p'}}\right)_{x,H}\right]\right)$ est une fonction continue sur G/H à support contenu dans $\omega(\text{supp}\phi) \bigcap \omega(\text{supp}\psi)$.

b) Il existe un unique opérateur borné de $L^p(G)$, noté $i_q(T)$, tel que pour tout $\phi, \psi \in C_{oo}(G)$ on ait $\quad (i_q(T)[\phi], [\psi]) = \int_{G/H} \left(T\left[\left(\frac{\phi}{q^{1/p}}\right)_{x,H}\right], \left[\left(\frac{\psi}{q^{1/p'}}\right)_{x,H}\right]\right) d_q\dot{x}. \; (1)$

Remarques.

1) Rappelons que $\left[\left(\frac{\phi}{q^{1/p}}\right)_{x,H}\right]$ désigne ici la classe de $L^p(H)$ contenant $\left(\frac{\phi}{q^{1/p}}\right)_{x,H}$. Nous ferons essentiellement usage de cette précision dans la preuve du théorème 5 ci-dessous.

2) Abstraction faite du distinguo entre $\mathcal{L}^p(G)$ et $L^p(G)$, cette définition a déjà été introduite dans [4] p. 76. A cette référence nous montrons entre autres que i_q est une isométrie de $CV_p(H)$ dans $CV_p(G)$ (cf [7] pour des résulats complémentaires).

La nécessité d'étendre les conditions de validité de la relation (1) nous conduit à introduire la définition suivante.

Définition 2. Soient $f \in \mathcal{L}^p(G)$ et A une partie de G/H. On dit que " A est associée à f " si A est négligeable et telle que pour tout $x \in G$ avec $\dot{x} \notin A$ on ait $\left(\frac{f}{q^{1/p}}\right)_{x,H} \in \mathcal{L}^p(H)$.

Remarque. De tels ensembles existent toujours (cf par exemple [10] p. 69 à 71 et p. 165 lignes 6 et 7 à partir du bas de la page).

Une *très légère modification* apportée aux arguments développés à ces deux dernières références fournit l'assertion suivante.

Proposition 2. Soient $f \in \mathcal{L}^p(G)$ et A *une* partie de G/H associée à f. Pour tout $x \in G$ posons $\ell(\dot{x}) = \int_H \frac{|f(xh)|^p}{q(xh)} dh$ si $\dot{x} \notin A$ et $\ell(\dot{x}) = 0$ sinon. On a $\ell \in \mathcal{L}^1(G/H, d_q\dot{x})$ et $\int_{G/H} \ell(\dot{x}) d_q\dot{x} = \|f\|_p^p$.

Proposition 3. Soient $T \in CV_p(H)$, $f \in \mathcal{L}^p(G)$, $g \in \mathcal{L}^{p'}(G)$, A, B *des* parties de G/H avec A associée à f et B à g. Pour tout $x \in G$ avec $\dot{x} \notin A \bigcup B$, on pose $\lambda(\dot{x}) = \left(T\left[\left(\frac{f}{q^{1/p}}\right)_{x,H}\right], \left[\left(\frac{g}{q^{1/p'}}\right)_{x,H}\right]\right)$ et $\lambda(\dot{x}) = 0$ sinon. Dans ces conditions

$$\lambda \in \mathcal{L}^1(G/H, d_q\dot{x}) \quad \text{et} \quad \int_{G/H} \lambda(\dot{x}) d_q\dot{x} = (i_q(T)[f], [g]). \tag{2}$$

Démonstration. Soient $(f_n)_{n=1}^\infty$ $(g_n)_{n=1}^\infty$ des suites de $C_{oo}(G)$ avec $\|f - f_n\|_p < 2^{-n}$ et $\|g - g_n\|_{p'} < 2^{-n}$ pour tout $n \in \mathbb{N}$. Pour tout $x \in G$ $n \in \mathbb{N}$ posons $\lambda_n(\dot{x}) = \left(T\left[\left(\frac{f_n}{q^{1/p}}\right)_{x,H}\right], \left[\left(\frac{g_n}{q^{1/p'}}\right)_{x,H}\right]\right)$. Vu la proposition 1, $\lambda_n \in C_{oo}(G/H)$. Pour tout $\dot{x} \in G/H - A \cup B$ on a

$$|\lambda_n(\dot{x}) - \lambda(\dot{x})| \leqq |||T|||_p \left\| \left(\frac{f_n}{q^{1/p}}\right)_{x,H} \right\|_p \left\| \left(\frac{g_n}{q^{1/p'}}\right)_{x,H} - \left(\frac{g}{q^{1/p'}}\right)_{x,H} \right\|_{p'} +$$

$$+ |||T|||_p \left\| \left(\frac{f_n}{q^{1/p}}\right)_{x,H} - \left(\frac{f}{q^{1/p}}\right)_{x,H} \right\|_p \left\| \left(\frac{g}{q^{1/p'}}\right)_{x,H} \right\|_{p'}.$$

D'où, grâce à la proposition 2, $\int_{G/H}^* |\lambda_n(\dot{x}) - \lambda(\dot{x})| d_q\dot{x} \leqq |||T|||_p (\|f_n\|_p \|g_n - g\|_{p'} + \|f_n - f\|_p \|g\|_{p'})$ où $\int_{G/H}^* m(\dot{x}) d_q\dot{x}$ désigne l'intégrale supérieure de la fonction m par rapport à la mesure $d_q\dot{x}$.

On obtient pour tout $n \in \mathbb{N}$

$$\int_{G/H}^* |\lambda_n(\dot{x}) - \lambda(\dot{x})| d_q\dot{x} \leqq 2^{-n} |||T|||_p (1 + \|f\|_p + \|g\|_{p'}). \tag{3}$$

Il en résulte que $\lambda \in \mathcal{L}^1(G/H, d_q\dot{x})$.

De $|(i_q(T)[f],[g]) - \int_{G/H} \lambda(\dot{x})d_q\dot{x}| \leqq |(i_q(T)[f],[g]) - (i_q(T)[f_n],[g_n])|+$
$+ |(i_q(T)[f_n],[g_n]) - \int_{G/H} \lambda(\dot{x})d_q\dot{x}|$ (3) et de $\int_{G/H} \lambda_n(\dot{x})d_q\dot{x} = (i_q(T)[f_n],[g_n])$ on tire pour tout $n \in \mathbb{N}$ $|(i_q(T)[f],[g]) - \int_{G/H} \lambda(\dot{x})d_q\dot{x}| \leqq 2^{-n+1}|||T|||_p(1 + ||f||_p + ||g||_{p'})$.

3. i_q est un morphisme d'algèbre.

Proposition 4. Pour tout $k \in C_{oo}(G)$ $f \in \mathcal{L}^p(H)$ on a $k *_H f \in C(G)$, $(k *_H f)_{x,H} = k_{x,H} * f \in \mathcal{L}^p(H)$ pour tout $x \in G$, $q^{1/p}(k *_H f) \in \mathcal{L}^p(G)$ et $||q^{1/p}(k *_H f)||_p \leqq ||f||_p||T_H|k|||_p$ où $T_H|k|(\dot{x}) = \int_H |k|(xh)|dh$.

Démonstration.

(a) Montrons la continuité de $k *_H f$ sur G.

Soient $\epsilon > 0$, $x \in G$, V un voisinage compact de e dans G et $n \in C_{oo}(G)$ avec $n \geqq 0$ $n = 1$ sur $V\text{supp}k$.

Posons $M = \max\left\{T_H n(\dot{y}) \mid \dot{y} \in G/H\right\} \cdot \sup\left\{\Delta_H(h)^{p'-1} \mid h \in (x^{-1}\text{supp}n) \cap H\right\}$.

Il existe V', voisinage ouvert de e dans G avec $V' \subset V$ et $|k(y^{-1}z) - k(z)|^{p'} < \dfrac{\epsilon^{p'}}{(1+M)(1+||f||_p)^{p'}}$ pour tout $y \in V', z \in G$. Avec ces choix, on obtient alors pour tout $y \in V' |k *_H f(y^{-1}x) - k *_H f(x)| < \epsilon$.

(b) Pour tout $x \in G$ $(k *_H f)_{x,H} = k_{x,H} * f$ ainsi $(k *_H f)_{x,H} \in \mathcal{L}^p(H)$ et $||(k *_H f)_{x,H}||_p \leqq ||k_{x,H}||_1||f||_p$.

(c) On a $\int_G^* q(x)|(k *_H f)(x)|^p dx \leqq \int_{G/H}^* ||k_{x,H}||_1^p||f||_p^p d_q\dot{x} = ||f||_p^p \int_{G/H}^* \left\{T_H|k|(\dot{x})\right\}^p d_q\dot{x}$ avec $T_H|k| \in C_{oo}(G/H)$.

Théorème 5. Pour tout $S, T \in CV_p(H)$ on a $i_q(ST) = i_q(S)i_q(T)$.

Démonstration.

(a) Pour tout $S \in CV_p(H)$ $k \in C_{oo}(G)$ $f, g \in \mathcal{L}^p(H)$ avec $S[f] = [g]$ on a $i_q(S)[q^{1/p}(k *_H f)] = [q^{1/p}(k *_H g)]$.

Il suffit de vérifier que pour tout $\psi \in C_{oo}(G)$, $(i_q(S)[q^{1/p}(k *_H f)], [\psi]) = ([q^{1/p}(k *_H g)], [\psi])$.

Soient $A \subset G/H$ associée à $q^{1/p}(k *_H f)$ et $\lambda_1(\dot{x}) = \left(S\left[(k *_H f)_{x,H}\right], \left[\left(\frac{\psi}{q^{1/p'}}\right)_{x,H}\right]\right)$ pour tout $x \in G$ avec $\dot{x} \notin A$ et $\lambda_1(\dot{x}) = 0$ si $\dot{x} \in A$. On sait que $\int_{G/H} \lambda_1(\dot{x}) d_q\dot{x} = (i_q(S)[q^{1/p}(k *_H f)], [\psi])$. Pour tout $x \in G$ avec $\dot{x} \notin A$, on a $S[(k *_H f)_{x,H}] = [k_{x,H}] * S[f] = [(k *_H g)_{x,H}]$.

Soient B un ensemble associé à $q^{1/p}(k *_H g)$ et $\lambda_2(\dot{x}) = \left(\left[(k *_H g)_{x,H}\right], \left[\left(\frac{\psi}{q^{1/p'}}\right)_{x,H}\right]\right)$ pour tout $x \in G$ avec $\dot{x} \notin B$ et $\lambda_2(\dot{x}) = 0$ si $\dot{x} \in B$. De $\int_{G/H} \lambda_1(\dot{x}) d_q\dot{x} = \int_{G/H} \lambda_2(\dot{x}) d_q\dot{x}$ on tire $(i_q(S)[q^{1/p}(k *_H f)], [\psi]) = ([q^{1/p}(k *_H g)], [\psi])$.

(b) Pour tout $k \in C_{oo}(G)$ $f \in \mathcal{L}^p(H)$ $S, T \in CV_p(H)$, on a $i_q(ST)[q^{1/p}(k *_H f)] = i_q(S)i_q(T)[q^{1/p}(k *_H f)]$.

Soient $g, \ell, m \in \mathcal{L}^p(H)$ avec $[g] = (ST)[f]$ $[\ell] = T[f]$, $[m] = S[\ell]$. Vu (a) $i_q(ST)[q^{1/p}(k *_H f)] = [q^{1/p}(k *_H g)]$ $i_q(S)\left(i_q(T)[q^{1/p}(k *_H f)]\right) = i_q(S)[q^{1/p}(k *_H \ell)] = [q^{1/p}(k *_H m)]$. De $[m] = (ST)[f]$ on tire $i_q(ST)[q^{1/p}(k *_H f)] = [q^{1/p}(k *_H m)]$, donc (b).

(c) Pour tout $\phi \in \mathcal{L}^p(G)$ $\epsilon > 0$ il existe $k \in C_{oo}(G)$ et $f \in C_{oo}(H)$ avec $||\phi - q^{1/p}(k *_H f)||_p < \epsilon$.

Il existe $\phi' \in C_{oo}(G)$ avec $||\phi - \phi'||_p < \frac{\epsilon}{2}$. Posons $k = \frac{\phi'}{q^{1/p}}$. Soit U un voisinage compact de e dans H . Il existe V voisinage ouvert de e dans H avec $V \subset U$ tel que pour tout $h \in V$ $||k - k_{h^{-1}}\Delta_H(h^{-1})||_p < \frac{\epsilon}{2(1 + \sup\{q(x)^{1/p} | x \in (\mathrm{supp}\phi')U\})}$.

Choisissons $f \in C_{oo}(H)$ avec $f \geqq 0$, $\int_H f(h)dh = 1$ et $\mathrm{supp} f \subset V$. On obtient alors $||q^{1/p}k - q^{1/p}(k *_H f)||_p \leqq \frac{\epsilon}{2}$ et par suite $||\phi - q^{1/p}(k *_H f)||_p < \epsilon$.

(d) Soient $S, T \in CV_p(H)$, $\phi \in \mathcal{L}^p(G)$ et $\epsilon > 0$. Il existe $k \in C_{oo}(G)$ $f \in C_{oo}(H)$ avec $||\phi - q^{1/p}(k *_H f)||_p < \frac{\epsilon}{2(1 + |||S|||_p|||T|||_p)}$. On a $||(i_q(ST) - i_q(S)i_q(T))([\phi])||_p \leqq$
$\leqq ||(i_q(ST) - i_q(S)i_q(T))([\phi] - [q^{1/p}(k *_H f)])||_p < \epsilon.$

Remarques.

1) Pour G abélien, ce théorème est une conséquence immédiate du point 2) de la proposition 11 ci-dessous.

2) Pour une autre approche qui passe par la mise au point d'une théorie de la p -induction et notamment un analogue "L^p" du théorème d'induction par étages, se reporter à [2] (p. 630 proposition 5). Il nous a semblé utile de proposer une preuve *directement dérivée* de la définition 1.

4. Quelques autres propriétés de i_q .

Dorénavant, nous commettons l'abus de langage qui consiste à poser pour $T \in CV_p(G)$ $\phi \in C_{oo}(G)$ $T\phi = T[\phi]$. Nous adoptons au surplus toutes les notations et conventions de [6] p. 96 et 97.

Proposition 6. Pour tout $S \in PM_p(H)$ on a $i_q(S) \in PM_p(G)$ et $L_{i_q(S)}(u) = L_S(\mathrm{Res}_H u)$ pour tout $u \in A_p(G)$.

Démonstration. Il existe $T \in PM_p(G)$ avec $L_T(u) = L_S(\mathrm{Res}_H u)$ pour tout $u \in A(G)$. Soient $\phi, \psi \in C_{oo}(G)$ on a (cf [6] p.9 lignes 1 et 2 à partir du bas)

$$L_T\left(\omega_{\lambda_G^{p'}}(\psi,\phi)\right) = \overline{(T\phi,\psi)} = L_S\left(\int_{G/H}\omega_{\lambda_G^{p'}}\left(\left(\frac{\psi}{q^{1/p'}}\right)_{y,H}, \left(\frac{\phi}{q^{1/p}}\right)_{y,H}\right)d_q\dot{y}\right) =$$

$$\int_{G/H}\overline{\left(S\left(\frac{\phi}{q^{1/p}}\right)_{y,H}, \left(\frac{\psi}{q^{1/p'}}\right)_{y,H}\right)}d_q\dot{y} = \overline{(i_q(S)\phi,\psi)}, \quad \text{d'où} \quad T = i_q(S)\,.$$

Remarque. Nous avons déjà utilisé cette assertion dans [6] (cf p. 102 ligne 10 à partir du haut).

Proposition 7 et définition 3. Les mesures de Haar invariantes à gauche de G et de H étant fixées, l'application i_q ne dépend pas du choix de la fonction continue strictement positive q satisfaisant à $q(xh) = q(x)\Delta_H(h)\Delta_G(h^{-1})$ pour tout $x \in G$ $h \in H$. On pose $i = i_q$.

Démonstration. Soient $\phi, \psi \in C_{oo}(G)$, $S \in CV_p(H)$ et $\alpha \in C_{oo}(H)$ avec $\alpha \geqq 0$ $\int_H \alpha(h)dh = 1$. Il est facile de se convaincre que $|(i_q(S\lambda_H^p(\alpha^*))\phi,\psi) - (i_q(S)\phi,\psi)| \leqq \leqq |||S|||_p\|\psi\|_{p'}\int_H \alpha(h)\|\phi_{h^{-1}}\Delta_G(h^{-1})^{1/p} - \phi\|_p dh$. Par ailleurs $(i_q(S\lambda_H^p(\alpha^*))\phi,\psi) =$

$$= \int_H\left\{S\left(\alpha\Delta_H^{1/p'}\right)\right\}(h)\Delta_H(h)^{-1/p'}\int_{G/H}\overline{\left(\lambda_H^{p'}(h)\left(\frac{\psi}{q^{1/p'}}\right)_{x,H}, \left(\frac{\phi}{q^{1/p}}\right)_{x,H}\right)}d_q\dot{x}dh =$$

$= \left(S\alpha^{1/p'}, \Delta_H^{1/p'}\, Res_H\omega_{\lambda_G^{p'}}(\psi,\phi)\right)$. Il suffit alors de choisir le support de α suffisamment petit pour pouvoir conclure.

Remarque. Il est intéressant de rapprocher cette proposition de la remarque du milieu de la page 165 de [10].

Proposition 8. Pour tout $T \in PM_p(G)$ et $u \in A_p(G)$ on a $uT = M_u(T)$ (nous adoptons les définitions de uT et $M_u(T)$ rappelées dans [6] en haut de la page 97).

Démonstration. Soient $\phi, \psi \in C_{oo}(G)$ et $u = \bar{k} * \check{\ell}$ avec $k \in L^p(G)$ $\ell \in L^{p'}(G)$. De $u\omega_{\lambda_G^{p'}}(\psi, \phi) = \int_G \overline{f(t)} * (g(t))\check{}dt$ où $f(t) = \tau_p\{_{t^{-1}}(\check{k})\phi\}$, $g(t) = \tau_{p'}\{_{t^{-1}}(\check{\ell})\psi\}$ et $\int_G \|f(t)\|_p\|g(t)\|_{p'}dt \leqq \|\phi\|_p\|\psi\|_{p'}\|k\|_p\|\ell\|_{p'}$ on tire $u\omega_{\lambda_G^{p'}}(\psi, \phi) \in A_p(G)$ $\|u\omega_{\lambda_G^{p'}}(\psi, \phi)\|_{A_p(G)} \leqq \|\phi\|_p\|\psi\|_{p'}\|k\|_p\|\ell\|_{p'}$ et $L_T\left(u\omega_{\lambda_G^{p'}}\psi, \phi)\right) =$
$= \int_G L_T(\overline{f(t)} * (g(t))\check{})dt = \int_G \overline{(T_{t^{-1}}(\check{k})\phi, {}_{t^{-1}}(\check{\ell})\psi)}dt = \overline{(M_u(T)\phi, \psi)}$. En utilisant le fait que $A_p(G)$ est complet, on obtient $L_T(u\omega_{\lambda_G^{p'}}(\psi, \phi)) = \overline{(M_u(T)\phi, \psi)}$ pour tout $u \in A_p(G)$. De $L_{uT}(\overline{\tau_p\phi} * (\tau_{p'}\psi)\check{}) = \overline{((uT)\phi, \psi)} = L_T(u\omega_{\lambda_G^{p'}}(\psi, \phi))$ on déduit finalement $M_u(T) = uT$.

Remarque. Cette assertion est mentionnée dans [6] (cf p. 97 ligne 9 à partir du haut).

Grâce à la proposition 6, pour tout $u \in A_p(G)$ et $S \in PM_p(H)$ on a $ui(S) = i(Res_H uS)$. Nous allons étendre cette assertion à $S \in CV_p(H)$.

Proposition 9. Pour tout $u \in A_p(G)\, S \in CV_p(H)$ on a $M_u(i(S)) = i(M_{Res_H u}(S))$.

Démonstration. Soient $\phi, \psi \in C_{oo}(G)$ et $S \in CV_p(H)$.

(I) Pour tout $k, \ell \in C_{oo}(G)$ $\alpha \in C_{oo}(H)$ avec $\alpha \geqq 0$ $\int_H \alpha(h)dh = 1$ on a

$$|(M_{k,\ell} \circ i \circ S \circ \lambda_H^p(\alpha^*)\phi, \psi) - (M_{k,\ell} \circ i \circ S\phi, \psi)| \leqq$$
$$\leqq |||S|||_p\|\psi\|_{p'}\|\ell\|_{p'}\{\|\phi\|_p \sup_{h \in \mathrm{supp}\alpha} \|k - {}_{h^{-1}}k\|_p +$$
$$+ \|k\|_p \sup_{h \in \mathrm{supp}\alpha} \|\phi_{h^{-1}}\Delta_G(h^{-1})^{1/p} - \phi\|_p\}. \tag{4}$$

Vu le théorème 5 l'expression à estimer n'excède pas $|||S|||_p\|\ell\|_{p'}\|\psi\|_{p'}I$ où $I = \left(\int_G \|i(\lambda_H^p(\alpha^*))_{t^{-1}}(\check{k})\phi - {}_{t^{-1}}(\check{k})\phi\|_p^p dt\right)^{1/p}$. Or, $I \leqq \|\lambda_1\|_p + \|\lambda_2\|_p$ avec $\lambda_1(t) = \int_H \alpha(h)\|\,[\phi\, {}_{t^{-1}}(\check{k})]_{h^{-1}}\Delta_G(h^{-1})^{1/p} - [\phi\, {}_{t^{-1}}(\check{k})_h]_{h^{-1}}\Delta_G(h^{-1})^{1/p}\|_p dh$ et $\lambda_2(t) = \int_H \alpha(h)\|\,[\phi\, {}_{t^{-1}}(\check{k})_h]_{h^{-1}}\Delta_G(h^{-1})^{1/p} - \phi\, {}_{t^{-1}}(\check{k})\|_p dh$. Les inégalités

$$\|\lambda_1\|_p \leqq \|\phi\|_p \left(\int_H \alpha(h)\|k - {}_{h^{-1}}k\|_p^p dh\right)^{1/p} \quad \text{et}$$

$$\|\lambda_2\|_p \leqq \|k\|_p \left(\int_H \alpha(h)\|\phi_{h^{-1}}\Delta_G(h)^{-1/p} - \phi\|_p^p dh\right)^{1/p} \quad \text{donnent (I).}$$

(II) Soient $u \in A_p(G), \epsilon > 0$ et U un voisinage ouvert de e dans H. Il existe alors V voisinage ouvert de e dans H avec $V \subset U$ et $|M_u o i o S o \lambda_H^p(\alpha^*)\phi, \psi) - (M_u o i o S\phi, \psi)| < \epsilon$ pour tout $\alpha \in C_{oo}(H)$ avec $\alpha \geqq 0$ $\int_H \alpha(h)dh = 1$ et $\operatorname{supp} \alpha \subset V$.

Il existe $(k_n)_{n=1}^{\infty}$ $(\ell_n)_{n=1}^{\infty}$, suites de $C_{oo}(G)$, avec $\sum_{n=1}^{\infty} \|k_n\|_p \|\ell_n\|_{p'} < \infty$ et $u = \sum_{n=1}^{\infty} \bar{k}_n * \check{\ell}_n$. Il existe N entier positif avec $\sum_{n=1+N}^{\infty} \|k_n\|_p\|\ell_n\|_{p'} < \dfrac{\epsilon}{4(1 + |||S|||_p\|\phi\|_p\|\psi\|_{p'})}$. Compte tenu de (I), il suffit de choisir V, voisinage ouvert de e dans H, avec $V \subset U$ et $\|k_n - {}_{h^{-1}}k_n\|_p < \dfrac{\epsilon}{2^{n+2}(1 + \|\phi\|_p\|\psi\|_{p'}\|\ell_n\|_{p'}|||S|||_p)}$ $\|\phi - \phi_{h^{-1}}\Delta_G(h^{-1})^{1/p}\|_p < \dfrac{\epsilon}{2^{n+2}(1 + |||S|||_p\|\psi\|_{p'}\|\ell_n\|_{p'}\|k_n\|_p)}$ pour tout $h \in V$ et $1 \leqq n \leqq N$.

(III) Pour tout $k, \ell, \alpha \in C_{oo}(H)$ avec $\alpha \geqq 0$ $\displaystyle\int_H \alpha(h)dh = 1$,
$|(i \circ M_{k,l} \circ S \circ \lambda_H^p(\alpha^*)\phi, \psi) - (i \circ M_{k,l} \circ S\phi, \psi)|$ se majore comme le membre de gauche de (4).

Posons $A = M_{k,\ell}(S - S\lambda_H^p(\alpha^*))$, $a = \left(\dfrac{\phi}{q^{1/p}}\right)_{y,H}$ et $b = \left(\dfrac{\psi}{q^{1/p'}}\right)_{y,H}$ pour tout $y \in G$. On a $|(i(A)\phi, \psi)| = \left|\displaystyle\int_{G/H} (Aa, b)d\dot{y}\right|$. Vu [5] p.5, proposition 1.(b) pour tout $y \in G$

$$|(Aa, b)| \leqq |||S|||_p\|b\|_{p'}\|\ell\|_{p'}\left\{\|a\|_p \left(\int_H \alpha(h)\|k - {}_{h^{-1}}k\|_p^p dh\right)^{1/p} + \right.$$

$$\left. + \|k\|_p \left(\int_H \alpha(h)\|a - a_{h^{-1}}\Delta_H(h^{-1})^{1/p}\|_p^p dh\right)^{1/p}\right\}.$$ En tenant compte de

$\displaystyle\int_{G/H}\int_H \alpha(h)\|a - a_{h^{-1}}\Delta_H(h^{-1})^{1/p}\|_p^p dh d\dot{y} = \int_H \alpha(h)\|\phi_{h^{-1}}\Delta_G(h^{-1})^{1/p} - \phi\|_p^p dh$ on obtient (III).

(IV) Soient $u \in A_p(H)$, $\epsilon > 0$ et U un voisinage ouvert de e dans H. Il existe V voisinage ouvert de e dans H tel que pour tout $\alpha \in C_{oo}(H)$ avec $\alpha \geqq 0$ $\int_H \alpha(h)dh = 1$ $\mathrm{supp}\alpha \subset V$ on ait $|(i \circ M_u \circ S \circ \lambda_H^p(\alpha^*)\phi, \psi) - (i \circ M_u \circ S\phi, \psi)| < \epsilon$.

On déduit (IV) de (III) en procédant comme pour la preuve de (II).

(V) Pour tout $u \in A_p(G)$, $\alpha \in C_{oo}(H)$ on a $(M_u \circ i \circ S \circ \lambda_H^p(\bar{\alpha}^*)\phi, \psi) =$
$= \left(S\alpha\Delta_H^{1/p'}, \Delta_H^{-1/p'} Res_H(u\omega_{\lambda_G^{p'}}(\psi, \phi))\right)$.

On se ramène au cas de $u = \bar{k} * \check{\ell}$ avec $k, \ell \in C_{oo}(G)$. Fixons $t \in G$ et posons $T = S\lambda_H^p(\bar{\alpha}^*)a = {}_{t^{-1}}(\check{k})\phi$ $b = {}_{t^{-1}}(\check{\ell})\psi$. On a $(i(T)a, b) = \int_H d(h)\left(\int_{G/H} \overline{c^* * f(h)}d_q\dot{x}\right)dh$

où $c = \left(\frac{a}{q^{1/p}}\right)_{x,H}$ $d = S(\Delta_H^{1/p'}\alpha)$ $f = \left(\frac{b}{q^{1/p'}}\right)_{x,H}$. Par suite $\int_G (i(T)a, b)dt =$
$= \int_H d(h)(\Delta_H(h^{-1})\Delta_G(h))^{1/p'}\overline{\left(\int_G a^* * b(h)dt\right)}dh$ d'où, en tenant compte de

$\int_G a^* * b(h)dt = \Delta_G(h^{-1})^{1/p'}(\lambda_G^{p'}(h)\phi, \psi)(\bar{k} * \check{\ell})(h)$, la relation (V).

(VI) Montrons la proposition. Supposons tout d'abord $u \in A_p(G) \cap C_{oo}(G)$. Vu (V)

$(M_u(i(S\lambda_H^p(\bar{\alpha}^*)))\phi, \psi) = \left(S\alpha\Delta_H^{1/p'}, \Delta_H^{-1/p'} Res_H\left(u\omega_{\lambda_G^{p'}}(\psi, \phi)\right)\right)$. Posons $v = Res_H u$.
Vu (V), appliqué au cas $H = G$, on a $(i\,(M_v\,(S\lambda_H^p(\bar{\alpha}^*)))\,\phi, \psi) =$

$$= \int_{G/H} \left(S\alpha\Delta_H^{1/p'}, \Delta_H^{-1/p'}\omega_{\lambda_H^{p'}}\left(\left(\frac{\psi}{q^{1/p'}}\right)_{x,H}, \left(\frac{\phi}{q^{1/p}}\right)_{x,H}\right)v\right)d_q\dot{x} =$$

$= \left(S\alpha\Delta_H^{1/p'}, \Delta_H^{-1/p'}\left(Res_H\left(\omega_{\lambda_G^{p'}}(\psi, \phi)\right)\right)v\right)$. L'assertion résulte alors de (II) et (IV).

Soit $u \in A_p(G)$ quelconque. Soit $\epsilon > 0$. Il existe $w \in A_p(G) \cap C_{oo}(G)$ avec $\| u - w \|_{A_p(G)} < \frac{\epsilon}{2(1 + |||S|||_p)}$. On obtient finalement

$$|||M_u(i(S)) - i\,(M_{Res_H u}(S))\,|||_p \leqq |||M_u(i(S)) - M_w(i(S))|||_p +$$

$$+ |||i\,(M_{Res_H w}(S)) - i\,(M_{Res_H u}(S))\,|||_p \leqq 2\,\| u - w \|_{A_p(G)}\,|||S|||_p.$$

Cette proposition nous permet de compléter la preuve du point (4) du théorème suivant (cf [7] théorème 3.).

Théorème 10. Soient G un groupe localement compact, H un sous-groupe fermé de G normal dans G et $p > 1$. Il existe alors R application linéaire de $CV_p(G)$ *sur* $CV_p(H)$ satisfaisant aux propriétés suivantes :

(1) $|||R(T)|||_p \leqq |||T|||_p$ pour tout $T \in CV_p(G)$,
(2) $R(i(S)) = S$ pour tout $S \in CV_p(H)$,
(3) $R(PM_p(G)) = PM_p(H)$.
(4) $R(M_u(T)) = M_{Res_H u}(R(T))$ pour tout $u \in A_p(G)$ et $T \in CV_p(G)$.
(5) $R(cv_p(G)) = cv_p(H)$.

Posons $W_H = \{T \mid T \in CV_p(G) \ \ \mathrm{supp} T \subset H\}$. On sait que i est une bijection de $CV_p(H)$ sur W_H. Vu la proposition 9, $i^{-1}(M_u(T)) = M_{Res_H u}(i^{-1}(T))$. Par ailleurs, soit P le projecteur de $CV_p(G)$ sur W_H *du* Théorème 2 de [7]. Alors $R = i^{-1} \circ P$ satisfait bien à la condition (4) de l'énoncé du théorème 10 ci-dessus.

5. Retour au cas des groupes abéliens.

Supposons G abélien et choisissons $q = 1$. Posons $H^\perp = \{\chi | \chi \in \hat{G}, \chi(h) = 1$ pour tout $h \in H\}$ et $\gamma(\chi) = \dot{\chi} = \chi H^\perp$ pour tout $\chi \in \hat{G}$. Contrairement à l'usage, introduisons les applications suivantes : $\tau(\gamma(\chi)) = Res_H \chi$, $\sigma(h)(\gamma(\chi)) = \chi(h)$ pour tout $\chi \in \hat{G}$ $h \in H$, $\rho(\chi')(\omega(x)) = \chi'(x)$, $\epsilon(x)(\chi) = \chi(x)$ pour tout $\chi' \in H^\perp$ $\chi \in \hat{G}$ $x \in G$. Ainsi τ, σ, ρ et ϵ sont des isomorphismes bicontinus de $\hat{G}/H^\perp$ sur $\hat{H}$, de H sur $(\hat{G}/H^\perp)\hat{}$, de $H^\perp$ sur $(G/H)\hat{}$ et de G sur $\hat{\hat{G}}$.

Soient $d\chi$ la mesure sur $\hat{G}$ duale de dx, $d\chi''$ celle sur $(G/H)\hat{}$ duale de $d\dot{x}$ et $d\chi'$ la mesure sur $H^\perp$ image via ρ^{-1} de $d\chi''$. Soit $d\dot{\chi}$ l'unique mesure sur $\hat{G}/H^\perp$ avec $d\chi = d\dot{\chi} d\chi'$, alors l'image par τ de $d\dot{\chi}$ est la mesure duale $d\nu$ de dh (cf [8] p. 244 (31.46) (c)).

Proposition 11. Avec toutes les notations qui précèdent, les assertions suivantes sont vérifiées :

1) pour tout $T \in CV_2(G)$ $u \in A_2(G)$ $g \in L^1(\hat{G})$ avec $\hat{g} \circ \epsilon = u$ on a $(uT)\hat{} = = \bar{g} * \hat{T}$.
2) Pour tout $T \in CV_p(H)$ on a $i(T)\hat{} = \hat{T} \circ \tau \circ \gamma$.
3) Pour tout $u \in A_2(G)$ $T \in CV_2(H)$ $g \in L^1(\hat{G})$ avec $\hat{g} \circ \epsilon = u$ on a $(u(i(T)))\hat{} = = \bar{g} *_{\hat{G}} (\hat{T} \circ \tau \circ \gamma)$ et $(i(Res_H uT))\hat{} = \Big((T_{H^\perp}\bar{g}) *_{\hat{G}/H^\perp} \hat{T} \circ \tau\Big) \circ \gamma$.

Démonstration.

(I) Preuve de 1).

Soient $\phi, \psi \in C_{oo}(G)$ et $k, \ell \in L^2(G)$ avec $u = \bar{k} * \check{\ell}$. Compte tenu de la proposition 8, on a $((uT)\phi, \psi) = \int_G (\hat{\tau}, (\tilde{\alpha} * \beta)\hat{}\,)\, dt$ où $\alpha = {}_{t^{-1}}(\check{k})\phi$ et $\beta = {}_{t^{-1}}(\check{\ell})\psi$.

Or pour tout $x \in G$ $\int_G (\tilde{\alpha} * \beta)(x) dt = \int_G \overline{\phi(z)}\psi(zx) \left(\int_G \overline{k(z^{-1}t)}\ell(x^{-1}z^{-1}t)dt \right) dz =$ $(\bar{k} * \check{\ell})(x)(\tilde{\phi} * \psi)(x)$ donc $((uT)\phi, \psi) = \int_{\hat{G}} \hat{T}(\chi)\overline{(\bar{k} * \check{\ell}\, \tilde{\phi} * \psi)\hat{}(\chi)} d\chi = \left(((k * \tilde{\ell})\hat{} * \hat{T})\hat{\phi}, \hat{\psi} \right)$; or $(k * \tilde{\ell})\hat{} = \bar{g}$, d'où $(uT)\hat{} = \bar{g} * \hat{T}$.

(II) Preuve de 2).

Soient de nouveau $\phi, \psi \in C_{oo}(G)$. On a $(i(T)\phi, \psi) = \int_{G/H} \left(\hat{T}(\phi_{x,H})\hat{}, (\psi_{x,H})\hat{} \right) d\dot{x} =$

$= \int_{\hat{H}} \hat{T}(\nu) \left(\int_{G/H} (\phi_{x,H})\hat{}(\nu)\overline{(\psi_{x,H})\hat{}(\nu)} d\dot{x} \right) d\nu$. Or pour tout $\nu \in \hat{H}$

$$\int_{G/H} (\phi_{x,H})\hat{}(\nu)\overline{(\psi_{x,H})\hat{}(\nu)} d\dot{x} = \int_{G/H} \left\{ \int_H \phi_{x,H}(h)\overline{\nu(h)} \left(\int_H \overline{\psi_{x,H}(h'h)}\nu(h'h)dh' \right) dh \right\} d\dot{x} =$$

$$= \int_H \nu(h') \left\{ \int_{G/H} \left(\int_H \phi_{x,H}(h)\overline{\psi_{x,H}(h'h)}dh \right) d\dot{x} \right\} dh' = \int_H \overline{\nu(h')}\phi * \tilde{\psi}(h')dh' =$$

$= (Res_H(\phi * \tilde{\psi}))\hat{}\,(\nu)$.

Par suite $(i(T)\phi, \psi) = \int_{\hat{G}/H^\perp} \hat{T}(\tau(\dot{\chi}))(Res_H(\phi * \tilde{\psi}))\hat{}(\tau(\dot{\chi}))d\dot{\chi}$. Pour tout $\chi \in \hat{G}$ $(Res_H(\phi * \tilde{\psi}))\hat{}(\tau(\gamma(\chi))) = \int_H (\phi * \tilde{\psi})(h)\overline{\tau(\gamma(\chi))(h)}dh = \int_H (\phi * \tilde{\psi})(h)\overline{\chi(h)}dh$. Grâce à la *formule de Poisson* $\int_H (\phi * \tilde{\psi})(h)\overline{\chi(h)}dh = \int_{H^\perp} (\phi * \tilde{\psi})\hat{}(\chi\chi')d\chi'$, c'est-à dire $(Res_H(\phi * \tilde{\psi}))\hat{} \circ \tau = T_{H^\perp}\left((\phi * \tilde{\psi})\hat{} \right)$. On aboutit ainsi à $(i(T)\phi, \psi) =$

$$= \int_{\hat{G}/H^\perp} \hat{T}(\tau(\dot{\chi})) \left\{ T_{H^\perp}(\phi * \tilde{\psi})\hat{} \right\}(\dot{\chi})d\dot{\chi} = \int_{\hat{G}} (\hat{T} \circ \tau \circ \gamma)(\chi)(\phi * \tilde{\psi})\hat{}(\chi)d\chi =$$

$$= \int_{\hat{G}} (\hat{T} \circ \tau \circ \gamma)(\chi)\hat{\phi}(\chi)\overline{\hat{\psi}(\chi)}d\chi.$$

(III) Preuve de 3).

En appliquant successivement 1) et 2), on obtient bien $(u(i(T)))\hat{} = \bar{g} * (i(T))\hat{} =$ $= \bar{g} * (\hat{T} \circ \tau \circ \gamma)$.

Pour tout $h \in H$ $\hat{g}(\epsilon(h)) = (T_{H^\perp}g)\hat{}(\sigma(h))$. Posons $f = (T_{H^\perp}g) \circ \tau^{-1}$ on a $f \in L^1(\hat{H})$. Soit δ l'application de H sur $\hat{\hat{H}}$ définie par $\delta(h)(\nu) = \nu(h)$ pour tout $\nu \in \hat{H}$. On a $\hat{f} \circ \delta = Res_H u$. En effet $\hat{f}(\delta(h)) = \int_{\hat{G}/H^\perp} f(\tau(\dot{\chi}))\overline{\delta(h)(\tau(\dot{\chi}))}d\dot{\chi} =$

$$= \int_{\hat{G}/H^\perp} f(\tau(\dot{\chi}))\overline{\sigma(h)(\dot{\chi})}d\dot{\chi} = \int_{\hat{G}/H^\perp} (T_{H^\perp}g)(\dot{\chi})\overline{\sigma(h)(\dot{\chi})}d\dot{\chi} = (T_{H^\perp}g)\hat{}(\sigma(h)) = u(h).$$

Vu 1) $(Res_H uT)\hat{} = \bar{f} *_{\hat{H}} \hat{T}$, ainsi $\{i((Res_H u)T)\}\hat{} = (\bar{f} *_{\hat{H}} \hat{T}) \circ \tau \circ \gamma =$

$$= \left(\overline{(f \circ \tau)} *_{\hat{G}/H^\perp} (\hat{T} \circ \tau)\right) \circ \gamma = \left((T_{H^\perp}\bar{g}) *_{\hat{G}/H^\perp} \hat{T} \circ \tau\right) \circ \gamma.$$

Remarques.

1) L'assertion 1) est déjà mentionnée dans [5] (p. 7) et [6] (pages 97 et 100).

2) La relation 2) était une des motivations d'une partie de [4] ! Elle permet, en choisissant $G = \mathbb{R}$ et $H = \mathbb{Z}$, de retrouver le Théorème de de Leeuw auquel il est fait allusion dans l'introduction ([9] p. 377, Theorem 4.5.). Rappelons que pour G abélien le résultat correspondant est dû à Saeki ([11] p. 411 Lemma 3.1.). On pourra aussi consulter [1] p. 105.

3) Pour $p = 2$ et G abélien, l'assertion 3) fournit une vérification directe de la proposition 9.

Bibliographie.

[1] Anker, J.-Ph., Aspects de la p-induction en analyse harmonique, Thèse de doctorat, Payot Lausanne (1982).

[2] Anker, J.-Ph., Applications de la p-induction en analyse harmonique, Comment. Math. Helvetici, 58 (1983) p. 622-645.

[3] Bourbaki, N., Eléments de mathématiques, Livre VI, Intégration, chapitres I à IV, deuxième édition revue et augmentée (1965) Hermann, Paris.

[4] Derighetti, A., Relations entre les convoluteurs d'un groupe localement compact et ceux d'un sous-groupe fermé, Bull. Sc. Math., 2ème série, 106 (1982) p. 69-84.

[5] Derighetti, A., A propos des convoluteurs d'un groupe quotient, Bull. Sc. Math., 2ème série, 107 (1983) p. 3-23.

[6] Derighetti, A., Quelques observations concernant les ensembles de Ditkin d'un groupe localement compact, Mh. Math. 101 (1986) p. 95-113.

[7] Derighetti, A., Convoluteurs et projecteurs, (à paraître).

[8] Hewitt, E. and Ross, K.A., Abstract Harmonic Analysis, vol. II, Springer-Verlag, Berlin;Heidelberg;
New York (1970).

[9] De Leeuw, K., On L_p Multipliers, Annals of Math. 81 (1965) p. 364-379.

[10] Reiter, H., Classical Harmonic Analysis and Locally Compact Groups, Clarendon Press, Oxford (1968).

[11] Saeki, S., Translation invariant operators on groups, Tôhoku Math. J. 22 (1970) p. 409-419.

CHARACTERIZATION OF THE TYPE OF SOME GENERALIZATIONS OF THE CAUCHY DISTRIBUTION

Jean-Louis DUNAU and Henri SENATEUR

INSA TOULOUSE Dépt. de Mathématiques
Laboratoire de Statistique et Probabilités UA CNRS 745
Université Paul Sabatier
TOULOUSE FRANCE

Abstract Let G be a closed connected semisimple subgroup of $SL(n,\mathbb{R})$, and $B = G/P$ where P is a parabolic subgroup of G. We define and characterize the type of a certain probability measure on B. The particular case where $G = SL(n,\mathbb{R})$ furnishes a characterization of various generalizations of the Cauchy distribution.

1. INTRODUCTION

An elementary property of the Cauchy distribution on $\mathbb{R}$, $\gamma(dx) = dx/\pi(1+x^2)$, is the following : let G be the group of homographies of $\mathbb{R}$: $x \to (ax+b)/(cx+d)$, where $ad-bc \neq 0$ and let H be the subgroup of affinities : $x \to ax+b$, where $a \neq 0$; if μ is a probability measure on $\mathbb{R}$, let us define the type of μ as the set $H\mu$ of probabilities on $\mathbb{R}$ which are images of μ by the affinities ; thus, the type of γ (the Cauchy type) is the set of probabilities $\gamma_{b,a}(dx) = a\, dx/\pi(a^2+(x+b)^2)$, where $a > 0$; then the Cauchy type is invariant by homography, that is $G\gamma=H\gamma$.

Conversely, Knight [9] proved that if μ is a probability with no atoms such that $G\mu = H\mu$, then μ is in the Cauchy type.

Knight and Meyer [10] have given a generalization to $\mathbb{R}^n$ of this result : a characterization of the type of the usual Cauchy measure on $\mathbb{R}^n$:

$$\frac{\Gamma((n+1)/2)dx_1\ldots dx_n}{\pi^{(n+1)/2}(1+\|x\|^2)^{(n+1)/2}} .$$

Other generalizations are possible. For instance, let $G = Mob(n)$ the conformal group of $\mathbb{R}^n$ (i.e. the group generated by inversions and symmetries of $\mathbb{R}^n$) and H the subgroup of similarities-translations, then we obtain [4] a characterization of the type of

$$\frac{2^{n-1}\Gamma((n+1)/2)dx_1\ldots dx_n}{\pi^{(n+1)/2}(1+\|x\|^2)^n} .$$

Another example is the characterization of the type of the Cauchy-Hua measure on real symmetric matrices (then G is the symplectic group, see [5]). Note that, in every case, to obtain the characterization theorem, we impose an additional condition which μ needs to satisfy (corresponding to the condition "μ is a probability with no atoms" in the n=1 case).

The aim of the present paper is to describe a more general framework which includes the above examples. We will consider a closed connected semisimple subgroup G of $SL(n,\mathbb{R})$ and B=G/P where P is a parabolic subgroup of G. For any probability measure μ on B, we will define the H-type of μ as the set $H\mu$ of the probabilities on B which are images of μ by a suitable subgroup H of G. We will define the Cauchy measure on B as the unique probability on B invariant by a compact subgroup of G. Theorem 1 (section 3) and Theorem 2 (section 4) are our main results and give the following characterization of the H-type of Cauchy measure : if μ is a probability on B (satisfying an additional condition) such that $G\mu=H\mu$, then μ is in the H-type of the Cauchy measure. Section 3 is devoted to the particular case of a minimal parabolic subgroup. The general case is in section 4. In section 5, as an illustration, we will consider $G=SL(n,\mathbb{R})$; in this case B is the set of flags on $\mathbb{R}^n$, and so we obtain a characterization of the type of the Cauchy distribution on the flags.

The authors are grateful to Professors Y. Guivarc'h and A. Raugi for some assistance and fruitful conversations concerning this paper. Their paper [7] is our most important reference.

2. PRELIMINARIES AND NOTATIONS

The background is the same as exposed in Guivarc'h et Raugi [7] ; for more details, see Helgason [8] ; we retain the notations of [7].

Let G be a closed connected subgroup of the special linear group $SL(n,\mathbb{R})$. If $M_n(\mathbb{R})$ denotes the set of nxn real matrices, then

$$\mathfrak{g} = \{X \in M_n(\mathbb{R}) \ ; \ \forall t \in \mathbb{R}, \exp(tX) \in G\}$$

is the Lie algebra of G. For any X in $\mathfrak{g}$, the linear transformation of $\mathfrak{g}$: $Y \to [X,Y] = XY - YX$ is denoted by adX. The linear mapping $X \to \mathrm{ad}\, X$, denoted by ad, is the adjoint representation of $\mathfrak{g}$. The Killing form is the bilinear form on $\mathfrak{g}$: $(X,Y) \to \mathrm{Tr}\,(\mathrm{ad}\, X . \mathrm{ad}\, Y)$ (where Tr denotes the trace of an endomorphism). Throughout the paper, we will suppose that G is *semisimple* (i.e. the Killing form is nondegenerate).

Consider a Cartan decomposition $\mathfrak{g} = \mathcal{K} \oplus \mathcal{P}$ of $\mathfrak{g}$ ($\mathcal{K}$ is a subalgebra, $\mathcal{P}$ a subspace, $[\mathcal{K},\mathcal{K}] \subset \mathcal{K}$, $[\mathcal{K},\mathcal{P}] \subset \mathcal{P}$ $[\mathcal{P},\mathcal{P}] \subset \mathcal{K}$, the restriction of the Killing form to $\mathcal{K}$ (resp. to $\mathcal{P}$) is negative definite (resp. positive definite), $\mathcal{K}$ and $\mathcal{P}$ are orthogonal with respect to the Killing form). Note that, if for any X in $\mathfrak{g}$ ${}^tX$ is also in $\mathfrak{g}$, then the decomposition of a matrix as a sum of a skew-symmetric and a symmetric matrix is a Cartan decomposition.

Let $\mathcal{A}$ be a maximal abelian (i.e. $[X,Y] = 0$ for any X and Y in $\mathcal{A}$) subalgebra of $\mathcal{P}$. If α is a linear form on $\mathcal{A}$, let us consider

$$\mathfrak{g}_\alpha = \{X \in \mathfrak{g} \ ; \ \forall H \in \mathcal{A}, \mathrm{ad}\, H(X) = \alpha(H)X\} ;$$

then α is a *root* if $\mathfrak{g}_\alpha \neq \{0\}$. Let Δ be the set of roots ; then $\mathfrak{g} = \bigoplus_{\alpha\in\Delta} \mathfrak{g}_\alpha$. Note that 0 is a root and $\mathfrak{g}_0 = \mathcal{M} \oplus \mathcal{A}$, where

$$\mathcal{M} = \{X \in \mathcal{K} ; \ \forall H \in \mathcal{A} \quad HX = XH\}.$$

Consider $\mathcal{A}' = \{H \in \mathcal{A} ; \ \forall \alpha \in \Delta\setminus\{0\} \ \alpha(H) \neq 0\}$. The *Weyl chambers* are the connected components of $\mathcal{A}'$. Let us choose a Weyl chamber W_0 and introduce :

$$\Delta_+ = \{\alpha \in \Delta \setminus \{0\} \ ; \ \forall H \in W_0 \quad \alpha(H) > 0\}$$

$$\Delta_- = \{\alpha \in \Delta \setminus \{0\} \ ; \ \forall H \in W_0 \quad \alpha(H) < 0\}.$$

A root α in Δ_- is simple if α is not the sum of two roots in Δ_- ; let Σ be the set of the simple roots in Δ_- ; any α in Δ_- is a linear combination of simple roots, with positive integer coefficients.

Consider $\mathcal{N} = \bigoplus_{\alpha\in\Delta_-} \mathfrak{g}_\alpha$ and $\tilde{\mathcal{N}} = \bigoplus_{\alpha\in\Delta_+} \mathfrak{g}_\alpha$. Then :

$$\mathfrak{g} = \mathcal{N} \oplus \mathcal{A} \oplus \mathcal{K} = \tilde{\mathcal{N}} \oplus \mathcal{A} \oplus \mathcal{K} \qquad \text{(Iwasawa decompositions)}$$

and

$$\mathfrak{g} = \mathcal{N} \oplus \mathcal{A} \oplus \tilde{\mathcal{N}} \oplus \mathcal{M} \qquad \text{(Bruhat decomposition)}.$$

Let N, $\tilde{N}$, A, K be the connected Lie subgroups of G with respective Lie Algebras $\mathcal{N}$, $\tilde{\mathcal{N}}$, $\mathcal{A}$, $\mathcal{K}$. Then N and $\tilde{N}$ are nilpotent subgroups, A is abelian, K is maximal compact,

$$G = NAK = KAN = \tilde{N}AK = KA\tilde{N}$$

are Iwasawa decompositions of G, and

$$G = K \,.\, \exp \overline{W_0} \,.\, K$$

(where $\overline{W}_0$ is the closure of W_0 in $\mathcal{A}$) is a polar decomposition of G. For any h in G, Ad h is the mapping $\mathfrak{g} \to \mathfrak{g}$ defined by : $\mathrm{Ad}\, h(X) = h X h^{-1}$; the mapping $h \to \mathrm{Ad}\, h$, denoted by Ad, is the adjoint representation of G. Then, for any X in $\mathfrak{g}$, $\mathrm{Ad}(\exp X) = \exp(\mathrm{ad}\, X)$. Thus, if α is a root, there exists a homomorphism $\psi_\alpha : A \to]0,+\infty[$ such that, for any a in A and any X in $\mathfrak{g}_\alpha$,

$$a X a^{-1} = \psi_\alpha(a) X$$

(indeed, consider H in $\mathcal{A}$ such that $a = \exp H$; then $\mathrm{ad}\, H(X) = \alpha(H) X$ and $a X a^{-1} = \mathrm{Ad}\, a\,(X) = (\mathrm{Ad}\,(\exp H))\,(X) = (\exp\,(\mathrm{ad}\, H))(X) = (\exp \alpha\,(H))\,(X) = \psi_\alpha(a)\, X)$.
The dual of $\mathcal{A}$ is generated by Σ, so a in A is determined by the $\psi_\alpha(a)$ $(\alpha\in\Sigma)$

Let $M = \{m \in K \ ; \ \forall a \in A \ ma=am\}$ be the centralizer of A in K and let $M' = \{m \in K \ ; \ m A m^{-1} = A\}$ be the normalizer of A in K. Then $\mathcal{M}$ is the Lie algebra of both of the closed subgroups M and M'. The finite group M'/M is called the *Weyl group* ; any element of M' permutes the Weyl chambers ; if an element of M'/M leaves a Weyl chamber invariant, then this element is the identity ; thus the Weyl group operates simply and transitively on the Weyl chambers.

The partition of G :

$$G = \bigcup_{m \in M'/M} N\, m\, \tilde{N}\, A\, M$$

is the Bruhat decomposition of G ; $N \tilde{N} A M$ is an open submanifold of G ; the other submanifolds $N\, m\, \tilde{N}\, A\, M$ are of lower dimension. The mapping $(u,\tilde{u},a,w) \to u\tilde{u}aw$ is an isomorphism of $N \times \tilde{N} \times A \times M$ onto $N\tilde{N}AM$.

Now we introduce $B = G/\tilde{N}\, A\, M$ and we will consider probability measures on B.

3. THE CAUCHY TYPE ON $B = G/\tilde{N}AM$

Since G operates transitively on B, for any measure μ on B, we can define the image measure $f\mu$ of μ by an element f in G. For any closed subgroup H of G such that $AN \subset H \subset MAN$ we define the H-type of μ as the set of images of μ by the elements of H.

If g is in G, $\bar{g}$ denotes the corresponding coset of $B=G/\tilde{N}\, A\, M$. Note that in each coset of B there is at most one element of N. Defining $B_N = \{\bar{u} \in B : u \in N\}$, we will identify N and B_N and thus imbed N in B. We will regard a probability measure on N (or on B_N) as a probability measure on B such that $\mu(B \setminus B_N) = 0$. Then, for any f in G, we can define the image measure $f\mu$ of μ by f ; if, in addition, $\mu(f^{-1}(B\setminus B_N)) = 0$, $f\mu$ is also a probability measure on N. Note that B_N is the part $N\tilde{N}AM$ of the Bruhat decomposition of G.

The compact subgroup K operates transitively on the homogeneous space $B = KA\tilde{N}/MA\tilde{N}$, and so there exists on B a unique probability measure which is invariant by K ; we will call this probability measure on B the Cauchy measure on B. Note that $G\gamma = NAK\gamma = NA\gamma \subset H\gamma$; hence the H-type of γ (the Cauchy H-type) is invariant by G.

Proposition 1 : Let μ be a probability measure on B such that, for any f in G, $\mu(f^{-1}(B\setminus B_N)) = 0$. Then the set $X_\mu = \{f \in G ; f\mu=\mu\}$ is compact.

Proof Consider a sequence $(g_n)_n$ in X_μ. Using polar decomposition, we may write $g_n = h_n a_n k_n$, where h_n and k_n are in K and a_n is in $\exp \overline{W_0}$. Then, for any α in Δ_-, $\psi_\alpha(a_n)$ is in $]0,1]$.

By the compactness of K and $[0,1]$, there exists a subsequence $\phi(n)$ such that, as $n \to \infty$, $(h_{\phi(n)})_n$ converges to h in K, $(k_{\phi(n)})_n$ converges to k in K and $(\psi_\alpha(a_{\phi(n)}))_n$ converges to λ_α in $[0,1]$, for any α in Δ_-. If $u = \exp \sum_{\alpha \in \Delta_-} X_\alpha$ is in N (X_α is in $\mathfrak{g}_\alpha$), we consider $\tau(\bar{u}) = \overline{\exp \sum_{\alpha \in \Delta_-} \lambda_\alpha X_\alpha}$. Thus we define a mapping $\tau : B_N \to B_N$. As $n \to \infty$, $a_{\phi(n)}$ converges to τ on B_N.

Now, we will prove that $\lambda_\alpha > 0$ for any α in Σ. Suppose there exists α_1 in Σ such that $\lambda_{\alpha_1} = 0$.

Consider the Weyl chamber W_1 such that

$$\forall H \in W_1 \begin{cases} \alpha_1(H) > 0 \\ \alpha(H) < 0 & \text{if } \alpha \in \Sigma,\ \alpha \neq \alpha_1 \end{cases}$$

and let m_1 be in M' such that its corresponding element in the Weyl group M'/M sends W_0 to W_1.

Then for any u in N, $m_1\tau(\bar{u})m_1^{-1} = \tau(\bar{u})$ and $m_1\tau(\bar{u})$ is not in B_N (by Bruhat decomposition, $N\tilde{m}_1$ and $NA\tilde{N}M$ are disjoints). Hence $m_1 \ \tau(B_N) \cap B_N \neq \emptyset$, $\tau(B_N) \subset m_1^{-1}(B\backslash B_N)$ and

$$h^{-1} \ \mu(\tau(B_N)) = 0$$

On the other hand, consider the function $h\tau k$ which is defined on $k^{-1}(B_N)$; by hypothesis $\mu(B\backslash k^{-1}(B_N)) = 0$ and so the image measure $h\tau k\mu$ is defined. The subsequence $(g_{\phi(n)})_n$ converges to $h\tau k$ μ-almost everywhere, hence $g_{\phi(n)}\mu$ converges weakly to $h\tau k\,\mu$ and $\mu = h\tau k\ \mu$. It follows that

$$h^{-1}\mu(\tau(B_N)) = 1$$

and so we obtain a contradiction.

Since $\lambda_\alpha > 0$ for any α in Σ, then $\lambda_\alpha > 0$ for any α in Δ_-, and there exists a in A such that $\psi_\alpha(a) = \lambda_\alpha$ and $a(\bar{u}) = \tau(\bar{u})$ for any u in N. Then the subsequence $(g_{\phi(n)})_n$ converges to hak as $n\to\infty$, $g_{\phi(n)}\mu$ converges weakly to $hak\mu$, and hak is in X_μ. This completes the proof.

<u>Théorem 1</u> Let μ be a probability measure on B_N (or on N) such that $\mu\ (f^{-1}(B\backslash B_N)) = 0$ for any f in G. Let H be a closed subgroup of G such that $AN \subset H \subset MAN$.
If for any g in G, $g\mu$ is in the H-type of μ (i.e. $G\mu = H\mu$) then, μ is in the Cauchy H-type (i.e. $\mu \in H\gamma$).

<u>Proof</u>. By Proposition 1, X_μ is a compact subgroup of G. Then, there exists g in G such that $X_\mu \subset g^{-1} K g$. Consider the probability measure $\nu = g\mu$. Then $\nu \in H\mu$, $G\nu = H\nu$, $X_\nu \subset K$. Then $K = X_\nu.(K\cap H)$ (indeed, for k in K, there exists h in H such that $k\nu=h\nu$ and $k = h\, h^{-1}k$ is in $(K\cap H)\ X_\nu$).

Consider m_s in M' such that its corresponding element in the Weyl group M'/M sends W_0 to $-W_0$. Then $(m_s^{-1} N\, m_s) \cap N = \{e\}$, $N\, m_s\, A\tilde{N}M = m_s\, A\tilde{N}M$ and $H\bar{m}_s = \bar{m}_s$.

Then $B = G\,\bar{m}_s = KAN\,\bar{m}_s = K\,\bar{m}_s = X_\nu\ (K \cap H)\ \bar{m}_s = X_\nu\,\bar{m}_s$. Thus the compact group X_ν operates transitively on B, ν and γ are two probability measures on B, invariant by X_ν. Hence $\nu=\gamma$ and μ is in the H-type of γ. This completes the proof.

4. THE GENERALIZED CAUCHY-TYPE

We will generalize Theorem 1 by replacing $MA\tilde{N}$ by a closed subgroup $\tilde{P}_\theta$ including $MA\tilde{N}$. Then we will define the type of a probability measure on $B_\theta = G/\tilde{P}_\theta$ and characterize the Cauchy type (the type of the unique K-invariant probability on B_θ).

We consider parabolic subgroups of G. More information on this subject can be found in Borel [1, section 11] ; see also Guivarc'h et Raugi [7], Dani and McCrudden [2].

Let θ be a subset of Σ (the set of simple roots in Δ_-) and define :

$$A_\theta = \{a \in A\ ;\ \forall\alpha \in \theta\quad \psi_\alpha(a) = 1\}.$$

Let $K_\theta = \{k \in K ; \forall a \in A_\theta \quad ka = ak\}$ be the centralizer of A_θ in K. Then $\tilde{P}_\theta = K_\theta A\tilde{N} = \tilde{N}AK_\theta$ is a standard parabolic subgroup of G (a parabolic subgroup is a closed subgroup of G including some subgroup $MA\tilde{N}$ or equivalently, a conjugate of a standard parabolic subgroup $\tilde{P}_\theta$, for some set $\theta \subset \Sigma$). Note that the mapping $\theta \to \tilde{P}_\theta$ is increasing. When $\theta=\Sigma$, we obtain $A_\Sigma=\emptyset$, $K_\Sigma=K$, $\tilde{P}_\Sigma=G$; when $\theta=\emptyset$, we get $A_\emptyset=A$, $K_\emptyset=M$, $\tilde{P}_\emptyset = MA\tilde{N}$: this is the case studied in section 3 ; $MA\tilde{N}$ is called a minimal parabolic subgroup.

Let $\langle\theta\rangle$ be the set of linear combinations of roots in θ, with positive integer coefficients. Then the Lie algebra of $\tilde{P}_\theta$ is $\bigoplus_{\alpha \in \langle\theta\rangle \cup \Delta_+} \mathfrak{g}_\alpha$. Moreover :

$$G = \bigcup_{m \in M'/M'_\theta} Nm\,\tilde{P}_\theta$$

where $M'_\theta = M' \cap K_\theta$. Let N_θ be the connected Lie subgroup of G with Lie algebra $\bigoplus_{\alpha \in {}^c\langle\theta\rangle \cap \Delta_-} \mathfrak{g}_\alpha$. Then $N\tilde{P}_\theta = N_\theta \tilde{P}_\theta$ and the mapping $(u,p) \to up$ is an isomorphism of $N \times \tilde{P}_\theta$ onto $N_\theta \tilde{P}_\theta$.

Consider the set $B_\theta = G/\tilde{P}_\theta$. If μ is a probability measure on B_θ, $f\mu$ denotes the image measure of μ by f in G. For any closed subgroup H of G such that $AN \subset H \subset K_\theta AN$, we define the H-type of μ as the set of images of μ by the elements of H.

For any g in G, $\bar{g}$ denotes the corresponding coset of B_θ. In each coset of B_θ there is at most one element of N_θ. Thus we identify N_θ and $B_N^\theta = \{\bar{u} \in B_\theta ; u \in N_\theta\}$, and so imbed N_θ in B_θ. We regard a probability measure on N_θ (or on B_N^θ) as a probability measure on B_θ such that $\mu\,(B_\theta \setminus B_N^\theta) = 0$.

The compact group K operates transitively on the homogeneous space $B_\theta = KA\tilde{N}/K_\theta A\tilde{N}$. Then we define the Cauchy measure γ on B_θ as the unique probability measure γ on B_θ which is invariant by K. Notice that $G\gamma = NAK\gamma = NA\gamma \subset H\gamma$; hence the H-type of γ (the Cauchy H-type) is invariant by G. Conversely, we obtain the following generalization of Theorem 1 :

<u>Theorem 2</u> Let μ be a probability measure on B_N^θ (or on N_θ) such that $\mu(f^{-1}(B_\theta \setminus B_N^\theta)) = 0$ for any f in G. Let H be a closed subgroup of G such that $AN \subset H \subset K_\theta AN$.
If for any g in G, $g\mu$ is in the H-type of μ (i.e. $G\mu=H\mu$), then μ is in the Cauchy H-type (i.e. $\mu \in H\gamma$).

<u>Proof</u> : Consider the set $X_\mu^\theta = \{f \in G ; f\mu=\mu\}$. Then X_μ^θ is a compact subgroup of G (the proof is analogous with that of Proposition 1). Hence X_μ^θ is in a conjugate of a maximal compact subgroup K, and so there exists g in G such that $X_\mu^\theta \subset g^{-1}\,K\,g$.
As in the proof of Theorem 1, consider the probability measure $\nu = g\mu$; then $\nu \in H\mu$, $G\nu=H\nu$, $X_\nu^\theta \subset K$, $K = X_\nu^\theta\,(K \cap H)$. Consider m_s (defined in the proof of Theorem 1) ; then $N\bar{m}_s=\bar{m}_s$, $H\bar{m}_s=\bar{m}_s$ and $B_\theta = K\bar{m}_s = X_\nu^\theta\,(K \cap H)\bar{m}_s = X_\nu^\theta \bar{m}_s$. By the same argument as in Theorem 1, we obtain $\nu=\gamma$, and μ is in the H-type of γ.

5. EXAMPLES : THE CAUCHY MEASURE ON THE FLAGS IN $\mathbb{R}^n$.

Throughout this section we will consider the case where $G = SL(n, \mathbb{R})$. Then B or B_θ are sets of flags. So we will obtain a characterization of the type of the Cauchy measure on the flags. At the end of the section we will give an explicit expression of the Cauchy measure.

In this section, $G = SL(n, \mathbb{R})$, $\mathfrak{g}$ is the set of matrices X in $M_n(\mathbb{R})$ such that $TrX = 0$. Let $\mathcal{P}$ (resp. $\mathcal{K}$) be the subset of symmetric (resp. skew-symmetric) matrices in $\mathfrak{g}$. Let $\mathcal{A}$ be the set of diagonal matrices in $\mathfrak{g}$. Then the roots are the linear forms α_{ij} $(1 \leq i \leq n,\ 1 \leq j \leq n)$ on $\mathcal{A}$ defined by :

$$X = \begin{pmatrix} \lambda_1 & & \\ & \ddots & \\ & & \lambda_n \end{pmatrix} \to \alpha_{ij}(x) = \lambda_i - \lambda_j$$

Note that, if $\alpha = \alpha_{ij}$ and $i \neq j$, $\mathfrak{g}_\alpha$ is the one-dimensional subspace:

$$\mathfrak{g}_\alpha = \{X = (x_{k\ell}) \in \mathfrak{g}\ ;\ x_{k\ell} = 0 \text{ if } (k, \ell) \neq (i,j)\}\ ;$$

$\mathfrak{g}_0 = \mathcal{A}$ and $\mathcal{M} = \{0\}$.

Consider the Weyl chamber W_0 :

$$W_0 = \{ X = \begin{pmatrix} \lambda_1 & & \\ & \ddots & \\ & & \lambda_n \end{pmatrix} \in \mathfrak{g}\ ;\ \lambda_1 < \dots < \lambda_n\}.$$

Then $\Delta_+ = \{\alpha_{ij};\ 1 \leq j < i \leq n\}$, $\Delta_- = \{\alpha_{ij}\ ;\ 1 \leq i < j \leq n\}$, and the set of simple roots is $\Sigma = \{\alpha_{i,i+1};\ 1 \leq i < n\}$.

Then $\mathcal{N}$ (resp $\tilde{\mathcal{N}}$) is the set of upper (resp. lower) triangular matrices, with diagonal entries equal to zero.

Thus $K = SO(n, \mathbb{R})$, A is the group of nxn diagonal matrices, with stricly positive diagonal entries, N (resp. $\tilde{N}$) is the group of nxn upper (resp. lower) real triangular matrices, with diagonal entries equal to 1 ; M is the subgroup of $SL(n, \mathbb{R})$ consisting of all diagonal matrices with diagonal entries equal to +1 or -1 ; M' is the subgroup of $SL(n, \mathbb{R})$ defined as follows : $g=(g_{ij})$ is in M' if and only if there exists a permutation s of $\{1,\dots,n\}$ such that, for any i in $\{1,\dots,n\}$, $g_{i,s(i)} = \pm 1$ and $g_{ij} = 0$ if $j \neq s(i)$. Note that AN is the subgroup of $SL(n, \mathbb{R})$ consisting of upper triangular matrices with strictly positive diagonal entries, and MAN the subgroup of upper triangular matrices in $SL(n, \mathbb{R})$.

Recall that $(E_1,\dots,E_{n-1})$ is a full flag in $\mathbb{R}^n$ if and only if : for any k in $\{1,\dots,n-1\}$, E_k is a subspace of $\mathbb{R}^n$, $\dim E_k = k$ and $E_1 \subset \dots \subset E_{n-1}$.
If g is a matrix in $SL(n, \mathbb{R})$ with columns $g_1,\dots,g_n$, let $\overline{g}$ denote the full flag :

$$\overline{g} = (\text{vect}\ (g_n),\ \text{vect}\ (g_{n-1},g_n),\dots,\ \text{vect}\ (g_2,\dots,g_n)).$$

Note that $\overline{g} = \overline{f}$ if and only if $f^{-1}g$ is in $\tilde{N}AM$.
Thus we identify $B=G/\tilde{N}AM$ and the set of the full flags in $\mathbb{R}^n$. Then B_N is the set of the full flags b in $\mathbb{R}^n$ such that there exists a matrix u in N so that $b = \overline{u}$.
For any matrix g in $SL(n, \mathbb{R})$, note that $\overline{g}$ is in B_N if and only if the principal minors of g are not zéro (the principal minors of a nxn matrix $g = (g_{ij})$ are the n determinants $d_1(g), \dots, d_n(g)$ where

$$d_k(g) = \begin{vmatrix} g_{k,k} & \cdots & g_{k,n} \\ \vdots & & \vdots \\ g_{n,k} & \cdots & g_{n,n} \end{vmatrix} \quad).$$

There is on $N \equiv B_N$ a unique K-invariant probability measure γ ; we will call it the Cauchy measure on N. For any matrix u in N :

$$u = \begin{pmatrix} 1 & u_{12} & \cdot & \cdot & \cdot & u_{1n} \\ & 1 & & & & \vdots \\ & & \cdot & & & u_{n-1,1} \\ & & & \cdot & & \\ & & & & \cdot & \\ & & & & & 1 \end{pmatrix}$$

let $u_1, \dots, u_n$ denote the columns of u. We shall see later that γ is the probability measure on $\mathbb{R}^{n(n-1)/2}$:

$$\frac{c_n \, du_{12} \dots du_{1n} \dots du_{n-1,1}}{\|u_n\|^2 \cdot \|u_{n-1} \wedge u_n\|^2 \dots \|u_2 \wedge \dots \wedge u_n\|^2}$$

The set $AN\gamma = MAN\gamma$ is the Cauchy type on N. Theorem 1 furnishes a characterization of the Cauchy type on N.

Now we will replace $MA\tilde{N}$ by a parabolic subgroup of $G = SL(n, \mathbb{R})$. Let α_i denote the simple root $\alpha_{i,i+1}$ and consider a subset θ of Σ. If k-1 is the cardinal of $^c\theta$ (the complement of θ with respect to Σ), let us define the positive integers $n_1, \dots, n_k$ such that :

$$\begin{cases} n = n_1 + \dots + n_k \\ ^c\theta = \{\alpha_{n_k}, \alpha_{n_{k-1}+n_k}, \dots, \alpha_{n_2+\dots+n_k}\}. \end{cases}$$

Then A_θ is the subgroup of diagonal matrices a in $SL(n, \mathbb{R})$ such that there exist $\lambda_1>0, \dots, \lambda_k>0$ so that :

$a =$ 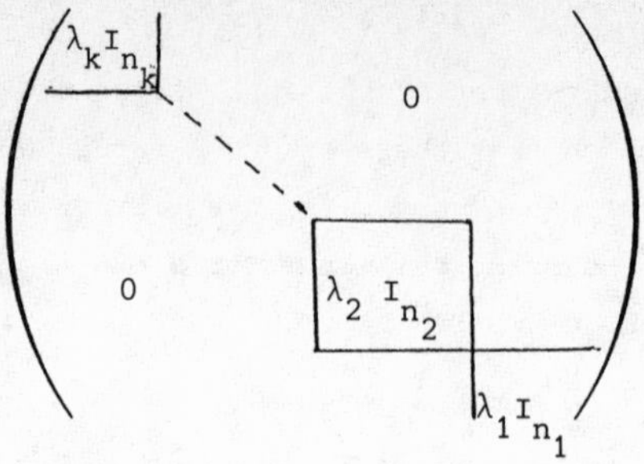

where I_k is the kxk identity matrix ; K_θ is the subgroup of all g in $SO(n, \mathbb{R})$ so that :

$g =$ 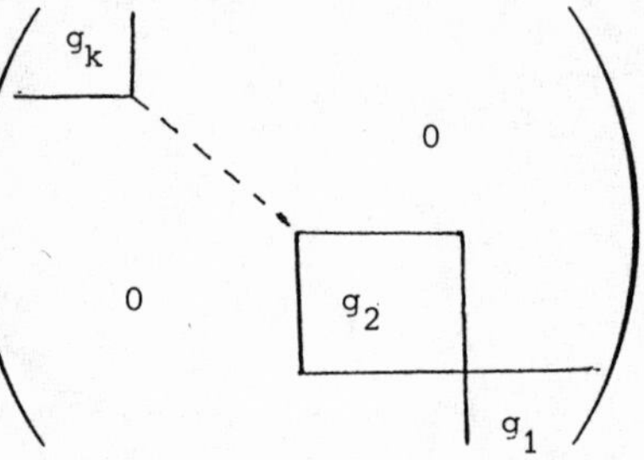

where g_i is in $O(n_i, \mathbb{R})$; the parabolic subgroup $\tilde{P}_\theta = K_\theta A\tilde{N}$ is the set of all matrices in $SL(n, \mathbb{R})$ of the form :

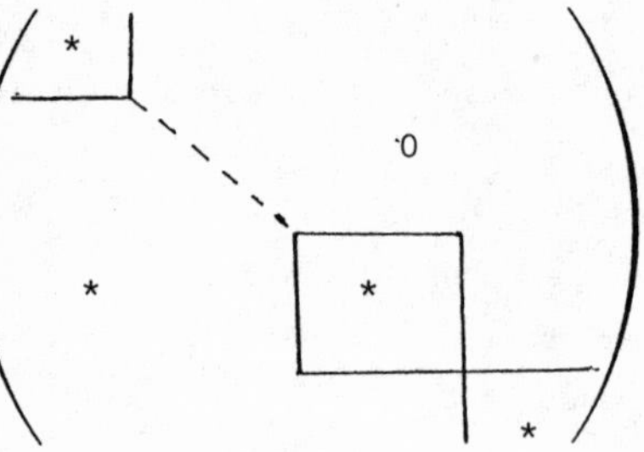

Then N_θ (satisfying to : $N\tilde{P}_\theta = N_\theta \tilde{P}_\theta$) is the subgroup of all matrices in $SL(n, \mathbb{R})$ of the form

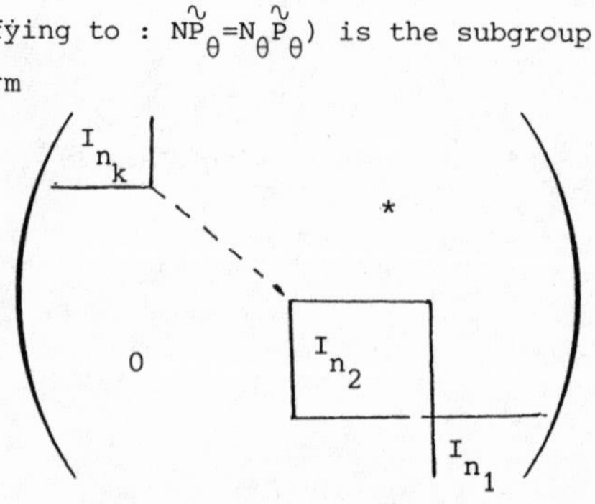

Recall that $(E_1,\dots,E_{k-1})$ is a flag in $\mathbb{R}^n$ if and only if :

$$\begin{cases} \forall i \in \{1,\dots,k-1\} \ E_i \text{ is a subspace of } \mathbb{R}^n \\ \{0\} = E_0 \subset E_1 \subset \dots \subset E_{k-1} \subset E_k = \mathbb{R}^n \\ \forall i \in \{1,\dots,k\} \quad n_i = \dim E_i - \dim E_{i-1} > 0 \end{cases}$$

More precisely, $(E_1,\dots,E_{k-1})$ is called a flag of the type $(n_1,\dots,n_k)$. Let g be a matrix in $SL(n, \mathbb{R})$ with columns $g_1,\dots,g_n$; let E_i denote the subspace of $\mathbb{R}^n$ generated by the $n_1+\dots+n_i$ vectors $g_n, g_{n-1},\dots,g_{1+n_{i+1}+\dots+n_k}$; then $\bar{g} = (E_1,\dots,E_k)$ is a flag of the type $(n_1,\dots, n_k)$; for f and g in $SL(n, \mathbb{R})$, observe that $\bar{f} = \bar{g}$ if and only if $f^{-1} g \in \tilde{P}_\theta$. Thus we identify $B_\theta = G/\tilde{P}_\theta$ and the set of the flags of the type $(n_1,\dots,n_k)$. Then B_N^θ is the set of flags d such that there exists a matrix u in N_θ so that $d=\bar{u}$.
The unique K-invariant probability measure γ on $N_\theta \equiv B_N^\theta$ is called the Cauchy measure.

Using results of Furstenberg and Tzkoni [6], we will obtain the density of γ. Let $p = (n^2-n_1^2\dots-n_k^2)/2$; identify N_θ with $\mathbb{R}^p$. For any g in $G = SL(n, \mathbb{R})$ and any b in B_θ , the measure $g^{-1}\gamma$ is absolutely continuous with respect to the Lebesgue measure on $\mathbb{R}^p$ and :

$$\frac{dg^{-1}\gamma}{d\gamma}(b) = \sigma_{B_\theta}(g,b),$$

where σ_{B_θ} is a certain cocycle on $G \times B_\theta$, which is calculated in [6]. Let f be the density of γ with respect to the Lebesgue measure on $\mathbb{R}^p$ (f is defined on $N_\theta = \mathbb{R}^p$). Then, for any g in G and any x in N_θ,

$$\sigma_{B_\theta}(g,\bar{x}) = f(g(x)). \ |(\text{Jac } g)(x)|/f(x).$$

Taking g=u in N_θ and x=I (then $\bar{x}$ is the canonical flag of the type $(n_1,\dots,n_k)$), we obtain $f(u) = f(I)\ \sigma_{B_\theta}(u,\bar{I})$. Let u be in N_θ :

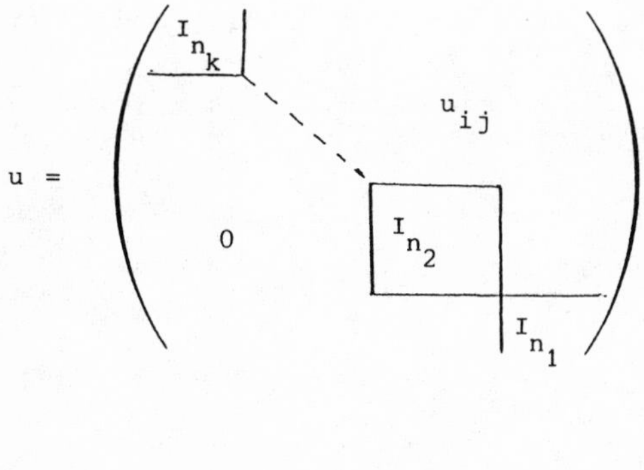

and let $u_1, \ldots, u_n$ denote the columns of u. Then γ is the probability measure on $\mathbb{R}^p$:

$$\frac{C \; du_{1,n_k+1} \cdots \; du_{1,n} \; \cdots \; du_{n-n_1,n}}{\| u_{1+n_2+\ldots n_k} \wedge \cdots \wedge u_n \|^{n_1+n_2} \cdots \| u_{1+n_k} \wedge \cdots \wedge u_n \|^{n_{k-1}+n_k}} .$$

Observe that, in the case where $G = SL(n+1, \mathbb{R})$, $n_1=1$, $n_2=n$, we obtain the usual Cauchy measure on $\mathbb{R}^n$:

$$\frac{C \; dx_1 \ldots dx_n}{(1+x_1^2+\ldots + x_n^2)^{(n+1)/2}} \quad ;$$

its type has been characterized in [10] and [3] .

The authors wish to thank J. Tapia (Université de Toulouse) for various hints concerning this paper.

References

[1] Borel, A. : Introduction aux groupes arithmétiques. Hermann, Paris, 1969.

[2] Dani, S.G. and McCrudden, M. : Parabolic subgroups and factor compactness of measures on semisimple groups, in Probability measures on groups VIII, Oberwolfach 1985, Lecture notes in Mathematics 1210, 34-40, Springer, 1986.

[3] Dunau, J.L. and Sénateur, H. : An elementary proof of the Knight-Meyer characterization of the Cauchy distribution. J. Multivar. Anal. 22, 74-78 (1987).

[4] Dunau, J.L. et Sénateur, H. : Une caractérisation du type de la loi de Cauchy-conforme sur $\mathbb{R}^n$. Probab. Th. Rel. Fields, 77, 129-135 (1988).

[5] Dunau, J.L. and Sénateur, H. : A characterization of the type of the Cauchy-Hua measure on real symmetric matrices. J. Theor. Probab. 1, 263-270 (1988).

[6] Furstenberg, H. and Tzkoni, I. : Spherical functions and integral geometry. Israel J. Math., 10, 327-338 (1971).

[7] Guivarc'h, Y. et Raugi, A.: Frontière de Furstenberg, propriétés de contraction et théorèmes de convergence. Z. Wahrsch. Verw. Gebiete, 69, 187-242 (1985).

[8] Helgason, S. : Differential geometry and symmetric spaces. Academic Press, New-York, 1962.

[9] Knight, F.B. : A characterization of the Cauchy type. Pro. Amer. Math. Soc. 55, 130-135 (1976).

[10] Knight, F.B. et Meyer, P.A. : Une caractérisation de la loi de Cauchy. Z. Wahrsch. Verw. Gebiete, 34, 129-134 (1976).

OPERATORS, STOCHASTIC PROCESSES, AND LIE GROUPS

by

Philip Feinsilver
Southern Illinois University
Carbondale, Illinois, USA

and

René Schott
Université de Nancy I
Nancy, FRANCE

I. Introduction

This study arose from the naïve inquiry: what is a process on a Lie group? We mean by this question -- how might you explain to someone -- mathematician, probabilist -- so as to illustrate clearly features new to the more general setting, extending from $\mathbb{R}^n$ to G.

First let us see how we put the usual theory on $\mathbb{R}$ into context. We start with the viewpoint that characteristic functions should be thought of as the mean values of processes (or random elements) on U(1). So we are thinking of $\mathbb{R}$ as a Lie algebra rather than as a basic group. In fact, we think of e^{itX}, say, as $e^{t\xi}$ with $t \in \mathbb{R}$, $\xi \in u(1)$. What is interesting is the dual nature of this expression. Think of e^{tx} with t a real parameter and x in some (Lie) algebra of operators (e.g. matrices). One could randomize either the t or the x. Notice that in the usual theory on $\mathbb{R}$ one is randomizing the x. More generally, one can look at functionals on operator algebras and get an operator-theoretic setting for the theory of stochastic processes. This is the approach of "quantum probability" -- see [PH], for example. Here we take the route of randomizing the t. Thus, we present here a class of examples of processes on Lie groups derived by the technique of "randomizing the Lie algebra" by putting in random variables as coefficients. In other words, if we have a Lie algebra $\mathfrak{g}$ associated to a group G, then with $\mathcal{M}$ denoting the space of real random variables, we map $\mathfrak{g} \to \mathcal{M} \otimes \mathfrak{g}$ and corresponding group elements become group-valued random variables with distribution induced by the coefficients.

It is perhaps surprising that even though the theoretical approach to processes on Lie groups sounds a lot like -- or can be put in a setting so as to sound like a "natural generalization of" -- the $\mathbb{R}^n$ theory, the results show that the Lie group theory is really different, intrinsically involving stochastic integrals. It provides a nice way of showing the existence and construction of stochastic integrals of certain types -- the types appearing in each specific case reflecting directly the group law.

To summarize the point of view here, we will take the abstract notion of a stochastic process on a Lie group as a measure on a function space and flesh it out with some specific constructions so we can see "what these processes actually look like." General constructions comprise Section II. Section III presents specific examples and discussion. The Appendix contains some remarks on computational techniques that are used in realizing the explicit constructions.

II. Constructions

We start with a process $w(t)$, a stochastically continuous, time-homogeneous process with independent increments on $\mathbb{R}^d$. Denote by $L(D)$, $D = (\partial/\partial x^1,\ldots,\partial/\partial x^d)$, the generator of the process.

G denotes a connected, simply-connected Lie group with Lie algebra $\mathfrak{g}$ having linear basis $\xi_1,\xi_2,\ldots,\xi_N$, $N \geq d$. Although the theory applies to the general case, we consider in this study exponential groups, or exponential subgroups of G -- i.e. we consider elements of G having the form $g(x^1,x^2,\ldots,x^N) = e^{x^1\xi_1}\cdots e^{x^N\xi_N}$, e denoting the exponential map. Briefly, we write

$$(2.1)\qquad g(x^1,\ldots,x^N) = \prod_1^N e^{x^k\xi_k} \qquad \text{(no summation convention)}$$

where the ordering in the multiplicaation is understood. (We remark that it is well-known that nilpotent groups, e.g., are exponential groups).

We would like to show that a corresponding process $X(t)$ exists on G and how to calculate its expected values (analogous to characteristic functions in the case of $\mathbb{R}$).

Note: We can assume that $d = N$ by filling in empty positions with zeros. This will be useful later. Typically we start with $\xi_1,\ldots,\xi_d$ where these generate $\mathfrak{g}$ as a Lie algebra.

Let $n > 0$. Let $\Delta w_n(j) = w(j/n)-w((j-1)/n)$, $w(0) = 0$, with the process being defined over $[0,1]$ as a generic time interval. Now map into G:

(2.2) $$X_{nj} = g(\Delta w_n(j))$$

according to (2.1).

Let U be a neighborhood of $0 \in \mathbb{R}^n$ corresponding to a neighborhood of the identity in G. Let F_{nj} denote the distribution of $\Delta w_n(j)$ and put $m_{nj} = \int_U x\, dF_{nj}(x)$, the (vector-valued) infinitesimal means. Now define the Lévy measure functions and covariance functions as in the limit theory, following the procedure in ([F], esp. pp. 80-83):

(2.3)
$$m_n(t) = \prod_{j=1}^{[nt]} g(m_{nj})$$
$$A_{rs,n}(t,U) = \sum_{j=1}^{[nt]} \int_U (x^r - m^r_{nj})(x^s - m^s_{nj})\, dF_{nj}(x)$$
$$M_n(t,dx) = \sum_{j=1}^{[nt]} F_{nj}(dx)$$

where we use directly the coordinates $(x^1,\ldots,x^N)$ as coordinates on the group, for $x \in U$ in particular. We consider the processes determined by

(2.4) $$X_n(t) = \prod_{j=1}^{[nt]} X_{nj}.$$

The desired process X(t) on G corresponding to the given process w(t) on $\mathbb{R}^n$ is the limit of these processes $X_n(t)$.

By the limit theory for the $\mathbb{R}^n$ case, we already know that the covariance functions and Lévy measure functions converge appropriately as required by the limit theorem for G (ibid. p. 82). The only problem arises in checking convergence of the mean functions $m_n(t)$. For the case of bounded operators ξ_j we have directly the estimates

(2.5)
$$\| e^{m^k_{nj}\xi_k} \| \le e^{|m^k_{nj}|\,\|\xi_k\|}$$
$$\| g(m_{nj}) \| \le \exp\left[\sum_k |m^k_{nj}|\,\|\xi_k\|\right]$$

so that since the m_{nj} are uniformly infinitesimal, one has compactness of $\{m_n(t)\}$, as a set of functions, with convergence to a continuous limit m(t). In general, one can invoke the method of ([F], pp. 118-119, an argument originally due to Stroock-Varadhan) to conclude that the $m_n(t)$ converge uniformly to a continuous function m(t) on the group. Thus, the limit theorem applies, and by construction, the generator of the process on G is $L(\xi_1,\ldots,\xi_N)$, where ξ_j replaces $\partial/\partial x^j$ in the original generator. In particular, Hunt's theory [H] is recovered.

Now consider the case where the components $w^k(t)$ are independent. The generator $L = \sum_{k=1}^{N} L_k$, with L_k the generator of the process $w^k(t)$ on $\mathbb{R}$. Then we have, $\langle\ \rangle$ denoting expected value,

$$\langle e^{\Delta w^k_{nj}\,\xi_k} \rangle = e^{\Delta t\, L_k(\xi_k)} \tag{2.6}$$

with $\Delta t = 1/n$, as long as one has a functional calculus for the ξ_k. So (2.6) makes sense for bounded operators when the moment generating function of Δw^k exists, or, via the spectral theorem, for unitary representations of G. By independence we have (ordered products):

$$\langle g(\Delta w_{nj})\rangle = \langle X_{nj}\rangle = \prod_{j=1}^{N} e^{\Delta t\, L_k(\xi_k)} \tag{2.7}$$

$$\langle X_n(t)\rangle = \left[\prod_{1}^{N} e^{\frac{1}{n}L_k(\xi_k)}\right]^{[nt]}. \tag{2.8}$$

$$X_n(t) = \prod_{j=1}^{[nt]} \prod_{k} e^{\Delta w^k_{nj}\,\xi_k} \tag{2.9}$$

can be calculated from the group law. We define components $W^k_n(t)$ by the relation $X_n(t) = g(W^1_n(t),\ldots,W^N_n(t))$. As we will see below, these W^k_n are typically discrete approximations to stochastic integrals mixing the components $w^k(t)$. In the formulation (2.8)-(2.9), we see that appropriate theorems on convergence of product integrals, on the group, (2.9), and for operators, (2.8), will give us the existence of the process as well as the expected values $\langle X(t)\rangle$. For this approach, see [S]. Notice that for the right-hand side of (2.8) one is looking precisely at a case of the Trotter product formula. Thus, we have, using either the limit theory invoked above or the theory of product integrals/Trotter product formula:

$$\langle X(t)\rangle = e^{tL(\xi)}, \tag{2.10}$$

$$X(t) = g(W^1(t),\ldots,W^N(t)) \tag{2.11}$$

with $W^k(t)$ the coordinates (components) of the limiting process. Here again, we are using the correspondence between the coordinates on the group and those on $\mathbb{R}^n$ to assert that in fact the processes $W^k(t)$ exist as a consequence of the existence of $X(t)$. The identification of $L(\xi_1,\ldots,\xi_N)$ as the generator follows, as noted above, from the probability limit theorem, independently of the Trotter formula. For unitary representations one can invoke the Lévy continuity theorems of [Bo], e.g. for amenable groups.

For further discussion of the connections with the operator-theoretic viewpoint see the "Polish paper" [FS]. Here we focus on providing some specific illustrations of this theory. Let us make a remark here. Notice that for the processes $w^k(t)$ it is not required that moment generating functions exist. However, if they do not, as for the real case, one must use unitary representations to guarantee the existence of the mean values $\langle X(t)\rangle$. But the applicability of the limit theory and the existence of the process $X(t)$ do not require any special considerations.

We summarize this section by stating

(2.12) *Theorem.* Let $w(t)$ be a process with stationary independent increments on $\mathbb{R}^d$, $w(0) = 0$. Let G be a connected, simply-connected Lie group with $\xi_1,\xi_2,\ldots,\xi_N$, $N \geq d$, a given, ordered basis for the Lie algebra $\mathfrak{g}$. Then one can associate to $w(t)$, via exponential coordinates, canonically a process $X(t)$ with coordinates $W(t)$

$$X(t) = g(W(t)) = e^{W^1(t)\xi_1}\cdots e^{W^N(t)\xi_N} \tag{2.13}$$

which one calls the w-process on G. The form of the processes $W^k(t)$ depends only on the group law of G. If the process $w(t)$ has independent coordinates $w^k(t)$ with generators $L_k(\partial/\partial x^k)$, then the process $W(t)$ has generator $L(\xi_1,\ldots,\xi_N) = \sum_{k=1}^{d} L_k(\xi_k)$.

III. Examples

To generate a process we typically start with $\xi_1,\ldots,\xi_d$, $d \leq N$, generating $\mathfrak{g}$ as a Lie algebra. We pick $w^1,\ldots,w^d$ independent, non-zero processes, filling in $w^{d+1} = \cdots = w^N = 0$. Then we employ the technique implicit in (2.8)-(2.9). See the Appendix for remarks on computational techniques. (Note: We typically write a, b, c, etc. for the variables x_1, x_2, x_3, etc.)

1. *Nilpotent groups*, e.g. the Heisenberg group, with basis for the Lie algebra δ, ξ, h, with $[\delta,\xi] = h$, h central. A standard realization is ξ = multiplication by x, $\delta = h\,\frac{d}{dx}$, h = multiplication by a scalar, h. A typical group element is

(3.1) $$g(a,b,c) = e^{a\xi} e^{b\delta} e^{ch} \text{ with}$$

(3.2) $$g(A,B,C)g(a,b,c) = g(A+a,\ B+b,\ C+c+Ba).$$

Inductively, we have

(3.3) $$g(A_{k-1},B_{k-1},C_{k-1})g(a_k,b_k,c_k) = g(A_k,B_k,C_k)$$

with $A_k = \sum_1^k a_j$, $B_k = \sum_1^k b_j$, $C_k = \sum_1^k (c_j+B_{j-1}a_j)$. We put $a_j = \Delta w_n^1(j)$, $b_j = \Delta w_n^2(j)$, $c_j = 0$, since δ, ξ generate the Lie algebra. Thus, with $t_n = [nt]/n$:

(3.4) $$X_n(t) = g\left[w^1(t_n), w^2(t_n), \sum_{j=1}^{nt_n} w^2\left[\frac{j-1}{n}\right]\Delta w_n^1(j)\right]$$

(3.5) $$X_n(t) = g(W_n^1(t), W_n^2(t), W_n^3(t))$$

with $W_n^3(t)$ an approximation to the Itô integral $\int_0^t w^2 dw^1$. Thus,

(3.6) $$X(t) = g(w^1(t), w^2(t), \int_0^t w^2 dw^1).$$

Notice that the form of the W^k processes depends only on the group law, not on the w^k processes.

Brownian motion on the Heisenberg group satisfies

(3.7) $$\langle B(t)\rangle = \exp\left[\frac{t}{2}(\delta^2 + \xi^2)\right].$$

In the standard realization this is the (quantum) harmonic oscillator process.

The *Poisson process*, corresponding to the standard real process $N(t)$ with $\langle e^{aN(t)}\rangle = e^{t(e^a-1)}$, satisfies, $\mathcal{N}(t)$ denoting the process on the Heisenberg group:

(3.8) $$\langle \mathcal{N}(t)\rangle = \exp\left[t(e^\delta + e^\xi - 2)\right].$$

If we let $\alpha = e^\xi$, $\beta = e^\delta$, $q = e^h$, we have $\beta\alpha = q\alpha\beta$, the multiplication in the group algebra. (Note that in general, the Poisson process on a group G involves the group algebra.) Introducing the Gaussian q-binomial coefficients (see [An], [F2]):

(3.9) $$\begin{bmatrix} n \\ k \end{bmatrix}_q = \frac{(1-q^n)\cdots(1-q^{n-k+1})}{(1-q)\cdots(1-q^k)}$$

one can show by induction that

$$(3.10)\qquad (\alpha+\beta)^n = \sum_{k=0}^{n} \begin{bmatrix} n \\ k \end{bmatrix}_q \alpha^{n-k}\beta^k. \qquad \text{So}$$

$$(3.11)\qquad e^{t(\alpha+\beta)} = \sum_0^\infty \frac{t^n}{n!} \sum_0^n \begin{bmatrix} n \\ k \end{bmatrix}_q \alpha^{n-k}\beta^k.$$

Thus, in the standard realization, for bounded continuous functions $f(x)$, with $N^3(t) = \int_0^t N^2 dN^1$,

$$(3.12)\qquad \langle e^{\xi N^1(t)} e^{\delta N^2(t)} e^{hN^3(t)} f(x)\rangle \ \langle e^{xN^1(t)} e^{hN^3(t)} f(x+hN^2(t))\rangle$$

$$(3.13)\qquad \langle e^{xN^1(t)} e^{hN^3(t)} f(x+hN^2(t))\rangle = \sum_0^\infty \frac{t^n}{n!} \sum_0^n \begin{bmatrix} n \\ k \end{bmatrix}_q e^{(n-k)x} f(x+kh).$$

Notice that the Lie group technique yields the joint distribution of N^1, N^2, and N^3 quite efficiently.

As another example in the nilpotent case, consider the Lie algebra with basis ξ_1, ξ_2, ξ_3, ξ_4 satisfying $[\xi_4,\xi_1] = \xi_2$, $[\xi_4,\xi_2] = \xi_3$, remaining commutators zero. We have $g(a,b,c,d) = e^{a\xi_1} e^{b\xi_2} e^{c\xi_3} e^{d\xi_4}$:

$$(3.14)\qquad g(A,B,C,D)g(a,b,c,d) = g(A+a, B+b+aD, C+c+bD + \frac{aD^2}{2}, D+d).$$

Starting with $w = (w^1,0,0,w^4)$, taking increments Δw on the right as in (3.3)-(3.4), we find

$$(3.15)\qquad X(t) = g\left[w^1(t),\ \int_0^t w^4 dw^1,\ \frac{1}{2}\int_0^t (w^4)^2 dw^1,\ w^4(t)\right]$$

$$(3.16)\qquad \langle X(t)\rangle = e^{t(L_1(\xi_1)+L_4(\xi_4))}.$$

A realization of this algebra is given by $\xi_4 = \frac{d}{dx}$, ξ_1 = multiplication by $\frac{1}{2}x^2$.

Other nilpotent groups for which one may do explicit calculations include groups of type H ([K]) and groups with Lie algebras like the latter example, generated by $\frac{d}{dx}$, $P(x)$, where $P(x)$ acts as multiplication by a polynomial.

2. *Solvable groups* are handled similarly. For example, the finite-difference algebra [F2], with commutation relations

$$(3.17)\qquad [S,\xi] = T, \quad [T,\xi] = hT, \quad [S,T] = 0$$

where $h \neq 0$ is a given scalar. A typical realization of this algebra is given by the action on functions $f(x)$:

(3.18) $$Sf(x) = \frac{f(x+h) - f(x)}{h}, \quad \xi f(x) = xf(x), \quad Tf(x) = f(x+h).$$

With $g(a,b,c) = e^{a\xi}e^{bS}e^{cT}$ one has the group law

(3.19) $$g(a,b,c)g(A,B,C) = g\left[a+A, b+B, C+ce^{Ah}+b\left(\frac{e^{Ah}-1}{h}\right)\right].$$

Starting with $w = (w^1,w^2,0)$, taking increments on the left, i.e. multiplying with time increasing from right to left, we find

(3.20) $$X(t) = g\left[w^1(t),\ w^2(t),\ \frac{1}{h}\int_0^t (e^{hw^1}-1)dw^2\right].$$

The *oscillator algebra*, corresponding to the so-called diamond group, given by the commutation relations

(3.21) $$[H,T] = H, \quad [T,K] = K, \quad [H,K] = Z$$

with Z central, has the realization $T = x\frac{d}{dx}$, $H = \frac{d}{dx}$, $K = x$, $Z = 1$. With $g(a,b,c,d) = e^{aK}e^{bH}e^{cZ}e^{dT}$, we have

(3.22) $$g(A,B,C,D)g(a,b,c,d) = g(A+ae^D, B+be^{-D}, c+C+aBe^D, d+D),$$

and starting with $w = (w^1,w^2,0,w^4)$ we find

(3.23) $$X(t) = g\left[\int_0^t e^{w^4}dw^1,\ \int_0^t e^{-w^4}dw^2,\ \int_0^t\left[\int_0^s e^{-w^4}dw^2\right]e^{w^4(s)}dw^1, w^4(t)\right].$$

In general, for solvable groups, the technique of semisimple splitting [Au], which has been utilized effectively in the study of random walks, allows one to embed a (non-exponential) solvable group into the semidirect product of a nilpotent group and a torus. Using a representation induced by a representation of the torus reduces to the nilpotent part and allows one to use the techniques presented here.

3. For *semisimple groups*, using the Cartan splitting of the Lie algebra $L^+ \oplus H \oplus L^-$, $L^\pm$ denoting positive/negative root spaces respectively, and H the Cartan subalgebra, from a representation induced from the Borel subalgebra $H \oplus L^+$, e.g. on Harish-Chandra modules, again one is left with a nilpotent part which can be handled as above.

4. Another interesting illustration is afforded by the Heisenberg group. Let $\gamma(t)$ be a process on $\mathbb{R}$ with moment generating function satisfying

(3.24) $$\langle e^{x\gamma(t)}\rangle = e^{tV(x)}$$

with V continuous on $\mathbb{R}$. Then taking $w(t) = (\gamma(t), b(t), 0)$, with $b(t)$ an independent Brownian motion, gives

(3.25) $$\langle e^{x\gamma(t)} e^{\delta b(t)} e^{h\int_0^t b\,d\gamma} \rangle = e^{t\left[\frac{h^2}{2}\frac{d^2}{dx^2} + V(x)\right]},$$

with $\delta = h\frac{d}{dx}$. Letting h = 1, apply to a suitable function f(x) to get

(3.26) $$\langle e^{\delta b(t)} e^{\int_0^t (x+b)\,d\gamma} \rangle f(x) = \langle e^{\int_0^t V(x+b(s))ds} f(x+b(t))\rangle$$

which, combined with (3.25), yields the Feynman-Kac formula.

We also remark that in the products $\prod X_{nj}$ one can group multiplying by increments from either the left or the right. E.g., for the Heisenberg case if we adjoin increments from the left, but keeping the time-ordering left-to-right,

(3.27) $$X(t) = \left[w^1(t), w^2(t), \bar{\int_0^t} w^1 dw^2\right]$$

where the bar over the integral indicates the "belated" Itô integral, where the increments are taken prior to the integrand.

5. We conclude by remarking that one should be able to extend, at least in some cases, to processes on homogeneous spaces. For background references in this regard, see e.g. [Z], [OM], [Be].

Appendix. Remarks on (basic) Lie theory.

1. To get an explicit realization of a Lie algebra, we employ the "right dual representation" which is a realization in terms of vector fields acting on functions of commuting variables. Namely, if $\xi_1,\ldots,\xi_N$ are the basis for a Lie algebra, look at $g(a_1,\ldots,a_N) = e^{a_1\xi_1}\cdots e^{a_N\xi_N}$. Using the "easy version" of the Campbell-Baker-Hausdorff formula $e^{a\xi}\eta e^{-a\xi} = e^{a(\mathrm{ad}\,\xi)}\eta$, one can calculate $g(a_1,\ldots,a_N)\xi_j$, where ξ_j is commuted to the jth spot. This gives an expression ξ_j^* as a first-order partial differential ooperator acting on functions of the a_j's with leading part $\frac{\partial}{\partial a_j}$.

Example. For the finite-difference algebra (see [F2]):

(A1) $$e^{bS}\xi e^{-bS} = \xi + bT, \qquad e^{cT}\xi e^{-cT} = \xi + chT$$

(A2) $$g(a,b,c)T = \frac{\partial}{\partial c}g, \qquad g(a,b,c)S = \frac{\partial}{\partial b}g, \quad \text{and}$$

(A3) $$g(a,b,c)\xi = e^{a\xi}e^{bS}(\xi + chT)e^{cT} = \left[\frac{\partial}{\partial a} + (b+ch)\frac{\partial}{\partial c}\right]g.$$

(A4) $$\xi^* = \frac{\partial}{\partial a} + (b+ch)\frac{\partial}{\partial c}\,, \quad S^* = \frac{\partial}{\partial b}\,, \quad T^* = \frac{\partial}{\partial c}\,.$$

2. To calculate the group law using the right dual representation, one has

(A5) $$g(a_1,\ldots,a_N)g(A_1,\ldots,A_N) = e^{A_1\xi_1^*}\cdots e^{A_N\xi_N^*}g(a_1,\ldots,a_N).$$

This reduces to calculating expressions of the type

(A6) $$E(A) = e^{A\xi^*}f(a_1,\ldots,a_N).$$

This may be calculated via the method of characteristics (Hamilton's equations), by solving the first-order partial differential equation $\frac{\partial E}{\partial A} = \xi^* E$.

Example. For the finite-difference group law:

(A7) $$g(a,b,c)g(A,B,C) = e^{A\xi^*}e^{BS^*}e^{CT^*}g(a,b,c)$$

(A8) $$g(a,b,c)g(A,B,C) = e^{A\xi^*}g(a,b+B,c+C).$$

$E(A) = e^{A\xi^*}g(a,b+B,c+C)$ satisfies $\frac{\partial E}{\partial A} = \frac{\partial E}{\partial a} + (b+ch)\frac{\partial E}{\partial c}$. One finds $E = g\left(a+A,b+B,ce^{Ah} + b\left(\frac{e^{Ah}-1}{h}\right) + C\right)$, giving the group law (3.19).

(*Note.* Additional references listed here are related importantly to our discussion, illustrating aspects of or discussing the theory of processes on groups.)

REFERENCES

[An] G. Andrews, The theory of partitions, Encyclopedia of Math. and Appl., Addison-Wesley, 1976.

[Au] L. Auslander, "An exposition of the structure of solvmanifolds," Bull. A.M.S., 79, 2(1973) 227-261.

[Be] M. Berger, "Les espaces symétriques non compacts," Ann. Ex. Norm. LXXIV, Fasc. 2 (1957) 85-177.

[Bo] Ph. Bougerol, "Extension du théorème de continuité de Paul Lévy aux groupes moyennables," Springer Lecture Notes in Math., 1064 (1984) 10-22.

[F] Ph. Feinsilver, "Processes with independent increments on a Lie group," Trans. A.M.S., 242(1978) 73-121.

[F2] Ph. Feinsilver, "Discrete analogues of the Heisenberg-Weyl algebra," Mh. Math., 104(1987) 89-108.

[FS] Ph. Feinsilver and R. Schott, "An operator approach to processes on Lie groups," Conference Proceedings "Probability Theory on Vector Spaces," Łańcut, Poland, 1987.

[G1] Y. Guivarc'h, "Croissance polynomiale et périodes des fonctions harmoniques," Bull. Math. Soc. France, 101(1973) 333-379.

[G2] Y. Guivarc'h, "Sur la loi des grands nombres et le rayon spectral d'une marche aléatoire," Astérisque, 74(1980) 47-98.

[GKR] Y. Guivarc'h, M. S. Keane, and B. Roynette, Marches aléatoires sur les groupes de Lie, Springer Lect. Notes in Math., 624(1977) esp. pp. 101, 142, 150.

[H] H. Heyer, Probability measures on locally compact groups, Ergeb. der Mat., 94, Springer-Verlag, 1977.

[K] A. Kaplan, "Fundamental solutions for a class of hypoelliptic PDE generated by composition of quadratic forms," Trans. A.M.S., 258(1980) 147-153.

[OM] T. Oshima and T. Matsuki, "Orbits on affine symmetric spaces under the action of the isotropy subgroups," J. Math. Soc. Japan, 32(1980) 399-414.

[PH] K. R. Parthasarathy and R. L. Hudson, "Quantum Itô's formula," Comm. Math. Phys., 93(1984) 301-323.

[R] A. Raugi, "Théorème de la limite centrale pour un produit semi-direct d'un groupe de Lie résoluble simplement connexe de type rigide par un groupe compact," Springer Lect. Notes in Math., 706(1979) 257-324.

[S] E. Siebert, "Holomorphic convolution semigroups on topological groups," Springer Lect. Notes in Math., 1064(1984) 421-449.

[Z] H. Zeuner, "Die Existenz einer Lévy-Abbildung auf einem homogenen Raum," Math. Z., 189(1985) 529-554.

Rectification à l'article de L. Gallardo et O. Gebuhrer "Lois de probabilités infiniment divisibles sur les hypergroupes commutatifs discrets dénombrables" (Probability Measures on Groups VII Lecture Notes 1064).

Léonard Gallardo
Département de Mathématiques et Informatique
Université de Bretagne Occidentale
6, avenue Victor Le Gorgeu
29287 BREST (France)

Le but de cette note est d'apporter une rectification à un théorème énoncé avec une hypothèse insuffisante dans l'article mentionné dans le titre (c.f. référence [1]).

Soit X un hypergroupe commutatif discret infini dénombrable et soit μ une mesure de probabilité infiniment divisible sur X i.e. pour tout entier n>0, il existe une probabilité μ_n telle que

(1) $(\mu_n)^{*n} = \mu$,

ce qui équivaut à

(1') $(\hat{\mu}_n(\chi))^n = \hat{\mu}(\chi) \qquad (\forall\ \chi\in\hat{X})$,

où $\hat{X}$ désigne le dual de X et $\hat{\mu}$ la transformée de Fourier de la mesure μ.

Comme me l'a fait remarquer H. Heyer, l'hypothèse "X n'a pas de sous hypergroupe compact non trivial" est insuffisante pour pouvoir affirmer que la suite μ_n converge étroitement vers δ_e. Autrement dit on n'a pas forcément

(2) $\lim_{n\to+\infty} \hat{\mu}_n(\chi) = 1$

(prendre par exemple X = $\mathbb{Q}$ avec la topologie discrète).

Il faut rechercher la cause de ce fait dans la topologie de $\hat{X}$ qui empêche parfois l'existence d'une "bonne" détermination de l'argument de $\hat{\mu}(\chi)$. Nous allons montrer qu'il suffit de supposer que $\hat{X}$ est connexe par arcs pour que (2) soit vérifiée. Cette hypothèse est plus forte que celle faite initialement car elle implique que X n'a pas de sous hypergroupe compact non trivial.

Hypothèse

Dans toute la suite l'hypergroupe X est supposé vérifier la condition C1 ou la condition C2 (cf [1], p. 119). On suppose de plus $\hat{X}$ connexe par arcs.

Lemme 1 : $\hat{\mu}$ (et donc $\hat{\mu}_n$) ne s'annule jamais sur $\hat{X}$.

Démonstration : Pour tout entier n>0, on a

$$|\hat{\mu}_n| = |\hat{\mu}|^{\frac{1}{n}},$$

d'où

$$|\hat{\mu}_n|^2 = \left(|\hat{\mu}|^2\right)^{\frac{1}{n}} \quad \text{et} \quad \lim_n |\hat{\mu}_n(x)|^2 = 0 \text{ ou } 1.$$

Mais comme $\hat{\mu}$ est continue et $\hat{\mu}(\mathbb{1}) = 1$, il existe un voisinage $V_{\mathbb{1}}$ de $\mathbb{1}$ dans $\hat{X}$ tel que $|\hat{\mu}(x)| \neq 0$ ($\forall\ x \in V_{\mathbb{1}}$). Ceci implique

$$\lim_{n\to+\infty} |\hat{\mu}_n(x)| = 1 \quad (\forall\ x \in V_{\mathbb{1}}).$$

Or $|\hat{\mu}_n|^2 = \widehat{\mu_n * \mu_n^-}$. Le théorème de continuité de Paul Lévy prouve alors que $\mu_n * \mu_n^- \longrightarrow \nu$ étroitement où ν est une probabilité idempotente ($\hat{\nu} = 0$ ou 1). Mais $\hat{X}$ est connexe donc $\hat{\nu} \equiv 1$, ce qui prouve que $\nu = \delta_e$. En particulier on a

$$\lim_n |\hat{\mu}_n(x)| = 1 \quad (\forall\ x \in \hat{X}),$$

donc $\hat{\mu}(x) \neq 0$ quel que soit $x \in \hat{X}$.

Lemme 2 : Pour tout $x \in \hat{X}$, il existe un nombre réel $\theta(x)$ tel que pour tout entier n>0,

$$\hat{\mu}_n(x) = |\hat{\mu}(x)|^{\frac{1}{n}} e^{i\frac{\theta(x)}{n}}$$

Démonstration : Fixons $x \in \hat{X}$ et soit $t \longrightarrow x_t$ une application continue de [0,1] dans $\hat{X}$ telle que $x_0 = \mathbb{1}$ et $x_1 = x$. Les applications

$$t \longrightarrow \hat{\mu}(x_t) \quad \text{et} \quad t \longrightarrow \hat{\mu}_n(x_t)$$

sont alors continues de [0,1] dans $\mathbb{C} - \{0\}$ d'après le lemme 1. Il existe alors des applications continues

$t \longrightarrow \theta(\chi_t)$ et $t \longrightarrow \theta_n(\chi_t)$ de [0,1] dans $\mathbb{R}$ telles que

$$\hat{\mu}(\chi_t) = \left|\hat{\mu}(\chi_t)\right| e^{i\theta(\chi_t)}$$
$$(t\in[0,1])$$
$$\hat{\mu}_n(\chi_t) = \left|\hat{\mu}(\chi_t)\right|^{\frac{1}{n}} e^{i\theta_n(\chi_t)}$$

avec $\theta(\chi_o) = \theta_n(\chi_o) = 0$ $(\forall\ n\in\mathbb{N}^*)$ et la fonction $\theta_n(\chi_t)$ est déterminée de manière unique par cette condition. Mais $\hat{\mu}_n(\chi_t)$ est une racine $n^{ième}$ de $\hat{\mu}(\chi_t)$ donc

$$\theta_n(\chi_t) = \frac{\theta(\chi_t)}{n} + \frac{2k_n(\chi_t)\pi}{n}$$

où $k_n(\chi_t)$ est un entier de [0, n-1]. Pour tout n>0 fixé, l'application

$$t \longrightarrow k_n(\chi_t) = \frac{1}{2\pi}\,[n\theta_n(\chi_t) - \theta(\chi_t)]$$

est donc continue de [0,1] dans $\mathbb{R}$, à valeurs entières et satisfait $k_n(\chi_o) = 0$. Ceci impose $k_n(\chi_t) \equiv 0$ donc

$$\theta_n(\chi_t) = \frac{\theta(\chi_t)}{n}$$

et le lemme 2 en résulte ainsi que l'assertion (2).

Les énoncés de la Proposition (4.2) et du Théorème 2) doivent donc être modifiés comme suit

(4.2) <u>Proposition</u> : Soit μ une probabilité I.D. sur un hypergroupe X vérifiant C1 ou C2 et tel que $\hat{X}$ est connexe par arcs. Soit μ_n une suite de probabilités telles que $(\mu_n)^{*n} = \mu$ pour tout entier n>0. Alors μ_n converge étroitement vers δ_e.

<u>Théorème 2</u> : Soit X un hypergroupe commutatif, discret, infini dénombrable tel que $\hat{X}$ soit connexe par arcs et vérifiant l'une des deux conditions suivantes :

C1 : Le support de la mesure de Plancherel $\hat{\sigma}$ contient le point $\mathbb{1}$.

C2 : Il existe un voisinage V de $\mathbb{1}$ tel que $\lim_{x\to\infty} \chi(x) = 0$ pour tout $\chi\in V-\{\mathbb{1}\}$.

Alors μ est infiniment divisible si et seulement s'il existe une (unique) mesure positive ν sur X avec $\nu(e)=0$ telle que

$$\hat{\mu}(\chi) = \exp\left(\sum_{x\in X} (\overline{\chi(x)} - 1)\,\nu(x)\right) \quad (\forall\chi\in\hat{X}).$$

Remarques :

1) Si on fait des hypothèses sur μ, on peut affaiblir les hypothèses topologiques concernant $\hat{X}$. Par exemple si μ est adaptée, il suffit de supposer que $\hat{X}$ possède un voisinage de $\mathbb{1}$ connexe par arcs et le théorème 2 est valable.

2) Si μ est symétrique, la connexité par arcs de $\hat{X}$ n'est pas nécessaire, l'ancienne hypothèse "X n'a pas de sous hypergroupe compact non trivial" est suffisante.

3) On pourra aussi remplacer la connexité par arcs par la connexité si la topologie de $\hat{X}$ est telle que pour toute fonction continue
$h : \hat{X} \longrightarrow \mathbb{C} - \{0\}$, il existe une détermination continue de arg h.

Référence :

[1] L. Gallardo et O. Gebuhrer : Lois de probabilité infiniment divisibles sur les hypergroupes commutatifs discrets dénombrables.
Probability Measures on Groups VII (Proceedings d'Oberwolfach)
Lecture Notes n° 1064 (1984), p. 116-129.

Convergence-of-Types Theorem for Simply Connected Nilpotent Lie Groups

W. Hazod and S. Nobel

The history of the convergence-of-types theorem is rather long. For probabilities on $\mathbb{R}$ see e.g. [5] VIII 2. Lemma 1, [7] II § 10 Theorem 1 and 2. For a simple proof see [14]. It can be stated as follows:
Let μ_n, μ, υ be non-degenerate probabilities on $\mathbb{R}$.
Let $\tau_n: x \longmapsto a_n x + b_n$ be affine transformations on $\mathbb{R}$. Assume $\mu_n \longrightarrow \mu$ and $\tau_n(\mu_n) \longrightarrow \upsilon$. Then the sequence $\{\tau_n\}$ is relatively compact.

Results of this type are the main tool for the investigation of limit behaviour of suitably normalized sums of independent random variables, especially for the investigation of stable, semistable and self-decomposable laws. For operator-normalized random variables on finite-dimensional vector spaces a convergence-of-types theorem was proved first in [6], and independently in [2], [21] and in [18] (called compactness lemma) in connection with operator-stable laws. For similar investigations in connection with selfdecomposability and semistability see e.g. [19], [12]. For generalizations to probabilities on vector spaces see e.g. [20], [13], [15], see also [8] and the literature cited there.

In the case of non-abelian groups only few examples are known: The groups of Euclidean motions [1], the Heisenberg groups [3] resp. the diamond groups [4].

It is known that nilpotent simply connected Lie groups play an essential rôle, see [9,10] resp. the survey on stability [8]. Hence the main result of this paper, a convergence-of-types theorem for simply connected nilpotent Lie groups is the essential tool linking the general theory of (semi-) stable laws on Lie groups with limit theorems. These applications will be treated in [17].

Since the groups under consideration are non-abelian affine transformations must be handeled with care. Therefore our results are stated twice: for transformations with and without shifts.

The class of measures we have to restrict to are called full or S-full. (For the vector space-case see e.g. [18], [15]).
In § 1 we investigate full and S-full probabilities and their properties. In § 2 we present as a main result the convergence-of-types theorem for full resp. S-full measures. And in § 3 we describe fullness by compactness properties of the invariance groups.

<u>Notations</u>. G, H, N will denote simply connected nilpotent Lie groups with Lie algebras $\mathfrak{G}$, $\mathfrak{H}$, $\mathfrak{N}$ respectively. Hom(G,H), Hom($\mathfrak{G}$,$\mathfrak{H}$), Aut(G), Aut($\mathfrak{G}$) will denote the sets of continuous homomorphisms and automorphisms of the Lie groups resp. the Lie algebras.

$\exp_G := \exp : \mathfrak{G} \longrightarrow G$ is the exponential map. exp and its inverse $\log : G \longrightarrow \mathfrak{G}$ are C^∞-isomorphisms. The differential of $\tau \in \mathrm{Hom}(G,H)$ resp. $\tau \in \mathrm{Aut}(G)$ is denoted by $\mathring{\tau}$. (Note that $\tau(\exp_G(X)) = \exp_H(\mathring{\tau}X)$.)

Let M(G) resp. $M^1(G)$ be the set of bounded measures resp. of probability measures on G. exp and log define isomorphisms on function and measure spaces:

For $f \in C_o(G)$ let $\mathring{f} := f \circ \exp \in C_o(\mathfrak{G})$, for $\mu \in M(G)$ let $\mathring{\mu} \in M(\mathfrak{G})$ be defined by $\langle f,\mu\rangle = \langle \mathring{f},\mathring{\mu}\rangle$, $f \in C_o(G)$.

Automorphisms resp. homomorphisms of G operate on functions and measures in the usual way: let e.g. $\tau \in \mathrm{Hom}(G,H)$, $f \in C_o(H)$, $\mu \in M(G)$.
Then $\tau(f) := f \circ \tau$, $\langle \tau(\mu),f\rangle = \langle \mu,\tau(f)\rangle$. Per definition we have $(\tau(f))^\circ = \mathring{\tau}(\mathring{f})$ and $(\tau(\mu))^\circ = \mathring{\tau}(\mathring{\mu})$.

Let $N \subseteq G$ be a normal subgroup, then let $\pi_N : G \longrightarrow G/_N$ be the canonical projection. Similar, if $\mathfrak{N} \subseteq \mathfrak{G}$ is an ideal, then let $\pi_{\mathfrak{N}} : \mathfrak{G} \longrightarrow \mathfrak{G}/_{\mathfrak{N}}$ be the canonical projection. There is a 1-1-correspondence between ideals $\mathfrak{N} \subseteq \mathfrak{G}$ and closed connected normal subgroups $N \subseteq G$.

For a subgroup $A \subseteq G$ let N(A) be the normalizer of A in G, for a subalgebra $\mathfrak{A} \subseteq \mathfrak{G}$ let $\mathfrak{N}(\mathfrak{A})$ be the normalizer of $\mathfrak{A}$ in $\mathfrak{G}$.

We assume $\mathfrak{G}$ to be step r+1 nilpotent, i.e. the descending central series is of the form

$$\mathfrak{G} = \mathfrak{G}^{(o)} \supseteq \mathfrak{G}^{(1)} = [\mathfrak{G},\mathfrak{G}] \supseteq \ldots \supseteq \mathfrak{G}^{(r)} \supseteq \mathfrak{G}^{(r+1)} = \{0\}.$$

$\mathfrak{Z}(\mathfrak{G})$ resp. Z(G) denotes the centre of $\mathfrak{G}$ resp. of G.

§ 1 Full probabilities.

Following [15] we consider two concepts of fullness: A probability measure μ on $\mathbb{R}$ is called full iff $\mu \neq \varepsilon_o$, S-full iff μ is non-degenerate i.e. $\mu \neq \varepsilon_x$, $x \in \mathbb{R}$. Equivalently, μ is S-full iff $\mu * \tilde{\mu}$ is full.
A probability measure on $\mathbb{R}^d$ is called full if it is not concentrated on a proper subspace. Hence μ is full iff for any projection $\pi : \mathbb{R}^d \longrightarrow \mathbb{R}$ the measure $\pi(\mu)$ is full on $\mathbb{R}$. Similar, μ is called S-full on $\mathbb{R}^d$ iff μ is not concentrated on a coset of a proper subspace, hence iff μ is not concentrated on a hyperplane. Equivalently, μ is S-full on $\mathbb{R}^d$ iff $\pi(\mu)$ is S-full on $\mathbb{R}$ for any projection $\pi : \mathbb{R}^d \longrightarrow \mathbb{R}$, resp. iff $\mu * \tilde{\mu}$ is full.

The difference between these concepts of fullness is carefully worked out in [15]. S-full measures are called non-degenerate in [2], non-singular in [6] resp. full in [18] and subsequent papers.

We now define fullness in the non-commutative case in a natural way and show that this property is closely related to the commutative structure.

<u>1.1</u> <u>Definition</u>. Let $\mu \in M^1(G)$. Then we denote by $S(\mu)$ the support of μ, by $G(\mu)$ the closed subgroup generated by $S(\mu)$. Further we denote by $CG(\mu)$ resp. $NG(\mu)$ the closed connected resp. closed connected normal subgroup generated by $S(\mu)$.

Obviously we have $S(\mu) \subseteq G(\mu) \subseteq CG(\mu) \subseteq NG(\mu)$.

<u>1.2</u> <u>Definition</u>. Let $\mu \in M^1(G)$.

μ is called full if $CG(\mu) = G$, i.e. if μ is not concentrated on a proper closed connected subgroup of G.

μ is called S-full if μ is not concentrated on a (right or left) coset of a proper closed connected subgroup. (We see later that it is not necessary to distinguish between right and left cosets).

The following lemma is folklore (see e.g. [16] Theorem 2.1).
We sketch a proof to make the paper more self-contained.

<u>1.3</u> <u>Lemma</u>. Let $\mathscr{H} \subsetneqq \mathscr{G}$ be a proper subalgebra. Then there exists a proper ideal $\mathscr{N} \subsetneqq \mathscr{G}$, such that $\mathscr{H} \subseteq \mathscr{N} \neq \mathscr{G}$. $\mathscr{N}$ can be chosen such that codim $\mathscr{N} = 1$, i.e. $\mathscr{G}/_{\mathscr{N}} \cong \mathbb{R}$. $\mathscr{N}$ is called a hyperplane-subalgebra then.

⟦ $\mathscr{H}$ is nilpotent. According to [11] XIV Prop. 1.1. we have $\mathscr{H} \subsetneq \mathscr{N}(\mathscr{H})$ if $\mathscr{H} \neq \mathscr{G}$.

So we obtain an ascending chain of subalgebras

$\mathscr{H} := \mathscr{H}_o \subset \mathscr{H}_1 = \mathscr{N}(\mathscr{H}_o) \subset \mathscr{N}(\mathscr{H}_1) \ldots$. The last element $\mathscr{H}_r \neq \mathscr{G}$ of this chain is a proper ideal containing $\mathscr{H}$.

The last assertion follows since any ideal in a nilpotent Lie algebra is contained in an ideal of codimension 1.⟧

<u>1.4</u> <u>Corollary.</u> Let H be a proper closed connected subgroup of G. Then there exists a proper closed connected normal subgroup $N \subsetneq G$, such that $H \subseteq N \neq G$. N can be chosen such that $G/_N \cong \mathbb{R}$.

In this case we call N a hyperplane-subgroup.

With this notations we can reformulate the definition of fullness:

<u>1.5</u> <u>Corollary.</u> Let $\mu \in M^1(G)$. Then

(i) μ is full iff μ is not concentrated on a hyperplane-subgroup, equivalently

(i^*) μ is full on G iff $\pi(\mu)$ is full on $\mathbb{R}$ for any surjective continuous homomorphism $\pi : G \longrightarrow \mathbb{R}$. ($\pi$ is open in this case).

(ii) μ is S-full iff μ is not concentrated on a coset of a hyperplane-subgroup, equivalently

(ii^*) μ is S-full on G iff $\pi(\mu)$ is S-full on $\mathbb{R}$ for any surjective continuous homomorphism $\pi : G \longrightarrow \mathbb{R}$.

<u>1.6</u> <u>Corollary.</u> Let $\mu \in M^1(G)$, $k \in \mathbb{N}$. Then

(i) μ is full [S-full] iff μ^k is full [S-full],

(ii) μ is S-full iff $\tilde{\mu}$ is S-full,

(iii) μ is S-full iff $\mu * \tilde{\mu}$ is full [iff $\tilde{\mu} * \mu$ is full].

⟦(i): μ is not full [S-full] iff for some hyperplane-subgroup N we have $\pi_N(\mu) = \varepsilon_o$ [resp. $\pi_N(\mu) = \varepsilon_x$, $x \in \mathbb{R}$]. This is the case iff $\pi_N(\mu^k) = \varepsilon_o$ [resp. $\pi_N(\mu^k) = \varepsilon_{kx}$].

(ii), (iii): Assume $\pi_N(\mu) = \varepsilon_x$, $x \in \mathbb{R}$, for some hyperplane subgroup N. Then $\pi_N(\tilde{\mu}) = \varepsilon_{-x}$ and $\pi_N(\mu * \tilde{\mu}) = \varepsilon_x * \varepsilon_{-x} = \varepsilon_o$, and vice versa.⟧

1.7 Proposition. Let $\mu \in M^1(G)$ and let $\upsilon = \exp(\mu - \varepsilon_e)$ be the corresponding Poisson measure.
Then we have: μ is full iff υ is full. This is the case iff υ is S-full.

(Since $e \in S(\upsilon)$, υ is full iff υ is S-full).

⟦ Follows immediately from the representation

$\upsilon = e^{-1}(\varepsilon_e + \sum_{k=1}^{\infty} \upsilon^k/k!)$ and from 1.6 (i), or directly from 1.5 (ii*).⟧

The following proposition shows a first connection between fullness of $\mu \in M^1(G)$ and of the corresponding measure $\mathring{\mu}$ on the tangent space $\mathcal{G}$:

1.8 Proposition. Let $\mu \in M^1(G)$ and $\mathring{\mu} \in M^1(\mathcal{G})$. Then we have:

(i) μ is full on G iff $\pi_{\mathcal{N}}(\mathring{\mu})$ is full on $\mathbb{R}$ for any hyperplane-algebra $\mathcal{N} \subset \mathcal{G}$.

Especially, if $\mathring{\mu}$ is full on the vector space $\mathcal{G}$, then μ is full on the group G.

(ii) μ is S-full on G iff $\pi_{\mathcal{N}}(\mathring{\mu})$ is S-full on $\mathbb{R}$ for any hyperplane-algebra $\mathcal{N} \subset \mathcal{G}$.

Especially, if $\mathring{\mu}$ is S-full on the vector space $\mathcal{G}$, then μ is S-full on the group G.

⟦ The first assertion is an immediate consequence of 1.5. To prove the second assertion we note that for any closed connected normal subgroup $N \subset G$ and the corresponding ideal $\mathcal{N} \subset \mathcal{G}$ of codimension 1 and for any $X \in \mathcal{G}$ we have $\log(\exp(X)\cdot N) = X + \mathcal{N}$.
Now assertion (ii) follows immediately.⟧

Now we are able to show that the concepts of fullness are essentially connected with the underlying vector space structure of the tangent space $\mathcal{G}$. We need two lemmata.

1.9 Lemma. Let $\mu \in M^1(G)$, $\mathring{\mu} \in M^1(\mathcal{G})$.

(i) μ is full on G iff $\pi_{[G,G]}(\mu)$ is full on (the abelian Lie group) $G/_{[G,G]} \cong \mathcal{G}/_{[\mathcal{G},\mathcal{G}]}$.

(ii) Equivalently, $\mathring{\mu}$ is concentrated on a proper ideal in $\mathcal{G}$ iff $\pi_{[\mathcal{G},\mathcal{G}]}(\mathring{\mu})$ is concentrated on a proper subspace of $\mathcal{G}/_{[\mathcal{G},\mathcal{G}]}$.

⟦ Let $\mathfrak{I}$ be a hyperplane algebra. Then $[\mathcal{G},\mathcal{G}] \subseteq \mathfrak{I}$. If $\mathring{\mu}$ is concentrated on $\mathfrak{I}$ then $\pi_{[\mathcal{G},\mathcal{G}]}(\mathring{\mu})$ is concentrated on the proper subspace $\mathfrak{I}/_{[\mathcal{G},\mathcal{G}]}$. Conversely, let $\pi_{[\mathcal{G},\mathcal{G}]}(\mathring{\mu})$ be concentrated on the proper subspace $\mathcal{V}$ of $\mathcal{G}/_{[\mathcal{G},\mathcal{G}]}$. Then $\mathring{\mu}$ is concentrated on the ideal $\pi^{-1}_{[\mathcal{G},\mathcal{G}]}(\mathcal{V}) \subset \mathcal{G}$.⟧

<u>1.10</u> <u>Lemma.</u> Let $A \subseteq G$ be a subsemigroup, let $\mathcal{A} := \log(A)$ and let $\mathcal{V}$ be the linear subspace generated by $\mathcal{A}$. Then $\mathcal{V}$ is a Lie subalgebra of $\mathcal{G}$.

<u>Proof:</u> Let $X, Y \in \mathcal{A}$, $x = \exp X$, $y = \exp Y$, $k \in \mathbb{N}$. We have $x^k \cdot y^k \in A$, therefore $Z(k) := \log(x^k y^k) \in \mathcal{A}$. The Campbell-Hausdorff formula yields for any $U,V \in \mathcal{G}$ the following representation:

$\log(\exp U \cdot \exp V) = U + V + \sum_1^r c_j\ h_j(U,V)$, where $c_j \in \mathbb{R}\setminus\{0\}$ and h_j are homogeneous $[\cdot,\cdot]$- polynomials of degree $j+1$. Therefore for $k \in \mathbb{N}$

$$Z(k) = \log(x^k y^k) = kX + kY + \sum_1^r c_j\ k^{j+1}\ h_j(X,Y).$$

Hence $\frac{1}{k^{r+1}} Z(k) \in \mathcal{V}$, and therefore

$$h_r(X,Y) = \lim_{k\to\infty} \frac{1}{c_r} \frac{1}{k^{r+1}} Z(k) \in \mathcal{V}.$$

We obtain $\sum_1^{r-1} c_j\ h_j(X,Y) \in \mathcal{V}$, and repeating the arguments above, we finally get $h_1(X,Y) = [X,Y] \in \mathcal{V}$.

Thus we have proved $[\mathcal{A},\mathcal{A}] \subseteq \mathcal{V}$, whence $[\mathcal{V},\mathcal{V}] \subseteq \mathcal{V}$ follows . □

<u>1.11</u> <u>Theorem</u>. Let $\mu \in M^1(G)$, $\mathring{\mu} \in M^1(\mathcal{G})$.

(i) If $\mathring{\mu}$ is full on $\mathcal{G}$ then μ is full on G.

(ii) Assume in addition that $S(\mu)$ is a semigroup. Then the converse is true: If μ is full on G then $\mathring{\mu}$ is full on the vector space $\mathcal{G}$.

<u>Proof:</u> (i) is proved in 1.8 (i).

(ii): Assume $S(\mu)$ to be a semigroup. Let $\mathcal{V}$ be the subspace of $\mathcal{G}$ generated by $S(\mathring{\mu})$. According to 1.10 $\mathcal{V}$ is a subalgebra.
If $\mathring{\mu}$ is not full on $\mathcal{G}$ then $\mathcal{V}$ is a proper subalgebra on which $\mathring{\mu}$ is

concentrated, hence μ is concentrated on the corresponding connected subgroup V of G. Therefore μ is not full. □

1.12 Lemma. Let $\mu, \upsilon \in M^1(G)$, let $\mathring{\mu}, \mathring{\upsilon}$ the corresponding measures on $\mathcal{G}$. Then $\pi_{[\mathcal{G},\mathcal{G}]}((\mu * \upsilon)^\circ) = \pi_{[\mathcal{G},\mathcal{G}]}(\mathring{\mu} * \mathring{\upsilon})$.

On the left * denotes convolution with respect to the group operation on G, on the right with respect to addition on $\mathcal{G}$.

Hence $\mathring{\mu}$ is concentrated on a coset of a hyperplane-algebra $\mathcal{N}$ iff $(\mu * \tilde{\mu})^\circ$ is concentrated on $\mathcal{N}$.

⟦ Obvious, since $G/_{[G,G]}$ is algebraically and topologically isomorphic to the vector space $\mathcal{G}/_{[\mathcal{G},\mathcal{G}]}$.⟧

In order to prove results similar to 1.11 for S-fullness we need the following lemma:

1.13 Lemma. Let $A \subseteq G$ be a monoid, $x_o \in G$ and $X_o := \log x_o$. Let further $\mathcal{A} := \log A$ and $\mathcal{B} := \log(x_o A)$. Assume $\mathcal{G} = \langle\mathcal{A}\rangle$ to be the vector space generated by $\mathcal{A}$. Let $\mathcal{V} := \langle\mathcal{B} - \mathcal{B}\rangle$ be the vector space generated by $\mathcal{B} - \mathcal{B}$.

Then $\mathcal{V} = \mathcal{G}$.

Proof: Since A is a monoid we have $x_o \in x_o A$, hence $X_o \in \mathcal{B}$. Therefore $\langle\mathcal{B} - X_o\rangle = \langle\mathcal{B} - \mathcal{B}\rangle$.

For any $U,V \in \mathcal{G}$ the Campbell-Hausdorff formula can be written as follows: $\log(\exp U \exp V) = U + \sum_1^r \beta_j(U,V)$, where the β_j are homogeneous of degree j in the variable V. (I.e. $\beta_j(U,V)$ is the sum of all brackets of order k, $j \le k \le r$, such that V appears exactly j-times).

Repeating the arguments of the proof of 1.10 we obtain for $y \in A$, $Y = \log y \in \mathcal{A}$, $k \in \mathbb{N}$:

$\beta_r(X_o,Y) = \lim_{k\to\infty} \frac{1}{k^r} (\log(x_o y^k) - X_o) \in \mathcal{V}$ and successively

$\beta_{r-1}(X_o,Y) \in \mathcal{V}$, ... finally $\beta_1(X_o,Y) \in \mathcal{V}$.

Note that $Y \longmapsto \beta_1(X_o,Y)$ is linear. Let $X \in \mathcal{G}^{(r)} \subseteq \mathcal{Z}(\mathcal{G})$.

Since $\langle\mathcal{A}\rangle = \mathfrak{g}$, there exists a representation

$X = \sum \alpha_i Y_i$, $\alpha_i \in \mathbb{R}$, $Y_i \in \mathcal{A}$. Therefore $\beta_1(X_o, X) = \sum \alpha_i \beta_1(X_o, Y_i) \in \mathcal{V}$.

On the other hand $\beta_1(X_o, X) = X$, since $X \in \mathfrak{g}^{(r)}$ and therefore all bracket products of higher order vanish.

We obtain $\mathfrak{g}^{(r)} \subseteq \mathcal{V}$.

Repeating these arguments, we obtain for $X \in \mathfrak{g}^{(r-1)}$: $\beta_2(X_o, X) \in \mathcal{V}$, and on the other hand $\beta_2(X_o, X) = X + \alpha [X_o, X]$, some $\alpha \in \mathbb{R}$.

Since $[X_o, X] \in \mathfrak{g}^{(r)} \subseteq \mathcal{V}$, we obtain $X \in \mathcal{V}$, hence $\mathfrak{g}^{(r-1)} \subseteq \mathcal{V}$, ..., finally $\mathfrak{g} \subseteq \mathcal{V}$ as asserted. □

<u>1.14</u> <u>Theorem.</u> Let $\mu, \upsilon \in M^1(G)$, $\mathring{\mu} \in M^1(\mathfrak{g})$.

(i) If $\mathring{\mu}$ is S-full on $\mathfrak{g}$ then μ is S-full on G.

(ii) Assume $\mu = \varepsilon_{x_o} * \upsilon$, where $S(\upsilon)$ is a monoid in G. Then the converse is true: If μ is S-full on G then $\mathring{\mu}$ is S-full on the vector space $\mathfrak{g}$.

<u>Proof:</u> (i) is proved in 1.8 (ii).

(ii): Assume μ to be S-full. This is the case iff υ is S-full and since $e \in S(\upsilon)$ iff υ is full.

According to 1.10 the vector space $\langle S(\mathring{\upsilon})\rangle$ generated by $S(\mathring{\upsilon})$ equals $\mathfrak{g}$.

On the other hand according to 1.13 the vector space generated by $S(\mathring{\mu}) - S(\mathring{\mu})$ is $\mathfrak{g}$, i.e. $\mathring{\mu} * (\mathring{\mu})^{\sim}$ is full on $\mathfrak{g}$. Therefore $\mathring{\mu}$ is S-full on $\mathfrak{g}$. □

It is well-known that S-full measures on a finite-dimensional vector space $\mathbb{E}$ form a subsemigroup of $M^1(\mathbb{E})$ which is open with respect to weak topology ([18,19]).

Let $\mathcal{F}(G)$ resp. $\mathcal{F}_S(G)$ be the subsets of full resp. S-full measures on G. Then we have:

<u>1.15</u> <u>Proposition.</u> $\mathcal{F}(G)$ and $\mathcal{F}_S(G)$ are open subsets of $M^1(G)$ w.r.t. weak topology. Moreover $\mathcal{F}_S(G)$ is a convolution semigroup.

<u>Proof:</u> 1.9 yields (i) $\mathcal{F}(G) = \pi^{-1}_{[G,G]}(\mathcal{F}(G/_{[G,G]}))$ and

(ii) $\mathcal{F}_S(G) = \pi^{-1}_{[G,G]}(\mathcal{F}_S(G/_{[G,G]}))$.

(The second assertion follows since $\mu \in \mathcal{F}_S(G)$ iff $\tilde{\mu} * \mu \in \mathcal{F}(G)$).
Since $G/_{[G,G]} \cong \mathfrak{G}/_{[\mathfrak{G},\mathfrak{G}]}$, since $\pi_{[G,G]}$ is a continuous homomorphism and since $\mathcal{F}_S(\mathfrak{G}/_{[\mathfrak{G},\mathfrak{G}]})$ is an open subsemigroup on the vector space $\mathfrak{G}/_{[\mathfrak{G},\mathfrak{G}]}$, we obtain: $\mathcal{F}_S(G)$ is an open subsemigroup of $M^1(G)$.
Let $\phi : M^1(G) \ni \mu \longmapsto \exp(\mu - \varepsilon_e) \in M^1(G)$.
Lemma 1.7 yields $\mathcal{F}(G) = \phi^{-1}(\mathcal{F}_S(G))$. Since ϕ is continuous $\mathcal{F}(G)$ is open (but in general not a convolution semigroup). □

<u>1.16 Remark.</u> A continuous convolution semigroup $(\mu_t)_{t\geq 0}$ is called full [S-full] iff every μ_t, $t > 0$, is full [S-full]. For vector spaces the fullness [S-fullness] of the convolution semigroup $(\mu_t)_{t\geq 0}$ is equivalent to the fullness [S-fullness] of μ_1. Lemma 1.9 yields that this holds also true in the group case.

§ 2 The Convergence-of-Types Theorem

We start with a version of the compactness lemma resp. convergence-of-types theorem for finite-dimensional vector spaces. This is well-known (see e.g. [18], [2], [6], [19], [12], [15], [21]). For the sake of completeness we give a short proof:

<u>2.1 Lemma.</u> Let $\mathbb{E}$, $\mathbb{F}$ be finite-dimensional vector spaces. Let $\mathrm{Hom}_V(\mathbb{E},\mathbb{F})$ denote the set of vector space homomorphisms.
Let $\lambda_n, \lambda \in M^1(\mathbb{E}), \mu \in M^1(\mathbb{F})$ and $\tau_n \in \mathrm{Hom}_V(\mathbb{E},\mathbb{F})$, $n \in \mathbb{N}$. Suppose $\sup_{n\in\mathbb{N}} \|\tau_n\| = \infty$. Suppose further $\lambda_n \xrightarrow[n \to \infty]{} \lambda$ and $\tau_n\lambda_n \xrightarrow[n \to \infty]{} \mu$.
Then λ is concentrated on a proper subspace of $\mathbb{E}$, i.e. λ is not full.

⌈ W.l.o.g. assume $\|\tau_n\| \xrightarrow[n\to\infty]{} \infty$. Let $\sigma_n := \frac{1}{\|\tau_n\|} \tau_n$.
Then $\|\sigma_n\| = 1$, $\sigma_n \in \mathrm{Hom}_V(\mathbb{E},\mathbb{F})$. Hence the sequence is relatively compact. Assume w.l.o.g. $\sigma_n \xrightarrow[n\to\infty]{} \sigma \in \mathrm{Hom}_V(\mathbb{E},\mathbb{F})$, $\|\sigma\| = 1$.
Continuity implies $\sigma_n\lambda_n \xrightarrow[n\to\infty]{} \sigma\lambda$.
But $\sigma_n = (\frac{1}{\|\tau_n\|} \cdot \mathrm{id}_{\mathbb{F}}) \circ \tau_n$.
Therefore, $\frac{1}{\|\tau_n\|} \xrightarrow[n\to\infty]{} 0$ and $\tau_n\lambda_n \xrightarrow[n\to\infty]{} \mu$ imply $\sigma_n\lambda_n \xrightarrow[n\to\infty]{} \varepsilon_o$.

Hence $\sigma(\lambda) = \varepsilon_o$, i.e. λ is concentrated on the proper subspace $\ker(\sigma)$.∎

Note that in the case of vector spaces, (i.e. abelian Lie algebras) we have: $\alpha \in \mathbb{R}$, $\tau \in \mathrm{Hom}_V(\mathbb{E},\mathbb{F}) \Longrightarrow \alpha\cdot\tau \in \mathrm{Hom}_V(\mathbb{E},\mathbb{F})$. This is of course not true for homomorphisms of non-abelian Lie algebras. Hence the proof of 2.1 does not work for arbitrary nilpotent Lie groups.

Let G, H be simply connected nilpotent Lie groups with Lie algebras $\mathcal{G}$ and $\mathcal{H}$ respectively. Let Hom(G,H) resp. $\mathrm{Hom}(\mathcal{G},\mathcal{H})$ denote the set of Lie group resp. Lie algebra homomorphisms. $\mathrm{Hom}(\mathcal{G},\mathcal{H})$ inherits the topology of the set of linear operators $\mathrm{Hom}_V(\mathcal{G},\mathcal{H})$ and by the isomorphism $\tau \longleftrightarrow \mathring{\tau}$ the topology on Hom(G,H) is defined. ($\mathrm{Hom}(\mathcal{G},\mathcal{H})$ is closed in $\mathrm{Hom}_V(\mathcal{G},\mathcal{H})$). Note that in the case G = H the usual topology on Aut(G) and the topology defined above coincide.

<u>2.2</u> <u>Theorem.</u> (Convergence-of-Types Theorem resp. Compactness Lemma for Homomorphisms).

Let λ_n, $\lambda \in M^1(G)$, $n \in \mathbb{N}$, $\mu \in M^1(H)$. Let $(\tau_n)_{n\in\mathbb{N}}$ be a sequence in Hom(G,H).

Assume $\lambda_n \xrightarrow[n\to\infty]{} \lambda$ and $\tau_n\lambda_n \xrightarrow[n\to\infty]{} \mu$.

(i) If λ is full then $\{\tau_n\}$ is relatively compact in Hom(G,H). For any limit point τ of $\{\tau_n\}$ we have $\tau\lambda = \mu$.

(ii) Let G = H and $\tau_n \in \mathrm{Aut}(G)$, $n \in \mathbb{N}$. If μ is full then $\{\tau_n^{-1}\}$ is relatively compact in Hom(G,G). For any limit point σ of $\{\tau_n^{-1}\}$ we have $\sigma\mu = \lambda$.

(iii) Let G = H and $\tau_n \in \mathrm{Aut}(G)$, $n \in \mathbb{N}$. If λ and μ are full then $\{\tau_n\}$ (and hence $\{\tau_n^{-1}\}$) is relatively compact in Aut(G). For any limit point τ of $\{\tau_n\}$ we have $\tau\lambda = \mu$ and $\tau^{-1}\mu = \lambda$.

<u>Prove:</u> If we replace λ_n, λ etc. by the Poisson measures $\lambda_n^* := \exp(\lambda_n - \varepsilon_e)$, $\lambda^* := \exp(\lambda - \varepsilon_e)$ etc., we have $\lambda_n^* \longrightarrow \lambda^*$, $\tau_n(\lambda_n^*) \longrightarrow \mu^*$. Proposition 1.7 allows us to assume that the supports of the measures involved are semigroups (indeed monoids). Hence w.l.o.g. $\lambda_n = \lambda_n^*$ etc. .

For the corresponding measures on the Lie algebras we obtain

$\mathring{\lambda}_n \xrightarrow[n\to\infty]{} \mathring{\lambda}$ in $M^1(\mathcal{G})$, $\mathring{\tau}_n\mathring{\lambda}_n \xrightarrow[n\to\infty]{} \mathring{\mu}$ in $M^1(\mathcal{H})$, and according to 1.11 λ is full on G iff $\mathring{\lambda}$ is full on $\mathcal{G}$.

Lemma 2.1 applies and yields the relative compactness of $\{\mathring{\tau}_n\}$ in $\mathrm{Hom}_V(\mathcal{G},\mathcal{H})$, hence in $\mathrm{Hom}(\mathcal{G},\mathcal{H})$ and hence of $\{\tau_n\}$ in Hom(G,H).

Let w.l.o.g. $\tau_n \xrightarrow[n\to\infty]{} \tau \in \mathrm{Hom}(G,H)$. Continuity of the action of Hom(G,H) implies $\tau_n\lambda_n \xrightarrow[n\to\infty]{} \tau(\lambda)$. Therefore $\tau(\lambda) = \mu$ as asserted.

To prove (ii) replace λ_n by $\mu_n := \tau_n(\lambda_n)$, λ by μ, τ_n by τ_n^{-1} and μ by λ. Now apply (i). Combining (i) and (ii) we obtain (iii). □

If we replace fullness by S-fullness we obtain similar results for affine transformations:

2.3 Theorem. (Convergence-of-Types Theorem resp. Compactness Lemma for affine transformations).

Let λ_n, $\lambda \in M^1(G)$, $n \in \mathbb{N}$, $\mu \in M^1(H)$. Let $(\tau_n)_{n\in\mathbb{N}}$ be a sequence in Hom(G,H) and $(x_n)_{n\in\mathbb{N}}$ a sequence in G. Put $y_n := \tau_n(x_n)$.

Assume $\lambda_n \xrightarrow[n\to\infty]{} \lambda$ and $\tau_n(\lambda_n * \varepsilon_{x_n}) = \tau_n(\lambda_n) * \varepsilon_{y_n} \xrightarrow[n\to\infty]{} \mu$.

(i) If λ is S-full then $\{\tau_n\}$ and $\{y_n\}$ are relatively compact in Hom(G,H) resp. in H. Let (τ,y) be a limit point of $\{(\tau_n,y_n)\}$. Then $\tau(\lambda) * \varepsilon_y = \mu$.

(ii) Let G = H and let $\tau_n \in \mathrm{Aut}(G)$.
If μ is S-full then $\{\tau_n^{-1}\}$ and $\{x_n\}$ are relatively compact in Hom(G,G) resp. G.
Let (σ,x) be a limit point of $\{(\tau_n^{-1},x_n)\}$.
Then $\sigma(\mu) * \varepsilon_{x^{-1}} = \lambda$.

(iii) Let G = H and $\tau_n \in \mathrm{Aut}(G)$. If λ and μ are S-full, then $\{\tau_n\}$ and $\{x_n\}$ are relatively compact in Aut(G) resp. G.
Let (τ,x) be a limit point of $\{(\tau_n,x_n)\}$.
Then $\tau(\lambda * \varepsilon_x) = \mu$, $\tau^{-1}(\mu) * \varepsilon_{x^{-1}} = \lambda$.

Proof: Put $\upsilon_n := \lambda_n * \tilde{\lambda}_n$, $\upsilon := \lambda * \tilde{\lambda}$, $\rho := \mu * \tilde{\mu}$.

Then $\upsilon_n \xrightarrow[n\to\infty]{} \upsilon$ and $\tau_n \upsilon_n \xrightarrow[n\to\infty]{} \rho$. We apply Theorem 2.2 and obtain

in (i): $\{\tau_n\}$ is relatively compact in Hom(G,H),

in (ii): $\{\tau_n^{-1}\}$ is relatively compact in Hom(G,G), and

in (iii): $\{\tau_n\}$ is relatively compact in Aut(G).

By assumption $\{\lambda_n\}$ and $\{\tau_n(\lambda_n) * \varepsilon_{y_n}\}$ are relatively compact.

The relative-compactness of $\{\tau_n\}$ implies the relative-compactness of $\{\tau_n(\lambda_n)\}$. Hence $\{y_n\}$ is relatively compact as asserted in (i).

The continuity of the action of affine transformations on G implies $\tau(\lambda) * \varepsilon_y = \mu$ for any limit point (τ, y) of $\{(\tau_n, y_n)\}$.

To prove (ii) put $\mu_n := \tau_n(\lambda_n * \varepsilon_{x_n})$, $y_n := \tau_n(x_n)$.

Then $\mu_n \xrightarrow[n\to\infty]{} \mu$, $\tau_n^{-1}(\mu_n * \varepsilon_{y_n^{-1}}) = \lambda_n \xrightarrow[n\to\infty]{} \lambda$.

The assertion follows from (i).

(iii) is proved combining (i) and (ii). □

§ 3 Compactness of the invariance group.

For finite-dimensional vector spaces it is well-known that probabilities are full iff the invariance groups are compact.
(This is not true for infinite-dimensional vector spaces). For general convolution structures fullness was defined by the compactness of the invariance group [8]. We want to show that for simply connected nilpotent Lie groups this leads to an equivalent definition.

3.1 Definition. Let $\mu \in M^1(G)$.

(i) $\mathfrak{I}(\mu) := \{\tau \in \mathrm{Aut}(G): \tau\mu = \mu\}$ is called the invariance group.

(ii) $\mathfrak{I}_s(\mu) := \{\tau \in \mathrm{Aut}(G): \tau\mu = \mu * \varepsilon_x$ for some $x \in G\}$ is called the shift-invariance group.

(iii) $\mathfrak{I}_a(\mu) := \{(\tau, x) \in \mathrm{Aut}(G) \times G: \tau\mu = \mu * \varepsilon_x\}$ is called the affine invariance group.

Obviously we have $\mathfrak{I}(\mu) \subseteq \mathfrak{I}_S(\mu) \subseteq \mathfrak{I}(\mu * \tilde{\mu})$ and $\mathfrak{I}(\mu)$, $\mathfrak{I}_S(\mu)$ resp. $\mathfrak{I}_a(\mu)$ are closed in Aut(G) resp. Aut(G) $\times$ G.

<u>3.2 Proposition.</u> (i) If μ is full then $\mathfrak{I}(\mu)$ is compact.

(ii) If μ is S-full then $\mathfrak{I}_S(\mu)$ and $\mathfrak{I}_a(\mu)$ are compact in Aut(G) resp. Aut(G) $\times$ G.

⟦ We apply theorem 2.2 (iii) to prove (i) resp. theorem 2.3 (iii) to prove (ii):

(i): Let $(\tau_n)_{n\in\mathbb{N}}$ be a sequence in $\mathfrak{I}(\mu)$.
Theorem 2.2 (iii) applies to $\lambda_n = \mu$, $\lambda = \mu$, $\tau_n\lambda_n = \tau_n\mu = \mu$, $n \in \mathbb{N}$.
Hence $\{\tau_n\}$ is relatively compact in Aut(G) and for every limit point τ we have $\tau\mu = \mu$.

(ii): Let $(\tau_n,x_n)_{n\in\mathbb{N}}$ be a sequence in $\mathfrak{I}_a(\mu)$. Theorem 2.3 (iii) applies to $\lambda_n = \mu$, $\lambda = \mu$, $\tau_n(\lambda_n) * \varepsilon_{x_n^{-1}} = \tau_n\mu * \varepsilon_{x_n^{-1}} = \mu$, $n \in \mathbb{N}$.
Hence $\{(\tau_n,x_n)\}$ is relatively compact in Aut(G) $\times$ G and for every limit point (τ,x) we have $\tau(\mu) * \varepsilon_{x^{-1}} = \mu$.⟧

At the conference at Oberwolfach it was posed as an open problem if fullness can be characterized in this way. Independently Mr. McCrudden and Mr. Siebert informed the first named author that the converse of proposition 3.2 holds true:

<u>3.3 Proposition.</u> If μ is not full then $\mathfrak{I}(\mu)$ is not compact. Indeed, $\mathfrak{I}(\mu)$ contains a non-relatively compact one-parameter subgroup. Hence the connected component $\mathfrak{I}(\mu)_o$ is not compact.

<u>Proof:</u> <u>1.</u> Let $\mathscr{V}$ be a hyperplane-algebra in $\mathfrak{g}$ on which $\mathring{\mu}$ is concentrated. Then there exist non-trivial Y, Z, such that $Z \in \mathcal{Z}(\mathfrak{g}) \cap \mathscr{V}$ and $Y \notin \mathscr{V}$.

⟦ a) Assume there exists $Y \in \mathcal{Z}(\mathfrak{g})$, $Y \notin \mathscr{V}$. Since $\mathscr{V}$ is nilpotent there exists a non-trivial $Z \in \mathcal{Z}(\mathscr{V})$. We have $[Z,Y] = 0$, and $[Z,\mathscr{V}] = 0$, hence $Z \in \mathcal{Z}(\mathfrak{g})$.

b) Assume now $\mathcal{Z}(\mathfrak{g}) \subseteq \mathscr{V}$, $Y \in \mathfrak{g}\setminus\mathscr{V}$.
Since $\mathcal{Z}(\mathfrak{g}) \neq \{0\}$ $\mathcal{Z}(\mathfrak{g}) \cap \mathscr{V} \neq \{0\}$ as asserted.⟧

<u>2.</u> The set $\{\mathring{\tau} \in \mathrm{Aut}(\mathcal{G}) : \mathring{\tau}|_{\mathcal{V}} = \mathrm{id}_{\mathcal{V}}\}$ contains a non-relatively compact one-parameter subgroup.

⟦ Fix X, Z as in <u>1.</u> Define for $t \in \mathbb{R}$ $\mathring{\tau}_t$ in the following way:

$\mathring{\tau}_t|_{\mathcal{V}} = \mathrm{id}_{\mathcal{V}}$, $\mathring{\tau}_t Y = Y + tZ$, $t \in \mathbb{R}$.

Obviously we have: $\mathring{\tau}_t[U,V] = [\mathring{\tau}_t U, \mathring{\tau}_t V]$ for $U,V \in \mathcal{G}$ and $\mathring{\tau}_t \in \mathrm{Gl}(\mathcal{G})$, hence $\mathring{\tau}_t \in \mathrm{Aut}(\mathcal{G})$. Further $\mathring{\tau}_t \mathring{\tau}_s = \mathring{\tau}_{t+s}$, $t,s \in \mathbb{R}$ and the orbit $\{\mathring{\tau}_t Y\}_{t \in \mathbb{R}}$ is not relatively compact.⟧

<u>3.</u> $\tau_t \in \mathfrak{I}(\mu)$.

⟦ Since $\mathring{\mu}$ is concentrated on $\mathcal{V}$ and $\mathring{\tau}_t|_{\mathcal{V}} = \mathrm{id}_{\mathcal{V}}$, we have $\mathring{\tau}_t \mathring{\mu} = \mathring{\mu}$. Hence the assertion.⟧ □

Combining 3.2 and 3.3 we obtain:

<u>3.4 Theorem.</u> Let $\mu \in M^1(G)$. Then we have

(i) μ is full iff $\mathfrak{I}(\mu)$ is compact. This is the case iff the connected component $\mathfrak{I}(\mu)_o$ is compact.

(ii) μ is S-full iff $\mathfrak{I}_S(\mu)$ is compact.
This is the case iff $(\mathfrak{I}_S(\mu))_o$ is compact.

(iii) μ is S-full iff $\mathfrak{I}_a(\mu)$ is compact.

<u>Proof:</u> (i) If μ is full then $\mathfrak{I}(\mu)$ is compact according to 3.2. Hence $(\mathfrak{I}(\mu))_o$ is compact.
Conversely, if μ is not full then according to 3.3 $(\mathfrak{I}(\mu))_o$ is not compact. Hence $\mathfrak{I}(\mu)$ is not compact.

(ii) Let μ be S-full. Then $\mathfrak{I}_S(\mu)$ and $\mathfrak{I}_a(\mu)$ are compact according to 3.2. Hence $(\mathfrak{I}_S(\mu))_o$ is compact.
Conversely, let μ be not S-full. Let $x_o \in S(\mu)$. Then $\upsilon := \mu * \varepsilon_{x_o^{-1}}$ is not full. Hence according to 3.3 $(\mathfrak{I}(\upsilon))_o$ is not compact. But $\mathfrak{I}(\upsilon) \subseteq \mathfrak{I}_S(\mu)$. Hence $(\mathfrak{I}_S(\mu))_o$ is not compact.

(iii) Compactness of $\mathfrak{I}_S(\mu)$ implies compactness of $\mathfrak{I}_a(\mu)$ as proved in 3.2 and vice versa. □

3.5 Concluding remarks.

In connection with self-decomposable measures the following subsemigroups of End(G) := Hom(G,G) are important:

$\mathcal{A}(\mu) := \{\tau \in \mathrm{End}(G) : \tau(\mu) = \mu\}$,

$\mathcal{A}_S(\mu) := \{\tau \in \mathrm{End}(G) : \tau(\mu) = \mu * \varepsilon_x$ for some $x \in G\}$.

$\mathcal{D}(\mu) := \{\tau \in \mathrm{End}(G) : \mu = \tau(\mu) * \upsilon_\tau$ for some $\upsilon_\tau \in M^1(G)\}$

(called the decomposability semigroup of μ).

Obviously $\mathfrak{I}(\mu) \subseteq \mathcal{A}(\mu)$ and $\mathfrak{I}_S(\mu) \subseteq \mathcal{A}_S(\mu) \subseteq \mathcal{D}(\mu)$.

(i) μ is full iff $\mathcal{A}(\mu)$ is compact. Then $\mathcal{A}(\mu) = \mathfrak{I}(\mu)$.

μ is S-full iff $\mathcal{A}_S(\mu)$ is compact. Then $\mathcal{A}_S(\mu) = \mathfrak{I}_S(\mu)$.

μ is S-full iff $\mathcal{D}(\mu)$ is compact. (For $G = \mathbb{R}^d$ see [19]).

⟦ Let μ be full [S-full]. Theorem 2.2. [resp. 2.3] implies the compactness of $\mathcal{A}(\mu)$ [resp. of $\mathcal{A}_S(\mu)$].

Let again μ be S-full. Then for any sequence $(\tau_n)_{n\in\mathbb{N}}$ in $\mathcal{D}(\mu)$ $\{\tau_n(\mu)\}$ (and $\{\upsilon_{\tau_n}\}$) is shift-compact.

Then Theorem 2.3 implies the relative-compactness of $\{\tau_n\}$.

Conversely, let μ be not full [not S-full]. Then according to 3.4 $\mathfrak{I}(\mu)$ [resp. $\mathfrak{I}_S(\mu)$] is not compact. Hence $\mathcal{A}(\mu)$ [resp. $\mathcal{A}_S(\mu)$ and $\mathcal{D}(\mu)$] is not compact.

Let μ be full and $\tau \in \mathcal{A}(\mu)$. Then $\mathring{\tau} \in \mathcal{A}(\mathring{\mu})$ on $\mathfrak{G}$. Proposition 1.2 in [19] yields $\mathring{\tau} \in \mathfrak{I}(\mathring{\mu})$, therefore $\tau \in \mathfrak{I}(\mu)$.

Since $\mathcal{A}_S(\mu) \subseteq \mathcal{A}(\mu * \tilde{\mu})$ the assertion $\mathcal{A}_S(\mu) = \mathfrak{I}_S(\mu)$ follows if μ is S-full.⟧

(ii) Fullness of probabilities can be characterized by the validity of the convergence-of-types theorem:

(*) μ is full iff for any sequence $(\tau_n)_{n\in\mathbb{N}}$ in End(G), $\lambda, \mu_n \in M^1(G)$, the conditions $\mu_n \xrightarrow[n\to\infty]{} \mu$, $\tau_n\mu_n \xrightarrow[n\to\infty]{} \lambda$ imply the relative-compactness of $(\tau_n)_{n\in\mathbb{N}}$ in End(G).

(**) μ is S-full iff for any sequence $(\tau_n)_{n\in\mathbb{N}}$ in End(G), $(x_n)_{n\in\mathbb{N}}$ in G, $\lambda, \mu_n \in M^1(G)$ the conditions $\mu_n \xrightarrow[n\to\infty]{} \mu$, $\tau_n(\mu_n) * \varepsilon_{x_n} \xrightarrow[n\to\infty]{} \lambda$ imply the relative-compactness of $(\tau_n)_{n\in\mathbb{N}}$ in End(G).

⟦ Theorem 2.2. [resp. 2.3] shows that (*) [resp. (**)] holds for full [resp. S-full] measures.

Conversely, the weaker conditions

(+) $\tau_n \in \mathrm{Aut}(G)$, $\mu_n = \mu = \lambda$, $\tau_n\mu = \mu \Longrightarrow \{\tau_n\}$ is relatively compact

resp.

(++) $\tau_n \in \mathrm{Aut}(G)$, $\mu_n = \mu = \lambda$, $\tau_n\mu * \varepsilon_{x_n} = \mu \Longrightarrow \{\tau_n\}$ is relatively compact

imply the fullness [resp. S-fullness] of μ according to 3.4.⟧

REFERENCES

[1] Baldi, P.: Lois stables sur les deplacements de $\mathbb{R}^d$. In: Probability measures on groups. Proceedings Oberwolfach (1978). Lecture Notes in Math. 706, 1 - 9. Springer (1979).

[2] Billingsley, P.: Convergence of types in k-spaces. Z. Wahrscheinlichkeitstheorie verw. Geb. 5, 175 - 179 (1966).

[3] Drisch, T., Gallardo, L.: Stable laws on the Heisenberg group. In: Probability measures on groups. Proceedings Oberwolfach (1983). Lecture Notes Math. 1064, 56 - 79 (1984).

[4] Drisch, T., Gallardo, L.: Stable laws on the diamond group. Unpublished manuscript.

[5] Feller, W.: An Introduction to Probability Theory and its Applications Vol. II. New York: Wiley (1966).

[6] Fisz, M.: A generalization of a theorem of Khintchin. Studia Math. 14, 310 - 313 (1954).

[7] Gnedenko, B.W., Kolmogorov, A.N.: Limit distributions for sums of independent random variables. Cambridge: Addison-Wesley (1954).

[8] Hazod, W.: Stable probability measures on groups and on vector spaces. A survey. In: Probability measures on groups VIII. Proceedings, Oberwolfach (1985). Lecture Notes Math. 1210, 304 - 352 (1986).

[9] Hazod, W., Siebert, E.: Continuous automorphism groups on a locally compact group contracting modulo a compact subgroup

and applications to stable convolution semigroups.
Semigroup Forum 33, 111 - 143 (1986).

[10] Hazod, W., Siebert. E.: Automorphisms on a Lie group contracting modulo a compact subgroup and applications to semistable convolution semigroups. J. of Theoretical Probability 1, 211 - 226 (1988).

[11] Hochschild, G.: The structure of Lie groups.
San Francisco-London-Amsterdam: Holden Day Inc. (1965).

[12] Jajte, R.: Semistable probability measures on $\mathbb{R}^N$.
Studia Math. 61, 29 - 39 (1977).

[13] Jurek, Z.J.: Convergence of types, self-decomposability and stability of measures on linear spaces.
In: Probability in Banach Spaces III. Proceedings Medford (1980). Lecture Notes Math. 860, 257 - 267 (1981).

[14] Letta, G.: Eine Bemerkung zum Konvergenzsatz für Verteilungstypen.
Z. Wahrscheinlichkeitstheorie verw. Geb. 2, 310 - 313 (1964).

[15] Linde, W., Siegel, G.: On the convergence of types for Radon probability measures in Banach spaces. Probability on Banach Spaces. Sønderborg. Proceedings. Lecture Notes Math.

[16] McCrudden, M.: On the Supports of Absolutely Continuous Gauss Measures on Connected Lie Groups. Mh. Math. 98, 295 - 310 (1984).

[17] Nobel, S.: Ph. D. Thesis University Dortmund. In preparation.

[18] Sharpe, M.: Operator stable probability measures on vector groups. Trans. Amer. Math. Soc. 136, 51 - 65 (1969).

[19] Urbanik, K.: Lévy's probability measures on Euclidean spaces.
Studia Math. 44, 119 - 148 (1972).

[20] Urbanik, K.: Lévy's probability measures on Banach spaces.
Studia Math. 63, 238 - 308 (1978).

[21] Weissmann, I.: On Convergence of Types and Processes in Euclidean Spaces. Z. Wahrscheinlichkeitstheorie verw. Geb. 37, 35 - 41 (1976).

W. Hazod, S. Nobel
Universität Dortmund, Postfach 500 500, D-4600 Dortmund 50, Germany

MD-SEMIGROUPS, DECOMPOSITION OF POINT PROCESSES, CENTRAL LIMIT THEOREMS FOR CERTAIN T_2-SEMIGROUPS

He Yuanjiang
Department of Mathematics
Zhongshan University
Guangzhou, The People's Republic of China

Abstract. ZH-semigroups and MD-semigroups (multiple Delphic semigroups) are defined. It is shown that MD-semigroups possess the same fundamental properties as D.G. Kendall's Delphic semigroups and a convolution semigroup of point processes is an MD-semigroup. Sufficient conditions under which the central limit theorems for ZH-semigroups take place are obtained, another "straight" proof of the central limit theorem for point processes is given, the decomposition and the classification of positive generalized renewal sequences are discussed.

Chapter 1. MD-semigroups and decomposition of point processes

D.G. Kendall [9] has defined Delphic semigroups and developed a full theory. Using this theory Liang Zhishun [12] has proved that many classes of point processes are Delphic semigroups respectively. In this chapter we follow the works of [9], [12]. First we define the MD-semigroups and show that they possess the same fundamental properties as Delphic semigroups, which are special cases of MD-semigroups. Then we prove that the convolution semigroup of all random point processes defined on a complete separable metric space is an MD-semigroup, so it possesses the same properties of decomposition and classification as Delphic semigroups.

Section 1. MD-semigroups

1.1 Definition. Suppose that a semigroup G satisfies the following conditions:

(a) G is a Hausdorff abelian monoid, that is, G is an abelian semigroup with an identity e and G carries a Hausdorff topology such that the mapping $(u,v) \to uv$ is continuous from $G\times G$ to G;

(b) for each $u\in G$, $(v\in G:v|u)$, the set of all factors of u, is compact;

(c) there exists a sequence (D_k) of continuous homomorphisms, $D_k : G \to (\mathbb{R}^+,+)$, k=1,2,..., where $(\mathbb{R}^+,+)$ is the additive semigroup of nonnegative reals, such that u=e if (and only if) $D_k(u) = 0$ for all k.

Then G will be called a ZH-semigroup.

1.2 Definition. If a semigroup G satisfies (a), (c) of 1.1, then $(u_{ij}) = (u_{ij} \in G : 1 \leq j \leq n_i, i=1,2,\ldots)$ will be called a multiple null triangular array or an MN-array if $\lim_{i\to\infty} \max_j D_k(u_{ij}) = 0$ for each natural number k. An element u of a semigroup S will be called infinitely divisible (i.d.) if for each natural number n there exists $v \in S$ such that $v^n = u$.

1.3 Definition. A ZH-semigroup G will be called a multiple Delphic semigroup or an MD-semigroup if it has the following central limit property:

1.4 RCLT. If $(u_{ij} : 1 \leq j \leq i, i=1,2,\ldots)$ is an MN-array and $u = \prod_{1 \leq j \leq i} u_{ij}$ for all i, then u is i.d.

1.5 Remark. If G is a Delphic semigroup of [9], and $D_1 = \Delta$, $D_k(g) = 0$ for all $g \in G$, $k=2,3,\ldots$, then G is also an MD-semigroup. So a Delphic semigroup is a special case of MD-semigroups.

The next three theorems show that an MD-semigroup possesses the same fundamental properties as Delphic semigroups. By Theorem 1.7 we can classify all elements of an MD-semigroup as follows:

a) the "simple" or "indecomposable" elements which are not identity e and have no factor but themselves and e;

b) the "decomposable" elements which are not simple and have at least a simple factor;

c) the elements which are i.d. and have no simple factor (the set of all these elements is called the class "I_0").

If G is an MD-semigroup, then we have the following three theorems.

1.6 Theorem. If $u \in G$ and u is i.d., then u has the decompositions

$$u = \prod_{1 \leq j \leq i} u_{ij}, \quad i=1,2,\ldots,$$

where $(u_{ij} : 1 \leq j \leq i, i=1,2,\ldots)$ is an MN-array.

1.7 Theorem. If $u \in G$ and u has no simple factor, then u is i.d.

1.8 Theorem. For each $u \in G$ there exists a representation

$$u = w \prod_{i \geq 1} v_i,$$

where w belongs to the class I_0 and each v_i is simple or equals e.

The proof of Theorem 1.6 is simple and is the same as that of [9], Theorem 1, so we omit it.

To prove Theorem 1.7, by RCLT 1.4 it is sufficient to prove the

1.7' Theorem. If $g \in G$ and g has no simple factor, then there exists an MN-array $(g_{ij} : 1 \leq j \leq n_i,\ i=1,2,\ldots)$ such that

$$u = \prod_{i \leq j \leq n_i} g_{ij}$$

for $i=1,2,\ldots$.

For ZH-semigroups G we first prove the following five lemmata.

1.9 Lemma. If a and b are two elements of G such that $a|b$, then the set $(u : a|u|b)$ is compact.

The proof of this lemma is the same as that of [9], Lemma 2.

1.10 Lemma. If g_r, $r=1,2,\ldots$, are elements of G such that $g_s|g_r$ when $r \leq s$, then $g^* = \lim_{n\to\infty} g_n$ exists and $g^*|g_r$ for all r.

Proof. The proof is similar to the proof of [9], Lemma 1. For each fixed r, $(u : u|g_r)$ is compact, so (g_n) has at least one cluster point g^* such that $g^*|g_r$. (g_n) will be convergent if and only if the cluster point is unique, i.e. if g^{**} is another cluster point, we must have $g^{**} = g^*$.

Since $g^*|g_r|g_1$ for all r and $(g : g^*|g|g_1)$ is compact, $g^*|g^{**}$ and (by symmetry) $g^{**}|g^*$, we have $D_k(g^{**}) = D_k(g^*)$ for all k. Suppose that $g^{**} = g^*d$, $d \in G$, then $D_k(d) = 0$ for all k, so $d=e$ and $g^* = g^{**}$. #

1.11 Lemma. If $g \in G$ and g has no simple factor and $D_1(g) \neq 0$, then there exists $u \in G$ such that $u|g$ and $0<D_1(u)<D_1(g)$.

Proof. Let $g_1 = g$, $a_1 = D_1(g_1)$, $E_1 = (v \in G : D_1(v) = a_1)$,

$E_2 = (v \in E_1 : v|g_1)$, $a_2 = \inf(D_2(v) : v \in E_2)$. Since E_2 is compact, there exists $g_2 \in E_2$ such that $D_2(g_2) = a_2$. Clearly, $D_1(g_2) = a_1$.

Let $E_3 = (v \in E_1 : v|g_2)$, $a_3 = \inf(D_3(v) : v \in E_3)$, then there exists $g_3 \in E_3$ such that $D_3(g_3) = a_3$. Since $g_3|g_2$, $D_2(g_3) \leq D_2(g_2) = a_2$, but $g_3 \in E_3 \quad E_2$, $D_2(g_3) \geq a_2$, so we must have $D_2(g_3) = a_2$. Clearly, $D_1(g_3) = a_1$.

Continuing this procedure, we obtain a sequence (E_n, a_n, g_n) such that $g_{n+1}|g_n$ for $n=1,2,\ldots$ and $D_k(g_n)=a_k$ for $k \leq n$. By Lemma 1.10 (g_n)

converges to some g^* and $g^*|g_n$ for $n=1,2,\ldots$, and by the continuity of D_k, $D_k(g^*) = a_k$ for all k, g^* is not simple and $g^*\in\bigcap_{i\geq 1} E_i$. Let g^* admit a decomposition of the form $g^* = uv$, $u\neq e$, $v\neq e$. If $D_1(u) = a_1$, then $u\in\bigcap_{i\geq 1} E_i$, so $D_k(u)\geq a_k$ for all k. But $u|g^*$, $D_k(u)\leq D_k(g^*)=a_k$ for all k, hence $D_k(u)=a_k$ for all k and we should have $D_k(v)=0$ for all k, so $v=e$ by (c) of 1.1. But this contradicts the original hypothesis on v. Hence $D_1(u)\neq a_1$. Similarly, $D_1(v)\neq a_1$ and so $0<D_1(u)<D_1(g)$ and $u|g^*|g$. #

<u>1.12 Lemma.</u> The same assumptions as in Lemma 1.11. For any $c>0$, there exists $u\in G$ such that $u|g$ and $0<D_1(u)\leq c$.

<u>Proof.</u> Using Lemma 1.11 repeatedly, we have for $n\geq D_1(g)/c$ the decomposition $g=u_1u_2\ldots u_n$ with $D_1(u_j)\neq 0$ für $j=1,2,\ldots,n$. Since $D_1(g) = D_1(u_1)+D_1(u_2)+\ldots+D_1(u_n)$, there exists $i\in(1,2,\ldots,n)$ such that $D_1(u_i)\leq D_1(g)/n\leq c$. Let $u=u_i$. Then $u|g$ and $0<D_1(u)\leq c$. #

<u>1.13 Lemma.</u> The same assumptions as in Lemma 1.11. For any $b\in(0,D_1(g))$, there exists u^* such that $u^*|g$ and $D_1(u^*)=b$.

<u>Proof.</u> Let $E=(u : u|g \text{ and } D_1(u)\leq b)$,

$$a = \sup(D_1(u) := u\in E). \qquad (*)$$

Then $a\leq b$. since E is nonempty and compact, there exists $u^*\in E$ such that $D_1(u^*)=a$. If $a<b$, let $g=u^*v^*$, by Lemma 1.12 there would exist a decomposition $v^* = v_0w_0$ such that $0<D_1(v_0)\leq b-a$.

Since $u^*v_0|g$ and $D_1(u^*v_0)\leq b$, $u^*v_0\in E$. Thus the inequality $D_1(u^*v_0)>a$ would contradict (*). So we must have $a=b$ and $D_1(u^*)=b$. #

<u>1.14 Remark.</u> It is easily seen that the above lemma still holds if instead of D_1 we take D_1+D_2, $D_1+D_2+D_3,\ldots$ respectively, because for each $i\geq 2$, $D_1+D_2+\ldots+D_i$, $D_1,D_2,\ldots$ is also a sequence of homomorphisms satisfying the condition (c) of Definition 1.1.

Now we turn to the proof of Theorem 1.7'. For each natural number i, let $n_i\geq((D_1+D_2+\ldots+D_i)(g))i$. Then $b := (D_1+D_2+\ldots+D_i)(g)/n_i\leq 1/i$. By Lemma 1.13 and Remark 1.14 let

$$g = g_{i1}h_1 \text{ such that } (D_1+D_2+\ldots+D_i)(g_{i1}) = b,$$

$$h_1 = g_{i2}h_2 \text{ such that } (D_1+D_2+\ldots+D_i)(g_{i2}) = b,$$

...

$h_{n_i-1} = g_{in_i}$. Then $(D_1+D_2+\ldots+D_i)(g_{in_i}) = b$.

So $g = \prod_{1\leq j\leq n_i} g_{ij}$, $(D_1+D_2+\ldots+D_i)(g_{ij})\leq 1/i$, $j=1,2,\ldots,n_i$,

$D_k(g_{ij})\leq 1/i$, $j=1,2,\ldots,n_i$, $k=1,2,\ldots,i$.

Hence for fixed k, $\lim_{i\to\infty} \max_j D_k(g_{ij})=0$ and $(g_{ij} : 1\leq j\leq n_i, i=1,2,\ldots)$ will be the MN-array which we desire. #

The proof of Theorem 1.8 differs only slightly from the proof of [9], Theorem 3. First we impose a partial ordering on Φ, the class of certain functions, then using Zorn's lemma we obtain a maximal element ϕ. But here we should convert "Δ" into "D_k for $k=1,2,\ldots$" before we make some appropriate modifications. For example, in [9] in order to prove that the set $(v : v\in \mathfrak{V}$ and $\phi(v)>0)$ is at most countable, the inequality $\Sigma\Delta(v_j)\leq\Sigma\phi(v_j)\Delta(v_j)\leq\Delta(u)<\infty$ is used. Instead of this, we will write here "for any fixed k and any finite subset E^* of $E = (v:v\in \mathfrak{V}$ and $\phi(v)>0)$, $\sum_{v_j\in E^*} D_k(v_j)\leq \sum_{v_j\in E^*} \phi(v_j)D_k(v_j)\leq D_k(u)<\infty$, hence $E_k := (v\in E : D_k(v)>0)$ is at most countable, and so is $E := \bigcup_{k\geq 1} E_k$". #

Section 2. Decomposition of random point processes

Suppose that (X,ρ_X) is a complete separable metric space, $\mathcal{B}$ is the ring consisting of all bounded Borel subsets of X, $(N,\mathcal{N})$ is the measurable space of all locally finite counting measures on (X,ρ_X), $\mathbb{P}$ is the convolution semigroup of all probability measures on $(N,\mathcal{N})$, every $P\in\mathbb{P}$ denotes a point process.

In [12] Liang Zhishun has proved the

2.1 Lemma. For all fixed $P\in\mathbb{P}$, $F = (Q : Q\in\mathbb{P}, Q|P)$, the set of all factors of P, is compact.

Proof. F is closed by [14], 3.2.9. By [14], 3.2.7. for each $\varepsilon>0$ and each bounded closed subset B X there exist an $n = n(B,\varepsilon)$ and a compact set C B such that

a) $\sup_{Q\in F} Q(\mu\in N : \mu(B)\geq n)\leq P(\mu\in N : \mu(B)\geq n)<\varepsilon$;

b) $\sup_{Q\in F} Q(\mu\in N : \mu(B\setminus C)>0)\leq P(\mu\in N : \mu(B\setminus C)>0)<\varepsilon$.

By [14], 3.2.7. F ist relatively compact, so F is compact. #

<u>2.2 Theorem.</u> Let $x_0 \in X$ be a fixed point, $B_k = (x : \rho(x,x_0) \leq k)$. Define $f_k(x) = \max(0, 1-\rho(x,B_k))$, $\mu f_k = \int_X f_k(x)\mu(dx)$, $k=1,2,\dots$. If we define (D_k) by $D_k(P) = -\log \int_N \exp(-\mu f_k) P(d\mu)$, $k=1,2,\dots$, then $\mathbb{P}$ is an MD-semigroup.

<u>Proof.</u> We observe that $P_e \in \mathbb{P}$ defined by $P_e(\mu(X) = 0) = 1$ is an identity. By [14], 3.1.2. we can introduce a metric d in $\mathbb{P}$, so that $\lim_{n\to\infty} d(P_n,P) = 0$ if and only if the sequence (P_n) converges weakly to P.

By [14], 3.1.10. the convolution is continous, so $\mathbb{P}$ is an abelian Hausdorff monoid. Condition (b) of 1.1 follows from Lemma 2.1, condition (c) of 1.1 is trivial, hence $\mathbb{P}$ is a ZH-semigroup. By the following Definition 2.3, Proposition 2.4, and Lemma 2.5. $\mathbb{P}$ has the property RCLT 1.4, so $\mathbb{P}$ is an MD-semigroup.

<u>2.3 Definition.</u> $(P_{ij} \in \mathbb{P} : 1 \leq j \leq i,\ i=1,2,\dots)$ will be called an infinitesimal triangular array if for all $A \in \mathcal{B}$,

$$\lim_{i\to\infty} \max_j P_{ij}(\mu(A) > 0) = 0$$

<u>2.4. Proposition.</u> ([14], Proposition 3.4.1.) A random point process is infinitely divisible if and only if there exists an infinitesimal triangular array $(P_{ij} : 1 \leq j \leq i,\ i=1,2,\dots)$ sucht that

$$\lim_{i\to\infty} \prod_{1\leq j\leq i} P_{ij} = P.$$

<u>2.5 Lemma.</u> $(P_{ij} : 1 \leq j \leq i,\ i=1,2,\dots)$ is an infinitesimal triangular array if and only if it is an MN-array with the (D_k) in Theorem 2.2.

<u>Proof.</u> Since $D_k(P_{ij}) = -\log \int_N \exp(-\mu f_k)\, P_{ij}(d\mu)$

$$\leq -\log \int_N \exp(-\mu(B_{k+1})) P_{ij}(d\mu) \leq -\log(1 - P_{ij}(\mu(B_{k+1}) > 0)),$$

so $\lim_{i\to\infty} \max_j P_{ij}(\mu(B_{k+1}) > 0) = 0 \Rightarrow \lim_{i\to\infty} \max_j D_k(P_{ij}) = 0.$

Since $\int_N \exp(-\mu(B_k)) P_{ij}(d\mu) = \int_{\mu(B_k)=0} \exp(-\mu(B_k)) P_{ij}(d\mu)$

$$+ \int_{\mu(B_k)>0} \exp(-\mu(B_k)) P_{ij}(d\mu) \leq P_{ij}(\mu(B_k) = 0) + P_{ij}(\mu(B_k) > 0) e^{-1},$$

$$P_{ij}(\mu(B_k) > 0) \leq \left(1 - \int_N \exp(-\mu(B_k)) P_{ij}(d\mu)\right) / (1 - e^{-1}),$$

so, for any $A \in \mathcal{B}$, there exists k such that $A \subset B_k$, and therefore

$$\lim_{i\to\infty} \max_j D_k(P_{ij}) = 0 \Rightarrow \lim_{i\to\infty} \min_j \int_N \exp(-\mu f_k) P_{ij}(d\mu) = 1$$

$$\Rightarrow \lim_{i\to\infty} \min_j \int_N \exp(-\mu(B_k)) P_{ij}(d\mu) = 1 \Rightarrow \lim_{i\to\infty} \max_j P_{ij}(\mu(B_k)>0) = 0$$

$$\Rightarrow \lim_{i\to\infty} \max_j P_{ij}(\mu(A)>0) = 0. \ \#$$

2.6 Corollary. For fixed k the convolution semigroup S_k of all k-dimensional nonnegative integral random vectors is an MD-semigroup. The convolution semigroup S of all countably-infinite-dimensional non-negative integral random vectors is an MD-semigroup.

Proof. S_k or S is the convolution semigroup of random point processes defined on spaces $X_k = (1,2,\dots,k)$ or $X = (1,2,\dots)$ respectively. #

Chapter 2. Central limit theorems for ZH-semigroups and their application

Section 3. Central limit theorems for ZH-semigroups

To verify the central limit properties is not easy; it usually depends on special properties of the semigroups. Now we show some sufficient conditions for the property of RCLT 1.4.

For a ZH-semigroup G we list the following properties:

3.1 CLT. If $(u_{ij} : 1 \leq j \leq i, i=1,2,\dots)$ is an MN-array and

$$u = \lim_{i\to\infty} \prod_{i \leq j \leq i} u_{ij},$$

then u is i.d.

3.2 T. If u has no simple factor, then u is i.d.

3.3 H. u=v if and only if $D_k(u) = D_k(v)$ for all k.

3.4 C. If (u_n) and (v_n) are two sequences in G, $\lim_{n\to\infty} u_n = u$ and $v_n | u_n$ for all n, then there exists a subsequence of (v_n) which converges to $v \in G$.

3.5 C'. If $(u_i, i \in I)$ and $(v_i, i \in I)$ are two nets from the same directed set I into G, $\lim u_i = u$ and $v_i | u_i$ for all $i \in I$, then there exists a subnet of (v_i) which converges to $v \in G$.

<u>3.6 Remark.</u> Either in 3.4 C or in 3.5 C' $v|u$. We only verify this property in the case of 3.4 C. Let $v_n w_n = u_n$ for all n, (v_{n_k}) and (w_{n_k}) be two convergent subsequences, $\lim_{k\to\infty} v_{n_k} = v$, $\lim_{k\to\infty} w_{n_k} = w$. Then $vw = u$, $v|u$.

<u>3.7 Proposition.</u> For a ZH-semigroup G, CLT => RCLT => T.

<u>Proof.</u> The first implication is trivial, the second one can be verified by virtue of Theorem 1.7.

<u>3.8 Theorem.</u> For a ZH-semigroup G, H => RCLT.

Before proving Theorem 3.8 we show a lemma.

Suppose that E^n is an n-dimensional Euclidean space, the length of each $x = (x_1, x_2, \ldots, x_n) \in E^n$ being given by $|x| = (x_1^2 + x_2^2 + \ldots + x_n^2)^{\frac{1}{2}}$. A subset $A = (a_1, a_2, \ldots, a_{rk})$ of E^n will be called a partition of $T \in E^n$ if $\sum_{1 \leq s \leq rk} a_s = T$. Moreover $||A|| := \max_{1 \leq s \leq rk} |a_s|$.

We collect the rk elements of A in k sets A_t, t=1,2,...,k, each A_t consisting of r elements. Let $B(k,r,A,E^n)$ be the minimum, over all possible choices of $(A_1, A_2, \ldots, A_k)$, of $\max_t (|\sum_{a_s \in A_t} a_s - T/k|)$. Let

$$B(k,r,\varepsilon,E^n) = \sup(B(k,r,A,E^n) : ||A|| \leq \varepsilon).$$

Since $B(k,r,\varepsilon,E^n) \leq K(k) B(2,r,\varepsilon,E^n)$, by [5], Theorem 1 and $B(2,r,\varepsilon,E^n) \leq (R_n)^{\frac{1}{2}} \varepsilon$ by [5], Theorem 3, we have the next lemma immediately.

<u>3.9 Lemma.</u> $B(k,r,\varepsilon,E^n) \leq K(k)(R_n)^{\frac{1}{2}} \varepsilon$, where $K(k)$ and R_n are positive and only depending on k and n respectively.

<u>Proof</u> of Theorem 3.8. Let $(u(i,j) : 1 \leq j \leq i,\ i=1,2,\ldots)$ be an MN-array, $u = \prod_{1 \leq j \leq i} u(i,j)$, i=1,2,... . For each fixed $l \in \mathbb{N}$ there exists a subarray $(v_{ij} : 1 \leq j \leq il,\ i=1,2,\ldots)$, within $v_{ij} = u(il,j)$. Moreover, $u = \prod_{1 \leq j \leq il} v_{ij}$, i=1,2,... . For each $n \in \mathbb{N}$, let

$$T = (D_1(u), D_2(u), \ldots, D_n(u)) \in E^n,$$

$$A_j(i) = (D_1(v_{ij}), D_2(v_{ij}), \ldots, D_n(v_{ij})),\ j=1,2,\ldots,il.$$

Then $A(i)=(A_1(i), A_2(i),\dots,A_{il}(i))$ is a partition of T. Let $\varepsilon = (nK(l)(R_n)^{\frac{1}{2}})^{-1}$. Then $||A(i)||\leq\varepsilon$ and $B(l,i,\varepsilon,E^n)\leq 1/n$ if

$$\max_{1\leq j\leq il} D_k(v_{ij})\leq(n^{3/2}K(l)(R_n)^{\frac{1}{2}})^{-1} \text{ for } k=1,2,\dots,n.$$

So for sufficiently large i, rearranging $(v_{ij} : j=1,2,\dots,il)$, we can get $(v_{ij}(t) : j=1,2,\dots,i,\ t=1,2,\dots,l)$ such that

$$\Big|\sum_{1\leq j\leq i} D_k(v_{ij}(t))-D_k(u)/l\Big|\leq 1/n,\ k=1,2,\dots,n,\ t=1,2,\dots,l. \qquad (*)$$

Rearranging the MN-array $(v_{ij} : 1\leq j\leq il,\ i=1,2,\dots)$, we can get

$$(v_{ij}(t) : 1\leq j\leq i,\ 1\leq t\leq l,\ i=1,2,\dots)$$

such that for any fixed $n\in\mathbb{N}$, for sufficiently large i, the inequalities (*) hold.

Suppose that $u_i(t) = \prod_{1\leq j\leq i} v_{ij}(t)$ for all i. Then $u = \prod_{1\leq t\leq l} u_i(t)$ for all i and

$$\lim_{i\to\infty} |D_k(u_i(t))-D_k(u)/l| = 0,\ t=1,2,\dots,l,\ k=1,2,\dots .$$

Suppose that $(u_{i_s}(t),\ s\in E,\ )$, $t=1,2,\dots,l$, defined on the same directed set E, is a subnet of the net $(u_i(t) : i\in\mathbb{N},\leq)$, $t=1,2,\dots,l$, such that $\lim u_{i_s}(t) = u(t)$, $t=1,2,\dots,l$. Then $u = u(1)u(2)\dots u(l)$, $D_k(u(1)) = D_k(u(2)) = \dots = D_k(u(l))$, $k=1,2,\dots$, hence $u(1) = u(2) = \dots = u(l)$, $u = (u(1))^l$. Since l is arbitrary, u is i.d. #

<u>3.10 Corollary.</u> For a ZH-semigroup G, H => T.

<u>Proof.</u> By virtue of 3.7, 3.8.

<u>3.11 Theorem.</u> For a ZH-semigroup G, H and C => CLT, H and C' => CLT.

<u>Proof.</u> We only have to modify the proof of 3.8 slightly, i.e. instead of inequalities (*) we write

$$\Big|\sum_{1\leq j\leq i} D_k(v_{ij}(t))-\sum_{1\leq j\leq il} D_k(u(il,j))/l\Big|\leq 1/n,\ k=1,2,\dots,n,\ t=1,2,\dots,l.$$

Hence we have convergent subsequences $(u_{i_s}(t))$ of $(u_i(t))$, $t=1,2,\dots,l$, or convergent subnets $(u_{i_s}(t), s\in E,<)$ of nets $(u_i(t), i\in \mathbb{N},\leq)$, $t=1,2,\dots,l$, respectively. #

Now we turn to some semigroups consisting of real sequences.

Let Q be the set of all positive sequences, the elements of Q being denoted by $u = (u_n)$, $v = (v_n)$, $w = (w_n),\dots$. In order to make Q into a semigroup, we define the multiplication termwise, that is, $u = vw$ if and only if $u_n = v_n w_n$ for all n. The distance between u and v is defined to be $\rho(u,v) = \sum_{n\geq 1} 2^{-n}|u_n-v_n|(1+|u_n-v_n|)^{-1}$. Then Q is an abelian Hausdorff monoid.

<u>3.12 Theorem.</u> Let G be a subsemigroup with identity $e = (1,1,\dots)$ of Q and satisfy the following conditions:

(a) for each fixed $n\in\mathbb{N}$; $u_n\in(0,1]$ for all $u\in G$, or $u_n\in[1,\infty)$ for all $u\in G$;

(b) G is closed in Q, that is, if $u(t)\in G$ for all $t\in\mathbb{N}$ and $\lim_{t\to\infty} u(t) = u\in Q$, then $u\in G$.

If we define (D_k) by $D_k : u \mapsto |\log u_k|$, $k=1,2,\dots$, then G is a ZH-semigroup with properties H and C; moreover, G is an MD-semigroup with property CTL.

<u>Proof.</u> G is a metric space. By virtue of Remark 3.6, property C implies (b) of Definition 1.1. Now we only verify that G has property C. Suppose that $u(t)$, $v(t)$ are sequences in G, $v(t)|u(t)$ for all t, $\lim_{t\to\infty} u(t) = u\in G$, then for each fixed n, $v_n(t)\in[\frac{1}{2}u_n,1]$ or $v_n(t)\in[1,2u_n]$ for sufficiently large t, hence we have a subsequence $(v(t_s))$ which converges to $v\in G$. #

<u>Section 4.</u> "Straight" proof of central limit theorem for random point processes

In the proof of Theorem 2.2 we have shown that $\mathbb{P}$ is a ZH-semigroup, and by virtue of Proposition 2.4, the central limit theorem for point processes, $\mathbb{P}$ has the property of RCLT 1.4, so $\mathbb{P}$ is an MD-semigroup. Now we turn to another "straight" proof of Proposition 2.4. Since $\mathbb{P}$ is a ZH-semigroup, by virtue of Theorem 3.11 we only have to verify that it has the property C and find out the desired (D_k) with the property of 3.3 H.

<u>4.1 Theorem.</u> $\mathbb{P}$ has property C.

<u>Proof.</u> Let (P_n), (Q_n) be two sequences in $\mathbb{P}$, $\lim\limits_{n\to\infty} P_n = P$, $Q_n | P_n$ for all n. Since (P_n) is relatively compact, by [14], Proposition 3.2.7, for each bounded closed subset B of X and each $\varepsilon>0$, there exist $k = k(B,\varepsilon)$ and a compact set $C\subset B$ such that

a) $\sup\limits_n Q_n(\mu(B)\geq k) \leq \sup\limits_n P_n(\mu(B)\geq k) < \varepsilon$,

b) $\sup\limits_n Q_n(\mu(B\setminus C)>0) \leq \sup\limits_n (\mu(B\setminus C)>0) < \varepsilon$.

Hence (Q_n) is relatively compact. #

We shall invoke the next

<u>4.2. Proposition.</u> ([14], Proposition 1.3.2) Let R* be a semiring generating . If the distributions P, Q on satisfy

$$P_{H_1,H_2,\ldots,H_m} = Q_{H_1,H_2,\ldots,H_m}$$

for all finite sequences $H_1,H_2,\ldots,H_m$ of disjoint sets in R*, then P=Q.

First we construct a semiring R generating . Let $z\in X$ be a fixed point, $G(t) = (x\in X : \rho(x,z)<t)$, $t=0,1,2,\ldots$, be a countable base of the space (X,ρ_X) consisting of bounded open sets and containing all G(t).

Let $\mathbb{E}$ be the class consisting of all sets each of which has a representation $\;_{1\leq i\leq s}\;_{1\leq j\leq s_i} u_{ij}$, where all u_{ij} belong to .

Let R be the class consisting of all sets each of which has a representation $\;_{1\leq i\leq n} A_i B_i^c$, where all A_i, B_i belong to $\mathbb{E}$, and B_i^c is the complement of B_i.

<u>4.3 Lemma.</u> R is a countable class and is a ring generating $\mathcal{B}$.

<u>Proof.</u> We only verify that R is a ring.

(a) It is obvious that $A_1, A_2 \in \mathbb{E} \Rightarrow A_1 \cup A_2 \in \mathbb{E}$, $A_1 \cap A_2 \in \mathbb{E}$.

(b) If $r_1, r_2 \in R$, i.e. $r_1 = \bigcup\limits_{1\leq i\leq n} A_i B_i^c$, $r_2 = \bigcup\limits_{1\leq j\leq m} S_j T_j^c$, where all A_i, B_i, S_j, T_j belong to $\mathbb{E}$, then

$r_1 \cup r_2 \in R$,

$$r_1 \cap r_2^c = (\bigcup_{1\le i\le n} A_iB_i^c) \cap (\bigcup_{1\le j\le m} S_jT_j^c)^c = (\bigcup_{1\le i\le n} A_iB_i^c) \cap (\bigcap_{1\le j\le m} (S_j^c \cup T_j)).$$

Without loss of generality suppose that $m=1$, $S_1 = S$, $T_1 = T$. Then

$$r_1 \cap r_2^c = (\bigcup_{1\le i\le n} A_iB_i^c) \cap (S^c \cup T) = (\bigcup_{1\le i\le n} A_iB_i^cS^c) \cup (\bigcup_{1\le i\le n} A_iB_i^cT)$$

$$= (\bigcup_{1\le i\le n} A_i(B_i \cup S)^c) \cup (\bigcup_{1\le i\le n} A_iTB_i^c) \in R. \quad \#$$

For each natural number k and each $r = \bigcup_{1\le i\le n} A_iB_i^c \in R$, let

$$g_{ik}^r(x) = \min(1,\ k\rho(x,A_i^c)),\ h_{ik}^r(x) = \min(1,\ k\rho(x,B_i^c)),$$

$$J_k^r(x) = \max_{1\le i\le n} (g_{ik}^r(x) - g_{ik}^r(x)\ h_{ik}^r(x)).$$

Them $\lim_{k\to\infty} J_k^r(x) = I_r(x) = 0$ when $x \notin r$, $= 1$ when $x \in r$. The set of functions

$$F^* = (J_k^r :\ r \in R,\ k=1,2,\dots)$$

is countable, and the set of functions

$$F = (t_1f(1)+t_2f(2)+\dots+t_nf(n) :\ f(k) \in F^*,\ t_k \text{ are nonnegative rational numbers, } 1\le k\le n,\ n=1,2,\dots)$$

is also countable. Let $F = (f_1, f_2, f_3, \dots)$. Then we have

<u>4.4 Theorem.</u> Suppose that the sequence (D_k) is defined by

$$D_k(P) = -\log \int_N \exp(-\mu f_k P(d\mu)),\ k=1,2,\dots\ .$$

Then $\mathbb{P}$ is a ZH-semigroup with property H.

<u>Proof.</u> We only verify that $\mathbb{P}$ has property H. Let $P, Q \in \mathbb{P}$, $D_k(P) = D_k(Q)$ for all k. By virtue of Proposition 4.2 it is sufficient to verify that

$$P_{H_1,H_2,\dots,H_m} = Q_{H_1,H_2,\dots,H_m}$$

for disjoint sets $H_1, H_2, \dots, H_m$ in R. It is equivalent to verify that

$$\sum_{j_1,j_2,\ldots,j_m \geq 0} P_{H_1,H_2,\ldots,H_m}(j_1,j_2,\ldots,j_m) h_1^{j_1} h_2^{j_2} \ldots h_m^{j_m}$$

$$= \sum_{j_1,j_2,\ldots,j_m \geq 0} Q_{H_1,H_2,\ldots,H_m}(j_1,j_2,\ldots,j_m) h_1^{j_1} h_2^{j_2} \ldots h_m^{j_m}$$

for all $h_i \in (0,1]$, i.e.

$$\int_N \exp(-\mu(c_1 I_{H_1} + c_2 I_{H_2} + \ldots + c_m I_{H_m})P(d\mu)$$

$$= \int_N \exp(-\mu(c_1 I_{H_1} + c_2 I_{H_2} + \ldots + c_m I_{H_m})Q(d\mu), \qquad (*)$$

where $c_i = -\log h_i$ for $i=1,2,\ldots,m$.

Let (g_k) be a sequence in F and

$$\lim_{k\to\infty} g_k(x) = c_1 I_{H_1}(x) + c_2 I_{H_2}(x) + \ldots + c_m I_{H_m}(x) \quad \text{for all } x \in X.$$

Since

$$\int_N \exp(-\mu g_k)P(d\mu) = \int_N \exp(-\mu g_k)Q(d\mu) \qquad \text{for } k=1,2,\ldots,$$

taking limits we have equation (*). #

4.5 Lemma. A triangular array $(P_{ij} \in \mathbb{P} : 1 \leq j \leq i,\ i=1,2,\ldots)$ is infinitesimal if and only if it is an MN-array with the (D_k) defined in Theorem 4.4.

Proof. Since every $f \in F$ is nonnegative continuous bounded and has a bounded support, and $J_1^{G(t)} \in F$ for each $G(t) \in \mathcal{U}$, the proof of this lemma is similar to that of Lemma 2.5. #

By virtue of Lemma 4.5, Theorem 4.1, Theorem 4.4, and Theorem 3.11, we again obtain Proposition 2.4, the CLT for point processes.

Section 5. Decomposition of positive generalized renewal sequences.

For the semigroup R^+ consisting of all positive renewal sequences, applying Theorem 3.12 we can again obtain some results which have been given by Kendall [8] and [9], Davidson [4], but it is more interesting that we can apply it to the semigroup GR^+ consisting of all positive generalized renewal sequences. In this section we shall show that the properties of decomposition and classification of GR^+ is very similar to R^+, and shall prove the central limit theorem for generalized renewal sequences.

Recall that a real sequence $(u_n, n\geq 1)$ will be called a generalized renewal sequence [11] if there exists a nonnegative real sequence $(f_n, n\geq 1)$ such that

$$u_1 = f_1,\ u_n = \sum_{1\leq r\leq n-1} f_r u_{n-r} + f_n,\ n=2.3,\ldots . \qquad (*)$$

Moreover, if $\sum_{n\geq 1} f_n \leq 1$, then (u_n) will be called a renewal sequence.

It is obvious from (*) that

$$u_n \geq u_1^n. \qquad (**)$$

Let R(GR) denote the set consisting of all (generalized) renewal sequences. Then $R\subset GR$, and by (**)

$$R^+ = (u\in R : u_n>0 \text{ for all } n) = (u\in R : u_1>0),$$

$$GR^+ = (u\in GR : u_n>0 \text{ for all } n) = (u\in GR : u_1>0).$$

The relation between R and GR has been given in [11] by the next two propositions.

<u>5.1 Proposition.</u> ([11] Proposition 1) A generalized renewal sequence u is a renewal sequence if and only if $u_n\leq 1$ for all n.

<u>5.2 Proposition.</u> ([11] Proposition 2) $(u_n)\in GR$ if and only if for each $N\in \mathbb{N}$, there exist $(v_n)\in R$ and constant $c>0$ such that $u_n = v_n c^n$ for $n=1,2,\ldots,N$.

Recall that R is closed under multiplication and is topologically closed (see [10], p. 422 or [8], p. 50). As a generalization of these properties of R, the next two propositions have been verified in [13].

<u>5.3 Proposition.</u> GR is closed under the termwise multiplication.

<u>Proof.</u> We give another proof different from [13]. Let (u_n), (u_n^*) belong to GR. For each fixed $N\geq 1$ there exist (v_n), (v_n^*) belonging to R and constants $c, c^*>0$ such that $u_n = v_n c^n$, $u_n^* = v_n^*(c^*)^n$ for $n=1,2,\ldots,N$. Then $(v_n v_n^*, n\geq 1) = (v_n)(v_n^*)\in R$ and $u_n u_n^* = v_n v_n^*(cc^*)^n$ for $n=1,2,\ldots,N$; hence $(u_n u_n^*)\in GR$. #

<u>5.4 Proposition.</u> GR is closed under the termwise limit.

<u>Proof.</u> Let $(u_n(1))$, $(u_n(2)),\ldots$ belong to GR, $\lim_{n\to\infty} u_n(k) = u_n$ for

all n, $f_1(k) = u_1(k)$, $f_n(k) = u_n(k) - \sum_{1\leq r\leq n-1} f_r(k)u_{n-r}(k)$, $n\geq 2$, $k\geq 1$.

Then the limit $f_n = \lim_{k\to\infty} f_n(k)$ exists for each $n\geq 1$ and $f_n\geq 0$ for each n.

Furthermore, $f_1 = u_1$, $f_n = u_n - \sum_{1\leq r\leq n-1} f_r u_{n-r}$ for $n\geq 2$; hence (u_n) belongs to GR. #

Let $R_0^+(GR_0^+)$ denote the set of all positive (generalized) renewal sequences which are infinitely divisible in $R^+(GR^+)$, or, equivalently, in R(GR). Let K := (u : u is a positive sequence, $u_2\geq u_1^2$, $u_n u_{n+2}\geq u_{n+1}^2$ for all $n\geq 1$).

BK := ($u\in K$: $u_n\leq 1$ for all $n\geq 1$).

From [8], Theorem 1 and from the fact that $u\in BK$ if and only if $u\in K$ and u is bounded we have the next

<u>5.5' Proposition.</u> $BK = R_0^+$.

<u>5.5 Proposition.</u> ([7], Theorem 6) $K = GR_0^+$.

The proof of Proposition 5.5 is similar to that of [8], Theorem 1; we omit it.

Let $c>0$, $p\geq 1$ and $k\in\mathbb{N}$. $v(\infty,c) := (c^n, n\geq 1)$,

$$v(k,1/p) := (p^{-\min(k,n)}, n\geq 1),$$

$$\tilde{v}(k,p) := (p^{\max(0,n-k)}, n\geq 1).$$

Then $v(\infty,c)\in GR_0^+$, $\tilde{v}(k,p)\in GR_0^+$ by 5.5, $v(k,1/p)\in R_0^+$ by 5.5'.

<u>5.6 Proposition.</u> Every $u\in GR_0^+$ has a representation

$$u = v(\infty,c) \prod_{k\geq 1} \tilde{v}(k,p_k).$$

<u>Proof.</u> Let $c = u_1$, $p_1 = u_2/u_1^2$, $p_k = u_{k+1}u_{k-1}/u_k^2$, k=2,3,... . #

<u>5.7 Proposition.</u> For each $k\geq 1$ and $p>1$, $\tilde{v}(k,p)$ has a factor which does not belong to GR_0^+.

<u>Proof.</u> By virtue of [3], Theorem 4, $v(k,1/p)$ has a factor u which belongs to $R^+\setminus R_0^+$, but $R_0^+ = GR_0^+\cap R^+$ by 5.5 and 5.5', hence $u\notin GR_0^+$. Since $\tilde{v}(k,p) = v(\infty,p)v(k,1/p)$, u is a factor of $\tilde{v}(k,p)$. #

<u>5.8 Remark.</u> Proposition 5.6 for R^+ has been verified first in [8], and its a little bit complex form for GR^+ has been verified in [7]. Proposition 5.7 has been verified first for R^+ in [3], and then for GR^+

in [1] with a long proof referring to [3].

Now we turn to the "arithmetic" of GR^+. Let

$$A = (v(\infty,c) : c>0), \quad R^* = (u \in GR^+ : u_1 = 1)$$

Then A is a group, R* is a semigroup.

5.9 Definition. Let S be an abelian semigroup, T be a subsemigroup, $u \in T$, $F(T,u) := (v \in T : vw = u$ for some $w \in T)$. Then every element of F(T,u) will be called a T-factor of u.

5.10 Proposition. A is a subsemigroup of GR^+, $F(GR^+,a) \subset A$ for each $a \in A$. GR^+ is a subsemigroup of GR, $F(GR,u) \subset GR^+$ for each $u \in GR^+$.

Proof. We only verify that $F(GR^+,a) \subset A$. Suppose that $a = uv$, $a_n = c^n$. Then $u_n v_n = c^n = (u_1 v_1)^n$, and $u_n \geq u_1^n$, $v_n \geq v_1^n$ by (**). Thus we have $u_n = u_1^n$ and $u \in A$. #

5.11 Remark. $u \in GR^+$ if and only if there exist $\tilde{u} \in A$, $u^* \in R^*$ such that $u = \tilde{u}u^*$. Furthermore, $\tilde{u}_n = n_1^n$, $u_n^* = u_n u_1^{-n}$ for all n.

5.12 Proposition. Let $u \in GR^+$. Then $u = vw$ if and only if $u^* = v^*w^*$ and $\tilde{u} = \tilde{v}\tilde{w}$. u is i.d. if and only if u* is i.d.

Proof. Using 5.11. #

Let u belong to $R^*(GR^+)$ and have decomposition $u = vw$, $v,w \in R^*(GR^+)$. If either v is equal to identity e (v belongs to A), or w is equal to e (w belongs to A), we shall call it pseudo decompotition, otherwise, true decomposition. If $u \in R^* \setminus (e)$ $(u \in GR^+ \setminus A)$ has at least one true decomposition, we call it $(GR^+$-)decomposable, otherwise, $(GR^+$-)indecomposable or $(GR^+$-)simple.

5.13 Proposition. Let $u \in GR^+$. Then u is GR^+-decomposable if and only if u* is decomposable.

Proof. From 5.12 and the fact that $v \in A$ if and only if $v^* = e$. #

From inequalities (**), Proposition 5.4, and Theorem 3.12 we obtain the next

5.14 Theorem. R* is an MD-semigroup with property CLT.

5.15 Theorem. Let $u \in R^*$. Then u has no simple factor if and only if $u = e$, i.e. the class I_0 of R* consists of one element e.

Proof. If u has no simple factor, then by 5.14 and 1.7 u is i.d. By Proposition 5.6 $u = v(\infty,c) \prod_{k\geq 1} \tilde{v}(k,p_k)$. Here $c = u_1 = 1$. For each k, if

$p_k>1$, then by 5.7 and 5.12 $\tilde{v}(k,p_k)$ has a factor w which belongs to R* but is not i.d. in R*. But w has no simple factor, so it is i.d.. The contradiction shows that p_k must be equal to 1. So u = e.

It is obvious that e belongs to I_0 of R* by virtue of 5.10. #

5.16 Theorem. If u∈R*, then $u = \prod_{i\geq 1} v(i)$, where each v(i) is in R* and is simple or equal to e.

Proof. From Theorem 5.14, Theorem 1.8, and Theorem 5.15.

5.17 Corollary. If $u \in GR^+$ and u has no GR^+-simple factor, then u belongs to A.

Proof. Let $u = u^*\tilde{u}$. Then u* has no simple factor by 5.13, so u* = e by 5.15. Hence $u = \tilde{u} \in A$. #

5.18 Corollary. If $u \in GR^+$, then $u = \tilde{u} \prod_{i\geq 1} v(i)$, where each v(i) is in R* and is simple or equal to e.

Proof. Since $u = \tilde{u}u^*$ and $u^* = \prod_{i\geq 1} v(i)$ by 5.16. #

Now we give the central limit theorem for GR.

5.19 Theorem. Let (u(i,j) : 1≦j≦i, i=1,2,...) be a triangular array in GR such that $\lim_{i\to\infty} \max_j(|u_k(i,j)-1|) = 0$ for all k. If

$\lim_{i\to\infty} \prod_{1\leq j\leq i} u(i,j) = u \in GR^+$, then u is i.d.

Proof. Without loss of generality suppose that (u(i,j)) is in GR^+. Since

$$\lim_{i\to\infty} \max_j(|u_k^*(i,j)-1|) = \lim_{i\to\infty} \max_j(|u_k(i,j)(u_1(i,j))^{-k}-1|) = 0,$$

$$\lim_{i\to\infty} \prod_{1\leq j\leq i} u_k^*(i,j) = \lim_{i\to\infty} \prod_{1\leq j\leq i} u_k(i,j)(u_1(i,j))^{-k} = u_k u_1^{-k} = u_k^*,$$

by virtue of Theorem 5.14 u* is i.d. Hence u is i.d. #

I am thankful to Professor Liang Zhishun for his instruction and help.

References

[1] Chen Zaifu: On the construction of class I_0 for positive infinitely divisible generalized renewal sequences. Natural Science Journal of Hainan University, Vol. 4, No. 4, December, 1986. (Chinese)

[2] Dai Yonglong: Random point processes. Publishing House of Zhongshan University, Guangzhou (1984). (Chinese)

[3] Davidson, R.: Arithmetic and other properties of certain Delphic semigroups: I, [15], 115-149. (Reprinted from Z. Wahrscheinlichkeitstheorie & verw. Geb. 10 (1968), 120-145)

[4] Davidson, R.: More Delphic theory and practice, [15], 183-200. (Reprinted from Z. Wahrscheinlichkeitstheorie & verw. Geb. 13 (1969), 191-203)

[5] Davidson, R.: Sorting vectors, [15], 201-207. (Reprinted from Proc. Cambridge Philos. Soc. 68 (1970), 153-157).

[6] He Yuanjiang: On Delphic semigroups. Chin. Ann. Math., 5A: 6 (1984), 691-696. (Chinese)

[7] Huang Zhirui: F-funtion clusters and their applications. Chin. Ann. Math., 5A: 3 (1984), 273-286. (Chinese)

[8] Kendall, D.G.: Renewal sequences and their arithmetic, [15], 47-72. (Reprinted from Symposium on Probability (Lecture Notes in Math. 31) (1967), 147-175)

[9] Kendall, D.G.: Delphic semigroups, infinitely divisible regenerative phenomena, and the arithmetic of p-functions, [15], 73-114. (Reprinted from Z. Wahrscheinlichkeitstheorie & verw. Geb. 9 (1968), 163-195)

[10] Kingman, J.F.C.: The stochastic theory of regenerative events. Z. Wahrscheinlichkeitstheorie & verw. Geb. 2 (1964), 180-224.

[11] Kingman, J.F.C.: Semi-p-functions. Trans. Amer. Math. Soc., Vol. 174 (1972), 257-273.

[12] Liang Zhishun: On Delphic semigroups in stochastic point processes. Chin. Ann. Math., 5A: 2 (1984), 127-132. (Chinese)

[13] Liang Zhishun; Huang Zhirui: Fundamental properties of generalized renewal sequences. Acta Scientiarum Naturalium Universitatis Sunyatseni, 1983, 1. (Chinese)

[14] Matthes, K.; Kerstan, J.; Mecke, J.: Infinitely divisible point processes. Wiley, New York (1978).

[15] Kendall, D.G.; Harding, E.F., eds.: Stochastic analysis. John Wiley & Sons, 1973.

INFINITE DIMENSIONAL ROTATION GROUP AND UNITARY GROUP

Takeyuki HIDA

Department of Mathematics

Faculty of Science, Nagoya University

Chikusa-ku, Nagoya, 464

JAPAN

§0. Introduction.

There have been several approaches to the infinite dimensional rotation group. Among others, the probabilistic approach has widely developed in connection with the white noise analysis. The main reason for this is that the white noise measure is kept invariant under the action of the rotation group and the group can even characterize the white noise measure.

As soon as the white noise is complexified, so is the rotation group. Thus, the infinite dimensional unitary group can naturally be introduced not only as the complexification of the rotation group itself, but also as a tool of the study of complex white noise.

In Section 1 and Section 2 we shall quickly give a review of a probabilistic interpretation of the infinite dimensional rotation group and the unitary group, and in addition, we shall present some supplementary remarks from the view point of the present approach.

Section 3 will be devoted to some new results which would show a good relationship between complex white noise and unitary representation theory of Lie groups. As is given in the concluding remarks, we have hope that our results for the case of a one-dimensional time would successfully be generalized to a multi-dimensional parameter case.

§1. Rotation group.

We start with the white noise measure μ on the space E^* of real generalized functions on R^1, where E^* is taken to be a member of a Gel'fand triple:

(1.1) $$E \subset L^2(R^1) \subset E^* .$$

A linear homeomorphism g of E is a rotation of E, if g preserves the $L^2(R^1)$-norm : $\|g\xi\| = \|\xi\|$ for every $\xi \in E$. The collection of such rotations forms a group which is called the infinite dimensional rotation group and is denoted by $O(E)$ or by O_∞ (see, e.g. [8]).

There is a subgroup, denoted by G_∞, of O_∞ which is isomorphic to the inductive limit of the finite dimensional rotation groups $SO(n)$. Another interesting subgroup, which is called the Lévy group and is denoted by $\mathcal{G}$, is important and is really infinite dimensional.

A one-parameter subgroup $\{g_t\}$ of O_∞ is often called a whisker if each g_t comes from a diffeomorphism of the parameter set $\bar{R} = R^1 \cup \{\infty\}$. It is defined in such a way that

(1.2) $$(g_t\xi)(u) = \xi(\psi_t(u))\sqrt{|\psi_t'(u)|}$$

with a suitable choice of a family $\{\psi_t(u), -\infty < t < \infty\}$ of functions of u satisfying

(1.3) $$\psi_t \circ \psi_s = \psi_{t+s} .$$

Such a g_t can, in general, not be approximated by finite dimensional rotations under the usual topology.

The most important and in fact the simplest example of a whisker is the shift $\{S_t;\ t \in R^1\}$ defined by

(1.4) $$(S_t\xi)(u) = \xi(u - t), \qquad t \in R^1.$$

Remind that u is the time variable. And we see that the shift stands for propagation of time.

It is known (see [4] Chapt.5) that there are two other simple and important whiskers and that together with the shift they form a three dimensional subgroup G_P of the O_∞ which is isomorphic to the group $PSL(2,R)$. The group G_P is particularly interesting in the probability theory; for one thing, G_P describes Lévy's projective invariance of Brownian motion. Note that the basic nuclear space should be taken suitably in this case as we shall see in Section 3 for the complex case.

As is easily seen, the operator g_t acting on the basic nuclear space E can be extended to a unitary operator, still denoted by the same symbol, on the complex Hilbert space $L^2{}_c(R^1) \simeq L^2(R^1) \oplus iL^2(R^1)$. Now it should be noted that this unitary representation of the group of diffeomorphisms of R^- defining G_P is irreducible. This property can illustrate, in particular, some intrinsic probabilistic interest for the so-called L^2-theory for second order stationary stochastic processes.

§2. Complex white noise and infinite dimensional unitary group.

Let the basic nuclear space E, the space E^* of generalized functions and the white noise measure μ be complexified to obtain $E_c = E + iE$, $E_c^* = E^* + iE^*$ and $\nu = \mu \times \mu$ on E_c^*, respectively. The infinite dimensional unitary group $U(E_c)$, sometimes denoted by U_∞, is a collection of operators g acting on E_c such that each g is a linear homeomorphism of E_c and it preserves the $L_c^2(R^1)$-norm : $\|g\xi\| = \|\xi\|$.

As in the case of the rotation group O_∞ we can find subgroups like the inductive limit of the $U(n)$ and a generalization of the Lévy group to the complex case, whiskers and the like. An essential extension of the rotation group may be introduced as multiplication by a factor of unit modulus:

(2.1) $$(g\xi)(u) = \exp[i\varphi(u)]\cdot\xi(u),$$

where φ is a real valued function such that $g\xi$ is again a member of E_c. If φ is taken in such a way that g given by (2.1) is a continuous, surjective mapping of E_c onto itself, then g is called a gauge transformation (see [2], Chapt.3). We are interested in a one-parameter group of gauge transformations which is a subgroup of the U_∞. Simplest examples have been introduced in [4], namely

i) $\{I_t;\ t \in R^1\}$. The I_t is defined by

(2.2) $$(I_t\xi)(u) = \exp[ict]\cdot\xi(u),\quad c \in R^1.$$

Obviously $\{I_t\}$ forms a one-parameter subgroup of U_∞.

<u>Remark</u>. In the book [4], this subgroup $\{I_t\}$ is called a gauge transformationon. In fact, I_t is a very particular example where φ in (2.1) is a real constant ct.

ii) The multiplication $\{\pi_t;\ t \in R^1\}$ is given by

(2.3) $$(\pi_t\xi)(u) = \exp[iut]\cdot\xi(u),\qquad t \in R^1.$$

Taking the shift $\{S_t\}$, which can now be viewed as a whisker in U_∞, into our consideration, we are given three dimensional subgroup H, generated by the three one-parameter subgroups $\{I_t\}$, $\{\pi_t\}$ and $\{S_t\}$. The group H is isomorphic to the (one-dimensional) Heisenberg group, and some quantum mechanical interpretation can be given using the space (L_c^2) of functionals of complex Brownian motion.

Living in the complexified space, we hold freedom to use the Fourier transform $\mathcal{F}$ acting on $L^2(R^1)$, by which the shift and the multiplication can be interchanged. We can further proceed to the fractional power of the Fourier transform; namely

iii) The Fourier-Mehler tansform $\{\mathcal{F}_\theta\ ;\ \theta \in R^1\}$. The $\mathcal{F}_\theta$ is defined by

(2.4) $$(\mathcal{F}_\theta\xi)(u) = \int K_\theta(u,v)\xi(v)dv,\qquad \theta \neq \frac{n}{2}\pi,$$

where the kernel K_θ is given by

(2.5) $$K_\theta(u,v) = \sqrt{\pi(1-\exp[2i\theta])}\cdot\exp\left(-i\frac{(u^2+v^2)}{2\tan\theta} + i\frac{uv}{\sin\theta}\right).$$

The following assertion is quite natural, but not trivial. (See [4] for details.) Let $\mathcal{S}_c$ be the complex Schwartz space.

<u>Proposition</u> 1. <u>The Fourier-Mehler transform</u> $\{\mathcal{F}_\theta;\ \theta \in R^1\}$ <u>is a one-parameter subgroup of</u> $U(\mathcal{S}_c)$. <u>It is periodic</u> :

$$\mathcal{F}_\theta = \mathcal{F}_{\theta'}\quad \text{for}\quad \theta \equiv \theta' \mod 2\pi.$$

<u>Both the Fourier transform and its inverse are imbedded in such a way</u>

that

$$\mathcal{F}_{\pi/2} = \mathcal{F} \text{ and } \mathcal{F}_{3\pi/2} = \mathcal{F}^{-1}.$$

The group H has now been extended to a four dimensional subgroup H' which is generated by H and $\{\mathcal{F}_\theta\}$, where the latter can play a special role. The relation among these four one-parameter subgroups may be expressed in terms of their infinitesimal generators. Let s, I, $i\pi$ and if be the generators of the $\{S_t\}$, $\{I_t\}$, $\{\pi_t\}$ and $\{\mathcal{F}_\theta\}$, respectively. Denote the Lie product by [,]. Then we have

Proposition 2. The commutation relations are

$$[\pi, s] = I \qquad [f, s] = \pi \qquad [f, \pi] = s$$

I commutes with others.

This shows, in particular, that the Fourier-Mehler transform can continuously change the shift to the multiplication, and so does conversely.

The group H' is a solvable Lie group. Its probabilistic roles are discussed in [4] Chapt.5, so we do not go into details here.

§3. Some connection with unitary representation of SL(2,R).

We now turn, in this section, to the case where the basic nuclear space E is taken to be the complex D_0 space given by

$$D_0 = \{\xi;\ \text{complex valued, } \xi \text{ and } w\xi \text{ are of } C^\infty\text{-class}\},$$

where $(w\xi)(u) = \xi(1/u)|u|^{-1}$.

As was briefly mentioned, in Section 1, for the whiskers in O_∞, we can also find three interesting whiskers which all together form a three dimensional subgroup, also denoted by G_P, of $U(D_0)$. The group G_P involves, in addition to the shift defined as in (1.4), a whisker $\{\tau_t;\ t \in R^1\}$ given by

(3.1) $$(\tau_t\xi)(u) = \xi(u\cdot e^t)e^{t/2}, \qquad t \in R^1,$$

which is called the dilation. Another whisker which is a member of the whiskers generating G_P is the special conformal transformation $\{\kappa_t;\ t \in R^1\}$. We define κ_t by

(3.2) $$\kappa_t = w\, S_t w, \qquad t \in R^1.$$

The action of κ_t may be expressed in the form

(3.3) $(\kappa_t\xi)(u) = \xi(\frac{u}{-tu+1})\frac{1}{|-tu+1|}$.

With this expression it can be proved that

$$\kappa_t \in U(D_o) \quad \text{for every} \quad t \in R^1,$$

and that κ_t is continuous in t. These facts mean that $\{\kappa_t\}$ is a whisker.

Remark. The κ_t is not a member of $U(\mathcal{S}_c)$.

The group G_P generated by the above three whiskers is isomorphic to PSL(2,R) as in the case of the rotation group, or one can say that G_P is isomorphic to $SO_o(2,1)$.

Our interest finally comes to a combination of G_P, that is the three whiskers mentioned above, and the gauge transformations. We still stick to one-parameter subgroups of $U(D_o)$, therefore we start with a possible expression of the form

(3.4) $(g_t\xi)(u) = \exp[if(t,u)]\cdot\xi(\psi_t(u))\sqrt{|\psi_t'(u)|}$, $t \in R^1$,

where f(t,u) is real-valued and where $\{\psi_t(u)\}$ is the same as in the expressions (1.2) and (1.3). The group property

$$g_t \cdot g_s = g_{t+s}$$

requires the following functional equation

(3.5) $f(s,u) + f(t, \psi_s(u)) = f(t + s, u)$.

It is known that, under suitable conditions, $\psi_t(u)$ has to be of the form

(3.6) $\psi_t(u) = k^{-1}(k(u) + t)$,

where k is a monotone map onto $\bar{R}$ and k^{-1} is the inverse map of k (see [4] Chapt. 5). With the expression (3.6) for the function ψ_τ the equation (3.5) for f(t,u) can easily be solved to obtain the following lemma.

Lemma. Suppose that f(t,u) is a C^∞-function on $R^1 \times \bar{R}$ for which the equation (3.5) holds, where ψ_t satisfies (3.6). Then, there is a smooth function h(u) on $\bar{R}$ such that f(t,u) is expressed in the form

(3.7) $$f(t, u) = h(\psi_t(u)) - h(u).$$

Proof. The equations (3.5) and (3.6) imply

$$f(s, u) + f(t, k^{-1}(k(u) + s)) = f(t + s, u).$$

Set $k(u) = v$, and set $f(t, k^{-1}(v)) = g(t, v)$ to obtain

$$g(s, v) + g(t, v + s) = g(t + s, v).$$

If the above equation is evaluated at $v = 0$, then we are given

$$g(s, 0) + g(t, s) = g(t + s, 0).$$

Change the variable s to $u = k^{-1}(s)$ to define

$$h(u) = g(k(u), 0).$$

Then, we have

$$g(t, k(u)) = g(t + k(u), 0) - g(k(u), 0).$$

Noting that $g(t, v) = f(t, k^{-1}(v))$, we finally obtain (3.7).

Using $k(u)$ and $h(u)$ in the formulae (3.6) and (3.7), the infinitesimal generator $\alpha = \frac{d}{dt} g_t\big|_{t=0}$ of $\{g_t\}$ is obtained. Namely, by evaluating

$$\frac{d}{dt}\{\exp[ih(\psi_t(u)) - h(u)]\xi(\psi_t(u))\sqrt{|\psi_t'(u)|}\}\Big|_{t=0}$$

we have

(3.8) $$\alpha = a(u)\frac{d}{du} + \frac{1}{2}a'(u) + ib(u),$$

where $a(u) = \frac{dk^{-1}}{du}(k(u))$ and $b(u) = a(u)h'(u)$.

We are now in a position to discover typical one-parameter subgroups of $U(D_o)$ expressed in the form (3.4) with $f(t, u)$ given by the formula (3.7). Our idea is that

1) we extend the group G_P, the roles of which has been well established, by using gauge transformations, and

2) the extended group should be finite dimensional.

Before we come to the study the problem in question, a remark is mentioned. Take a real-valued function $h(u)$ in the Schwartz space and define a gauge transformation

$$m_h : \xi(u) \text{ --------> } \exp[ih(u)]\cdot\xi(u).$$

Then we obviously obtain

$$m_h^{-1} G_P \, m_h \simeq G_P \,,$$

although the transformed group involves a member of the form (3.4) with f expressible as (3.7).

With this remark we shall proceed to the following steps to find a new one-parameter subgroups of U_∞.

i) First, the one-parameter group $\{I_t\}$, introduced in § 2, is taken up as a special group of the gauge transformations. Its generator is obviously iI, I being the identity.

ii) We then come to combination of a gauge transformation and each whisker in G_P. Each infinitesimal generator is obtained by using (3.8) and is listed below:

$$\begin{array}{lll} & \text{for the shift} & s - ih_1'(u)I \\ (3.9) & \text{for the dilation} & \tau + iuh_2'(u)I \\ & \text{for the special conformal transformation} & \kappa + iu^2h_3'(u)I. \end{array}$$

The function h(u) in the formula (3.7) is chosen differently in the above three cases. The last generator is particularly interested in the application to the unitary representation theory of Lie groups as is seen in the following main theorem.

<u>Theorem</u> <u>Let</u> $\{I_t\}$ <u>be the one-parameter subgroup of</u> U_∞ <u>given in Section</u> 2, <u>and let</u> G_P <u>be the group generated by the three whiskers</u> $\{S_t\}$, $\{\tau_t\}$ <u>and</u> $\{\kappa_t\}$. <u>The lowest dimensional subgroup of</u> U_∞ <u>obtained from</u> $\{I_t\}$ <u>and combinations of gauge transformations and whiskers in</u> G_P <u>is four dimensional and its Lie algebra is generated by</u> I (the identity), s, τ, and $\kappa^c \equiv \kappa + ic\pi$, <u>where</u> c <u>is a real number and where</u> π <u>is multiplication operator by</u> u. <u>The one-parameter subgroup</u> $\{\kappa_t^c\}$ <u>corresponding the last generator is given by</u>

$$(3.10) \qquad (\kappa_t^c \, \xi)(u) = \xi\left(\frac{u}{-tu+1}\right)|-tu+1|^{-1+ic} \,.$$

<u>Proof</u>. If we want to have a finite dimensional subgroup of U_∞, then as is easily seen from commutation relations, $h_1(u)$ in the above list (3.9) should be affine. In this cace the term $ih_1'(u)$ which is $i \cdot cI$ (c being a real const.) shall be covered by I, the generator of

$\{I_t\}$, so we are not given any new generator. Similarly, we are not able to find any real-valued function h_2 for the generator τ.

We finally come to the last case of κ. Again, noting the restriction that only finite dimensional group (the same for Lie algebra) is permitted, we observe the commutation relations among k^c, $\jmath$ and τ to see that $u^2h_3'(u)$ should be affine. Therefore appears a new member of the form

(3.11) $\kappa^c = \kappa + icuI$,

for which $h_3(u)$ is taken to be $\log|u|$. It is easy to see that the one-parameter group with the generator (3.11) is defined by (3.10).

Before we close this section concluding remarks are mentioned.

Remark. i) The expression (3.10) is nothing but the formula that gives the principal continuous series with spin zero of the unitary representation of the group SL(2,R).

ii) We have so far discussed the case where the (time-)parameter space is one dimensional. In the higher (say d-)dimensional case we are given a subgroup G_P which is isomoprphic to $SO_o(d+1,1)$, and we can play a similar game using gauge transformations to extend the G_P, where much closer connections with the unitary representation theory would be developed.

[REFERENCES]

[1] P. Lévy, Problèmes concrets d'analyse fonctionnelle. Gauthier-Villars, 1951.

[2] Ian J. R. Aitchison, An informal introduction to gauge field theories. Cambridge Univ. Press, 1982.

[3] T. Hida, A role of Fourier transform in the theory of infinite dimensional unitary group. J. Math. Kyoto Univ. 13-1 (1973), 203-212.

[4] ______ , Brownian motion. Iwanami Pub. Co. 1975 (in Japanese); English ed. Springer-Verlag, 1980; Russian ed. Nauka, 1987.

[5] ______ , Brownian functionals and the rotation group. Math. + Physics, ed. by L. Streit, World Scientific, 1985, 167-194.

[6] ______, K.-S.Lee and S.-S. Lee, Conformal invariance of white noise. Nagoya Math. J. 98 (1985), 167-194.

[7] N. Obata, A characterization of the Lévy Laplacian in terms of infinite dimensional rotation groups. To appear.

[8] H.Yoshizawa, Rotation group of Hilbert space and its application to Brownian motion. Proc. International Conference on Functional Analysis and Related Topics. 1969, Tokyo. 414-423.

A Note on the Semigroup of Analytic Mappings with a Common Fixed Point

by Göran Högnäs, Åbo Akademi *

1. Introduction

Products of random matrices have been widely studied in recent years and the structure of those semigroups of matrices that admit, e.g., tight convolution sequences of probability measures is well understood, cf. Bougerol (1987), Högnäs (1987). Roughly speaking, tight convolution sequences can exist only in group-like structures. In the same way, other probabilistic notions, such as recurrence or existence of various invariant measures, will more or less automatically impose a richer algebraic structure on an *a priori* rather general semigroup, see, e.g., Högnäs and Mukherjea (1980), Mukherjea (1987).

The purpose of this note is to start an investigation along the same lines for a semigroup of non-linear mappings whose structure is sufficiently close to the linear case to permit a very similar kind of reasoning. In fact, we will heavily rely on a representation of those mappings as infinite upper triangular matrices.

Let f be a function from $\Re^d$ to itself. Suppose that f has a fixed point which we will take to be the origin. We will say that f is *analytic* (at the origin) if its Maclaurin series converges and the sum coincides with f within a disk around the origin with positive radius. The Maclaurin series of f is a power series in d variables with no constant terms. (In the context of *formal* power series, of one variable, Henrici (1974) terms such series *nonunits* in the integral domain of formal power series. For the case $d > 1$, however, this terminology might be a little ambiguous.) We will assume that f is completely determined by its Maclaurin series.

The set A of all such analytic functions is a semigroup under composition of mappings. We will strive to show that tightness of convolution sequences and existence of invariant measures on A will force upon A essentially the same structural restrictions as one has in the case of matrix semigroups.

2. The semigroup of analytic functions as a limit of matrix semigroups

Let the function f belong to A. Consider the power series representation, i.e. the Maclaurin series, for f truncated to terms of order $\leq k$. Call this polynomial function $\pi_k(f)$. If g is another element of A we obtain, by virtue of the assumption of the origin as common fixed point of all elements of A,

$$\pi_1(f \circ g) = \pi_1(f) \circ \pi_1(g) = \pi_1(\pi_1(f) \circ \pi_1(g))$$

(This is the chain rule for the Jacobians.) In other words, the first order terms in the expansion of $f \circ g$ involves only the first order terms in f and g.

* Matematiska institutionen, Fänriksgatan 3, SF-20500 Åbo, Finland

The above observation applies to higher order terms, too:

$$\pi_k(f \circ g) = \pi_k(\pi_k(f) \circ \pi_k(g)), k = 1, 2, \ldots$$

Phrased in another way, the truncation to terms of order k or less is a *homomorphism* from the semigroup A to the semigroup of polynomial functions of order k where the operation is composition of functions followed by a truncation. This operation is clearly associative. Let us denote it by $\bullet_k$ or just $\bullet$ if we do not want to stress explicitly the dependence on k. Thus $\pi_k(f \circ g) = \pi_k(f) \bullet_k \pi_k(g)$.

Remark. In exactly the same way the truncation (projection) π_k is a homomorphism from the set P of formal power series in d variables without constant terms to the set of polynomial functions of order k or less. The operation in P corresponding to the composition of functions is the ***substitution*** of the power series g into the series f. The new series, which we will denote $f \circ g$, is well defined as a formal power series because of the absence of constant terms, cf. Henrici (1974).

Let us define a topology on A (and P) as follows: A sequence $\{f_n\}$ converges to an element $f \in A$ iff the coefficients of the polynomials $\pi_k(f_n - f)$ go to zero as $n \to \infty$, for each k. This topology on A is generated by a metric, d, say, which has to be non-complete, though, because a Cauchy sequence may approach a formal power series which does not necessarily converge outside of the origin.

The topology does, however, make A into a *topological* semigroup. This is because the composition of polynomial functions of degree at most k is a jointly continuous operation (depends continuously on the coefficients of the two factors) for every positive integer k. Furthermore, the homomorphism π_k discussed above is defined to be a *continuous* one.

The topology on A is completely determined by these finite-dimensional representations.

We will now, following Henrici (1974), introduce a useful matrix representation of the elements of the set P of formal power series without constant terms (of which our semigroup A of analytic mappings constitutes a subset).

In dimension $d = 2$ the matrix representation of $f \in P$ is written as follows (x, y are the variables and $u(x,y), v(x,y)$ the two components of f): The first two *rows* contain the coefficients in the power series for $u(x,y)$ and $v(x,y)$. The next four rows correspond to the power series (obtained by using the usual Cauchy product rule) for u^2, uv, vu, v^2; the next eight rows correspond to $u^3, u^2v, uvu, uv^2, vu^2, vuv, v^2u$, and v^3. The *columns* correspond to the coefficients of x, y, x^2, xy, yx, y^2, $x^3, x^2y, xyx, xy^2, \ldots$. The fact that the variables *commute* causes a certain arbitrariness in the assignment of values for the coefficient of xy and yx or u^2v, uvu or vu^2, say. Let us agree to "symmetrize" the matrix so that the columns (rows) corresponding to the same products are equal. Thus the fourth and fifth column (row) are equal, as are the eighth, ninth and eleventh. Modulo this additional requirement each formal power series or analytic mapping determines a matrix uniquely. On the other hand, the first d rows of a matrix fixes the formal power series.

As an example let us look at the matrix representation of the map

$$\begin{pmatrix} x \\ y \end{pmatrix} \mapsto \begin{pmatrix} \sin xy \\ y + \sin 2x + x^2 + x^2 y \end{pmatrix}:$$

$$\begin{pmatrix} 0 & 0 & 0 & \frac{1}{2} & \frac{1}{2} & 0 & 0 & 0 & 0 & 0 & \dots \\ 2 & 1 & 1 & 0 & 0 & 0 & -\frac{4}{3} & \frac{1}{3} & \frac{1}{3} & 0 & \dots \\ 0 & 0 & 0 & 0 & 0 & 0 & 0 & 0 & 0 & 0 & \dots \\ 0 & 0 & 0 & 0 & 0 & 0 & 0 & \frac{2}{3} & \frac{2}{3} & \frac{1}{3} & \dots \\ 0 & 0 & 0 & 0 & 0 & 0 & 0 & \frac{2}{3} & \frac{2}{3} & \frac{1}{3} & \dots \\ 0 & 0 & 4 & 2 & 2 & 1 & 4 & \frac{2}{3} & \frac{2}{3} & 0 & \dots \\ \dots & & & & & & & & & & \end{pmatrix}$$

The first 2×2 block is the Jacobian, the first 6×6 block corresponds to the truncation $\pi_2(f)$, the first 14×14 block to $\pi_3(f)$, etc.

To see that matrix multiplication really preserves the operation $\circ$ of P and A, let us look at the second and the fourth rows (the $v-$ and the $uv-$rows) and the eighth column (the x^2y-column) of the matrix corresponding to $f \circ g$. In the calculations below the elements of the matrix corresponding to f and g are termed $a_{.,.}$ and $b_{.,.}$. The row indices called $u, v, uu, uv, vu, \dots$ above are termed $1, 2, 11, 12, 21, \dots$ and the column indices $x, y, xx, xy, yx, yy, \dots$ analogously $1, 2, 11, 12, 21, 22, \dots$. The elements in the matrix for $f \circ g$ that we are looking for are denoted $c_{2,112}$ and $c_{12,112}$.

$c_{2,112} \;=\; a_{2,1}b_{1,112} + a_{2,2}b_{2,112} + a_{2,11}b_{11,112} + a_{2,12}b_{12,112} + a_{2,21}b_{21,112} + a_{2,22}b_{22,112} + a_{2,111}b_{111,112} + \dots + a_{2,222}b_{222,112}$

$$c_{12,112} = a_{12,11}b_{11,112} + a_{12,12}b_{12,112} + \dots a_{12,222}b_{222,112}$$

The conventions emanating from the commutativity of the variables, the equality of certain rows and columns, are, of course, preserved by matrix multiplication. Also, the interpretation of the rows below the "independent" d first rows is preserved: the $uv-$row is formed by Cauchy-multiplying the $u-$ and the $v-$row. Take for example $c_{12,112}$. A tedious computation shows that the sum of the 24 terms in the expression for $c_{1,11}c_{2,2} + c_{1,1}c_{2,12}$ is equal to the sum

$$\sum_k a_{12,k}b_{k,112} = c_{12,112}$$

where the summation extends over all the 12 two- and three-digit indices $11, 12, \dots, 111, 112, \dots$.

3. Argabright's conjecture

As a first application of our matrix representation we prove the so-called Argabright conjecture for P and A, equipped with the topology introduced above.

Let S be a locally compact second countable (LCCB) subsemigroup of P or A. The measure μ defined on the Borel sets of S is said to be r^**−invariant* if

$$\mu(Bx^{-1}) = \mu(B)$$

for each $x \in S$ and Borel set $B \subset S$, where Bx^{-1} is a notation for $\{s|sx \in B\}$. *Argabright's conjecture* (cf. Argabright (1966), Michael (1964), Mostert (1964)), usually formulated for a general LCCB semigroup S, states that the support of an r^*–invariant measure μ is necessarily a left group, i.e. isomorphic to a product of a locally compact set E and a locally compact group G with multiplication rule $(e,g)(e',g') = (e,gg')$. The measure μ can then be factored into a product measure on $E \times G$ with a Haar measure as second factor. For a *finite* measure μ the conjecture was proved by Mukherjea and Tserpes (1976); this reference also contains a detailed discussion of invariance properties of measures on semigroups including some useful alternative formulations of the conjecture.

Theorem. *Let S be a LCCB subsemigroup of A or P. Then the Argabright conjecture holds for S.*

Proof: Call the r^*-invariant measure μ. With no loss of generality we may restrict S to equal the support of μ. S is then a locally compact subset of A or P. In particular, we have at our disposal the continuous homomorphism π_k mapping S into the semigroup T_k of polynomial functions of degree at most k and with $\bullet_k$, composition followed by truncation to terms of degree $\leq k$, as semigroup operation. Denote by S_k the image of S under π_k. S is σ- compact and so are all the S_k's.

Let C be any Borel subset of S_k. Then $\mu_k(C) = \mu\{\pi_k^{-1}(C)\}$ defines a measure on S_k. For $y \in S_k$ ($y = \pi_k(x)$ for some $x \in S$) we have $\mu_k(Cy^{-1}) = \mu\{s|\pi_k(s)y \in C\} = \mu\{s|\pi_k(sx) \in C\} = \mu\{s|sx \in \pi_k^{-1}(C)\} = \mu((\pi_k^{-1}(C))x^{-1}) = \mu(\pi_k^{-1}(C)) = \mu_k(C)$. Hence the measure μ_k is r^*-invariant on S_k.

We are now in a position to use the same kind of argument as in Högnäs and Mukherjea (1980), p. 72f. For fixed but arbitrary k all elements of S_k have the same rank because of the r^*-invariance of μ_k. From this fact we draw the conclusion that the matrix semigroup $\pi_k(a)S_k\pi_k(b)$ is bicancellative for all $\pi_k(a), \pi_k(b) \in S_k$. Take $a, b \in S$ and consider the equation

$$asbatb = asbaub, \quad s,t,u \in S.$$

Operating with the homomorphism π_k on both sides and using the cancellativity of $\pi_k(a)S_k\pi_k(b)$ we conclude that $\pi_k(atb) = \pi_k(aub)$ for all k which implies that $atb = aub$. Consequently, aSb is bicancellative whence it follows that S is a left group.

4. Tightness of convolution sequences

Let us return to the whole semigroup A of mappings from $\Re^d$ to itself analytic at a common fixed point (the origin). Recall that we have imposed a special topology on A depending only on the coefficients in the power series representations. Suppose we have a probability measure μ on A such that it generates a tight convolution sequence $\{\mu^n\}$. What algebraic properties does the semigroup S generated by the support of μ necessarily possess? As the results in the previous sections suggest, such a semigroup must have a rather special structure. A partial result in that direction is the following

Proposition. *Let $d = 2$. Let S be a LCCB subsemigroup of A generated by the support of a probability measure μ. Suppose that the convolution sequence $\{\mu^n\}$ is tight. Then S admits a completely simple subsemigroup K with compact group factor and we have either*

(i) the Jacobians of the elements of K are non-singular and K is a compact group isomorphic to the set K_1 of Jacobians,

or

(ii) the Jacobians of the elements of K have the common rank 1 and any idempotent of K may be written, modulo a not necessarily linear change of coordinates,

$$\begin{pmatrix} x \\ y \end{pmatrix} \mapsto \begin{pmatrix} h(y) \\ y \end{pmatrix}$$

where $h(y)$ is a function analytic at the origin with zero constant and first degree terms in its power series representation,

or

(iii) $K = \{0\}$ and μ^n converges weakly to the point mass at 0.

For the proof we refer to Högnäs (1988).

Acknowledgements. I want to thank the organizers, especially Prof. Herbert Heyer and the personnel of the Geschäftstelle in Freiburg, for their helpfulness and hospitality. A travel grant from the Academy of Finland (Science Research Council) is gratefully acknowledged. Thanks are also due to my colleagues Paul Lindholm and Gunnar Söderbacka for valuable discussions and references.

References

Argabright, L. N. (1966), A note on invariant integrals on locally compact semigroups, *Proc. Amer. Math. Soc.* **17**, 377 - 382.

Henrici, P. (1974), *Applied and Computational Complex Analysis, Vol. 1* (John Wiley & Sons, New York-London-Sydney-Toronto).

Högnäs, G. (1988), Invariant measures and random walks on the semigroup of matrices. In Heinz Langer (ed.): Proceedings of the conference on Markov Processes and Stochastic Control, Gaußig, DDR, 11 - 15 January, 1988 (to appear).

Högnäs, G. and A. Mukherjea (1980), Recurrent random walks and invariant measures on semigroups of $n \times n$ matrices, *Math. Z.* **173**, 69 - 94.

Michael, J. H. (1964), Right invariant integrals on locally compact semigroups, *J. Austral. Math. Soc.* **4**, 273 - 286.

Mostert, P. S. (1964), Comments on the preceding paper of Michael's, *J. Austral. Math. Soc.* **4**, 287 - 288.

Mukherjea, A. and N. A. Tserpes (1976), *Measures on Topological Semigroups: Convolution Products and Random Walks* (Lecture Notes in Mathematics 547, Springer-Verlag, Berlin-Heidelberg-New York).

Localizations of Feller infinitesimal generators and uniqueness of corresponding killed processes

Jan Kisyński

Abstract. Let G be infinitesimal generator of a Feller transition semigroup on a compact C^{∞} manifold M with boundary. Assume that G is defined by means of a sufficiently smooth integrodifferential elliptic boundary system of Ventcel. Let U be an open subset of $M \setminus \partial M$. Then the operator $(\mathbb{1}_U G)|_{C_c^{\infty}(U)}$ uniquely determines the canonical cadlag Markov process corresponding to G before its first exit time from U. This statement is formulated and proved in rigorous measure theoretical language.

1. Introduction

1.1. The manifold M. Throughout this paper we denote by M a compact C^{∞} manifold with boundary ∂M, or without boundary. In the second case we admitt that $\partial M = \emptyset$.

1.2. Feller semigroups and generators. By a Feller generator on M we mean the strong infinitesimal generator of a Feller semigroup on M, i.e. of a one-parameter strongly continuous semigroup of non-negative linear contractions of the space $C(M)$.

1.3. Filtrations and Markov processes. Take into account the compact metrizable space $M_\Delta = M \cup \{\Delta\}$ such that $\Delta \notin M$ is a separated point of M_Δ and the topology induced on M by M_Δ coincides with original topology of M. Denote by W the space of all cadlag M_Δ-valued functions ω defined on R^+ (i.e. M_Δ-valued right-continuous functions on R^+, having left-side limits everywhere on $R^+ \setminus \{0\}$), such that the set $\omega^{-1}(\Delta)$ either is empty, or is equal to $[\zeta, \infty)$ for some $\zeta = \zeta(\omega) \in R^+$. For every $t \in R^+$ define the evaluation map $X_t : W \longrightarrow M$ such that $X_t(\omega) = \omega(t)$ for every $\omega \in W$. Denote by $(X_t)_{t \geqslant 0}$ the canonical process on W, i.e. the map $R^+ \times W \ni (t, \omega) \longrightarrow X_t(\omega) \in M_\Delta$. Denote by $\mathcal{B}(M_\Delta)$ the Borel σ-field of M_Δ and by $\mathbb{P}$ the set of all probability measures on $\mathcal{B}(M_\Delta)$. Define the σ-fields of subsets of W:

$$\mathring{\mathcal{F}} = \sigma\{X_s^{-1}(B) : s \in R^+,\ B \in \mathcal{B}(M_\Delta)\},$$

$$\mathring{\mathcal{F}}_t = \sigma\{X_s^{-1}(B) : 0 \leqslant s \leqslant t,\ B \in \mathcal{B}(M_\Delta)\},$$

According to one of fundamental existence theorems of the theory of Markov processes, see [2 ; p. 46], for each Feller generator G on M there is unique system $(P_\mu^G)_{\mu \in \mathbb{P}}$ of probability measures $\mathbb{P}_\mu^G$ on the σ-field $\mathring{\mathcal{F}}$ such that

(1.3.1) $$P_\mu^G\{X_0 \in B\} = \mu(B) \qquad \text{for every } B \in \mathcal{B}(M_\Delta) \text{ and } \mu \in \mathbb{P}$$

and

(1.3.2) $$E_\mu^G[\tilde{f}(X_{t+s}) \,|\, \mathring{\mathcal{F}}_t] = (e^{sG}f)(X_t)$$

$\mathbb{P}_\mu^G$-a.s., for every $\mu \in \mathbb{P}$, $t \in R^+$, $s \in R^+$, $f \in C(M)$ and $\tilde{f} \in C(M_\Delta)$ such that $\tilde{f}(\Delta) = 0$ and $\tilde{f} = f$ on M.

Given a Feller generator G on M, we define the σ-fields of subsets of the space W:

$$\mathcal{F}^G = \bigcap_{\mu\in\mathbb{P}} (\mathring{\mathcal{F}})^{P^G_\mu}, \quad \mathcal{F}^G_t = \bigcap_{h>0,\ \mu\in\mathbb{P}} (\mathring{\mathcal{F}}_{t+h})^{P^G_\mu}, \qquad t\in R^+,$$

where $(\mathcal{G})^P$ stands for the completion of a σ-field $\mathcal{G}$ with respect to the measure $P|_{\mathcal{G}}$, i.e. with respect to the restriction to $\mathcal{G}$ of a probability measure P defined on some σ-field containing $\mathcal{G}$. Then $(\mathcal{F}^G_t)_{t\geqslant 0}$ is a right-continuous filtration on the measurable space $(W,\mathcal{F}^G)$, and each first entry time of the process $(X_t)_{t\geqslant 0}$ to any Borel subset of M_Δ is an optional time of this filtration. Every measure P^G_μ has unique extension to a measure on $\mathcal{F}^G$, denoted again by P^G_μ. If ε_x is the unit mass concentrated at a point $x\in M_\Delta$, then we write P^G_x instead of $P^G_{\varepsilon_x}$.

1.4. Events strictly prior then an optional time. For every optional time T of a filtration $(\mathcal{F}^G_t)_{t\geqslant 0}$ we denote by $\mathcal{F}^G_{T-}$ the following σ-field of events strictly prior then T:

$$\mathcal{F}^G_{T-} = \sigma\{\mathcal{F}_{0-} \cup \mathcal{G}^G_{T-}\},$$

where

$$\mathcal{G}^G_{T-} = \{A\cap\{t<T\}:\ t\in R^+,\ A\in\mathcal{F}^G_t\}$$

and

$$\mathcal{F}_{0-} = \{X_0^{-1}(B)\ :\ B\in\mathcal{B}_u(M_\Delta)\},$$

$\mathcal{B}_u(M_\Delta)$ being the σ-field of universally measure sets over $\mathcal{B}(M_\Delta)$. Since $P^G_\mu\{X_0\in B\} = \mu(B)$ for every $\mu\in\mathbb{P}$ and $B\in\mathcal{B}_u(M_\Delta)$, we have $\mathcal{F}_{0-} = \bigcap_{\mu\in\mathbb{P}} (\mathring{\mathcal{F}}_0)^{P^G_\mu} \subset \mathcal{F}^G_0$.

1.5. The purpose of the present paper is to find a possibly wide class $\mathcal{G}$ of Feller generators on M such that the following local unique determination principle is true for $\mathcal{G}$:

if U is an arbitrary open subset of $M \setminus \partial M$ and T is the first exit time of the process $(X_t)_{t \geq 0}$ from U,

then, for G ranging over $\mathcal{G}$, the σ-fields $\mathcal{F}^G_{T-}$ and the restricted probability measures $\mathbb{P}^G_\mu \big|_{\mathcal{F}^G_{T-}}$, $\mu \in \mathbb{P}$, are uniquely determined by the operators $G_U = (\overline{\mathbb{1}_U G})\big|_{C_c^\infty(U)}$.

In the above $\mathbb{1}_U$ stands for multiplications by the indicator function of the set U, and $\overline{\mathbb{1}_U G}$ denotes the closure of $\mathbb{1}_U G$ treated as an operator from $C(M)$ into $C(U)$, the former space being equiped with the topology of uniform convergence on intersections of U with compact subsets of $M \setminus \partial M$.

Every Feller generator G on M satisfies the principle of non-negative maximum: if $f \in \mathcal{D}(G)$, $\mathring{x} \in M$ and $f(\mathring{x}) = \max\{f(x) : x \in M\} \geq 0$, then $Gf(\mathring{x}) \leq 0$. Using only this principle and the density of $\mathcal{D}(G)$ in $C(M)$, one can prove that, for each open $U \subset M \setminus \partial M$, the closure $\overline{\mathbb{1}_U G}$ exists, see [11; p.535], and for justification of our definition of G_U it remains to show that $C_c^\infty(U) \subset \mathcal{D}(\overline{\mathbb{1}_U G})$. The former follows easily from the Proposition III.2.9. of [3; p. 494], if we only assume that G is defined by means of a sufficiently smooth, (satisfying in particular the condition 7.2.1), elliptic integrodifferential boundary system of Ventcel. Under similar assumptions we are able to carry over a proof of the local unique determination principle. In this proof we rest upon construction of resolvent presented in Chapter III of [3], and the mentioned Proposition III.2.9. is one of our fundamental tools. After elimination the condition (7.2.1) the local unique determination principle works in a version, formulated not in terms of operators G_U, but in terms of restrictions of operators of Waldenfels occuring in concrete systems of Ventcel.

2. Definitions of integrodifferential operators of Ventcel and of Waldenfels.

We enclose a list of unavoidable definitions. Except of the first of them, all they are taken from [3].

2.1. By a Lévy map on M we mean a C^∞ map $v : M \times M \longrightarrow T(M)$ such that, for every x and y in M, $v(x,y) \in T_x(M)$, $v(x,x) = 0$ and $d_y v(x,y)\big|_{x=y} = id_{T_x(M)}$.

2.2. An operator of Lévy on M is a linear operator S of $C^2(M)$ into the space $b\mathcal{B}_{loc}(M \setminus \partial M)$ of Borel functions on M locally bounded on $M \setminus \partial M$ such that

$$Sf(x) = -a(x)f(x) + df(x)\cdot u(x) + \int_M [f(y) - f(x) - df(x)\cdot v(x,y)]\, s(x,dy)$$

for every $f \in C^2(M)$ and $x \in M$, where $a \in b\mathcal{B}_{loc}(M \setminus \partial M)$ is non-negative, u is a vector field on M such that $df(\cdot)\cdot u(\cdot) \in b\mathcal{B}_{loc}(M \setminus \partial M)$ whenever $f \in C^1(M)$, v is a Lévy map and s is a Lévy kernel, i.e. a non-negative Borel kernel on $M \times \mathcal{B}(M)$ such that $\int_M \varphi(\cdot,y)d(\cdot,dy) \in b\mathcal{B}_{loc}(M \setminus \partial M)$ for every non-negative function $\varphi \in C^2(M \times M)$ vanishing on the diagonal of $M \times M$. The values of $s(x,\{x\})$ do not influence the operator S and the usual convention is that $s(x,\{x\}) = 0$ for every $x \in M$.

2.3. An operator of Waldenfels on M is a linear operator W: $C^2(M) \longrightarrow b\mathcal{B}_{loc}(M \setminus \partial M)$, such that $W = P + S$, where $P : C^2(M) \longrightarrow b\mathcal{B}_{loc}(M \setminus \partial M)$ is a homogeneous differential operator of second order whose coefficient tensor field is non-negative definite at each point of M, and $S : C^2(M) \longrightarrow b\mathcal{B}_{loc}(M \setminus \partial M)$ is an operator of Lévy.

2.4. If $W = P + S$ is an operator of Waldenfels on M, then its absorbtion coefficient a, Lévy kernel s, convection vector field u and diffusion part P are uniquely determined by the formulas :

(2.4.1) $a(x) = - W\mathbb{1}(x)$ for every $x \in M$,

(2.4.2) $\int f(y)s(x,dy) = Wf(x)$ for every $x \in M$ and every $f \in C^2(M)$ such that $x \notin \operatorname{supp} f$,

(2.4.3) $df(x) \cdot u(x) = Wf(x) - \int [f(y) - df(x) \cdot v(x,y)] s(x,dy)$ for every $x \in M$ and every $f \in C^2(M)$ such that $f(x) = 0$ and $d^2f(x) = 0$,

(2.4.4) $Pf(x) = Wf(x) - \int f(y)s(x,dy)$ for every $x \in M$ and every $f \in C^2(M)$ such that $f(x) = 0$ and $df(x) = 0$.

It is easy to check that each operator of Waldenfels satisfies the principle of maximum:

(2.4.5) if $f \in C^2(M)$ attains non-negative maximum at a point $x \in M \setminus \partial M$, then $Wf(x) \leq 0$.

2.5. Let t be a non-negative Borel kernel on $\partial M \times \mathcal{B}(M)$. Define the order r of t so that $r \in \{0,1,2,\infty\}$,

$$r = 0 \Leftrightarrow \sup_{x \in \partial M} t(x,M) < \infty,$$

$$r \leq k \Leftrightarrow \begin{cases} \sup_{x \in \partial M} \int_M (F(x,y))^{k/2} t(x,dy) < \infty \quad \text{for each non-negative} \\ \text{function } F \in C^\infty(\partial M \times M) \text{ such that } F(x,x) = 0 \text{ for} \\ \text{every } x \in \partial M, \end{cases}$$

for $k = 1$ and for $k = 2$, and $r = \infty \Leftrightarrow r \notin \{0,1,2\}$.

A non-negative Borel kernel on $\partial M \times \mathcal{B}(M)$ of order 0, 1 or 2 will be called a Ventcel kernel on M.

2.6. To a Ventcel kernel t on M of order r, and to a Lévy map v if $r = 2$, we associate the operators $T_k : C^k(M) \longrightarrow b\mathcal{B}(\partial M)$ defined by:

$$T_0 f(x) = \int_M f(y)t(x,dy) \qquad \text{if} \quad r = 0,$$

$$T_1 f(x) = \int_M [f(y) - f(x)]t(x,dy) \qquad \text{if} \quad r \in \{0,1\},$$

$$T_2 f(x) = \int_M [f(y) - f(x) - df(x) \cdot v(x,y)]t(x,dy) \quad \text{if} \quad r \in \{0,1,2\}.$$

Correctness of two last definitions follows from suitable Taylor developments.

2.7. By an operator of Ventcel on M of order zero we shall mean an operator $\Gamma : C(M) \longrightarrow b\mathcal{B}(\partial M)$ of the form

(2.7.0) $\Gamma f(x) = T_0 f(x) - (a(x) + t(x,M))f(x), \quad f \in C(M), \quad x \in \partial M,$

where $a \in b\mathcal{B}(\partial M)$ is non-negative and t is a Ventcel kernel of order zero.

By an operator of Ventcel on M of order one we shall mean an operator $\Gamma : C^1(M) \longrightarrow b\mathcal{B}(\partial M)$ of the form

(2.7.1) $\Gamma f(x) = T_1 f(x) - a(x)f(x) + df(x) \cdot u(x), \quad f \in C^1(M), \quad x \in \partial M,$

where the Ventcel' kernel occuring in T_1 has order 0 or 1, the function $a \in b\mathcal{B}(\partial M)$ is non-negative, and u is a non-vanishing $b\mathcal{B}$ vector field on ∂M tangent to M, such that at each point $x \in \partial M$ the vector $u(x)$ is directed strictly to the interior of M.

By an operator of Ventcel on M of order two we shall mean an operator $\Gamma : C^2(M) \longrightarrow b\mathcal{B}(\partial M)$ of the form

(2.7.2) $\Gamma f(x) = T_2 f(x) - a(x)f(x) + df(x) \cdot (u(x) + w(x)) + (Q\gamma_0 f)(x),$

$$f \in C^2(M), \quad x \in \partial M, \quad w(x) = \int_M (1 - p(x))v(x,y)t(x,dy),$$

where t is a Ventcel' kernel (of order 0, 1 or 2, occuring also in T_2), v is a Lévy map (the same which occurs in T_2), $p : T(M)\big|_{\partial M} \longrightarrow T(\partial M)$ is a C^∞ map such that, for each $x \in \partial M$, $p\big|_{T_x(M)}$ is a projection of $T_x(M)$ onto $T_x(\partial M)$, $a \in b\mathcal{B}(\partial M)$ is non-negative, u is a $b\mathcal{B}$ vector field on ∂M tangent to M, such that at each point $x \in \partial M$ the vector $u(x)$ either is tangent to ∂M, or is directed stricty to the interior of M, $\gamma_0 : C^2(M) \longrightarrow C^2(\partial M)$ is the trace operator, and finally Q is a second order homogeneous elliptic differential operator on ∂M.

In local coordinates, the length of the vector $(1-p(x))v(x,y)$ and the distance from y to M have the same order of magnitude as $M \ni y \longrightarrow x \in \partial M$. Together with conditions defining the order of t, this implies absolute convergence of the $T_x(M)$-valued integral defining $w(x)$ and gives possibility of rewritting (2.7.2) in the form

$$(2.7.3) \qquad \Gamma f(x) = \int_M \left[f(y) - f(x) - df(x)\cdot p(x)v(x,y)\right] t(x,dy) -$$
$$- a(x)f(x) + df(x)\cdot u(x) + (Q\gamma_0 f)(x).$$

2.8. Given a second order Ventcel operator Γ, and given the maps v and p, one can successively determine a, t, u and Q from the equalities: $a(x) = -(\Gamma \mathbb{1})(x)$,

$$\int_M f(y)t(x,dy) = (\Gamma f)(x) \quad \text{for } f \in C^2(M) \text{ and } x \in \partial M \setminus \operatorname{supp} f,$$

$$df(x)\cdot u(x) = \Gamma f(x) - T_2 f(x) - df(x)\cdot w(x) \quad \text{for } x \in \partial M \text{ and}$$
$$f \in C^2(M) \text{ such that } d^2 f(x) = 0,$$

and from (2.7.2). Similar remarks concern operators of Ventcel of order 0 and 1.

2.9. In the sequel, by an operator of Ventcel on M we shall mean an operator of Ventcel of order 0, 1 or 2. Our assumptions about u in (2.7.1) and about Q in (2.7.2) coincide with assumptions used in Chapter III of [3] and are more restrictive then analogous conditions occuring in definitions of Chapter II of [3]. It is evident from (2.7.0), (2.7.1) and (2.7.) that every operator of Ventcel satisfies the principle of non-negative maximum at boundary:

(2.9.1) if $f \in C^2(M)$ attains non-negative maximum at a point $x \in \partial M$, then $\Gamma f(x) \leq 0$.

3. Hölderian elliptic boundary systems of Ventcel'.

3.1. A Lévy kernel s on M will be called hölderian if there are $\beta \in (0,1)$ and $\gamma \in (0,1)$ such that $\mathcal{K}_\beta f \in C^\gamma(M)$ for every $f \in C^\gamma(M)$, where

$$\mathcal{K}_\beta f(x) = \int_M f(y)\overline{x,y}^{2-\beta} s(x,dy), \qquad x \in M,$$

$\overline{x,y}$ being the geodesic distance from x to y in the sense of a C^∞ Riemann metric on an open C^∞ manifold M containing M. The definition of a hölderian Lévy kernel is independent of a choice of $\tilde{M}$ and the C^∞ Riemann metric on $\tilde{M}$.

3.2. If s is a hölderian Lévy kernel, and β, γ and the Riemann metric are such that $\mathcal{K}_\beta$ maps $C^\gamma(M)$ into itself, then

$$\sup_{x \in M} \int_M \overline{x,y}^{2-\beta} s(x,dy) < \infty,$$

whence, by density of $C^\gamma(M)$ in $C(M)$, $\mathcal{K}_\beta \in L(C(M),C(M))$. The closed graph theorem implies that $\mathcal{K}_\beta \in L(C^\gamma(M),C^\gamma(M))$. Consequently, by interpolation, $\mathcal{K}_\beta \in L(C^\alpha(M), C^\alpha(M))$ for every $\alpha \in (0,\gamma)$.

3.3. An operator of Waldenfels $W = P + S$ will be called elliptic hölderian if it has three properties: $W \in L(C^{2+\alpha}(M), C^\alpha(M))$ for some $\alpha \in (0,1)$, the Lévy kernel of S is hölderian, and P is elliptic. By interpolation, one can assume that $\alpha \in (0, \gamma \wedge \frac{1}{3}\beta)$, where β and γ are constants occuring in 3.1. As we shall see, if $\alpha \in (0, \gamma \wedge \frac{1}{3}\beta)$, then S is a compact operator of $C^{2+\alpha}(M)$ into $C^\alpha(M)$.

3.4. Example. Let ρ be the Prohorov metric in $\mathbb{P}$, corresponding to a geodesical distance in M defined by a C^∞ Riemann metric on $\tilde{M}$. This means that, for $P \in \mathbb{P}$ and $Q \in \mathbb{P}$, $\rho(P,Q) = \inf \{\varepsilon > 0;\ P(K) \leq Q(K^\varepsilon) + \varepsilon$ for every compact $K \subset M\}$, where $K^\varepsilon = \{x \in M : \inf\{\overline{x,y} : y \in K\} < \varepsilon\}$. See [5; p. 96]. Let k be a probabilistic Borel kernel on M satisfying the Lipschitz condition

$$\rho(k(x,\cdot), k(y,\cdot)) \leq L\, \overline{x,y}$$

for every $x \in M$ and $y \in M$, L being a finite constant independent of x and y. Let β be any positive number. Then the formula

$$s(x,B) = \int_B \overline{x,y}^{\,\beta-2}\, k(x,dy), \qquad x \in M, \qquad B \in \mathcal{B}(M),$$

defines a hölderian Lévy kernel on M. The proof follows easily from Theorem 1.2 of [5; p. 96].

3.5. An operator of Ventcel Γ of order $r \in \{0,1,2\}$ will be called hölderian if there is an $\alpha \in (0,1)$ such that the restriction of Γ to $C^{2+\alpha}(M)$ is continuous operator of $C^{2+\alpha}(M)$ into $C^{2-r+\alpha}(\partial M)$ and the operator T_r occuring in (2.7.r), after restriction to $C^{2+\alpha}(M)$, is a compact operator of $C^{2+\alpha}(M)$ into $C^{2-r+\alpha}(\partial M)$.

3.6. Example. Let $M = \{z \in \mathbb{C} : |z| \leq 1\}$. For any $s > 0$ define the kernel t_s on $\partial M \times \mathcal{B}(M)$ such that for every $\varphi \in \mathbb{R}$,

$t_s(e^{i\varphi}, \{e^{i\varphi}\}) = 0$ and

$$t_s(e^{i\varphi}, B) = \int_{-\pi}^{\pi} \Big[\int_0^1 \mathbb{1}_B(re^{i(\varphi - \psi)}) \frac{dr}{(1-r)^2 + (\psi^2)^s} \Big] d\psi$$

for every Borel set $B \subset M \setminus \{e^{i\varphi}\}$. If $s \geqslant 3$, then t_s is not a Ventcel kernel. If $0 < s < 3$, then t_s is a Ventcel kernel of order $[s]$ and, for every $\alpha \in (0,1)$, the operator $T_{[s]}$ with kernel t_s is a compact operator of $C^{2+\alpha}(M)$ into $C^{2-[s]+\alpha}(\partial M)$.

3.7. An integrodifferential system of Ventcel on M is a triple (W, Γ, δ) composed of an operator of Waldenfels W, an operator of Ventcel Γ, and a non-negative function δ defined on ∂M. We shall speak that a Ventcel system (W, Γ, δ) on M is elliptic hölderian if the four conditions are satisfied:

(3.7.1) W is elliptic hölderian,

(3.7.2) Γ is hölderian,

(3.7.3) $\delta \in C^{2-r+\alpha}(\partial M)$ for some $\alpha \in (0,1)$, where $r \in \{0,1,2\}$ is the order of Γ,

(3.7.4) if the order of Γ is zero, then δ is strictly positive everywhere on ∂M, and if the order of Γ is two, the Γ is strongly transversal at each point $x \in \partial M$, at which $\delta(x) = 0$.

According to [3 ; p. 488], by strong transversality of Γ at a point $x \in \partial M$ we mean that, in (2.7.4), either $t(x, M \setminus \partial M) = \infty$, or the vector $u(x)$ is non-zero and directed strictly to the interior of M.

4. Resolvents and semigroups corresponding to elliptic hölderian systems of Ventcel.

Elaborated analytical reasonings of Sato and Ueno [11] and of Bony, Courrege and Priouret [3] base on

(i) theory of boundary problems for second order elliptic differential equations in Hölder spaces,

(ii) compactness properties of integral parts of the operators of Waldenfels and of Ventcel,

(iii) maximum principles (2.4.5) and (2.9.1), and transversality condition (3.7.4),

and lead in Sections III.2.5 - III.2.8 of [3] to construction of resolvent of Feller semigroup corresponding to an elliptic system of Ventcel. The following sketch exhibits some important for us elements of this construction.

If (W, Γ, δ) is an elliptic hölderian Ventcel system on M, r = order of Γ and $\alpha \in (0,1)$ is sufficiently small, then, by (i) and (ii), for every $\lambda > 0$ the map

(4.1) $J_\lambda : C^{2+\alpha}(M) \ni f \longrightarrow ((\lambda - W)f, (\lambda\delta\gamma_0 - \Gamma)f) \in C^{\alpha}(M) \times C^{2-r+\alpha}(\partial M)$

is Fredholm of index zero. Put

(4.2) $$\mathcal{D} = \{f \in C^2(M) : \delta\gamma_0 Wf = \Gamma f\}.$$

According to Theorem XI of [3; p. 451], (iii) implies that

(4.3) if $f \in \mathcal{D}$ and f attains a non-negative maximum at a point $x \in M$, then $Wf(x) \leq 0$,

whence

(4.4) $\|\lambda f - Wf\|_{C(M)} \geq \lambda \|f\|_{C(M)}$ for every $f \in \mathcal{D}$ and $\lambda > 0$.

If $\lambda > 0$, $f \in C^{2+\alpha}(M)$ and $J_\lambda f = 0$, then, according to (4.1), $(\lambda - W)f = 0$ and $(\delta\gamma_0 W - \Gamma)f = (\lambda\delta\gamma_0 - \Gamma)f = 0$, so that $f \in \mathcal{D}$. But then $f = 0$, by (4.4). As a consequence, for every $\lambda > 0$, being Fredholm of index zero,

(4.5) J_λ is an isomorphism of $C^{2+\alpha}(M)$ onto $C^{\alpha}(M) \times C^{2-r+\alpha}(\partial M)$

Take now into account the operators

(4.6) $$\mathring{R}_\lambda : C^{2-r+\alpha}(M) \ni f \longrightarrow J^{-1}(f, \delta\gamma_0 f) \in C^{2+\alpha}(M), \qquad \lambda > 0.$$

For $f \in C^{2-r+\alpha}(M)$ and $\lambda > 0$, according to (4.1) and (4.6), we have $((\lambda - W)\mathring{R}_\lambda f, (\lambda\delta\gamma_0 - \Gamma)\mathring{R}_\lambda f) = (f, \delta\gamma_0 f)$, so that $(\lambda - W)\mathring{R}_\lambda f = f$ and $(\delta\gamma_0 W - \Gamma)\mathring{R}_\lambda f = (\lambda\delta\gamma_0 - \Gamma)\mathring{R}_\lambda f - \delta\gamma_0 f = 0$, whence

(4.7) $$\mathring{R}_\lambda C^{2-r+\alpha}(M) \subset \mathcal{D}$$

and

(4.8) $$(\lambda - W)\mathring{R}_\lambda = \mathrm{id}_{C^{2-r+\alpha}(M)} .$$

These two relations imply that

(4.9) for every $\lambda > 0$ the set $(\lambda - W)\mathcal{D}$ is dense in $C(M)$. Moreover, if $f \in C^{\infty}(M)$, $\lambda > 0$ and $g = \mathring{R}_{\lambda} f$ attains a negative minimum at the point $x \in M$, then, by (4.8), (4.7) and (4.3), $f(x) = (\lambda - W)g(x) \leq \lambda g(x) < 0$, whence

(4.10) if $\lambda > 0$, $f \in C^{\infty}(M)$ and $f \geq 0$, then $\mathring{R}_{\lambda} f \geq 0$.

In spite of complicated outline of reasonings, thus far each step was a rather easy consequence of (i) - (iii). But next step is decisive and really hard: the operator $\mathring{R}_{\lambda}$ can be expressed by Green operators and harmonic operators, and an analysis of two former takes one in [11; pp. 564-565] and in [3; pp. 488-492] to conclusion that

(4.11) $$\lim_{\lambda \to \infty} \| \lambda \mathring{R}_{\lambda} f - f \|_{C(M)} = 0 \quad \text{for every } f \in C^{\infty}(M).$$

Now, it is an immediate consequence of (4.7) and (4.11) that

(4.12) $\mathcal{D}$ is dense in $C(M)$.

According to Theorem 1.2 of [11; p.534], the conditions (4.4) and (4.12) imply that the operator $W|_{\mathcal{D}}$, restriction of W to the domain $\mathcal{D}$, is closeable from $C(M)$ to $C(M)$. Let G be its closure:

(4.13) $$G = \overline{W|_{\mathcal{D}}}.$$

Then, by (4.12), (4.9) and (4.4), by virtue of Hille-Yosida theorem, G is infinitesimal generator of a one-parameter strongly continuous semigroup $(N_t)_{t \geq 0}$ of linear contraction in $C(M)$. Moreover, it follows from (4.8) that

(4.14) $$(\lambda - G)^{-1} f = \mathring{R}_{\lambda} f \quad \text{for every } f \in C^{\infty}(M) \text{ and } \lambda > 0,$$

which, by (4.10), implies that the operators $(\lambda - G)^{-1}$, $\lambda > 0$, are non-negative. Consequently, the Hille approximation formula

$$N_t f = s - \lim_{n \to \infty} \left(\frac{n}{t}\right)^n \left(\frac{n}{t} - G\right)^{-n} f, \qquad f \in C(M), \quad t > 0,$$

shows that also the operators N_t are non-negative.

5. Killing at first exit time from an open set

This Section appoints the outline of proof of the local unique determination principle formulated in 1.5. For every Feller infinitesimal generator on M we denote by $(P^G_\mu)_{\mu \in \mathbb{P}}$ the corresponding Markov system of probability measures on the filtered measurable space $(W, \mathcal{F}^G, \mathcal{F}^G_t)$ described in 1.3. We identify $C(M)$ with $\{f \in C(M_\Delta) : f(\Delta) = 0\}$. If U is an open subset of M, then we define

$$C_o(U) = \{ f \in C(M) : f = 0 \text{ on } M \setminus U \}$$

and by a Feller semigroup on U we mean a one-parameter strongly continuous non-negative linear contractions of $C_o(U)$.

5.1. Theorem. Let G_1 and G_2 be Feller generators on M, U an open subset of M, and T the first exit time of the canonical process $(X_t)_{t \geqslant 0}$ from U. Then the two conditions are equivalent:

(5.1.1) $E_x^{G_1}[f(X_t); t < T] = E_x^{G_2}[f(X_t); t < T]$ for every $x \in U$, $f \in C_o(U)$ and $t \in R^+$,

(5.1.2) $\mathcal{F}^{G_1}_{T-} = \mathcal{F}^{G_2}_{T-}$ and $P^{G_1}_\mu(A) = P^{G_2}_\mu(A)$ for every $\mu \in \mathbb{P}$ and every $A \in \mathcal{F}^{G_1}_{T-}$.

The proof bases on monotone class arguments.

5.2. Theorem. Let G be a Feller generator on M, U an open subset of M, $(\mathring{N}_t)_{t\geq 0}$ a Feller semigroup on U, and G_o a pregenerator of $(\mathring{N}_t)_{t\geq 0}$. Denote by T the first exit time of the canonical process $(X_t)_{t\geq 0}$ from U. Suppose that

(5.2.1) for every $f_o \in \mathcal{D}(G_o)$ there is $f \in \mathcal{D}(G)$ such that for each $x \in U$ the three equalities are satisfied: $f(x) = f_o(x)$, $Gf(x) = G_o f_o(x)$ and $P_x^G\{T<\infty, f(X_T) \neq 0\} = 0$.

Then $E_x^G[f(X_t); t<T] = N_t(x,f)$ for every $x \in U$, $t \in R^+$ and $f \in C_o(U)$.

Proof. I. The formula

$$\hat{N}_t(x,f) = E_x^G[f(X_t); \ t<T] + f(\Delta)P_x^G\{T \leq t\},$$

where $t \in \mathbb{R}^+$, $x \in M$ and $f \in b\mathcal{B}_u(M_\Delta)$, defines a semigroup $(\hat{N}_t)_{t\geq 0}$ of Markov transition kernels on the measurable space $(M_\Delta, \mathcal{B}_u(M_\Delta))$. Moreover, we have

$$\hat{N}_t(x,f) = \hat{E}_x[f(X_t)] = \int_W f(X_t(\omega))\hat{P}_x(d\omega),$$

the probability measures $\hat{P}_x$, $x \in M$, on the σ-field $\mathcal{F}^*$ of universally measurable sets over $(W, \mathring{\mathcal{F}})$ being defined by

$$\hat{\mathbb{P}}_x(A) = P_x^G(\mathcal{K}^{-1}(A)), \qquad A \in \mathcal{F}^*,$$

where

$$(\mathcal{K}\omega)(t) = \begin{cases} \omega(t), & \text{if } t < T(\omega), \\ \Delta, & \text{if } t \geq T(\omega). \end{cases}$$

Since $\mathcal{K}^{-1}\{X_t \in B\} = \{X_t \in B, t < T\} \in \mathcal{F}^*$ for every $t \in \mathbb{R}^+$ and $B \in \mathcal{B}(M)$, and since $\mathcal{K}^{-1}\{X_t = \Delta\} = \{T \leq t\} \in \mathcal{F}^*$ for every $t \in \mathbb{R}^+$, it follows that $\mathcal{K} \in \mathcal{F}^*/\mathring{\mathcal{F}}$. Consequently $\mathcal{K} \in \mathcal{F}^*/\mathcal{F}^*$, so that the probability measures $\hat{\mathbb{P}}_x$ are well defined.

By righ continuity of every function ω in W and by Lebesque bounded convergence theorem, for every $f \in C(M_\Delta)$, $t \in \mathbb{R}^+$ and $x \in M_\Delta$ we have

$$\hat{N}_t(x,f) = \lim_{k \to \infty} \int_W f(\omega(\frac{[kt]+1}{k}))\hat{\mathbb{P}}_x(d\omega) = \lim_{k \to \infty} \hat{N}_{k^{-1}([kt]+1)}(x,f).$$

This implies that if $f \in C(M_\Delta)$ then the function

$$(5.2.2) \qquad (t,x) \longrightarrow \hat{N}_t(x,f)$$

is $\mathcal{B}(R^+) \times \mathcal{B}_u(M_\Delta)$ - measurable on $R^+ \times M_\Delta$. By monotone class theorem, the same measurability is preserved for every $f \in b\mathcal{B}(M_\Delta)$. Finally, $\mathcal{B}(R^+) \times \mathcal{B}_u(M_\Delta) \subset \mathcal{B}_u(R^+ \times M_\Delta)$, and so, if $f \in b\mathcal{B}_u(M_\Delta)$ then the function (5.2.2) is $\mathcal{B}_u(R^+ \times M_\Delta)$ - measurable on $R^+ \times M$. This measurability is needed for definition of the resolvent $\hat{R}_\lambda$ occuring in steps IV - VI of present proof.

II. Denote by $(N_t)_{t \geqslant 0}$ the Feller semigroup on M, generated by G. Let $(\widetilde{N}_t)_{t \geqslant 0}$ be the corresponding markovian Feller semigroup on M_Δ such that

$$\widetilde{N}_t(x,f) = E_x^G[f(X_t)] = \int_W f(X_t(\omega))P_x^G(d\omega)$$

for every $t \in \mathbb{R}^+$, $x \in M_\Delta$ and $f \in C(M_\Delta)$. Then

$$\widetilde{N}_t f = N_t(f - f(\Delta)\mathbb{1}) + f(\Delta)\mathbb{1}, \qquad t \in \mathbb{R}^+, \qquad f \in C(M_\Delta),$$

where $\mathbb{1}$ denotes the function on M_Δ equal identically to one. If $\widetilde{G}$ denotes the strong infinitesimal generator of $(\widetilde{N}_t)_{t \geqslant 0}$, then $\mathcal{D}(\widetilde{G}) = \mathcal{D}(G) + \mathbb{R}\mathbb{1}$ and $\widetilde{G}f = G(f - f(\Delta)\mathbb{1})$ for every $f \in \mathcal{D}(\widetilde{G})$.

III. Denote by $\widehat{C}$ the pointwise right-continuity space of the semigroup $(\widehat{N}_t)_{t \geqslant 0}$, i.e. let

$$\widehat{C} = \{f \in b\mathcal{B}_u(M_\Delta) : \lim_{t \downarrow 0} \widehat{N}_t(x,f) = f(x) \quad \text{for each } x \in M_\Delta\}.$$

Observe that

(5.2.3) if $f = \mathbb{1}_U h$, where $h \in C(M)$, then $f \in \widehat{C}$.

Indeed, if $h \in C(M_\Delta)$ and $f = \mathbb{1}_U h$, then $\widehat{N}_t(x,f) = \widehat{N}_t(x,h) = N_t(x,h) - E_x^G[h(X_t); T \leq t]$.

If $x \in U$, then $P_x^G\{T > 0\} = 1$, whence $\lim_{t \downarrow 0} E_x^G[b(X_t); T \leq t] = 0$, and so, $\lim_{t \downarrow 0} \widehat{N}_t(x,f) = \lim_{t \downarrow 0} N_t(x,h) = h(x) = f(x)$. If $x \in M_\Delta \setminus U$, then $P_x^G\{T = 0\} = 1$, so that $\widehat{N}_t(x,f) = 0$ for every $t > 0$ and hence $\lim_{t \downarrow 0} \widehat{N}_t(x,f) = 0 = f(x)$.

IV. Denote by $\hat{\mathcal{G}}$ the weak infinitesimal generator of the semigroup $(\hat{N}_t)_{t\geq 0}$. The theorem will follow when we show that

$$G_o \subset \hat{\mathcal{G}}. \tag{5.2.4}$$

Indeed, we have $(\lambda - \hat{\mathcal{G}})^{-1} = \hat{R}_\lambda \big|_{\hat{C}}$ for every $\lambda > 0$, where

$$\hat{R}_\lambda f(x) = \int_0^\infty e^{-\lambda t} \hat{N}_t(x,f)\,dt, \qquad f \in b\mathcal{B}_u(M_\Delta), \qquad x \in M_\Delta .$$

Since $\hat{C}$ is complete with respect to the topology of uniform convergence on M_Δ, it follows from the above formulas that the operator $\hat{\mathcal{G}}$ is closed with respect to this topology. So, the inclusion (5.2.4) implies that

$$\overline{G_o} \subset \hat{\mathcal{G}},$$

where $\overline{G_o}$, the closure of G_o, is the infinitesimal generator of the semigroup $(\mathring{N}_t)_{t\geq 0}$. Since, by (5.2.3), $C_o(U) \subset \hat{C}$, it follows that

$$\int_0^\infty e^{-\lambda t} \hat{N}_t(x,f)\,dt = (\lambda - \hat{\mathcal{G}})^{-1} f(x) = (\lambda - G_o)^{-1} f(x) = \tag{5.2.5}$$

$$= \int_0^\infty e^{-\lambda t} \mathring{N}_t(x,f)\,dt$$

for every $\lambda > 0$, $x \in M$ and $f \in C_o(U)$. But, for fixed x and f, the function $t \longrightarrow \mathring{N}_t(x,f)$ is continuous, $t \longrightarrow \hat{N}_t(x,f)$ is right--continuous, and (5.2.5) means that both they have the same Laplace transform. Consequently, these functions are identical. This means that $\hat{N}_t f = \mathring{N}_t f$ for every $t \in R^+$ and $f \in C_o(U)$, and the theorem follows.

V. The inclusion (5.2.4) is a consequence of (5.2.1) and of the following statement:

$$(5.2.6)\qquad \begin{cases} \text{if } \lambda > 0, \ f \in \mathcal{D}(G) \ \text{ and } \ P_x^G\{f(X_T) = 0\} = 1 \\ \text{for every } x \in U, \ \text{ then } \ \hat{R}_\lambda \mathbb{1}_U(\lambda - G)f = \mathbb{1}_U f. \end{cases}$$

Indeed, suppose that (5.2.6) is true, take any $f_o \in \mathcal{D}(G_o)$ and choose $f \in \mathcal{D}(G)$ with properties as in (5.2.1). Then $\mathbb{1}_U(\lambda - G)f \in \hat{C}$, by (5.2.3). Consequently. by (5.2.6), $f_o = \mathbb{1}_U f \in \hat{R}_\lambda \hat{C} = \mathcal{D}(\hat{\mathcal{G}})$ and $\hat{\mathcal{G}} f_o = \lambda f_o - (\lambda - \hat{\mathcal{G}})f_o = \lambda f_o - \hat{R}_\lambda^{-1} \mathbb{1}_U f = \lambda f_o - \mathbb{1}_U(\lambda - G)f =$
$= \mathbb{1}_U G f = G_o f_o$.

VI. In the proof of (5.2.6) we follow Itô and Mc Kean [8 ; Section 3.9]. Suppose that $f \in \mathcal{D}(G)$ and $P_x^G\{f(X_T) = 0\} = 1$ for every $x \in U$. Take any $\lambda > 0$ and put $g = \mathbb{1}_U(\quad - G)f$. We have to prove that $\hat{R}_\lambda g = \mathbb{1}_U f$.

Since $g(\Delta) = 0$, we have

$$\hat{R}_\lambda(x,g) = \int_0^\infty e^{-\lambda t} E_x^G[g(X_t);\ t < T]dt = E_x^G \int_0^T e^{-\lambda t} g(X_t)dt$$

for every $\lambda > 0$ and $x \in M$. But if $t \in [0, T(\omega))$, then $X_t(\omega) \in U$, so that $g(X_t(\omega)) = \lambda f(X_t(\omega)) - (Gf)(X_t(\omega))$. Moreover $f \in \mathcal{D}(G) \subset C(M)$, so that $Gf = \tilde{G}f$. Consequently

$$E_x^G \int_0^T e^{-\lambda t} g(X_t)dt = E_x^G \int_0^T e^{-\lambda t}(\lambda f - \tilde{G}f)(X_t)dt.$$

According to Theorem 5.1 of Dynkin [4], the right side of this equality is equal to

$$(5.2.7) \qquad f(x) - E_x^G\left[e^{-\lambda T} f(X_T)\right].$$

If $x \in U$ then, by assumption, $P_x^G\{f(X_T) = 0\} = 1$, so that (5.2.7) is equal to $f(x)$. If $x \in M_\Delta \setminus U$, then $P_x^G\{T = 0\} = 1$ and $E_x^G\left[e^{-\lambda T} f(X_T)\right] = E_x^G\left[f(X_0)\right] = f(x)$, so that (5.2.7) is equal to zero. Consequently (5.2.7) is equal to $\mathbb{1}_U(x) f(x)$ for each $x \in M_\Delta$. This proves the statement (5.2.6) and, at the same time, completes the whole proof of Theorem 5.2.

6. A connection between infinitesimal generator and jumps of sample paths.

6.1. Theorem. Let G be the strong infinitesimal generator of a Feller semigroup $(N_t)_{t \geq 0}$ on M. Suppose that $\varphi \in C(M)$, $\psi \in \mathcal{D}(G)$ and $\operatorname{supp}\varphi \cap \operatorname{supp}\psi = \emptyset$. Let $\tilde{\varphi} \in C(M_\Delta)$ and $\tilde{\psi} \in C(M_\Delta)$ be such that $\tilde{\varphi}(\Delta) = \tilde{\psi}(\Delta) = 0$, $\tilde{\varphi} = \varphi$ on M and $\tilde{\psi} = \psi$ on M. Then

$$E_x^G\left[\sum_{0 < u \leq t} \tilde{\varphi}(X_{u-})\tilde{\psi}(X_u)\right] = \int_0^t N_u(x, \varphi \cdot G\psi)\,du$$

for every $t \in (0,\infty)$ and every $x \in M$.

6.2. Comments. The proof can be found in [9]. According to [10], in the case of non-negative φ and ψ the remaining assumptions can be considerably weakened. Also in [10] some simple applications are shown and the relation to the paper of N.Ikeda and S. Watanabe [7] is explained.

7. The Proposition III.2.9 of Bony, Courrège and Priouret.

7.1. Let (W,Γ,δ) be a system of Ventcel on M. Assume that this system is elliptic and hölderian in the sense of Section 3.7. As we already know from Section 4, if the linear subspace D of $C^2(M)$ is defined by (4.2), i.e. if

$$D = \left\{ f \in C^2(M) : \delta\gamma_0 Wf = \Gamma f \right\},$$

then the operator $G = W|_D$ is a Feller generator on M.

Applicability of Theorem 5.2 in our further reasonings depends on the Proposition III.2.9 of [3], which in our situation takes following form.

7.2. Proposition. Let (W,Γ,δ) be an elliptic hölderian system of Ventcel on M. Let r be the order of Γ. Suppose that

(7.2.1) there is an $\alpha_0 \in (0,1)$ such that the map $f \longrightarrow \delta\gamma_0 Wf - \Gamma f$ is continuous from $C^{4-r+\alpha_0}(M)$ into $C^{2-r+\alpha_0}(\partial M)$.

Then, for every function $f \in \bigcup_{\alpha>0} C^{4-r+\alpha}(M)$, every compact $K \subset M \setminus \partial M$ and every $\varepsilon > 0$, there is a function $g \in D$ such that $g = f$ on K and $\| g - f \|_{C(M)} \leq \varepsilon$.

Note that the condition (7.2.1) is satisfied for every elliptic hölderian Ventcel system (W,Γ,δ) for which r = 2. If r = 0 or r = 1, then (7.2.1) is an additional smoothness assumption.

8. Local unique determination principle under additional assumptions.

8.1. Lemma. Let G_1 and G_2 be Feller infinitesimal generators on M. Let $U_1 \subset U_2 \subset \dots$ be an increasing sequence of open subsets of M and let $U_\infty = \bigcup_{n=1}^{\infty} U_n$. For each $n = 1,2,\dots,$ and also for $n = \infty$, denote by T_n the first exit time of the canonical process $(X_t)_{t\geq 0}$ from U_n. Under above assumption, if the equality

$$(8.1.n) \quad E_x^{G_1}\left[f(X_t);\ t<T_n\right] = E_x^{G_2}\left[f(X_t);\ t<T_n\right] \text{ for every } x\in U_n \text{ and } f\in C_0(U_n)$$

is true for each $n = 1,2,\dots,$ then it is also true for $n = \infty$.

Proof. It is sufficient to prove $(8.1.\infty)$ for $f\in C_c(U_\infty) = \bigcup_{n=1}^{\infty} C_c(U_n)$. For such f the equality $(8.1.\infty)$ follows from the fact that $\lim_{n\to\infty} T_n = T_\infty$ $P_x^{G_i}$ - a.s. on W, for $i = 1,2$ and every $x\in U_\infty$, by quasi-left-continuity of the process $(W, \mathcal{F}, \mathcal{F}_t, X_t, \theta_t, P_x^{G_i})$.

8.2. Assumptions. Let G_1 and G_2 be Feller infinitesimal generators on M determined by two elliptic hölderian Ventcel systems $(W_1, \Gamma_1, \delta_1)$ and $(W_2, \Gamma_2, \delta_2)$ satisfying the condition (7.2.1). Let s_1 and s_2 be the Lévy kernels of the operators W_1 and W_2. Let V be an open subset of $M\setminus\partial M$. Suppose that

(8.2.1) there is an open neighbourhood $\mathcal{O}$ of $\partial M\cup\partial V$ such that $s_1(x,\mathcal{O}) = s_2(x,\mathcal{O}) = 0$ for every $x\in V$.

Suppose moreover that

(8.2.2) $(W_1f)(x) = (W_2f)(x)$ for every $x \in V$ and every $f \in C_c^{\infty}(V)$.

Let T denote the first exit time of the canonical process $(X_t)_{t\geqslant 0}$ from V.

8.3. Lemma. Under Assumptions 8.2 the equality (5.1.1) is satisfied.

It follows ewasily from Proposition 7.2 that $C_c^{\infty}(U) \subset \mathfrak{D}(\overline{\mathbb{1}_U G_i})$ and $\overline{\mathbb{1}_U G_i}\big|_{C_c^{\infty}(U)} = W_i\big|_{C_c^{\infty}(U)}$. Consequently the Lemma means that the local unique determination principle, formulated in Section 1.5, is valid for the class of Feller infinitesimal generator determined by elliptic hölderian Ventcel systems under additional assumptions (7.2.1) and (8.2.1). Elimination of these additional assumptions will be the subject of our subsequent sections.

Proof. Choose an increasing sequence $U_1 \subset U_2 \subset \dots$ of open sets such that $V = \bigcup_{n=1}^{\infty} \overline{U}_n$ and that, for each n, the boundary ∂U_n of U_n is a C^{∞} submanifold of $M \setminus \partial M$. Without loss of generality we can assume that, for every n,

(8.3.1) $V \subset U_n \cup \mathcal{O}$, and so $\partial U_n \subset \mathcal{O}$.

For every fixed n, by (8.2.2) and (8.2.1), the equalities

(8.3.2) $W_{0,n}f_0 = W_1 f\big|_{\overline{U}_n} = W_2 f\big|_{\overline{U}_n}$ for every $f_0 \in C^2(\overline{U}_n)$ and every $f \in C^2(M)$ such that $f = f_0$ on $\overline{U}_n$ and $f = 0$ on $(M \setminus \overline{U}_n) \setminus \mathcal{O}$, define a hölderian elliptic operator of Waldenfels $W_{o,n}$ on the compact C^∞ manifold $\overline{U_n}$ with boundary ∂U_n. Let

$$D_{o,n} = \left\{ f \in C^2(\overline{U_n}) \; : \; f = W_o f = 0 \quad \text{on} \quad \partial U_n \right\}.$$

The Theorem XVI of [3; p. 480] then says that the operator

$$G_{o,n} = W_{o,n} \Big| D_{o,n}$$

is pregenerator of a Feller semigroup $(N_{n,t})_{t \geqslant 0}$ on the open set U_n. By Lemma 8.1, the Lemma 8.3 will follow when we show that

$$(8.3.3) \qquad E_x^{G_i}\left[f(X_t) ; \; t < T_n \right] = N_{n,t}(x,f)$$

for every $x \in U_n$, $f \in C_o(U_n)$ and $i = 1,2$.

In order to prove (8.3.3) we shall apply the Theorem 5.2. All what we have to do is to show that, for $i = 1$ as well as for $i = 2$, the condition (5.2.1) is satisfied for the Feller generator G_i on M and for the pregenerator $G_{o,n}$. Let i be fixed. If $f_o \in \mathcal{D}(G_{o,n}) = D_{o,n}$ then, by Proposition 7.2, there is a function $f \in \mathcal{D}(G_i)$ such that $f = f_o$ on $\overline{U_n}$ and $f = 0$ on $(M \setminus \overline{U_n}) \setminus \vartheta$. Then, by (8.3.2), $G_i f = G_{o,n} f_o$ on U_n. In order to check that (5.2.1) is satisfied, it is sufficient to prove that for such f we must have

$$P_x^{G_i}\left\{ T_n < \infty, \; f(X_{T_n}) = 0 \right\} = 1 \qquad \text{for every} \quad x \in U_n.$$

Since, by right continuity of sample paths and by (8.3.1), $X_{T_n} \in M \setminus U_n \subset (M \setminus \overline{U_n}) \cup \vartheta$, since $f = 0$ on ∂U_n and on $(M \setminus \overline{U_n}) \setminus \vartheta$ and since $P_x^{G_i}\{T_n > 0\} = 1$ for every $x \in U$, it is sufficient to prove that

(8.3.4) $$P_x^{G_i}\{0<T_n<\infty,\ X_{T_n}\in \mathcal{O}\setminus \overline{U}_n\} = 0 \quad \text{for every } x\in M.$$

We shall prove (8.3.4) applying the Theorem 6.1 and the Proposition 7.2. Let $K_1, K_2, \ldots$ be a sequence of compact sets such that

$$\mathcal{O}\setminus\overline{U}_n = \bigcup_{\nu=1}^{\infty} K_\nu$$

Then

$$\{0<T_n<\infty,\ X_{T_n}\in\mathcal{O}\setminus\overline{U}_n\} = \bigcup_{\mu,\nu=1}^{\infty} A_{\mu,\nu},$$

where

$$A_{\mu,\nu} = \{0<T_n\leq\mu,\ X_{T_n}\in K_\nu\}.$$

Using Proposition 7.2, for every $\nu = 1,2,\ldots$ we can construct non-negative functions $\varphi_\nu \in C(M)$ and

$\psi_\nu \in \{f\in C^2(M) : \delta\gamma_0 W_i f = \Gamma_i f\}\subset \mathcal{D}(G_i)$ such that

$\operatorname{supp}\varphi_\nu \cap \operatorname{supp}\psi_\nu = \emptyset$, $\operatorname{supp}\psi_\nu\subset\mathcal{O}$, $\varphi_\nu \geq 1$ on $\overline{U}_n$ and $\psi_\nu\geq 1$ on K_ν. By Theorem 6.1 we have

(8.3.5) $$P_x^{G_i}(A_{\mu,\nu})\leq E_x^{G_i}\Big[\sum_{0<u\leq\mu}\varphi_\nu(X_{u-})\psi_\nu(X_u)\Big] \leq \mu \sup_M(\varphi_\nu\cdot G\psi_\nu).$$

For $y\notin\operatorname{supp}\psi_\nu$ we have $(G_i\psi_\nu)(y) = (W_i\psi_\nu)(y) = s_i(y,\psi_\nu)\geq 0$. On the other hand, by (8.2.1), $s_i(y,\psi_\nu)\leq s_i(y,\mathcal{O})\cdot\sup_M \psi_\nu = 0$. Consequently $G_i\psi_\nu = 0$ outside of $\operatorname{supp}\psi_\nu$, whence

$\varphi_\nu\cdot G_i\psi_\nu = 0$ everywhere on M. This, together with (8.3.5), implies that $P_x^{G_i}(A_{\mu,\nu}) = 0$, whence (8.3.4) follows, completing the proof.

9. Approximation of hölderian Lévy kernels.

9.1. Assumptions. Let U and V be open subsets of M such that $\overline{V} \subset U \subset M \setminus \partial M$ and that the boundary ∂V of V is a C^∞ submanifold of $M \setminus \partial M$. Let s_1 and s_2 be Lévy kernels on M such that

(9.1.1) $\quad s_1(x,B) = s_2(x,B) \quad$ for every $x \in \overline{V}$ and every Borel $B \subset U$.

Assume that s_1 and s_2 are hölderian in the sense of definition 3.1. Let $v : M \times M \longrightarrow T(M)$ be a Lévy map. We imbed M into an open C^∞ manifold $\widetilde{M}$ and choose on $\widetilde{M}$ a Riemann metric. We denote by dy the element of corresponding riemannian volume. Let $\alpha \in (0, \min_{i=1,2} (\gamma_i \wedge \frac{1}{3}\beta_i))$, where γ_i and β_i are exponents occuring in the definition 3.1, corresponding to our fixed Riemann metric. Let $\varepsilon > 0$ be arbitrary.

9.2. Theorem. Under Assumptions 9.1 there are non-negative C^∞ functions Φ_1 and Φ_2 on $M \times M$ satisfying the three conditions:

(9.2.1) $\quad \Phi_1 = \Phi_2 \quad$ on $V \times V$,

(9.2.2) $\quad \Phi_1 = \Phi_2 = 0$ on $M \times \mathcal{O}$, where $\mathcal{O}$ is a neighbourhood of $\partial M \cup \partial V$,

(9.2.3) $\quad \| S_i^* - S_i \|_{L(C^{2+\alpha}(M), C^\alpha(M))} \leq \varepsilon$ for $i = 1,2$, the operators of Lévy S_i and S_i^* being defined by

$$(S_i f)(x) = \int_M \left[f(y) - f(x) - df(x) \cdot v(x,y) \right] s_i(x,dy)$$

and by

$$(S_i^* f)(x) = \int_M \left[f(y) - f(x) - df(x)v(x,y) \right] \Phi_i(x,y)dy.$$

The troublesome but elementary proof is omitted.

10. An application of the Trotter-Kato theorem about convergence of semigroups.

10.1. Lemma. Let U and V be open subsets of M such that $\bar{V} \subset U \subset M \setminus \partial M$ and that the boundary ∂V of V is a C^∞ submanifold of $M \setminus \partial M$. Let $(W_{i,o}, \Gamma_{i,o}, \delta_{i,o})$, $i = 1,2$, be two elliptic hölderian systems of Ventcel on M such that

(10.1.1) $(W_{1,o}f)(x) = (W_{2,o}f)(x)$ for every $x \in U$ and every $f \in C_c^\infty(U)$.

Then there are two sequencens $(W_{i,k}, \Gamma_{i,k}, \delta_{i,k})_{k=1,2,\ldots}$, $i = 1,2$, of elliptic hölderian systems of Ventcel on M satisfying the four conditions:

(10.1.2) $(W_{1,k}f)(x) = (W_{2,k}f)(x)$ for every $x \in V$, $f \in C_c^\infty(V)$ and $k = 1,2,\ldots$,

(10.1.3) for every $k = 1,2,\ldots$ there is an open neighbourhood $\mathcal{O}_k$ of $\partial V \cup \partial M$ such that $s_{1,k}(x, \mathcal{O}_k) = s_{2,k}(x, \mathcal{O}_k) = 0$ for every $x \in M$, where $s_{i,k}$ denotes the Lévy kernel of $W_{i,k}$,

(10.1.4) each of the systems $(W_{i,k}, \Gamma_{i,k}, \delta_{i,k})$, $i = 1,2$, $k = 1,2,\ldots$, satisfies the condition (7.2.1),

(10.1.5) $\lim_{k\to\infty} \sup_{0\leq u\leq t} \| N_{i,k,u}f - N_{i,o,u}f \|_{C(M)} = 0$ for every $f\in C(M)$, $t\in R^+$ and $i = 1,2$, $(N_{i,k,t})_{t\geq 0}$ being Feller semigroup on M generated by the operator $G_{i,k} = \overline{W_{i,k}|\mathcal{D}_{i,k}}$, where $\mathcal{D}_{i,k} = \{f\in C^2(M) : \delta_{i,k}\gamma_o W_{i,k}f = \Gamma_{i,k}f\}$.

10.2. Sketch of the proof. By an application of Theorem 9.2, and by smoothing of $\Gamma_{i,o}, \delta_{i,o}$ and of the differential parts of $W_{i,o}$ if the order of $\Gamma_{i,o}$ is less then 2, we can construct the two sequences $(W_{i,k}, \Gamma_{i,k}, \delta_{i,k})_{k=1,2,\ldots}$, $i = 1,2$, which have the properties (10.1.2) - (10.1.4) and are such that there is an $\alpha\in(0,1)$ such that, for every $\lambda>0$ and for $i = 1,2$,

(10.2.1) $$\lim_{k\to\infty} \| \mathcal{J}_{i,k,\lambda} - \mathcal{J}_{i,o,\lambda} \|_{L(C^{2+\alpha}(M), C^{\alpha}(M)\times C^{2-r_i+\alpha}(\partial M))} = 0$$

where r_i = order of $\Gamma_{i,k}$ is independent of k and $\mathcal{J}_{i,k}$, denotes the isomorphism (4.1) corresponding to the system $(W_{i,k}, \Gamma_{i,k}, \delta_{i,k})$. Let $(R_{i,k,\lambda})_{\lambda>0}$ be the resolvent of the semigroup $(N_{i,k,t})_{t\geq 0}$. According to (4.6),

$$R_{i,k,\lambda} f = \mathcal{J}^{-1}_{i,k,\lambda}(f,0) \text{ for every } f\in C^{\alpha}(M),$$

so that, by (10.2.1), $\lim_{k\to\infty} \| R_{i,k,\lambda} - R_{i,o,\lambda} \|_{L(C^{\alpha}(M), C^{2+\alpha}(M))} = 0$, and consequently

$$\lim_{k\to\infty} \| R_{i,k,\lambda} f - R_{i,o,\lambda} f \|_{C(M)} = 0 \quad \text{for every} \quad f \in C(M),$$
$$\lambda > 0 \quad \text{and} \quad i = 1,2.$$

By the Trotter-Kato theorem, the former limit relation implies (10.1.5).

11. A limit passage in the equalities (5.1.2).

11.1. Theorem. Let $G_{i,k}$, $i = 1,2$, $k = 0,1,\dots$, be Feller generators on M. Let U be an open subset of M_Δ, and T the first exit time of the process $(X_t)_{t\geq 0}$ from U. Suppose that the equalities

$$(11.1.k) \quad \mathcal{F}_{T-}^{G_{1,k}} = \mathcal{F}_{T-}^{G_{2,k}} \quad \text{and} \quad P_\mu^{G_{1,k}}(A) = P_\mu^{G_{2,k}}(A) \quad \text{for every } \mu \in \mathcal{P} \text{ and } A \in \mathcal{F}_{T-}^{G_{1,k}}$$

are true for $k = 1,2,\dots$. Suppose moreover that

$$\lim_{k\to\infty} \sup_{0\leq u\leq t} \| \exp(uG_{i,k})f - \exp(uG_{i,0})f \|_{C(M)} = 0$$

for every $t \in R^+$, $f \in C(M)$ and $i = 1,2$. Then the equalities (11.1.k) are true also for $k = 0$.

The proof is divided into three subsequent Sections.

11.2. Metrization of W. For arbitrarily fixed $s > 0$ denote by Λ_s the set of all strictly increasing Lipschitz functions λ on $[0,\infty)$ satisfying the four conditions: $\lambda(0) = 0$, $\lambda(s) = s$, $\lim_{t\to\infty} \lambda(t) = \infty$, and

$$\gamma(\lambda) = \operatorname*{ess\,sup}_{t\geqslant 0} \left| \log \frac{d\lambda(t)}{dt} \right| < \infty .$$

Take into account the space $D_{M_\Delta}[0,\infty)$ of all M_Δ-valued right-continuous functions on $[0,\infty)$, having left-side limits everywhere on $(0,\infty)$. Choose on M_Δ a metric r compatible with the topology of M_Δ and define the distance between elements ω and $\widetilde{\omega}$ of $D_{M_\Delta}[0,\infty)$ by

$$d_s(\omega,\widetilde{\omega}) = \inf_{\lambda\in\Lambda_s}\left[\gamma(\lambda)\vee\int_0^\infty e^{-u}d(\omega,\widetilde{\omega},\lambda,u)du\right],$$

where

$$d(\omega,\widetilde{\omega},\lambda,u) = \sup_{t\geqslant 0} r(\omega(t\wedge u),\widetilde{\omega}(\lambda(t)\wedge u)).$$

After omission of the condition that each function $\lambda\in\Lambda_s$ has to satisfy the equality $\lambda(s) = s$, the metric d_s would be identical with the metric d defined on page 117 of book of Ethiér and Kurtz [5]. The condition that $\lambda(s) = s$ for every $\lambda\in\Lambda_s$ implies that the evaluation $X_s : \omega \longrightarrow X_s(\omega) = \omega(s)$ is a continuous map of the metric space $(D_{M_\Delta}[0,\infty), d_s)$ into metric space (M_Δ, r). Since Δ is a separated point of M_Δ, W is a closed subset of $(D_{M_\Delta}[0,\infty), d_s)$. Comparison of $d_s(\omega,\widetilde{\omega})$ with Billingsley's distance d_0, see [1; (14.17), (14.18)], between restrictions of $\omega(s\cdot)$ and $\widetilde{\omega}(s\cdot)$ to $[0,1]$, and with Ethiér-Kurtz distance between $\theta_s\omega$ and $\theta_s\widetilde{\omega}$, shows that (W,d_s) is separable and complete. The Borel σ-field of (W,d_s) is identical with $\mathring{\mathcal{F}}$. See [1; Section 14] and [5; Chapter 3, Sections 5 - 7].

11.3. Weak convergence of $P_x^{G_{i,k}}$ to $P_x^{G_{i,o}}$ as $k \to \infty$. The convergence conditions imposed in Theorem 11.1 onto sequences of semigroups $(\exp(tG_{i,k}))_{k=0,1,\dots}$ imply that

$$\lim_{h \downarrow 0} \alpha_\varepsilon(h) = 0 \tag{11.3.1}$$

for every $\varepsilon > 0$, where

$$\alpha_\varepsilon(h) = \sup\{P_x^{G_{i,k}}\{r(x,X_t) > \varepsilon\} \ : x \in M_\Delta \ , t \in [0,h], \ i=1,2, \ k=0,1,..\}$$

Reasonings based on (11.3.1), similar to that presented in Section VI.5 of the book of Gichman and Skorochod [6], show that, for $i = 1,2$ and for every $x \in M_\Delta$ and every $s > 0$,

$$\lim_{k \to \infty} P_x^{G_{i,k}} = P_x^{G_{i,o}} \tag{11.3.2}$$

in sense of weak convergence of measures on (W, d_s).

11.4. Completion of the proof of Theorem 11.1. According to Theorem 5.1, it is sufficient to prove that

$$E_x^{G_{1,0}}[f(X_s); \ s < T] = E_x^{G_{2,0}}[f(X_s); \ s < T] \tag{11.4.1}$$

for every $x \in U$, every non-negative $f \in C_o(U)$ and every $s \geqslant 0$. If $s = 0$, then both the sides of (11.4.1) are equal to $f(x)$, because $P_x^{G_{i,0}}\{T > 0\} = 1$ for $x \in U$.

Hence_forth assume that $x \in U$, $0 \leqslant f \in C_o(U)$ and $s > 0$ are arbitrary but fixed.

Let D be the first contact time of the process $(X_t)_{t \geqslant 0}$ with the compact set $K = M_\Delta \setminus U$. By quasi-left-continuity of the processes $(W,\mathcal{F}^{G_{i,0}},\mathcal{F}_t^{G_{i,0}},X_t,\theta_t,P_x^{G_{i,0}})$, we have $P_x^{G_{i,0}}\{D = T\} = 1$ for $i = 1,2$. Consequently, instead of (11.4.1), it is sufficient to prove that

$$(11.4.2) \qquad E_x^{G_{1,0}}\left[f(X_s);\ s<D\right] = E_x^{G_{2,0}}\left[f(X_s);\ s<D\right].$$

For $n = 1,2,\ldots$ put $V_n = \left\{x\in M_\Delta\ :\ r(x,K)< \frac{1}{n}\right\}$. Let $\underline{D}_n$ be the first constact time of the process $(X_t)_{t\geqslant 0}$ with $\overline{V_n}$, and $\overline{D_n}$ the first entry time into V_n. Then $\underline{D}_1\leqslant\overline{D}_1\leqslant\underline{D}_2\leqslant\overline{D}_2\leqslant\ldots$ and $\lim_{n\to\infty}\underline{D}_n = \lim_{n\to\infty}\overline{D_n} = D$ everywhere on W. Moreover, each $\underline{D}_n$ is a lower semi-continuous function on the metric space (W,d_s), and each $\overline{D}_n$ is upper semi-continuous. Consequently the events $O_n = \left\{s + \frac{1}{n} < \underline{D}_n\right\}$ are open, and $F_n = \left\{s + \frac{1}{n}\leqslant \overline{D}_n\right\}$ are closed.

For each $k = 1,2,\ldots$ and $i = 1,2$ the random variable $f(X_s)\mathbb{1}_{O_n}$ is $\mathcal{F}_{\underline{D}_n-}^{G_{i,k}}$ - measurable. But $\underline{D}_n\leqslant D\leqslant T$ every where on W, so that $\mathcal{F}_{\underline{D}_n-}^{G_{i,k}}\subset\mathcal{F}_{T-}^{G_{i,k}}$, whence, by (11.1.k), $E_x^{G_{1,k}}\left[f(X_s)\mathbb{1}_{O_n}\right] = E_x^{G_{2,k}}\left[f(X_s)\mathbb{1}_{O_n}\right]$ for every $k = 1,2,\ldots$ and consequently

$$(11.4.3) \qquad a_n^1 = a_n^2,$$

where

$$a_n^i = \liminf_{k\to\infty} E_x^{G_{i,k}}\left[f(X_s)\mathbb{1}_{O_n}\right].$$

For each $n = 1,2,\ldots$ the random variable $f(X_s)\mathbb{1}_{O_n}$ is lower semi-continuous on (W,d_s), and $f(X_s)\mathbb{1}_{F_n}$ is upper semi-continuous. So, (11.3.2) and non-negativity of f imply that $E_x^{G_{i,0}}\left[f(X_s)\mathbb{1}_{O_n}\right]\leqslant$ $\leqslant a_n^i\leqslant\limsup_{k\to\infty} E_x^{G_{i,k}}\left[f(X_s)\mathbb{1}_{F_n}\right]\leqslant E_x^{G_{i,0}}\left[f(X_s)\mathbb{1}_{F_n}\right] \leqslant$ $\leqslant E_x^{G_{i,0}}\left[f(X_s);\ s<D\right]$. But $O_1\subset O_2\subset\ldots$ and $\bigcup_n O_n = \{s<D\}$, so

that $\lim_{n\to\infty} E_x^{G_i,0}\left[f(X_s)\mathbb{1}_{O_n}\right] = E_x^{G_i,0}\left[f(X_s);\ s<D\right]$ and consequently

$$(11.4.4)\qquad E_x^{G_i,0}\left[f(X_s);\ s<D\right] = \lim_{n\to\infty} a_n^i .$$

Now, (11.4.2) follows at once from (11.4.3) and from (11.4.4), which completes the proof.

12. The main result.

12.1. Theorem. Let $(W_i, \Gamma_i, \delta_i)$, $i = 1,2$, be two elliptic hölderian systems of Ventcel on a compact C^∞ manifold M with boundary ∂M. Let G_i, $i = 1,2$, be the corresponding Feller generators on M defined by $G_i = \overline{W_i|_{\mathcal{D}_i}}$, where $\mathcal{D}_i = \{f\in C^2(M) :$ $\Gamma_i f = \delta_i \gamma_0 W_i f\}$. Let U be an open subset of $M\setminus\partial M$ and T the first exit time of the canonical process $(X_t)_{t\geqslant 0}$ from U. Suppose that

$$(12.1.1)\qquad (W_1 f)(x) = (W_2 f)(x) \quad \text{for every } f\in C_c^\infty(U) \text{ and every } x\in U.$$

Then

$$(12.1.2)\qquad \mathcal{F}_{T-}^{G_1} = \mathcal{F}_{T-}^{G_2} \text{ and } P_\mu^{G_1}(A) = P_\mu^{G_2}(A) \text{ for every } \mu\in\mathbb{P} \text{ and every } A\in\mathcal{F}_{T-}^{G_1}.$$

12.2. Remarks. Note that the σ-fields $\mathcal{F}_{T-}^{G_i}$ in Theorem 12.1 correspond to the special filtrations $(\mathcal{F}_t^{G_i})_{t\geqslant 0}$ denoted in the book of Dynkin [4; p. 108] by $(\overline{\mathcal{N}}_{t+0})_{t\geqslant 0}$. These filtrations need not satisfy the condition that

(12.2.1) if $A \in \mathcal{F}$ and $P_\mu(A) = 0$ for every $\mu \in \mathbb{P}$, then $A \in \mathcal{F}_0$,

occuring as one of axioms in the definition of a standard Markov process in the book of Blumenthal and Getoor[2; p. 45]. Theorem 12.1 would be false for filtrations satisfying (12.2.1) or after replacing $\mathcal{F}^{G_i}_{T-}$ by $\mathcal{F}^{G_i}_{T}$.

Theorem 12.1 remains true for a compact C^∞ manifold M without boundary, if only the Ventcel systems are replaced by two single elliptic hölderian operators of Waldenfels W_i, $i = 1,2$, in which case $G_i = \overline{W_i|_{C^2(M)}}$.

If $\partial M = \emptyset$ or if $\partial M \neq \emptyset$ and both the Ventcel systems satisfy the condition (7.2.1), then $\overline{\mathbb{1} W_i|_{C^\infty_c(U)}} = \overline{\mathbb{1}_U G_i|_{C^\infty_c(U)}}$, the closure being taken in the sense discussed in Section 1.5.

12.3. Proof of Theorem 12.1. Let V be any open set such that $\overline{V} \subset U$ and that ∂V is a C^∞ submanifold of $M \setminus \partial M$. Write $(W_{i,0}, \Gamma_{i,0}, \delta_{i,0})$ instead of $(W_i, \Gamma_i, \delta_i)$. Then the assumptions of Lemma 10.1 are satisfied. Let $(W_{i,k}, \Gamma_{i,k}, \delta_{i,k})_{k=1,2,\ldots}$, $i = 1,2$, be two sequences of elliptic hölderian systems of Ventcel satisfying (10.1.2) - (10.1.5). Let $G_{i,k}$ be the corresponding Feller generators. Denote by S the first exit time of $(X_t)_{t \geqslant 0}$ from V. By (10.1.2) - (1o.1.4), by Lemma 8.3 and by Theorem 5.1, $\mathcal{F}^{G_{1,k}}_{S-} = \mathcal{F}^{G_{2,k}}_{S-}$ and $\mathbb{P}^{G_{1,k}}_\mu(A) = \mathbb{P}^{G_{2,k}}_\mu(A)$ for every $k = 1,2,\ldots$, $\mu \in \mathbb{P}$ and $A \in \mathcal{F}^{G_{1,k}}_{S-}$. By Theorem 11.1, these equalities together with (10.1.5) imply that

$$\mathcal{F}_{S-}^{G_{1},0} = \mathcal{F}_{S-}^{G_{2},0} \quad \text{and} \quad P_{\mu}^{G_{1},0}(A) = P_{\mu}^{G_{2},0}(A) \quad \text{for every} \quad \mu \in \mathbb{P} \quad \text{and} \quad A \in \mathcal{F}_{S-}^{G_{1},0}$$

By Theorem 5.1, the former is equivalent to condition that

$$(12.3.1) \qquad E_{x}^{G_{1},0}\left[f(X_t);\ t<S\right] = E_{x}^{G_{2},0}\left[f(X_t);\ t<S\right]$$

for every $x \in V$, $f \in C_o(V)$ and $t \in R^+$.

Let now $U_1 \subset U_2 \subset \cdots$ be a sequence of open sets such that $\bigcup_{n=1}^{\infty} U_n = U$ and that, for every $n = 1,2,\ldots$, $\overline{U_n} \subset U$ and ∂U_n is a C^{∞} submanifold of $M \setminus \partial M$. Let T_n be the first exit time of $(X_t)_{t \geq 0}$ from U_n. By (12.3.1),

$$E_{x}^{G_{1},0}\left[f(X_t);\ t<T_n\right] = E_{x}^{G_{2},0}\left[f(X_t);\ t<T_n\right]$$

for every $n = 1,2,\ldots$, $x \in U_n$, $f \in C_o(U_n)$ and $t \in R^+$. Now, the Lemma 8.1 implies that the equalities (5.1.1) are satisfied. By Theorem 5.1, equalities (5.1.1) are equivalent to (5.1.2). But (5.1.2) is exactly the same as (12.1.2). The proof is complete.

References

[1] P. Billingsley, Convergence of Probability Measures, John Wiley and Sons, Inc., 1968.

[2] R.M. Blumenthal and R.K. Getoor, Markov Processes and Potential Theory, Academic Press, 1968.

[3] J.-M. Bony, Ph. Courrége et P. Priouret, Semi-groupes de Feller sur une variété a bord compacte et problémes aux limites intégro--différentielles du second ordre donnant lieu au principe du

maximum, Ann. Inst. Fourier 18, 2 (1968), p. 369-521.

[4] E.B. Dynkin, Markov Processes, Vol. I, Springer-Verlag, 1965.

[5] S.N. Ethiér and T.G. Kurtz, Markov Processes, Characterization and Convergence, John Wiley and Sons, 1986.

[6] I.I. Gichman and A.V. Skorochod, Theory of Stochastic Processes, Vol. I (in russian), "Nauka", Moscow, 1971.

[7] N. Ikeda and S. Watanabe, On some relations between the harmonic measure and the Lévy measure for a certain class of Markov processes, J. Math. Kyoto Univ. 2 (1962), p. 79-95.

[8] K. Itô and H. Mc Kean, Diffusion processes and their sample paths, Springer-Verlag, 1965.

[9] J. Kisyński, On a formula of N. Ikeda and S. Watanabe concerning the Lévy kernel, p. 260-279 in "Probability Measures on Groups VII", Lecture Notes in Mathematics, Vol. 1064, Springer-Verlag, 1984.

[10] J. Kisyński, On jumps of paths of Markov processes, p.130-145 in "Probability Measures on Groups VIII", Lecture Notes in Mathematics, Vol. 1210, Springer-Verlag, 1986.

[11] K. Sato and T. Ueno, Multi-dimensional diffusion and the Markov process on the boundary, J. Math. Kyoto Univ. 4 (1965), p. 529-605.

[12] W. von Waldenfels, Fast positive Operatoren, Zeitschrift für Wahrscheinlichkeitstheorie 4 (1965), p. 159-174.

A DICHOTOMY THEOREM FOR RANDOM WALKS ON HYPERGROUPS

AJAY KUMAR and AJIT IQBAL SINGH*

Abstract. We define possible and recurrent elements for random walks on commutative hypergroups and show that for a large class of random walks, either no elements is recurrent or all possible elements are recurrent and they form a closed subhypergroup.

Chung and Fuchs [4] initiated the study of recurrence properties of random walks on the real line $\mathbb{R}$ i.e sums of independent identically distributed random variables taking values in $\mathbb{R}$. Loynes [13] adapted this study to products of identically distributed independent random elements i.e. functions from some fixed probability space $(\Omega', \mathcal{B}_{\Omega'}, P')$ taking values in a topological group. In this note we generalise his Dichotomy theorem to a large class of random walks on a commutative hypergroup K ([6], same as convos in [10] whose notation and terminology we follow here). The corresponding proof in the group case cannot be adapted in this situation but the technique of replacing Ω' by a concrete probability space of sequences in K on the lines of chapter 8[3] (c.f. [5]) is very useful. Motivation for our formulation of the problem comes from the example of space of orbits G_B where $G = \mathbb{R}^n$ and $B = SO(n)$. Earlier attempts of generalizing random walks or Markov chains for that matter (c.f. [9], [7]) presumably have their motivation from discrete hypergroups and consequently the two approaches diverge as soon as they lift away from groups. We thank K. R. Parthasarathy for useful discussions.

Let Y_n be a sequence of independent random elements i.e. functions on Ω' to K with common distribution μ. Let $\Omega = \{Y(t') = (Y_n(t'))_n : t' \in \Omega'\}$. Consider this as a subspace of $K^{\mathbb{N}}$ with the product topology and product measure $\mu^{\mathbb{N}}$ defined on the σ-algebra $\mathcal{B}_{\Omega}$ of subsets of Ω generated by

---

* née Ajit Kaur Chilana.

$\mathcal{S} = \{(\prod_j C_j) \cap \Omega : C_j \in \mathcal{B}_K$ for every j and $C_j = K$ for all but finitely many j$\}$, $\mathcal{B}_K$ being the σ-algebra of Borel subsets of K. For $S = (\prod_j C_j) \cap \Omega \in \mathcal{S}$

$P'\{t' : Y(t') \in S\} = \mu^{\mathbb{N}}(S)$ so that $\mu^{\mathbb{N}}(\Omega) = 1$ and $P'\{t' : Y(t') \in B\} = \mu^{\mathbb{N}}(B)$ for every $B \in \mathcal{B}_\Omega$. Thus $(\Omega, \mathcal{B}_\Omega, P = \mu^{\mathbb{N}}|_{\mathcal{B}_\Omega})$ is a probability space and we may consider $\{Y_n\}$ as defined on Ω and given by coordinate functions.

We next note that for $n \in \mathbb{N}$, the n-fold convolution product [14] defined on K^n to $M^+(K)$ with the cone topology is continuous and, therefore, for $j,n \in \mathbb{N}$, $j < n$ the functions $X_{j,n}$; $X_{j,j}$ and X_j are $M^+(K)$-valued random variables, where

$X_{j,n}(t) = Y_n(t) * Y_{n-1}(t) * \ldots\ldots * Y_j(t)$, for $t \in \Omega$, $X_{j,j} = Y_j$ and $X_j = X_{1,j}$, it being understood that the elements of K are identified with the corresponding point masses [10]. Further, X_j and $X_{j+1,n}$ are independent.

Let f be a continuous non-negative function on K with compact support, $C_f := \{x \in K : f(x) > 0\}$ the corresponding cozero set and $\nu \in M^+(K)$. Then $\int_K f d\nu > 0$ if and only if $\nu(C_f) > 0$ and this is true if and only if $(\mathrm{supp}\,\nu) \cap C_f \neq \emptyset$. Therefore the set $\{\nu \in M^+(K), \nu(C_f) > 0\} = \{\nu \in M^+(K) : \int_K f d\nu > 0\}$ is open in $M^+(K)$ with the cone topology. Let $\mathcal{C}$ be the class of such cozero sets and $\mathcal{B}_{\mathcal{C}}$ be the σ-algebra generated by $\mathcal{C}$. Then for each B in $\mathcal{B}_{\mathcal{C}}$, $\{t : X_{j,n}(t)(B) > 0\}$ and $\{t : X_j(t)(B) > 0\}$ are in $\mathcal{B}_\Omega$. Clearly if τ is the shift ([3], ch.8) then $X_{j+1,n} = X_{n-j} \circ \tau^j$ and $X_n = X_n \circ \sigma$ for each finite permutation σ of $N_n = \{j \in \mathbb{N}, j \leq n\}$. So $P\{t : X_{j+1,n}(t)(B) > 0\} = P\{t : X_{n-j}(t)(B) > 0\}$ and we have the following analogue of the <u>zero or one law</u> (c.f. [3], Theorem 8.2.1).

<u>Proposition 1</u>. Let $\{B_n\}$ be a sequence of sets in $\mathcal{B}_{\mathcal{C}}$ then $P\{t : X_n(t)(B_n) > 0 \text{ i.o.}\}$ is zero or one, where, i.o. stands for infinitely often.

<u>Remarks (i)</u>. Since K is locally compact Hausdorff, $\mathcal{B}_\Omega$ contains the class of σ-compact open subsets of K. Further it coincides with the σ-algebra generated by compact G_δ subsets of K and thus consists of Baire sets and their complements ([8], p.220). So by ([8], 51.A) for $B \in \mathcal{B}_\Omega$ either B or its complement is contained in a σ-compact set. If K is σ-compact and every open set in K is an F_σ, then every open set in K is σ-compact and, therefore, $\mathcal{B}_{\mathcal{C}} = \mathcal{B}_K$. On the other hand a

stronger form of the converse holds as shown below:

K has a base $\mathcal{N}_1$ of compact symmetric neighbourhoods at the identity e of K and $\mathcal{U} = \{S \subset K \times K : S \supset U_N = \bigcup_{c \in K} (c*N) \times (c*N)$ for some $N \in \mathcal{N}_1\}$ is a uniformity giving the topology of K. Also for $N \in \mathcal{N}_1$, $x \in K$, $U_N(x) = \{y \in K : (x,y) \in U_N\} = N * N * x$ is compact and thus K is uniformly locally compact. Therefore by ([11], 7.T) K is paracompact, K is the union of a disjoint open family $\{K_\alpha : \alpha \in \Lambda\}$ of σ-compact spaces and if K is connected then it is σ-compact. In fact for any fixed N in $\mathcal{N}_1$, if we put $M_x = \bigcup_{j \in \mathbb{N}} x * N^j$, $x \in K$ then distinct M_x's constitute such a family. If Λ is uncountable then $\Lambda = \Lambda_1 \cup \Lambda_2$ where Λ_1 and Λ_2 are both uncountable. Now $U = \bigcup_{\alpha \in \Lambda_1} K_\alpha$ is an open subset which is not in $\mathcal{B}_\mathcal{C}$. For, if it were, then either U or $V = \bigcup_{\alpha \in \Lambda_2} K_\alpha$ is σ-bounded. Since $\{K_\alpha : \alpha \in \Lambda\}$ is an open cover of every superset S of U or V which cannot have a countable subcover, we conclude that this is not so. Now suppose that $\mathcal{B}_\mathcal{C} = \mathcal{B}_K$. Then Λ is countable and thus K is σ-compact. Further using ([8], 50.D), K is a countable union of compact G_δ sets and thus $\mathcal{B}_\mathcal{C}$ is the class of Baire sets. So by ([8], 51.D) every compact Baire set is a G_δ and consequently {e} is the intersection of a countable family of open sets. But by a result of K.A. Ross (c.f. [16], Theorem A.3) in this case K is metrizable. Thus $\mathcal{B}_K = \mathcal{B}_\mathcal{C}$ if and only if K is σ-compact and metrizable if and only if K is second countable.

(ii) Since the involution $x \to \bar{x}$ in K is continuous and convolution of sets in K is also continuous ([10], 3.2(c)) there exists a base $\mathcal{N}$ of neighbourhoods at identity e of K such that $\mathcal{N} \subset \mathcal{C}$, each N in $\mathcal{N}$ is symmetric and for $N \in \mathcal{N}$ there exists $N' \in \mathcal{N}$ such that $N' * N' \subset N$. Using ([10] 3.1B and 4.2D), $A * c \in \mathcal{C}$ for each A in $\mathcal{C}$ and c in K and thus $\mathcal{N}_c = \{N_c = N * c, N \in \mathcal{N}\} \subset \mathcal{C}$ is a base of neighbourhoods at c. So the following definitions generalise their counterparts in groups ([13] § 4).

<u>Definitions</u>. An element c in K is called <u>possible</u> if for every neighbourhood N_c of c with $N \in \mathcal{N}$ there exist $k \in \mathbb{N}$ such that $P\{t : \text{supp } X_k(t) \cap N_c \neq \phi\} > 0$ or, equivalently, $P\{t : X_k(t)(N_c) > 0\} > 0$. $c \in K$ is called <u>recurrent</u>, if for every neighbourhood N_c of c with $N \in \mathcal{N}$, $P\{t : \text{supp } X_n(t) \cap N_c \neq \phi \text{ i.o.}\} = 1$ or, equivalently, $P\{t : X_n(t)(N_c) > 0 \text{ i.o.}\} = 1$.

Let P be the set of possible values and R the set of recurrent values. Clearly P and R are closed.

<u>Proposition 2(i)</u>. $P * P \subset P$, $e \in P$ if and only if $P \cap \bar{P} \neq \phi$.

(ii) $R * P \subset R$.

(iii) If $R \cap \bar{P} \neq \phi$, then e is recurrent and every possible value is recurrent.

(iv) If K is σ-compact (equivalently, if K is Lindelöf) then $P \neq \phi$.

(v) If K is compact then $P = R$ is a subhypergroup.

<u>Proof</u>. (i). Let $b, c \in P$ and $N, N' \in \mathcal{N}$ be such that $N' * N' \subset N$. There exist $n, m \in \mathbb{N}$ such that $P\{t : \text{supp } X_n(t) \cap (N'*b) \neq \phi\} > 0$ and $P\{t : \text{supp } X_m(t) \cap (N' * c) \neq \phi\} > 0$. So
$P\{t : \text{supp } X_{n+1,n+m}(t) \cap (N'*c) \neq \phi\} > 0$ and since X_n and $X_{n+1,n+m}$ are independent,

$P[\{t:\text{supp}X_n(t) \cap (N'*b) \neq \phi\} \cap \{t:\text{supp}X_{n+1,n+m}(t) \cap (N'*c) \neq \phi\}] > 0.$

Let $z \in b * c$, then $c \in \bar{b} * z$ ([10],4.1B) and, therefore,

$\{t:\text{supp}X_n(t) \cap (N'*b) \neq \phi\} \cap \{t:\text{supp}X_{n+1,n+m}(t) \cap (N' * c) \neq \phi\}$

$\subset \{t:\text{supp}X_n(t) \cap (N'*b) \neq \phi\} \cap \{t:\text{supp}X_{n+1,n+m}(t) \cap (N'*\bar{b}*z) \neq \phi\}$

$\subset \{t : \text{supp } X_{n+1,n+m}(t) \cap (N*z*\text{supp } \overline{X_n(t)}) \neq \phi\}$

$= \{t : \text{supp } X_{n+m}(t) \cap (N*z) \neq \phi\}.$

So $P\{t : \text{supp } X_{n+m}(t) \cap (N*z) \neq \phi\} > 0$. Hence z is possible.

(ii) Let $b \in R$, $c \in P$ and $z \in b*c$. Let N and N' be as in (i). Now b being recurrent $P\{t : X_n(t)(N'*b) > 0 \text{ i.o.}\} = 1$ and since $c \in P$ there exists $k \in \mathbb{N}$ such that $P\{t: X_k(t)\ (N'*c) > 0\} > 0$. Then by arguments similar to those in (i) above,

$P[\{t: \text{supp}X_{k+1,k+n}(t) \cap (N'*b) \neq \phi \text{ i.o.}\} \cap \{t:\text{supp}X_k(t) \cap (N'*c) \neq \phi\}]>0$

and the following inclusion holds

$\{t:\text{supp}X_{k+1,k+n}(t) \cap (N'*b) \neq \phi \text{ i.o.}\} \cap \{t:\text{supp}X_k(t) \cap (N'*c) \neq \phi\}$

$\subset \{t : \text{supp } X_{k+n}(t) \cap (N*z) \neq \phi \text{ i.o.}\}.$

So $P\{t : \text{supp } X_{k+n}(t) \cap (N*z) \neq \phi \text{ i.o.}\} > 0$.

Therefore by Proposition 1, $P\{t : \text{supp } X_{k+n}(t) \cap (N*z) \neq \phi \text{ i.o.}\} = 1$.

Hence z is recurrent.

(iii) follows from (i) and (ii) above.

(iv) If $P = \phi$, then for each c in K there is a $U_c \in \mathcal{N}_c$ such that $P\{t : \text{supp } X_k(t) \cap U_c \neq \phi\} = 0$ for each k. In particular $P\{t : Y_1(t) \in U_c\} = 0$. Since K is σ-compact, there exist a countable subcover $\mathcal{U}_1$ of $\mathcal{U} = \{U_c : c \in K\}$ and we have

$$1 = P\{t : Y_1(t) \in K\}$$
$$= P\{t : Y_1(t) \in \bigcup_{U \in \mathcal{U}_1} U\} \leq \sum_{U \in \mathcal{U}_1} P\{t: Y_1(t) \in U\} = 0,$$

a contradiction. Hence $P \neq \phi$.

(v) We first note that if $K = \bigcup_{j=1}^{p} K_j$, $K_j \in \mathcal{B}_{\mathcal{C}}$ then for at least one j, $P(t: X_k(t)(K_j) > 0 \text{ i.o.}) > 0$ or equivalently (in view of Proposition 1) $P(t: X_k(t)(K_j) > 0 \text{ i.o.}) = 1$. So it follows that $R \neq \phi$. Thus P is a nonempty compact subset of K satisfying $P * P \subset P$. Therefore by ([10], 10.2F) we have that $\overline{P} = P$. So $R \cap \overline{P} = R \neq \phi$ and therefore by (iii) above $e \in R = P$.

Let us consider the case when $P \neq \phi$. Then P is a subhypergroup if and only if $\overline{P} = P$. This clearly occurs if K is hermitian. It is also so if P is compact ([10], 10.2F). Another instance occurs if $\{Y_n\}$ is quasi-symmetrically distributed in the sense that $\mu \equiv \overline{\mu}$, where μ is the common distribution of Y_n's. Because in this case for any n and $B \in \mathcal{B}_{\mathcal{C}}$, $P\{t : \text{supp } X_n(t) \cap B \neq \phi\} = 0$ if and only if $P\{t : \text{supp } \overline{X_n(t)} \cap B \neq \phi\} = 0$ and consequently c is possible for (Y_n) if and only if $\overline{c}$ is so. Thus we have the following Dichotomy-theorem for hypergroups.

Theorem 3. If (Y_n) is quasi-symmetrically distributed (in particular, if K is hermitian) then either no element is recurrent or all possible elements are recurrent and they form a closed subhypergroup.

Remark. K is hermitian or compact for a large class of hypergroups ([1], [2], [10], [12], [14] which are not groups. On the other hand the conditions in the above Theorem is not necessary in the case of groups ([15], § 11).

References

1. W. R. Bloom and H. Heyer, The Fourier transform for probability measures on hypergroups, Rend. di Mat. 2 (1982), 315-334.
2. W. R. Bloom and H. Heyer, Convergence of convolution products of probability measures on hypergroups, Rend. di Mat 3 (1982), 547-563.
3. K. L. Chung, A Course in Probability Theory (1968) Academic Press.

4. K. L. Chung and W.H.J. Fuchs, On the distribution of values of sums of random variable, Mem. Amer. Math. Soc. (1951), 1-12.

5. J. L. Doob, Stochastic Processes. John Wiley and Sons Inc.(1953) New York.

6. C. F. Dunkl, The measure algebra of a locally compact hypergroup, Trans Amer. Math. Soc. 179(1973), 331-348.

7. L. Gallardo and O. Gebuhrer, Marches aleatoires et hypergroups, Expo. Math. 5(1987), 41-73.

8. P. R. Halmos, Measure Theory, Van Nostrand 1950, Springer Verlag 1974 East West Press (New Delhi, Madras).

9. H. Heyer, Probability theory on hypergroups, Probability Measures on groups VII. Lecture notes in Math. 1064 (1984) 481-550.

10. R. I. Jewett, Spaces with an abstract convolution of measures, Adv. in Math. 18(1975), 1-101.

11. J. L. Kelley, General Topology, (1955), Van Nostrand.

12. R. Lasser, Orthogonal Polynomials and hypergroups Rend. di Mat. (VII) 3,2 (1983), 185-209.

13. R. M. Loynes, Products of Independent Random elements in a topological group, Z. Wahrscheinlichkeitstheorie verw. Geb. 1(1963), 446-455.

14. K. A. Ross, Centers of hypergroups, Trans. Amer. Math. Soc. 243 (1978), 251-269.

15. K. Schmidt. Cocycles on Ergodic Transformation groups, Macmillan 1976.

16. R. C. Vrem, Lecunarity on compact hypergroup, Math. Zeit. 164(1978) 93-104.

Department of Mathematics
University of Delhi
Delhi-110 007, INDIA

A MODIFICATION OF STATIONARITY FOR STOCHASTIC PROCESSES INDUCED BY ORTHOGONAL POLYNOMIALS

R. LASSER
Institut für Medizinische Informatik und Systemforschung, MEDIS
der Gesellschaft für Strahlen- und Umweltforschung, GSF
Ingolstädter Landstr. 1, 8042 München-Neuherberg, FRG

For many applications it is desirable to have generalizations of the classical theory of (weakly) stationary processes to certain classes of nonstationary ones. In [3] we introduced a class of stochastic processes with a stationarity condition based on the notion of hypergroups, and we call a family of square integrable random variables $\{X_a: a\in K\}$, indexed by a hypergroup K, 'K-weakly stationary' whenever

(1) the means are constant, i.e., $EX_a = c$ for each $a \in K$,

(2) the covariance function $d(a,b) = E\left[(X_a - c)\overline{(X_b - c)}\right]$ is bounded, continuous and satisfies $d(a,b) = \int_K d(x,e)\,dp_a * p_{\bar{b}}(x)$.

Here we are mainly interested in that case where the parameter set K equals $\mathbb{N}_0$, bearing a polynomial hypergroup structure induced by a sequence of orthogonal polynomials $P_n(x)$ with a certain positivity property (P). In [2] we discuss the connection between hypergroups on $\mathbb{N}_0$ and orthogonal polynomials, and, some basic facts concerning the Haar weights h(n), the dual space D_s, the Plancherel measure π and Fourier coefficients $\check{f}(n)$. For the polynomial hypergroups $K = \mathbb{N}_0$ we shall use the notion of $(P_n(x))$-weakly stationary, to distinguish them from other hypergroup structures on $\mathbb{N}_0$.

An incentive to study K-weakly stationary processes are the common estimators of the mean

$$Y_n = \frac{1}{2n+1}\sum_{k=-n}^{n} X_k \quad , \quad n\in\mathbb{N}_0 \quad ,$$

where $(X_k)_{k\in\mathbb{Z}}$ is a weakly stationary stochastic process. In [3] we showed that $(Y_n)_{n\in\mathbb{N}}$ is K-weakly stationary with $K = \mathbb{N}_o$, having a polynomial hypergroup structure induced by the Jacobi polynomials $P_n^{(\frac{1}{2},-\frac{1}{2})}(x)$. However, this is only a special case of the following class of K-weakly stationary processes.

Let $P_n^{(\alpha,\beta)}(x)$ denote the Jacobi polynomials with parameters $\alpha,\beta>-1$, normalized such that $P_n^{(\alpha,\beta)}(1) = 1$. - Note: if $\alpha+\beta+1\geq 0$ and $\alpha\geq\beta>-1$, the polynomials $P_n^{(\alpha,\beta)}(x)$ induce a hypergroup structure on $\mathbb{N}_o$, see [2]. - Formula (2.7), (3.15) and Gegenbauer's formula (7.5) in [1] yield for $\gamma > -\frac{1}{2}$:

$$P_{2n}^{(\gamma,\gamma)}(\cos\tfrac{t}{2}) = \frac{(2n)!}{(2\gamma+1)_{2n}} \sum_{k=0}^{2n} \frac{\left(\gamma+\frac{1}{2}\right)_{2n-k} \left(\gamma+\frac{1}{2}\right)_k}{(2n-k)! \quad k!} \cos(n-k)t$$

$$= \frac{(2n)!}{(2\gamma+1)_{2n}} \sum_{k=-n}^{n} \frac{\left(\gamma+\frac{1}{2}\right)_{n+k} \left(\gamma+\frac{1}{2}\right)_{n-k}}{(n+k)! \quad (n+k)!} e^{-ikt} \quad .$$

Using (3.13) in [1] we have

$$P_n^{(\gamma,-\frac{1}{2})}(\cos t) = \frac{(2n)!}{(2\gamma+1)_{2n}} \sum_{k=-n}^{n} \frac{\left(\gamma+\frac{1}{2}\right)_{n+k} \left(\gamma+\frac{1}{2}\right)_{n-k}}{(n+k)!\ (n-k)!} e^{-ikt} \quad .$$

Now, let $(X_k)_{k\in\mathbb{Z}}$ be a weakly stationary stochastic process. Fix $\gamma>-\frac{1}{2}$ and define

$$Y_n^{\gamma} = \frac{(2n)!}{(2\gamma+1)_{2n}} \sum_{k=-n}^{n} \frac{\left(\gamma+\frac{1}{2}\right)_{n+k} \left(\gamma+\frac{1}{2}\right)_{n-k}}{(n+k)!\ (n-k)!} X_k \quad . \qquad (1)$$

Since Y_n^{γ} is a convex combination of the X_k's, the Y_n^{γ} may be viewed as unbiased estimates of the mean of $(X_k)_{k\in\mathbb{Z}}$. The stochastic process $(Y_n^{\gamma})_{n\in\mathbb{N}_0}$ is K-weakly stationary, where $K = \mathbb{N}_0$ is bearing the hypergroup structure induced by the polynomials $P_n^{(\gamma,-\frac{1}{2})}(x)$. To demonstrate this we write for $m,n\in\mathbb{N}_0$:

$$P_n^{(\gamma,-\frac{1}{2})}(x)P_m^{(\gamma,-\frac{1}{2})}(x) = \sum_{k=|n-m|}^{n+m} g(n,m,k)P_k^{(\gamma,-\frac{1}{2})}(x) \quad ,$$

and let μ denote the spectral measure of $(X_k)_{k\in\mathbb{Z}}$, that is,

$$d_x(m) = E\Big[(X_m-c)(\overline{X_o-c})\Big] = \int_{-\pi}^{\pi} e^{-imt}d\mu(t) \ .$$

For the covariance d_y^{γ} of $(Y_n^{\gamma})_{n\in\mathbb{N}_0}$ formula (1) yields:

$$d_y^{\gamma}(n,m) = \int_{-\pi}^{\pi} P_n^{(\gamma,-\frac{1}{2})}(\cos t)P_m^{(\gamma,-\frac{1}{2})}(\cos t)\, d\mu(t)$$

$$= \sum_{k=|n-m|}^{n+m} g(n,m,k) \int_{-\pi}^{\pi} P_k^{(\gamma,-\frac{1}{2})}(\cos t)\, d\mu(t)$$

$$= \sum_{k=|n-m|}^{n+m} g(n,m,k)d_y^{\gamma}(k,0) \quad .$$

In particular, $\gamma = \frac{1}{2}$ yields the classical unbiased estimator of the mean.

In [3] we generalized the spectral representation theorems of Bochner and Cramér, respectively, to K-weakly stationary processes, where K is a commutative hypergroup. Further, we investigated for polynomial hypergroups $K = \mathbb{N}_0$ subclasses which extend notions, such as moving average processes or autoregressive processes in an appropriate way. For autoregressive processes, however, (even for the order 1) a characterization seems to be rather intricate. The following theorem contains a complete description of autoregressive processes of order 1.

Let $(P_n)_{n\in\mathbb{N}_0}$ be an orthogonal sequence with property (P), and endow $K = \mathbb{N}_0$ with the corresponding hypergroup structure. Given a $(P_n(x))$-weakly stationary process $(X_n)_{n\in\mathbb{N}_0}$ on a probability space (Ω,Σ,P) with zero mean, let H denote the $L^2(P)$ closure of the linear span of $\{X_n : n\in\mathbb{N}_0\}$. Then the translation

$$p_m * X_n = \sum_{k=|n-m|}^{n+m} g(n,m,k)X_k \tag{2}$$

can easily be extended to a linear, norm-decreasing operator T_m on H, compare [3] . Let $(Z_n)_{n\in\mathbb{N}_0}$ be a white noise with respect to $(P_n)_{n\in\mathbb{N}_0}$, i.e., an uncorrelated sequence of random variables Z_n with $E(Z_n\bar{Z}_n) = g(n,n,0)$. A $(P_n(x))$-weakly stationary process $(X_n)_{n\in\mathbb{N}_0}$ is called autoregressive (of order q) with respect to $(P_n)_{n\in\mathbb{N}_0}$, if there exist $b_1,\ldots,b_q\in\mathbb{C}$ $(b_q \neq 0)$, such that

$$X_n + b_1p_1 * X_n + \ldots + b_qp_q * X_n = Z_n \tag{3}$$

for any $n\in\mathbb{N}_0$.

<u>Theorem</u>: Suppose $(X_n)_{n\in\mathbb{N}_0}$ is autoregressive of order 1 with respect to $(P_n)_{n\in\mathbb{N}_0}$, that is,

$$X = \alpha p_1 * X_n + Z_n \tag{4}$$

for all $n\in\mathbb{N}_0$ with a complex number $a\in\mathbb{C}$. If $|\alpha| < 1$, then X_n is a moving average process. Explicitely,

$$X_n = \sum_{k=0}^{\infty} (g_{k,\alpha})^{\vee}(k)p_k * Z_n h(k) \quad , \tag{5}$$

where $g_{k,\alpha}(x)$ are the continuous functions on D_s given by

$$g_{k,\alpha}(x) = \frac{\left[\alpha P_1(x)\right]^k}{1-\alpha P_1(x)} \quad . \tag{6}$$

Conversely, each process $(X_n)_{n\in\mathbb{N}_0}$ defined by (5) with $|\alpha| < 1$ satisfies equation (4).

To prove the theorem we show two auxiliary results. For $i \le m$ designate

$$\beta_{m,i} = \sum_{j=i}^{m} ((\alpha P_1)^j)^{\vee}(i)h(i)$$

and for $n, m \in \mathbb{N}_o$:

$$Y_{n,m} = \sum_{i=0}^{m} \beta_{m,i} p_i * Z_n \quad , \tag{7}$$

Lemma 1: The random variables $Y_{n,m}$ of (7) satisfy

$$Y_{n,m} = \sum_{k=0}^{m} \alpha^k p_1^k * Z_n \quad ,$$

where $p_1^{k}* Z_n = p_1*(p_1^{k-1}*Z_n)$ is defined recursively as in (2) with $p_1^{o}* Z_n = Z_n$. In particular, $(Y_{n,m})_{m\in\mathbb{N}_o}$ does converge in $L^2(P)$, provided $|\alpha| < 1$.

Proof. Note that the k-fold convolution product p_1^k may be written as

$$p_1^k = \sum_{j=0}^{k} a_{kj} p_j,$$

where p_j is the point measure in $j\in\mathbb{N}_o$ and $a_{kj} = (P_1^k)^{\vee}(j)h(j)$. Hence, for given $m\in\mathbb{N}_o$ we obtain

$$\sum_{k=0}^{m} \alpha^k p_1^k * Z_n = \sum_{k=0}^{m} \alpha^k \left[\sum_{j=0}^{k} a_{kj} p_j * Z_n \right] = \sum_{i=0}^{m} \gamma_{m,i} p_i * Z_n \quad ,$$

where the $\gamma_{m,i}$ are given by

$$\gamma_{m,i} = \sum_{j=i}^{m} \alpha^j a_{ji} \quad .$$

Since

$$\gamma_{m,i} = \sum_{j=i}^{m} \alpha^j a_{ji} = \sum_{j=i}^{m} \left[(\alpha P_1)^j\right]^{\vee}(i)h(i) = \beta_{m,i} ,$$

we have demonstrated that

$$Y_{n,m} = \sum_{k=0}^{m} \alpha^k p_1^{k} * Z_n .$$

Finally,

$$\|p_1^{k} * Z_n\|_2 \le \sup_{n \in \mathbb{N}_o} \|Z_n\|_2 \le 1$$

and $|\alpha| < 1$ yield convergence of $Y_{n,m}$ in $L^2(P)$ with $m \to \infty$.

Now observe that for $|\alpha| < 1$

$$\sum_{j=0}^{k} |(g_{k,\alpha})^{\vee}(k)|^2 h(k) < \infty$$

with $g_{k,\alpha}$ as defined in (6). In fact,

$$|(g_{k,\alpha})^{\vee}(k)|^2 \; h(k) \le \|g_{k,\alpha}\|_2^2 \le C \int_{D_S} |\alpha P_1(x)|^{2k} d\pi(x) \le \tilde{C}\alpha^{2k}$$

with C und $\tilde{C}$ being appropriate constants. Thus the X_n given in (5) are well defined elements of $L^2(P)$. Moreover, we show

<u>Lemma 2</u>: If $|\alpha| < 1$ then $X_n = \lim_{m \to \infty} Y_{n,m}$

Proof. Consider

$$X_{n,m} = \sum_{k=0}^{m} (g_{k,\alpha})^{\vee}(k) p_k * Z_n h(k) .$$

Since

$$(g_{k,\alpha})^{\vee}(k)h(k) - \beta_{m,k} = \sum_{j=m+1}^{\infty} \left[(\alpha P_1)^j\right]^{\vee}(k)h(k) = \left[\frac{\left[\alpha P_1\right]^{m+1}}{1-\alpha P_1}\right]^{\vee}(k)h(k),$$

we obtain

$$\|X_{n,m} - Y_{n,m}\|_2^2 \le \sum_{k=0}^{m} |(g_{k,\alpha})^{\vee}(k)h(k) - \beta_{m,k}|^2$$

$$\le C \sum_{k=0}^{m} \int_{D_S} |\alpha P_1(x)|^{2m+2}\, d\pi(x) \le \tilde{C}(m+1)|\alpha|^{2m+2} \quad ,$$

where C and $\tilde{C}$ are appropriate constants. Hence $\|X_{n,m} - Y_{n,m}\| \to 0$.

Now we can prove the theorem. First, assume $(X_n)_{n\in\mathbb{N}_0}$ to satisfy equation (4) with $|\alpha|<1$. Then, by Lemma 1 we get

$$X_n = \alpha^{m+1} p_1^{m+1} * X_n + \sum_{k=0}^{m} \alpha^k p_1^k * Z_n = \alpha^{m+1} p_1^{m+1} * X_n + Y_{n,m} .$$

Therefore, $\|X_n - Y_{n,m}\|_2 = |\alpha|^{m+1} \|p_1^{m+1} * X_n\|_2$. Since $(X_n)_{n\in\mathbb{N}_0}$ is $(P_n(x))$-weakly stationary, $\|X_n - Y_{n,m}\|_2 \to 0$ as $m \to \infty$. With Lemma 2 we have shown that X_n may be written as in (5). Conversely, if X_n is given by (5) we obtain

$$Y_{n,m} - \alpha p_1 * Y_{n,m} = Z_n - \alpha^{m+1} p_1^{m+1} * Z_n \quad ,$$

and the continuity of the translation operator yields

$$X_n - \alpha p_1 * X_n = Z_n .$$

References

1. Askey, R.: Orthogonal polynomials and special functions. SIAM: Philadelphia 1975
2. Lasser, R.: Orthogonal polynomials and hypergroups. Rend. Mat. 3, 185-209 (1983)
3. Lasser, R., Leitner, M.: Stochastic processes indexed by hypergroups (to appear in Journal of Theoretical Probability) .

LE PROBLEME DE LA CLASSIFICATION DES FAMILLES EXPONENTIELLES NATURELLES DE $\mathbb{R}^d$ AYANT UNE FONCTION VARIANCE QUADRATIQUE.

Gérard LETAC, Laboratoire de Statistique et Probabilités,
Université Paul Sabatier,
118, route de Narbonne,
31062 TOULOUSE Cedex.

Sommaire : Cet exposé décrit la fonction-variance V d'une famille exponentielle naturelle de $\mathbb{R}^d$ et examine le problème de la classification des fonctions-variances de la forme

$$V(m_1,\dots,m_d) = \sum_{i,j=1}^{d} A_{ij} m_i m_j + \sum_{j=1}^{d} B_j m_j + C,$$

où les A_{ij}, B_j et C sont des matrices réelles (d,d) symétriques.

Abstract : This lecture describes the variance function V of a natural exponential family on $\mathbb{R}^d$ and considers the problem of the classification of the variance functions of the following form

$$V(m_1,\dots,m_d) = \sum_{ij=1}^{d} A_{ij} m_i m_j + \sum_{j=1}^{d} B_j m_j + C,$$

where A_{ij}, B_j and C are symmetrical (d,d) real matrices.

Le présent exposé ne donne pas encore de solution complète du problème mentionné dans le sommaire, ce n'est pas non plus un "survey" d'un domaine encore à développer ; il est destiné à intéresser les probabilistes à un sujet qui appartenait entièrement aux statisticiens jusqu'à un passé récent. Toutefois, pour apaiser l'éditeur, il y aura quelques résultats (ou démonstrations) nouveaux (Th. 3.1, 4.1 et 5.1).

J'expliquerai d'abord ce que sont les familles exponentielles : les "naturelles", puis les "générales", et j'indiquerai quelques occasions où on les rencontre en probabilités. On parlera ensuite des familles à variance quadratique, de l'espoir de les classer, et on terminera par des exemples et des morceaux sporadiques de classification, petits assemblages partiels d'un grand puzzle.

§1. Les familles exponentielles. Un modèle statistique est une partie F de l'ensemble des probabilités sur un espace mesurable $(\Omega,\mathfrak{A})$. L'art du statisticien ayant à trouver la probabilité P sur $(\Omega,\mathfrak{A})$ qui rende le mieux compte du phénomène étudié est d'abord de délimiter la famille F où on cherche P, grâce à la géométrie du phénomène, puis d'estimer ensuite les paramètres qui caractérisent P au milieu de F. Les familles exponentielles sont de tels modèles, mais nous n'expliquerons pas ici quelles caractéristiques de phénomènes aléatoires y conduisent, en renvoyant à de grands traités (Centsov (1981), Barndorff-Nielsen (1978) et Brown (1986)).

Commençons par les familles exponentielles naturelles, pour lesquelles $(\Omega,\mathfrak{A})$ est un espace vectoriel E réel de dimension finie muni de ses boréliens. On notera E^* son dual et $(\theta,x) \mapsto \langle\theta,x\rangle$ l'application bilinéaire canonique de E^*xE dans $\mathbb{R}$.

Soit μ une mesure de Radon positive sur E, pas nécessairement bornée, soit

$$L_\mu(\theta) = \int_E \exp\langle\theta,x\rangle\,\mu(dx) \tag{1.1}$$

sa transformée de Laplace. D'après Hölder, $\left\{\theta\in E^* \; ; \; L_\mu(\theta)<\infty\right\}$ est un ensemble convexe et nous notons par $\Theta(\mu)$ son intérieur. D'après Hölder encore, $\theta \mapsto k_\mu(\theta) = \mathrm{Log}\, L_\mu(\theta)$ est une fonction convexe sur $\Theta(\mu)$. Désormais nous désignons par $\mathfrak{M}(E)$ l'ensemble des μ tels que $\Theta(\mu)$ soit non vide et tels que μ ne soit pas concentré sur un hyperplan affine. Dans ces conditions k_μ est strictement convexe, et réelle analytique sur $\Theta(\mu)$.

Fixons nous alors μ dans $\mathfrak{M}(E)$ et θ dans $\Theta(\mu)$. On définit la probabilité sur E :

$$P(\theta,\mu)(dx) = \exp(\langle\theta,x\rangle\ ) - k_\mu(\theta))\mu(dx) \tag{1.2}$$

Le modèle statistique sur E $F(\mu) = \{P(\theta,\mu) \; ; \; \theta\in\Theta(\mu)\}$ est appelé <u>famille exponentielle naturelle</u> engendrée par μ (certains auteurs disent "linéaire" à la place de "naturelle").

Parlons maintenant des familles exponentielles générales. C'est à elles qu'on se réfère le plus souvent dans la littérature statistique. Elles sont, elles, définies sur un espace abstrait $(\Omega,\mathfrak{A})$ muni d'une mesure ν positive, mais pas nécessairement bornée ; on considère alors une application mesurable $w \mapsto t(w)$ de Ω dans E espace vectoriel réel de dimension finie, telle que l'image $\mu=t\nu$ de ν par t soit un élément de E. On introduit alors la probabilité suivante sur $(\Omega,\mathfrak{A})$, où θ est dans $\Theta(\mu)$:

$$P(\theta,\nu,t)(d\omega) = \exp(\langle\theta,t(\omega)\rangle - k_\mu(\theta))\ \nu(d\omega) \qquad (1.3)$$

Le modèle statistique sur $(\Omega,\mathcal{A})$ $F(\nu,t) = \{P(\theta,\nu,t)\ ;\ \theta\in\Theta(\mu)\}$ est appelé <u>famille exponentielle générale</u> engendrée par (ν,t). Quant à $F(\mu)=F(t\nu)$, on l'appelle la famille <u>naturelle associée</u> à $F(\nu,t)$. Les familles générales sont aussi appelées familles de Pitman-Koopman-Darmois, les pères fondateurs des années trente. Un exemple classique suffit : introduisons le modèle statistique des lois gaussiennes sur $\mathbb{R}$ non concentrées en un point :

$$F = \left\{N_{m,\sigma^2}\ ;\ m\in\mathbb{R},\ \sigma>0\right\}, \qquad \text{avec} \quad N_{m,\sigma^2}(d\omega) = \exp\left(\frac{-(\omega-m)^2}{2\sigma^2}\right)\frac{d\omega}{\sigma\sqrt{2\pi}}\ .$$

Si σ est fixé, $\left\{N_{m,\sigma^2}\ ;\ m\in\mathbb{R}\right\}$ est une famille exponentielle naturelle avec $E=\mathbb{R}$, $\mu=N_{0,\sigma^2}$, $\Theta(\mu)=\mathbb{R}$ et $k_\mu(\theta) = \dfrac{\sigma^2\theta^2}{2}$.

En revanche, si m et σ sont inconnus, F est une famille générale avec $\Omega=\mathbb{R}$, $\nu(d\omega) = \dfrac{d\omega}{\sqrt{2\pi}}$, $E=\mathbb{R}^2$, $t(\omega) = (\omega,\ -\dfrac{\omega^2}{2})$. Enfin μ est la mesure

sur $\mathbb{R}^2$ concentrée sur la parabole de $\mathbb{R}^2\left\{(\omega,\ -\dfrac{\omega^2}{2})\ ;\ \omega\in\mathbb{R}\right\}$, image de ν par t, dont la transformée de Laplace est :

$$L_\mu(\theta_1,\theta_2) = \int_{\mathbb{R}^2}\exp(\theta_1x_1+\theta_2x_2)\ \mu(dx) = \int_{-\infty}^{+\infty}\exp(\theta_1\omega - \theta_2\frac{\omega^2}{2})\frac{d\omega}{\sqrt{2\pi}}\ .$$

Il est clair que

$$\Theta(\mu) = \{(\theta_1,\theta_2)\ ;\ \theta_2>0\}$$

et que

$$k_\mu(\theta_1,\theta_2) = \frac{\theta_1^2}{2\theta_2} - \frac{1}{2}\,\mathrm{Lg}\ \theta_2\ .$$

La mesure μ est donc bien un élément de $\mathcal{M}(\mathbb{R}^2)$ et la famille $F(\mu)$ de $\mathbb{R}^2$ est la famille naturelle associée à F.

Laissons le lecteur jouer avec la famille générale $\left\{N_{0,\sigma^2}, s^2>0\right\}$ et trouver la famille associée. En fait, il ne sera à peu près pas question des familles générales ici : à mon avis la grande majorité des propriétés intéressantes des familles générales, y compris les problèmes d'estimation, sont en fait ceux de la famille naturelle associée. Revenons donc à ces familles exponentielles naturelles, pour quelques concepts et propriétés.

Observons d'abord que si μ et μ' sont dans $\mathcal{M}(E)$, alors

$F(\mu)=F(\mu')$ si et seulement si il Dxiste (a,b) dans $E\times\mathbb{R}$ tel que $\mu'(dx) = \exp(\langle a,x\rangle + b)\,\mu(dx)$. Si F est une famille exponentielle naturelle sur E, notons par $\mathcal{B}_F$ l'ensemble des μ de $\mathcal{M}(E)$ tels que $F(\mu)=F$. Il est clair que $F\subset\mathcal{B}_F$.

Plus important : si μ est dans $\mathcal{M}(E)$ et θ est dans $\Theta(\mu)$, alors

$$k'_\mu(\theta) = \int_E x\, P(\theta,\mu)(dx), \qquad (1.4)$$

où $k'_\mu(\theta)$ désigne l'élément de E qui est la différentielle de k_μ en θ. A cause de (1.4), l'ensemble $k'_\mu(\Theta(\mu)) = M_F$ est appelé le <u>domaine des moyennes</u> de F. Il ne dépend que de F, et non d'un μ de $\mathcal{B}_F$. Or, k_μ étant strictement convexe et analytique sur $\Theta(\mu)$, $\theta\mapsto k'_\mu(\theta)$ est donc un difféomorphisme analytique de $\Theta(\mu)$ sur M_F et on note par $\psi_\mu : M_F \to \Theta(\mu)$ sa réciproque. D'où une nouvelle paramétrisation de F par son domaine des moyennes ; notant, pour m dans M_F :

$$P(m,F) = P(\psi_\mu(m),\mu),$$

alors $P(m,F)$ ne dépend que de F et m et non d'un μ de $\mathcal{B}_F$. Et on a $F = \{P(m,F)\ ;\ m\in M_F\}$.

En prenant θ dans $\Theta(\mu)$ et non (comme Barndorff-Nielsen) dans l'ensemble, légèrement plus grand peut être, $\{\theta; L_\mu(\theta)<\infty\}$, on s'est assuré que les moments exponentiels au voisinage de 0 de la probabilité $P(\theta,\mu)$ existent (si on veut, en mettant une norme sur E, on est sûr que pour $\varepsilon>0$ assez petit, $\int_E \exp(\varepsilon\|x\|)\, P(\theta,\mu)(dx)$ est fini. On peut donc parler de la <u>variance de P(m,F),</u> c'est-à-dire de la forme bilinéaire symétrique sur $E^*\times E^*$, définie par

$$V_F(m)(\eta,\zeta) = \int_E \langle\eta,x-m\rangle\,\langle\zeta,x-m\rangle\, P(m,F)(dx)$$

Il est facile de voir que $V_F(m)$ est définie positive, et que si on confond la forme bilinéaire $V_F(m)$ sur $E^*\times E^*$ avec l'application linéaire associée de E^* dans E, alors on a, pour m dans M_F :

$$V_F(m) = (\psi'_\mu(m))^{-1} \qquad (1.5)$$

Le point important est que la connaissance de V_F donne la connaissance de F, c'est-à-dire que si F et F' sont des familles exponentielles naturelles sur E telles que $M_F = M_{F'}$ et $V_F = V_{F'}$, alors $F=F'$. En fait, on a même un résultat plus fort : si J est un ouvert non vide contenu dans $M_F\cap M_{F'}$, tel que $V_F(m) = V_{F'}(m)$ pour tout m de J, alors $F=F'$. Grâce à l'analyticité de $m\mapsto V_F(m)$, c'est facile à démontrer.

La fonction $m\mapsto V_F(m)$ est appelée la <u>fonction variance de F.</u> Elle est aussi amusante à manipuler que des transformées de Fourier et joue un rôle important dans les problèmes de classification dont nous parlerons tout à l'heure.

§2. Les familles exponentielles naturelles et les probabilités.

Cette section est une parenthèse pour rappeler aux probabilistes que les familles exponentielles sont leur prose de Mr Jourdain et qu'ils les manient familièrement.

(a) Fonctions génératrices des moments. Dans les cours élémentaires de probabilités, on considère des variables aléatoires X sur $\mathbb{R}$ telles que $\mathbb{E}(\exp \theta X)$ soit fini pour $|\theta|$ petit et θ réel. Ceci a l'avantage de donner les moments de X par développement en série entière au voisinage de zéro ; c'est dire que la loi de X est dans $\mathcal{M}(\mathbb{R})$. Plus généralement, c'est un problème classique, étant donné une mesure de Radon μ positive sur $\mathbb{R}$, de se demander si les polynômes sont denses dans $L^2_{\mathbb{R}}(\mu)$, question bien naturelle quand on présente les polynômes d'Hermite et de Laguerre. Une réponse raisonnable est : oui, si μ est dans $\mathcal{M}(\mathbb{R})$ et 0 est dans $\Theta(\mu)$. La démonstration vaut d'être rappelée (Sz Nagy (1965))) : soit f orthogonale aux polynômes, θ dans $\frac{1}{2}\Theta(\mu)$. Alors $x \mapsto f(x)\exp(\theta x)$ est μ intégrable, $F(z) = \int_{-\infty}^{+\infty} f(x)\exp(zx)\mu(dx)$ est analytique dans la bande $\left\{z\in\mathbb{C} \ ; \ z\in \frac{1}{2}\Theta(\mu)\right\}$, et $F^{(n)}(0) = \int_{-\infty}^{+\infty} x^n f(x)\ \mu(dx) = 0$. Donc F est nulle dans cette bande, la transformée de Fourier $t \mapsto F(it)$ de la mesure bornée $f(x)\mu(dx)$ est nulle et donc f=0.

(b) Lois des grands nombres, grandes déviations. Chacun connait la démonstration de poche suivante de la loi des grands nombres pour des variables aléatoires de Bernouilli : si $(X_n)_{n=1}^{\infty}$ sont indépendantes, avec $P(X_n=1)=p$, $P(X_n=0) = 1-p=q$ et $0<p<1$, alors, pour s et $\varepsilon>0$ on a, d'après l'inégalité de Markov

$$a_n = P\left(\frac{X_1+\ldots+X_n}{n} \geqslant \mathbb{E}(X_1)+\varepsilon\right) = P(\exp s(X_1+\ldots+X_n) \geqslant \exp sn(p+\varepsilon))$$

$$\leqslant \exp(-sn(p+\varepsilon))(q+p\exp s)^n = (\varphi_\varepsilon(s))^n .$$

En prenant s assez proche de 0, on a $0<\varphi_\varepsilon(s)<1$, et donc $\sum_{n=1}^{\infty} a_n<+\infty$, ce qui permet d'appliquer Borel-Cantelli et de conclure à $\overline{\lim}\ \frac{1}{n}(X_1+\ldots+X_n)\leqslant p$ presque sûrement.

Il est clair que si on remplace la loi de Bernouilli par une loi μ des X telle que μ est dans $\mathcal{M}(\mathbb{R})$ avec 0 dans $\Theta(\mu)$, la même technique est applicable pour montrer que $\lim_{n\to\infty} \frac{1}{n}(X_1+\ldots+X_n)=m=\mathbb{E}(X_1)$ presque-sûrement, c'est-à-dire que le cadre normal de cette démonstration est celui des familles exponentielles naturelles. En fait cette méthode montre que $\overline{\lim_{n\to+\infty}} \sqrt{a_n} <1$, et le théorème des grandes déviations est plus précis. Il affirme que si m et m_1 sont dans M_F, avec $m<m_1$, et si la loi des X_1 est $P(m,F)$, alors

$$a_n^{\frac{1}{n}} = P(\frac{1}{n}(X_1+\ldots+X_n) \geqslant m_1)^{\frac{1}{n}} \underset{n\to+\infty}{\longrightarrow} \exp - \int_m^{m_1} \frac{m_1-t}{V_F(t)} dt.$$

(c) <u>Inégalité de Hoeffding</u> (1963). Elle affirme que si X est une variable aléatoire de [-1,+1], alors pour θ réel on a

$$\mathbb{E}(\exp\theta(X-\mathbb{E}(X))) \leqslant \exp \frac{\theta^2}{2} .$$

Sa démonstration est typique des familles naturelles : on suppose $\theta>0$; si $k(\theta)=\operatorname{Log}\mathbb{E}(\theta(X-\mathbb{E}(x))$ alors $k(0)=k'(0)=0$ et $0\leqslant k''(\theta)\leqslant 1$ car $k''(\theta)$ est la variance de la loi : $\exp(\theta x-k(\theta)x$ Loi de $(X-\mathbb{E}(X))$. Donc

$$k'(\theta) = \int_0^\theta k''(s)ds\leqslant\theta \qquad \text{et} \qquad k(\theta) = \int_0^\theta k'(s)ds\leqslant\frac{\theta^2}{2}.$$

(d) <u>Convergence vers la loi de Poisson.</u> Soit F la famille exponentielle de Poisson, c'est-à-dire engendrée par

$$\mu(dx) = \sum_{k=0}^{\infty} \delta_k(dx)/k!,$$

où δ_k est la masse de Dirac en k. On peut être frappé par le fait que, à la différence du théorème central limite, le théorème de convergence de la loi binomiale vers la loi de Poisson semble un résultat erratique, isolé. Il consiste en ceci : si $m>0$ et si * désigne la convolution, alors

$$P_n(m) = \left(\left(1-\frac{m}{n}\right)\delta_0 + \frac{m}{n}\delta_1\right)^{*n} \text{ converge fortement vers } P(m,F).$$

Prohorov(1953) est plus précis et montre que

$$n\|P_n(m)-P(m,F)\|=n \sum_{k=0}^{\infty} \left| \binom{n}{k}(1-\frac{m}{n})^{n-k}(\frac{m}{n})^k - \frac{m^k}{k!}e^{-m} \right|$$

tend vers la fonction (de Prohorov) $\phi(m)$ définie par

$$\phi(m) = me^{-m}\left(\frac{m^{a}}{a!} - \frac{m^{a-1}}{(a-1)!} + \frac{m^{A-1}}{(A-1)!} - \frac{m^{A}}{A!}\right)$$

où les entiers a et A sont les fonctions de m définies par

$$a+\sqrt{a}\leqslant m<a+1+\sqrt{a+1} \qquad \text{et} \qquad A-\sqrt{A}\leqslant m<A+1 - \sqrt{A+1}.$$

En fait, si on introduit ν dans $\mathcal{M}(\mathbb{R})$ telle que ν soit concentrée sur l'ensemble des entiers >0, avec $\nu(\{j\})>0$ pour j=0 et 1, et si $G=F(\nu)$, on peut montrer que si m>0, alors

$$P_m(m) = (P(\frac{m}{n},G))^{*n} \quad \text{satisfait} \quad \lim_{n\to\infty} n\|P_n(m)-P(m,F)\| = V_G''(0)\phi(m)$$

C'est une conséquence de l'article de A. Barbour (1987). (En fait, on peut montrer que $M_G=]0,b[$ avec $0<b\leqslant+\infty$, mais que V_G est la restriction à M_G d'une fonction analytique dans un ouvert contenant 0 : cette remarque est nécessaire pour donner un sens à $V_G''(0)$). Le résultat classique rappelé avant devient le cas particulier de l'unique famille naturelle G concentrée sur {0,1}, par exemple engendrée par $\nu=\delta_0+\delta_1$.

§3. Puissances et affinités des familles naturelles.

Avant d'aborder les problèmes intéressants, considérons quelques définitions et points techniques. Si μ est dans $\mathcal{M}(E)$, désignons (avec B. Jorgensen (1987)) par $\Lambda(\mu)$ l'ensemble des nombres p>0 tels qu'il existe μ_p dans $\mathcal{M}(E)$ avec $\Theta(\mu_p)=\Theta(\mu)$ et $pk_\mu= k_{\mu_p}$. Il y aurait de bonnes raisons (voir la fin du §5) d'appeler $\Lambda(\mu)$ l'ensemble de Wallach de μ. Il est facile de voir que $\Lambda(\mu)$ est un semi-groupe additif fermé, que μ_p est unique et que, si $F_p = F(\mu_p)$, alors

$$M_{F_p} = p\, M_F ,$$

$$P(pm,F_p)*P(p'm,F_{\mu'}) = P((p+p')m,F_{p+p'})$$

et

$$V_{F_p}(m) = p\, V_F(\frac{m}{p}) \qquad (3.1)$$

Bien entendu, $\Lambda(\mu)$ contient $\mathbb{N}\setminus\{0\}$. S'il est égal à $]0,+\infty[$, la famille F est formée de probabilités indéfiniment divisibles (Un problème soulevé à cette conférence par une conversation avec Paul Ressel est le suivant : si Λ est un semi-groupe additif fermé de $]0,+\infty[$, existe-t-il μ dans $\mathcal{M}(E)$ tel que $\Lambda=\Lambda(\mu)$?).

Considérons ensuite l'image d'une famille naturelle F par une affinité φ de E dans un espace E' de même dimension, c'est-à-dire que $\varphi(x) = a_\varphi(x) + b_\varphi$, avec b_φ dans E' et a_φ isomorphisme linéaire de E dans E'. On voit alors facilement que si μ est dans $\mathcal{M}(E)$, si $\varphi\mu$ est son image par φ et $F=F(\mu)$ alors

$$\varphi\, P(m,F) = P(\varphi(m),\ \varphi(F)),$$

et $\varphi(F)=F(\varphi\mu)$. On a en outre l'importante formule : si ${}^t a_\varphi$ est l'isomorphisme de $(E')^*$ dans E^* transposé de a_φ, si la fonction variance $V_F(m)$ est considérée comme à valeurs dans les applications linéaires symétriques de E^* dans E, et si m est dans $M_{\varphi(F)} = \varphi(M_F)$, alors :

$$V_{\varphi(F)}(m) = a_\varphi \circ V_F(\varphi^{-1}(m)) \circ {}^t a_\varphi \tag{3.2}$$

En quelque sorte, les affinités de E dans E' définissent des morphismes de familles exponentielles. Plus généralement, on peut se demander s'il existe d'autres fonctions $\varphi : E \to E'$ non affines, et deux familles naturelles F et F' sur E et E' tels que $P \mapsto \varphi(P)$ définisse une application de F dans F'. Pour $F=F'=\{N_{m,1}\ ;\ m \in \mathbb{R}\}$, le problème est posé dans Letac (1988). Le résultat suivant n'est pas très important. Il donne une solution du problème qui n'est que partielle, puisqu'elle suppose $E'=M_{F'}$.

> <u>Théorème 3.1.</u> : Soit F et F' deux familles exponentielles naturelles sur deux espaces réels E et E' telle que pour tout P de F son image φP est dans F'. Alors, si $M_{F'} = E'$, il existe b dans E' et un isomorphisme a de E sur E' tels que si μ est dans $\mathcal{B}_F$, $x \mapsto \varphi(x)-a(x)-b$ est nulle μ presque partout.

<u>Démonstration</u> : Celle-ci est assez longue ; nous la détaillerons pour dim E=1 avec des indications pour le cas général. On suppose donc d'abord $E=E'=\mathbb{R}$.

Soit μ dans $\mathcal{B}_F$ et μ' dans $\mathcal{B}_{F'}$. Alors il existe $g : \Theta(\mu) \to \Theta(\mu')$ telle que $\varphi P(\theta,\mu) = P(g(\theta),\mu')$.
Donc si θ est dans $\Theta(\mu)$ et $s+g(\theta)$ est dans $\Theta(\mu')$ on a

$$\int_{\mathbb{R}} \exp(s\varphi(x)+\theta x)\mu(dx) = \exp(k_\mu(\theta) + k_{\mu'}(s+g(\theta)) - k_{\mu'}(g(\theta))) \tag{3.3}$$

Considérons alors la mesure ν dans $\mathbb{R}^2$ qui est l'image de μ par $x \mapsto (x,\varphi(x))$: on veut montrer que ν est concentrée sur un sous espace affine de $\mathbb{R}^2$ $\{(x,y)\ ;\ y=ax+b\}$, car ceci entraînera $\varphi(x)-ax-b$ μ-presque partout. La Transformée de Laplace de μ donnée par (3.3) et est finie sur $A_1=\{(\theta,0)\ ;\ \theta\in\Theta(\mu)\}$ et $A_2=\{(\theta_0,s)\ ;\ s+g(\theta_0)\in\Theta(\mu)\}$ où θ_0 est fixé dans $\Theta(\mu)$; $L_\nu(\theta,s)$ est donc finie sur le convexe de $\mathbb{R}^2$ engendré par $A_1 \cup A_2$, lequel est d'intérieur non vide. Donc si ν n'est pas concentré sur un sous-espace affine, on a donc ν dans $\mathcal{M}(\mathbb{R}^2)$; nous montrons que cette hypothèse conduit à une contradiction,

D'après (3.3) on a

$$k_\nu(\theta,s) = k_\mu(\theta) + k_{\mu'}(s+g(\theta)) - k_{\mu'}(g(\theta)) \qquad (3.4)$$

On en déduit que $g : \Theta(\mu)\to\mathbb{R}$ est analytique réelle. En effet

$$(\theta,s) \mapsto k_\nu(\theta,s) - k_\mu(\theta) = k_{\mu'}(s+g(\theta)) - k_{\mu'}(g(\theta))$$

est analytique sur $\Theta(\nu)$, donc en faisant $s\to 0$, on obtient que $\theta \mapsto k'_{\mu'}(g(\theta))$ est analytique sur $\Theta(\mu')$, ainsi que

$$\theta\mapsto g(\theta) = \psi_{\mu'} \circ k'_{\mu'}(g(\theta)).$$

Notons pour simplifier :

$$A = k''_\mu(\theta) - k''_{\mu'}(g(\theta))\,(g'(\theta))^2 + (k'_{\mu'}(s+g(\theta))-k'_{\mu'}(\theta))\,g''(\theta)$$

$$B = k''_{\mu'}(s+g(\theta)).$$

Alors la matrice de covariance de $P((\theta,s)\nu)$ est

$$C=\begin{bmatrix} A + B(g'(\theta))^2 & Bg'(\theta) \\ Bg'(\theta) & B \end{bmatrix}$$

Comme $B>0$ et $\det C>0$, cela entraine $A>0$. Nous distinguons alors 2 cas :

- ou bien il existe θ_0 dans $\Theta(\mu)$ tel que $g''(\theta_0)\neq 0$. Dans ce cas, soit :

$$A_0 = k''_\mu(\theta_0) - k''_{\mu'}(g(\theta_0))(g'(\theta_0))^2.$$

A ce point, <u>utilisons l'hypothèse</u> $M_{F'} = \mathbb{R}$ pour dire que

$$s\mapsto\lambda(s) = k'_{\mu'}(s+g(\theta_0)) - k'_{\mu'}(\theta_0)$$

a pour image $\mathbb{R}$ tout entier. Donc pour $\theta=\theta_0$, $A=A_0+\lambda\, g''(\theta_0) >0$ pour tout λ de $\mathbb{R}$, ce qui est impossible.

- ou bien $g''(\theta)=0$ pour tout θ de $\Theta(\mu)$, et donc il existe a et b réels tels que $g(\theta)=a\theta+b$ si θ est dans $\Theta(\mu)$. Si $a=0$, (3.3) montre que ν est la mesure produit $\nu=\mu\otimes P(b,\mu')$, ce qui est incompatible avec le fait que ν soit la mesure image de μ par $x\mapsto(x,\varphi(x))$. Si $a\neq 0$, on considère la mesure μ'' définie par

$$\mu''(dy) = \exp\left(\frac{by}{a}\right)(H_a\mu')(dy),$$

où $H_a(y) = ay$. Le calcul à partir de cette définition et (3.3) donne alors

$$\int_{\mathbb{R}} \exp(sa\ \varphi(x) + \theta(x-a\varphi(x)))\mu(dx) = \exp(k_{\mu''}(s)+k_{\mu}(\theta)-k_{\mu''}(\theta)).$$

Si alors ν' est l'image de μ par l'application $x \mapsto (x-a\varphi(x),\ a\varphi(x))$, on voit donc que

$$k_{\nu'}(\theta,s) = k_{\mu}(\theta) - k_{\mu''}(\theta) + k_{\mu''}(s),$$

ce qui montre que ν' est une mesure produit et donne la contradiction.

Envisageons le cas général $E=E'=\mathbb{R}^d$. Il existe $g:\Theta(\mu)\to\Theta(\mu')\subset\mathbb{R}^d$ tel que $\varphi P(\theta,\mu) = P(g(\theta),\mu')$. Pour avoir la matrice de covariance de $P(\theta,s),\nu)$, on calcule la matrice hessienne de $k_{\mu'}(g(\theta))$:

$$^t g'(\theta)\ k''_{\mu'}(g(\theta))\ g'(\theta) + \sum_i \frac{\partial k_{\mu'}}{\partial \alpha_i}\ (g(\theta)g''_i(\theta),$$

les matrices (d,d) :

$$A=k''_{\mu'}(\theta) - {}^t g'(\theta)\ k''_{\mu'} g'(ac) = \sum_i \left(\frac{\partial k_{\mu'}}{\partial \alpha_i}\ (s+g(\theta)) - \frac{\partial k_{\mu'}}{\partial \alpha_i}(g(\theta))\right) g''_i(\theta),$$

$$B = k''_{\mu'}(s+g(\theta)) = {}^t B,$$

et la matrice de covariance C dans $\mathbb{R}^{2d}$ de $P(\theta,s),\nu)$:

$$C = \begin{bmatrix} A+g'(\theta)\ B\ {}^t g'(\theta) & g'(\theta)B \\ \\ {}^t B\ {}^t g'(\theta) & B \end{bmatrix}$$

Comme C est définie positive on en déduit que A est définie positive en formant $({}^t x, {}^t y)C\binom{x}{y}$, avec $y= -{}^t g'(\theta)x$. On utilise alors $M_{F'} = \mathbb{R}^d$ pour montrer que

$$s \mapsto \sum_C \left(\frac{\partial k_{\mu'}}{\partial \alpha_i}(s+g(\theta)) - \frac{\partial k_{\mu'}}{\partial \alpha_i}\ g(\theta))\right)\ {}^t x\ g''_i(\theta)\ x$$

a pour image $\mathbb{R}$ si un des $g''_i(\theta)$ n'est pas nul, ce qui contredit la définie-positivité de A. Chacun des g_i est donc affine, et on conclut comme dans le cas d=1. □

§4. Les familles naturelles dans $\mathbb{R}$ à fonction variance quadratique.

Il convient de citer maintenant les articles de Morris (1982) et

(1983) comme la source des questions que nous considérons dans toute la suite. Malgré un ou deux à-peu-près, il me semble que c'est une des lectures les plus intéressantes de la statistique des dernières années.

	F	μ	M_F	V_F
(1)	Normales de variance 1	$N_{0,1}$	$\mathbb{R}$	1
(2)	Poisson	$\sum_{k=0}^{\infty} \delta_k/k!$	$]0,+\infty[$	m
(3)	Bernouilli	$\delta_0+\delta_1$	$]0,1[$	$m-m^2$
(4)	Géométrique	$\sum_{k=0}^{\infty} \delta_k$	$]0,+\infty[$	$m+m^2$
(5)	Exponentielle	$\mathbb{1}_{]0,+\infty]}(x)\ dx$	$]0,+\infty[$	m^2
(6)	Cosinus-Hyperbolique	$(\mathrm{ch}\ \frac{\pi x}{2})^{-1} dx$	$\mathbb{R}$	$1+m^2$

Pour simplifier, nous dirons que la fonction variance d'une famille naturelle F sur $\mathbb{R}$ est <u>quadratique</u> si il existe un polynôme P de degré ≤ 2 tel que $V_F(m)=P(m)$ pour tout m de M_F. Dans les articles cités, Carl N. Morris observe que six familles exponentielles naturelles sur $\mathbb{R}$ très familières ont une fonction variance quadratique. Voici ces six familles décrites au Tableau ci-dessus, chacune par une mesure μ qui l'engendre.

L'application des formules 3.1 et 3.2 (notez que pour la famille (3) $\Lambda(\mu) = \mathbb{N}\setminus\{0\}$ et que les autres sont indéfiniment divisibles) permet de fabriquer à partir de ce tableau d'autres fonctions variances quadratiques. Morris montre alors qu'on obtient ainsi <u>toutes</u> les fonctions variances quadratiques (sur ce point, on peut consulter Letac-Mora (1989) pour une démonstration plus détaillée). Laissons de côté les déformations par affinité, assez triviales dans $\mathbb{R}$, et considérons le passage $F \to F_p$ pour les F du tableau. On obtient ainsi

(1) $F_p = \{N_{m,p}\ ;\ m \in \mathbb{R}\}$

(2) $F = F_p$

(3) si p entier >0

F_p est la famille des binomiales sur $\{0,\ldots,p\}$.

(4) F_p est la famille négative binomiale engendrée par

$$\sum_{k=0}^{\infty} p(p+1)\dots(p+k-1)\delta_k/k!$$

(5) F_p est une famille gamma engendrée par $x^{p-1}\mathbb{1}_{]0,+\infty[}(x)\,dx$

(6) F_p est engendrée par $\mu_p(dx) = |\Gamma(\frac{p}{2} + i\,\frac{x}{2})|\,dx$.

Il convient de rapprocher de ce résultat de Morris un théorème de Philip Feinsilver (1986) sur les fonctions variances quadratiques dont nous allons donner une autre démonstration.

Avant d'énoncer ce théorème, introduisons quelques notations. Soit F une famille exponentielle naturelle sur $\mathbb{R}$, m_0 dans M_F fixé et $\mu=P(m_0,F)$. Notons $\psi(m) = \psi_\mu(m)$ et $\psi_1(m) = k_\mu(\psi(m))$. Par conséquent

$$\psi(m) = \frac{1}{V_F(m)} \qquad \text{et} \qquad \psi'_1(m) = \frac{m}{V_F(m)} \tag{4.1}$$

Notons

$$f(x,m) = \exp(x\,\psi(m) - \psi_1(m)).$$

On a donc

$$P(m,F)\,(dx) = f(x,m)\,\mu(dx).$$

Puisque ψ et ψ_1 sont analytiques réelles au voisinage de m_0, il existe donc $r>0$ et des fonctions $P_n : \mathbb{R} \to \mathbb{R}$ telles que pour $|m-m_0| < r$ et pour tout x on ait

$$f(x,m) = \sum_{n=0}^{\infty} (m-m_0)^n\,P_n(x). \tag{4.2}$$

Si $J_n : \mathbb{R}\times M_F \to \mathbb{R}$ est défini par

$$J_n(x,m) = (f(x,m))^{-1}\left(\frac{\partial}{\partial m}\right)^n f(x,m),$$

on voit facilement que $J_0(x,m)=1$, que $P_n(x)=J_n(x,m_0)$ et que

$$J_{n+1}(x,m) = \frac{\partial}{\partial m} J_n(x,m) + \frac{x-m}{V_F(m)} J_n(x,m).$$

Ceci permet de bâtir facilement la démonstration par récurrence du fait que que les P_n sont des polynômes de degré n.

<u>Théorème 4.1.</u> (Feinsilver 1986). Soit F une famille exponentielle naturelle sur $\mathbb{R}$, m_0 dans M_F fixé, $\mu=P(m_0,F)$ et les $(P_n)_{n=0}^{\infty}$ définis comme ci-dessus. Alors les $(P_n)_{n=0}^{\infty}$ sont

une <u>famille orthogonale de $L^2(\mu)$</u> si et seulement si la fonction variance de F est quadratique.

<u>Démonstration</u> [⇒] Sans perte de généralité nous supposons $m_0=0$; soit $r>0$ tel que $f(x,m) = \sum_{n=0}^{\infty} m^n P_n(x)$ $\forall (m,x)$ dans $]-r,r[\times\mathbb{R}$. Le choix de μ entraine que $\psi(0)=0\in\Theta(\mu)$. Il existe donc r_1 dans $]0,r]$ tel que $|m|<r_1$ implique $\psi(m)$ dans $\frac{1}{2}\Theta(\mu)$ et donc si $|m|$ et $|m'|<r_1$, $f(x,m)\, f(x,m')$ est μ-intégrable. D'après l'hypothèse, on a donc pour $|m|<r_1$ et $|m'|<r_1$

$$\int_{\mathbb{R}} f(x,m)\, f(x,m')\, \mu(dx) = \sum_{n=0}^{\infty} (mm')^n \int_{\mathbb{R}} (P_n(x))^2\, \mu(dx) \qquad (4.3)$$

Notons $I =]-r_1^2, r_1^2[$; (4.3) implique qu'il existe $F_1 : I \to \mathbb{R}$ analytique réelle telle que

$$k_\mu(\psi(m) + \psi(m')) - \psi_1(m) - \psi_1(m') = F_1(mm'). \qquad (4.4)$$

Appliquons $\frac{\partial^2}{\partial m \partial m'}$ aux deux membres de (4.4) dans le carré

$C = \{(m,m') \; ; \; |m|<1, \text{ et } |m'|<r_1\}$. Il existe donc une fonction $F_2 : I\to\mathbb{R}$ analytique réelle telle que, pour (m,m') dans C on ait

$$k''_\mu(\psi(m)+\psi(m'))\psi'(m)\psi'(m') = F_2(mm'). \qquad (4.5)$$

Notons pour simplifier $V=V_F$.
Introduisons alors $g:C\to M_F$ définie par

$$\int_0^{g(m,m')} \frac{dt}{V(t)} = \int_0^{m} \frac{dt}{V(t)} + \int_0^{m'} \frac{dt}{V(t)},$$

c'est-à-dire que

$$\psi(g(m,m')) = \psi(m) + \psi(m')$$

et que

$$g(m,m') = k'_\mu(\psi(m)+\psi(m')).$$

Si on rappelle que $k''_\mu(\theta)$ est la variance de $P(\theta,\mu)$, on obtient que

$$k''_\mu(\psi(m) + \psi(m')) = V(g(m,m')). \qquad (4.6)$$

Prenant le logarithme du 1er membre de (4.5), et utilisant (4.1) et (4.6), on voit qu'il existe $F_3 : I\to\mathbb{R}$ tel que pour tous (m,m') de C on ait

$$\text{Log } V(g(m,m')) - \text{Log } V(m) - \text{Log } V(m') = F_3(mm') \tag{4.7}$$

Appliquons alors l'opérateur $m \frac{\partial}{\partial m} - m' \frac{\partial}{\partial m'}$ aux deux membres de (4.7), on obtient, dans C :

$$m \frac{V'(g)}{V(g)} \frac{\partial g}{\partial m} - \frac{m\, V'(m)}{V(m)} = m' \frac{V'(g)}{V(g)} \frac{\partial g}{\partial m'} - m' \frac{V'(m')}{V(m')} \tag{4.8}$$

On tire de la définition de g que $\frac{1}{V(g)} \frac{\partial g}{\partial m} = \frac{1}{V(m)}$, ce qui permet de simplifier (4.8) :

$$m \frac{V'(g) - V'(m)}{V(m)} = m' \frac{V'(g) - V'(m')}{V(m')} \tag{4.9}$$

Dans (4.9), faisons $m' \to 0$. Puisque $g(m,m') \to m$, que

$$\frac{V'(g(m,m')) - V'(m)}{m'} \to V''(m) \frac{\partial g}{\partial m'}(m,0)$$

et que $\frac{1}{V(m')} = \frac{1}{V(g)} \frac{\partial g}{\partial m'} \to \frac{1}{V(0)}$, (4.9) fournit

$$m\, V''(m) = V'(m) - V'(0),$$

qui entraine que la restriction de V à $]-r_1, r_1[$ est un polynôme du second degré. Puisque V est analytique dans M_F, le résultat est montré.

[⇐] Cette partie est bien détaillée dans l'article de Morris (1982) (Théorème 4). ■

Il y a alors plusieurs manières d'étendre le résultat de Morris. D'abord en étudiant les fonctions-variances cubiques sur $\mathbb{R}$, c'est-à-dire celles qui sont restriction à M_F d'un polynôme de degré 3. C'est ce qui est fait par Marianne Mora(1986), qui en donne une classification complète : plus de détails dans Letac-Mora(1989). On pourrait chercher à poursuivre, avec des polynômes de degré 4 ou plus. Cependant il y a à mon avis peu de chances de réussir une classification aussi détaillée que pour les degrés ≤ 3, avec calcul explicite des lois. Une obstruction à ce programme est dans le phénomène de réciprocité, décrit dans Letac-Mora(1989) et Letac(1986). C'est un procédé réflexif, défini sur une large classe de familles naturelles, qui à F associe une famille réciproque F', telle que $m^3 V_{F'}(1/m) = V_F(m)$. Si F est polynomial de degré ≤ 3, il en va de même pour F'; la remarque est aussi valable pour l'ensemble des fonctions variance de la forme $P+Q\sqrt{R}$ où P,Q et R sont des polynômes de degré ≤ 3, ≤ 2, ≤ 2, respectivement. De nombreuses familles courantes sont de ce genre, et leur classification est un problème ouvert dont la solution généralisera directement la classification de Mora- Morris.

Signalons dans ce domaine un beau résultat de Bar-Lev et Bshouty(1989), qui affirme que si $]a,b[= M_F$ est borné et si V_F est la restriction à $]a,b[$ d'une fonction méromorphe f dans un ouvert de $\mathbb{C}$ contenant $[a,b]$ telle que $f(a)=f(b)=0$, alors F n'est autre que l'affinité d'une binomiale, c'est-à-dire qu'il existe un entier $p>0$ tel que $V_F(m) = \frac{1}{p}(m-a)(b-m)$ sur $]a,b[$. On peut aussi facilement étudier les variances de la forme Am^a, avec $M_F \supset]0,+\infty[$; c'est "l'échelle de Tweedie" : détails dans Barlev et Enis(1987).

Cependant, on peut chercher à généraliser le résultat de Morris en dimensions supérieures. C'est ce qui nous occupe maintenant.

§5. La classification dans $\mathbb{R}^d$ des familles naturelles de fonction variance quadratique.

Le problème quadratique en dimensions supérieures se pose ainsi dans $\mathbb{R}^d$: quelles familles exponentielles naturelles dans $\mathbb{R}^d$ sont telles qu'il existe des matrices réelles (d,d) symétriques A_{ij}, B_j, C telles que, pour m dans M_F :

$$V_F(m) = \sum_{t,j=1}^{d} A_{ij} m_i m_j + \sum_{j=1}^{d} B_j m_j + C \ ?$$

Avec une approche différente, le problème est déjà mentionné dans Feinsilver(1985).

On peut présenter le problème sans coordonnées, en considérant E à la place de $\mathbb{R}^d$, $\mathbb{Q}(E^*)$ les formes quadratiques sur E^* et enfin

- une application bilinéaire symétrique A de ExE dans $\mathbb{Q}(E^*)$.
- une application bilinéaire B de E dans $\mathbb{Q}(E^*)$
- un élément C de $\mathbb{Q}(E^*)$,

et chercher la famille exponentielle naturelle F sur E - si elle existe -telle que pour tout m de M_F on ait

$$V_F(m) = A(m,m) + B(m) + C \tag{5.1}$$

Nous allons maintenant traiter complètement le cas A=0 (assez trivial : Théorème 5.1), donner des exemples pour $A\neq 0$ et formuler une conjecture pour le cas B=0 et C=0, qui ressemble au cas gamma (n° 5 dans $\mathbb{R}$).

Pour traiter les problèmes à plusieurs dimensions, empruntons le vocabulaire des représentations de groupe. On dira que μ de $\mathbb{M}(E)$ est <u>réductible</u> si E est décomposable en somme-directe $E_1 \oplus E_2$ (il n'y a pas de structure euclidienne ici) telle qu'il existe μ_j dans $\mathbb{M}(E_j)$ j=1,2 avec $\mu=\mu_1 \otimes \mu_2$; on dira alors que $F=F(\mu)$ et que V_F sont réductibles (notons qu'alors tous les éléments de $\mathcal{B}_F$ sont réductibles). Dans le cas opposé on dira que μ, F et V_F sont irréductibles. Il est bien clair que le problème de classification des fonctions variance sur E qui sont quadratiques, c'est-à-dire de la forme (5.1), est celui de la recherche des fonctions variances quadratiques irréductibles dans toutes dimensions. Par exemple si $V_F(m)=C$ (cas A=0 et B=0) et dimE>1, alors F est une famille normale de E de covariance C : elle est réductible, car C est diagonalisable. Voici maintenant le principal résultat de l'exposé :

Théorème 5.1. : Soit F une famille exponentielle de E telle qu'il existe une application linéaire B de E dans $\mathcal{Q}(E^*)$, et C dans $\mathcal{Q}(E^*)$ tels que pour tout m de M_F on ait

$$V_F(m) = B(m) + C. \tag{5.2}$$

Alors il existe une base $e = (e_1,\dots,e_d)$ de E telle que, en notant $m = \sum_{j=1}^{d} m_j e_j$, et $e^* = (e_1^*,\dots,e_d^*)$ la base duale de E^*, il existe des réels $(b_j, c_j)_{j=1}$ tels que la matrice représentative de $V_F(m)$ dans la base (e^*, e) soit diagonale et égale à :

$$[V_F(m)]^e_{e^*} = \begin{bmatrix} b_1 m_1 + c_1 & 0 & \dots & 0 \\ 0 & b_2 m_2 + c_2 & \dots & 0 \\ \dots & \dots & \dots & \dots \\ 0 & 0 & \dots & b_1 m_1 + c_d \end{bmatrix}$$

En outre si $I_j = \{t \in \mathbb{R} \; ; \; b_j t + c_j > 0\}$, I_j est non vide et $M_F = \prod_{j=1}^{d} I_j e_j$, et la famille F est un produit de lois normales et d'affinités de lois de Poisson à 1 dimension.

Corollaire 5.2. Si dim E>1, il n'y a pas de fonctions variances irréductibles de la forme (5.2).

Démonstration du Théorème : Au prix d'une translation éventuelle, nous supposons que 0 est dans M_F, et donc $V_F(0)=C$ est définie positive ; on munit alors E^* de la structure euclidienne induite par C, c'est-à-dire du produit scalaire sur E^* $\langle\alpha|\beta\rangle = \langle\alpha, C(\beta)\rangle$.
(Rappelons qu'un élément C de $\mathcal{Q}(E^*)$ est identifié à la forme bilinéaire symétrique polaire sur $E^* \times E^*$, elle même identifiée à une application linéaire symétrique $\beta \mapsto C(\beta)$ de E^* dans E). On identifie alors canoniquement E^* et E dans cette structure euclidienne, B(m) devient un endomorphisme symétrique de l'espace euclidien E, et $V_F(m)$ est un endomorphisme symétrique défini positif de E égal à, pour m dans M_F :

$$V_F(m) = B(m) + i_E,$$

où i_E est l'endomorphisme identique de E. Soit alors μ fixé dans $\mathcal{M}(E)$ tel que $F=F(\mu)$. Nous allons noter ψ au lieu de ψ_μ pour simplifier. On écrit alors $\psi'_m(x)$ pour sa différentielle en m évaluée en x de E, et $\psi''_m(x,y)$ sa différentielle seconde évaluée en x,y de E. Rappelons que

$$\psi''_m(x,y) = \psi''_m(y,x).$$

On a donc $\psi'_m(x) = (i_E+B(m))^{-1}(x)$ et, en dérivant par rapport à m :

$$\psi''_m(x,y) = -(i_E+B(m))^{-1}B(x)\ (i_E+B(m))^{-1}(y) \tag{5.2}$$

(Ici on a utilisé le fait que $m\mapsto B(m)$ est linéaire, ainsi que la formule de dérivation de l'inverse d'un endomorphisme :

$$\frac{d}{dt}(A(t))^{-1} = -(A(t))^{-1}\left(\frac{d}{dt}A(t)\right)(A(t))^{-1}.)$$

Introduisons ensuite $u = (i_E+B(m)^{-1}x$ et $v=(i_E+B(m))^{-1}y$.
Puisque l'expression (5.2) doit être symétrique en x et y, on en tire :

$$B(u+B(m)(u))(v) = B(v+B(m)(v)(u) \tag{5.3}$$

Faisant m=0 (qui est dans M_F) on obtient

$$B(u)(v) = B(v)u \qquad \forall(u,v) \text{ dans } E^2 \tag{5.4}$$

En utilisant la linéarité de B dans (5.3), on a d'après (5.4) :

$$B(B(m)(u))(v) = B(B(m)(v))(u)$$

$$B(v)B(m)(u) = B(u)B(m)(v)$$

$$B(v)B(u)(m) = B(u)B(v)(m).$$

Donc B(u) et B(v) commutent. Comme ce sont des endomorphismes symétriques, donc diagonalisables, il existe une base commune de diagonalisation $e=(e_1,\dots,e_d)$ de la famille $\{B(u)\ ;\ u\in E\}$. Il existe donc des réels $b_1,\dots,b_d$ tels que si m est dans M_F

$$[V_F(m)]_e^e = \begin{bmatrix} 1+b_1m_1 & \dots & 0 \\ \dots & \dots & \dots \\ 0 & \dots & 1+b_dm_d \end{bmatrix}$$

Introduisons alors $I_j = \{t;1+b_jt>0\}$: $I_j \neq \emptyset$ puisque $V_F(m)$ est définie-positive. $I_j = \mathbb{R}$ si $b_j = 0$ et I_j est une demi droite si $b_j \neq 0$.

Pour voir que $M_F = \prod_{j=1}^{d} I_j e_j$, considérons la famille exponentielle F_j sur $\mathbb{R}$, qui est une famille normale $\{N_{m1}, m\in\mathbb{R}\}$ si $b_j=0$, et qui est l'image d'une famille de Poisson standard (Ligne 2 du tableau de Morris) par $x\mapsto\varphi(x) = b_j x - \frac{1}{b_j}$ si $b_j\neq 0$. Par (3.2) on voit que $M_{F_j} = I_j$ et $V_{F_j}(m) = 1+b_j m$. Soit alors F' la famille produit sur E des F_j. Alors $V_F(m) = V_{F'}(m)$ sur un voisinage de 0. D'après la remarque de la fin du §1, on a F=F'. La forme indiquée dans l'énoncé, avec les c_j à la place de 1, apparait quand on fait une translation correspondant au cas où $0\notin M_F$. □

Nous donnons maintenant 7 exemples de fonctions variances quadratiques en dimension >1 qui soient irréductibles. Elles sont copiées des cas 3,4 et 5 du tableau de Morris du §4. On imite le cas 3 en partant d'un tétraèdre de E à d+1 sommets $e_0, e_1 \ldots e_d$ et en considérant la mesure $\mu = \sum_{j=0}^{d} \delta_{e_j}$. Prenant $e_0=0$ et prenant $e=(e_1,\ldots,e_d)$ pour base de E, il est plus simple ici de travailler dans $\mathbb{R}^d$, avec $e_j = {}^t(0,\ldots,1,0.0)$. On obtient ainsi une famille F concentrée sur $\mathbb{N}^d$, dont les puissances F_p, avec p entier, seront les lois multinomiales. Ensuite, pour passer au cas 4, on forme $\mu=(\delta_0 - \delta_{e_1} - \ldots - \delta_{e_d})^{*-1}$, et la famille $F(\mu)$ donnera un analogue des lois géométriques, la famille F_p, avec p>0 donnera un analogue des lois négatives binomiales de paramètre p. Pour énoncer le résultat concernant ces deux familles, introduisons la notation suivante : si $m\in E$, on note $m\otimes m$ l'élément de $\mathbb{Q}(E^*)$ défini par $\theta\mapsto(\langle\theta,m\rangle)^2$.

<u>Proposition 5.2.</u> : Soit $E=\mathbb{R}^d$ muni de sa base canonique $e=e_1,\ldots e_d)$

(1) Soit $\mu = \delta_0 + \sum_{j=1}^{d} \delta_{e_j}$ et $F=F(\mu)$. Alors $M_F=\{m; \forall j, m_j>0, \Sigma m_j<1\}$

$$P(m,F)(e_j) = m_j \quad , \quad P(m,F)(0) = 1 - \sum_{j=1}^{d} m_j ,$$

$$\text{et } [V_F(m)]_e^e = \begin{bmatrix} m_1 - m_1^2 & & - m_1 m_d \\ \ldots & \ldots\ldots & \ldots\ldots \\ m_d m_1 & & m_d - m_d^2 \end{bmatrix} = \mathrm{Diag}(m) - m\otimes m.$$

Si $p\in\Lambda(\mu) = \mathbb{N}\setminus\{0\}$ alors $M_{F_p} = \{m; \forall j\ m_j>0 \text{ et } \Sigma\, m_j<p\}$

$$P(m,F_p)(n_1e_1+\ldots+n_de_d) = \begin{pmatrix} & p & \\ n_0 & n_1 & n_d\end{pmatrix}\left(1 - \frac{\Sigma\, m_j}{p}\right)^{n_0} \sum_{j=1}^{n} (\frac{m_j}{p})^{n_j}$$

avec $n_0,\ldots n_d$ entiers ≥ 0 tels que $n_0+\ldots n_d = p$, et

$$\left[V_{F_p}(m)\right]_e^e = \mathrm{Diag}(m) - \frac{1}{p}\, m\otimes m. \qquad (5.5)$$

(2) Soit $\mu = \delta_0 + \sum_{n=1}^{\infty}\left(\sum_{j=1}^{d} \delta_{e_j}\right)^{*n}$ et $F=F(\mu)$. Alors $M_F=\{m;m_j>0\forall j\}$

et, en notant $S=m_1+\ldots+m_d$, si $n_1\ldots n_d$ sont dans $\mathbb{N}$:

$$P(m,S)(n_1e_1+\ldots+n_de_d) = \frac{1}{1+S}\left(\frac{m_1}{1+S}\right)^{n_1}\cdots\left(\frac{m_d}{1+S}\right)^{n_d}\binom{n_1+\ldots+n_d}{n_1,\ldots,n_d},$$

et $$[V_F(m)]_e^e = \mathrm{Diag} m + m\otimes m. \qquad (10.6)$$

Si $p\in\Lambda(\mu) =]0,+\infty[$, alors $M_{F_p} = M_F$ et

$$V_F(m) = \mathrm{Diag}(m) + \frac{1}{p}\, m\otimes m. \qquad (5.7)$$

Démonstration : Le (1) est assez trivial par les méthodes élémentaires. Traitons le (2) en détail à titre d'exemple. Adoptons les notations suivantes : si $\theta={}^t(\theta_1,\ldots,\theta_d)\in\mathbb{R}^d$, on note $e^\theta = {}^t(e^{\theta_1},\ldots,e^{\theta_d})$ et $\sigma = \Sigma\, e^{\theta_j}$.
Donc pour μ comme au (2) on a $L_\mu(\theta) = (1-\sigma)^{-1}$ sur $\Theta(\mu) = \{\theta;\theta_j<0\}$, $k_\mu(\theta) = -\mathrm{Log}(1-\sigma)$, et $m=k'_\mu(\theta) = \frac{e^\theta}{1-\sigma}$ dans $M_F=\{m;m_j>0\ \forall j\}$. Si m est dans M_F on note $\mathrm{Log}\, m = {}^t(\mathrm{Log}\, m_1,\ldots \mathrm{Log}\, m_d)$ et $S=\Sigma\, m_j$. Puisque $S = \frac{\sigma}{1-\sigma}$, on obtient donc $1-\sigma = 1/(1+S)$, $e^\theta=m/(1+S)$ et donc :

$$\psi_\mu(m) = \mathrm{Log}\left(\frac{m}{1+S}\right) \quad \text{si } m\in M_F.$$

Le calcul de $(\psi'_\mu(m))^{-1} = V_F(m)$ conduit alors à (5.6). Ensuite $k_\mu(\psi_\mu(m)) = \mathrm{Log}(1+S)$ et, puisque

$$P(m,F)(dx) = \exp(\langle\psi_\mu(m),x\rangle - k_\mu(\psi_\mu(m))\ \mu(dx).$$

on obtient, pour $n_1,\ldots,n_d$ dans $\mathbb{N}$:

$$P(m,F)(n_1e_1+\ldots+n_de_d) = \frac{1}{1+S}\left(\frac{m_1}{1+S}\right)^{n_1}\ldots\left(\frac{m_d}{1+S}\right)^{n_d}\mu(n_1,\ldots,n_d)$$

D'après la définition de μ, on a $\mu(n_1,\ldots,n_d) = \binom{n_1+\ldots+n_d}{n_1,\ldots,n_d}$ et on obtient l'expression annoncée. La formule (5.7) s'obtient à partir de (5.6) et de (3.2). On pourrait faire le calcul explicite de $P(m,F_p)$ en imitant la méthode précédente.

On voit donc que les formules (5.5) et (5.7) fournissent des fonctions variances quadratiques irréductibles dans l'espace de dimension d l'irréductibilité est évidente à partir de la définition des μ qui les engendrent). L'analogie avec le tableau de Morris, qui consiste à remplacer m par $m\otimes m$ dans les lignes 3 et 4, est évidente. Elle a ses limites, car il est facile de voir que pour la ligne 5, passer de m^2 à $m\otimes m$ ne donne qu'une forme quadratique positive non définie et que pour la ligne 6, passer de $1+m^2$ à $I+m\otimes m$ ne donne pas une fonction variance dans $E=\mathbb{R}^d$; on vérifie cette affirmation en observant que la dérivée de $m\mapsto(I+m\otimes m)^{-1}$ n'a pas la symétrie d'une Hessienne, c'est-à-dire que si $f_m(\theta) = (I+m\otimes m)^{-1}(\theta)$, alors $f'_{m_0}(\theta,\eta)$ n'est pas symétrique en θ et η.

En fait l'analogie avec le (5), le cas gamma $\frac{m^2}{p}$, existe mais est plus subtile. Je renvoie aux articles Letac(1988 aet b) pour plus de détails sur le rôle joué par l'invariance du cas gamma par le groupe des homothéties positives de $\mathbb{R}$, et comment cette idée conduit à une bonne généralisation. Je me contenterai d'indiquer ici brièvement les résultats, qui conduisent à 5 sortes de généralisations du cas gamma. Pour cela je rappelle qu'un cône symétrique est un cône C saillant ouvert convexe de l'espace euclidien E tel que

(1) $C = \{x\in E\ ;\ \langle x,y\rangle >0\ \forall y\in C\}$

(2) Si G(E) est le groupe des automorphismes de E et si G est l'ensemble des g de G(E) tels que la restriction de g à C soit une permutation de C, alors G agit transitivement sur C (c'est-à-dire que pour tous x et y de C il existe g dans G tel que gx=y).

Un cône symétrique est dit réductible s'il est le produit de deux cônes symétriques, et irréductible sinon. Il y a, à isomorphisme linéaire près, 5 types de cônes symétriques irréductibles :

(1) Le cône de révolution dans $\mathbb{R}^{m+1}$, avec $m\geq 2$, défini par

$C - \left\{(x_0,x_1,\ldots,x_m)\ ;\ x_0^2 - x_1^2 - \ldots - x_m^2>0 \text{ et } x_0>0\right\}$, de dimension $n=m+1$

(2) Les matrices (m,m) réelles définies positives (dimension $n=m(m+1)/2$)

(3) Les matrices (m,m) complexes hermitiennes définies-positives (dimension $n=m^2$)

(4) Les matrices (m,m) de quaternions à symétrie hermitiennes définies positives (dimension $n=2m^2-m$).

(5) Les matrices (3,3) d'octonions à symétrie hermitienne définies positives (dimension n=27).

A un cône symétrique irréductible on associe deux entiers r et d et une fonction $x \mapsto \det(x)$ de C dans $\mathbb{R}$ tels que

$$\int_C \exp(-\langle\theta,x\rangle)(\det x)^{p-\frac{n}{r}} dx = \Gamma_c(p)\ (\det \theta)^{-p}$$

si $p>(r-1)\frac{d}{2}$ (ici $\Gamma_c(p)$ est une certaine constante). Dans ces conditions $\mu_p(dx) = (\det x)^{p-\frac{n}{r}} \mathbb{1}_c(x) \dfrac{dx}{\Gamma_c(p)}$ engendre une famille exponentielle $F(\mu_p)$. Si on note

$\Lambda_c = \left\{\frac{d}{2},\ldots,\frac{d}{2}(r-1)\right\} \cup \left]\frac{d}{2}(r-1),+\infty\right[$, (appelé ensemble de Wallach) un important résultat de Gyndikin (1975) est que il existe μ_p dans $\mathcal{M}(E)$ tel que $\Theta(\mu_p) = -C$ et $L_\mu(\theta) = (\det \theta)^{-p}$ si et seulement si p est dans Λ_c. Si $p\in\left\{\frac{d}{2},\ldots,\frac{d}{2}(r-1)\right\}$, μ_p n'a plus l'expression explicite mentionnée mais est singulière par rapport à la mesure de Lebesgue.

Ces familles $F(\mu_p)$ pour p dans Λ_c, sont en fait très familières dans le cône n°2 : ce sont les lois de Wishart réelles. Les cas n°3 et n°4 existent dans la littérature, ce sont les lois de Wishart complexes (voir Goodman(1962) et quaternioniques (voir S. Andersson(1975) le cas n°1 est étudié dans Letac (1988b)).
Venons-en alors aux fonctions variances irréductibles associées : elles sont quadratiques. On obtient pour chaque cas :

(1) d=1, r=2

$$V_{F(\mu_p)}(m)(\theta) = \frac{1}{2p}\ (2(\theta_0 m_0+\theta_1 m_1+\ldots+\theta_m m_m)m +$$

$$(m_0^2 - m_1^2 - \ldots - m_m^2)\ {}^t(-\theta_0, \theta_1, \ldots, \theta_m)).$$

(2) d=1, r=m

$$V_{F(\mu_p)}(m)(\theta) = \frac{1}{p}\, m\theta m \text{ (produit ordinaire des matrices réelles)}$$

(3) d=2, r=m

$$V_{F(\mu_p)}(m)(\theta) = \frac{1}{p}\, m\theta m \text{ (produit de matrices complexes)}$$

(4) d=4, r=m

$$V_{F(\mu_p)}(m)(\theta) = \frac{1}{p}\, m\theta m \text{ (produit de matrices de quaternions)}$$

(5) d=8, r=3

$V_{F(\mu_p)}(m)(\theta) = \frac{1}{2p}(m(\theta m)+(m\theta)m)$ (produit de matrices (3,3) d'octonions).

Il y a quelques remarques à faire sur ces formules, en dépit de la ressemblance formelle des variances dans les cas 2 et 3, ils sont naturellement distincts si on explicite en termes d'applications linéaires sur des espaces vectoriels réels. En outre, en basses dimensions les cas (1) (2) (3) (4) ne sont pas toujours distincts. Ainsi m=2 dans les cas 2, 3 et 4 correspondent aux valeurs m=2 m=3 et m=5 dans le cas n°1, la correspondance se faisant par

$$\begin{bmatrix} x_0+x_m & z \\ \bar{z} & x_0-x_m \end{bmatrix} \to (x_0, z, x_m)$$

où $z \in \mathbb{R}$, $\mathbb{C}$ ou $\mathbb{H}$. Même la valeur m=9 du cas n°1 peut s'interpréter avec une matrice (2,2) définie positive d'octonions.

On trouvera des détails sur les cônes symétriques dans Faraut(1988) sur les cas (1) et (2) dans Letac(1988 a et b), sur les cas 3,4,5 dans la thèse à venir de M. Cazalis.

Je me hasarde alors à formuler une conjecture pour le cas B=C=0 : si une fonction variance quadratique irréductible est de la forme

$$V_F(m) = A(m,m),$$

où A est une application bilinéaire symétrique de ExE dans $\mathbb{Q}(E^*)$ alors il existe un cône symétrique irréductible C sur E telle que

V_F ait l'une des 5 formes ci-dessus.

Pour terminer, mentionnons une observation concernant les exemples concrets ci-dessus de fonctions variances quadratiques irréductibles. Si F est une famille naturelle, si $m_0 \in M_F$ est fixé notons $\mu = P(m_0, F)$ est

$$f(x,m) = \exp(\langle\psi_\mu(m),x\rangle - k_\mu(\psi_\mu(m)) = \frac{dP(x,m)}{dP(x,m_0)}$$

Enfin, pour h et h' dans E assez petits pour que m_0+h et m_0+h' soient dans M_F et $\psi_\mu(m_0+h)$ et $\psi_\mu(m_0+h')$ soient dans $\Theta(\mu)/2$ (rappelons que celui-ci contient 0), formons

$$g(h,h') = \int_E f(x,m_0+h)f(x,m_0+h')P(m_0,F)(dx)$$

on a vu au Th.4.1. que si $E=\mathbb{R}$ et V_F est quadratique alors il existe G définie dans un voisinage de 0 tel que $g(h,h')=G(hh')$.

Prenons maintenant F_p comme au (1) de la Proposition 5.2. On trouve :

$$g(h,h') = \left(1+\left(1-\sum_{j=1}^{d} m_{0j}\right)^{-1}\left(\sum_{j=1}^{d} h_j\right)\left(\sum_{j=1}^{d} h'_j\right)+\sum_{j=1}^{d}\frac{h_j h'_j}{m_{0j}}\right)^{p}$$

Si F_p est comme au (2) de la Prop. 5.2. :

$$g(h,h') = \left(1+\left(1+\sum_{j=1}^{d} m_{0j}\right)^{-1}\left(\sum_{j=1}^{d} h_j\right)\left(\sum_{j=1}^{d} h'_j\right)-\sum_{j=1}^{d}\frac{h_j h'_j}{m_{0j}}\right)^{-p}$$

Si F_p est associée au cône symétrique n°1 (cône de révolution) en notant $B(x,y) = x_0y_0 - x_1y_1 - \dots - x_ny_n$ et $B(x,x) = B(x)$

on obtient, en notant $r = a\sqrt{B(m_0)}$:

$$g(h,h') = \left(1-4B\left(\frac{m_0}{r},\frac{h}{r}\right)B\left(\frac{m_0}{r},\frac{h'}{r}\right)+2B\left(\frac{h}{r},\frac{h'}{r}\right)+B\left(\frac{h}{r}\right)B\left(\frac{h'}{r}\right)\right)^{-p}$$

Si F_p est associée au cône symétrique n°2 (lois de Wishart sur les matrices symétriques définies positives) on obtient

$$\left(\operatorname{Det}\left(I-\left(m_0^{-1/2}\, h\, m_0^{-1/2}\right)\left(m_0^{-1/2}\, h'\, m_0^{-1/2}\right)\right)\right)^{-p}.$$

Dans tous les cas, à l'exception du cône de révolution, il

existe une fonction bilinéaire de h et h', $B_1(h,h')$ à valeurs dans un espace E_1, et une fonction G définie dans un voisinage de 0 et E_1 telle que $g(h,h') = G(B_1(h,h'))$. Il y a peut être là le germe d'une caractérisation à la manière du Th. 4.1.

Références

Andersson,S.,(1975). Invariant normal models. Ann. Statist.,3 : 132-154.

Barbour,A.,(1987). Asymptotic expansions in the Poisson limit theorem. Ann. Prob.,15 : 748-766.

Barlev,S.K. and Bshouty,D.(1989). Rational variance functions. To appear in Ann.Statist.

Barlev,S.K. and Enis,P.,(1986). Reproducibility and natural exponential families with power variance functions. Ann. Statist. 14 : 1507-1522.

Barndorff-Nielsen,O.,(1978). Information and Exponential families in Statistical Theory. Wiley, New-York.

Brown,L.(1986). Foundations of Exponential Families. I.M.S. Lecture notes, Hayward.

Cencov,N.N.,(1981), Statistical decision rules and optimal inference, AMS Publications, Providence.

Faraut,J.,(1988), Algèbres de Jordan et cônes symétriques. Publications du département de Mathématiques de l'Université de Poitiers.

Feinsilver,Ph.,(1985). Bernouilli systems in several variables, in Probability on Groups VII, Lecture Notes in Math., vol.1064, Springer-Verlag, Berlin and New-York : 86-98.

Feinsilver,Ph.,(1986). Some classes of orthogonal polynomials associated with martingales. Proc. A.M.S.,98 : 298-302.

Goodman,N.R.,(1963). Statistical analysis based on a certain multivariate complex Gaussian distribution (an introduction). Ann. Math. Statist., : 152-177.

Gyndikin,S.G.,(1975). Invariant generalized functions in homogeneous domains. Funct. Anal. Appl.,9 : 50-52.

Hoeffding,W.,(1963). Probability inequalities for sums of bounded random variables. Amer. Stat. Ass. J.58 : 13-30.

Jorgensen,B.,(1987). Exponential dispersion models. J.R. Statist. Soc.49,n°2 : 127-162.

Letac,G.,(1986). La réciprocité des familles exponentielles naturelles sur ℝ. C.R. Acad. Sc. Paris 303 Série I,2 : 61-64.

Letac,G.,(1988). Problème 6573. Amer. Math Monthly 95 n°5 : 461.

Letac,G.,(1988a). A characterization of the Wishart exponential families by an invariance property. To appear in Journal of Theoretical Probability.

Letac,G.,(1988b). Les familles exponentielles statistiques invariantes par les groupes du cône et du paraboloïde de révolution. Soumis aux Ann. Statist..

Letac,G. and Mora,M.,(1989). Natural real exponential families with cubic variance functions. To appear in Ann. Statist..

Mora,M.,(1986). Classification des fonctions variance cubiques des familles exponentielles sur ℝ. C.R. Acad. Sc. Paris 302, Série I, 16 : 582-591.

Morris,C.N.,(1982). Natural exponential families with quadratic variance functions. Ann. Statist.10 : 65-80.

Morris,C.N.,(1983). Natural exponential families with quadratic variance functions : statistical theory. Ann. Statist.11 : 315-519.

Prohorov, J.V.,(1953). Asymptotic behaviour of the binomial distribution. Uspchi Mat. Nank. 8 : 135-142.

Sz-Nagy,B.,(1965). Introduction to real functions and orthogonal expansions. Oxford University Press, New York.

CONVOLUTION PRODUCTS OF NON-IDENTICAL DISTRIBUTIONS ON A COMPACT ABELIAN SEMIGROUP

by

Arunava Mukherjea
University of South Florida
Tampa, Florida 33620-5700

1. <u>Introduction</u> Convolution products of non-identical distributions on general algebraic structures have not been as well-studied as such products for identical distributions. The reason is, of course, that the situation in this case is more complicated and complete results are not easy to come by. Maximov [4,5] studied this problem for compact (and also finite) groups and Csiszar [2], Tortrat [7] for non-compact groups. Other contributions in this area include Heyer [3] and Center and Mukherjea [1].

This paper is the first step towards understanding how the weak convergence of convolution products of non-identical distributions takes place in the class of finite dimensional stochastic matrices which form a compact topological semigroup under multiplication and usual topology. Here we restrict ourselves to a smaller class, namely the class of compact abelian semigroups. As we will show, the abelian assumption leads to results somewhat similar to those for groups. The non-abelian case will be treated in another paper.

Our results are very easily understood in the simple

situation of [0,1], the unit interval under multiplication and usual topology, which is also isomorphic to the abelian semigroup of stochastic matrices $\{\begin{pmatrix} a & 1-a \\ 0 & 1 \end{pmatrix} : 0 \leq a \leq 1\}$. This example is discussed at length in the next section.

Necessary and sufficient conditions for the weak convergence of the convolution products $\mu_{k,n} = \mu_{k+1} \cdots \mu_n$, as n tends to infinity, for every $k \geq 1$, of a sequence of probability measures in terms of conditions on the individual μ_n's are still unknown even for compact groups (except in some special cases, for example see [1] for finite groups). Maximov in his interesting paper [4] considers, on a compact (second countable) group G, probability measures μ_n such that for each $k \geq 1$, $\mu_{k,n}$ converges weakly to ν_k, as n tends to infinity. It then follows that ν_k, as k tends to infinity, converges weakly to ω_H, the Haar probability measure on some compact subgroup H. Maximov's main results in [4] can then be stated as follows:

(1) For any open set $V \supset H$, $\sum_{n=1}^{\infty} [1 - \mu_n(V)] < \infty$.

(2) The subgroup g(F) generated by the set F of essential points, where $F = \{x \in G: \sum_{n=1}^{\infty} \mu_n(N(x)) = \infty$ for every open set N(x) containing x $\}$, is contained in H, and in case G is a finite group, g(F) = H.

In the case of a compact semigroup, Maximov's result (1) does not hold as our example [0,1] will show. The result (2) is also false in non-group situations; for example, for any probability measure μ on [0,1] (under multiplication), the convolution sequence μ^n always converges weakly to either the unit mass at 1 or the unit mass at 0 so that H is either the singleton 0 or the singleton 1, whereas g(F) is the closed semigroup generated by the support of μ. The question then is what happens in the non-group situation. Our paper answers this question for compact abelian semigroups. Our main result is the following:

<u>Theorem</u> Let G be a compact (second countable) Hausdorff topological semigroup and (μ_n) be a sequence of Borel probability measures on G. Let (n_i) be a given subsequence of positive integers. Then we have:

i) There exists a subsequence $(\rho_i) \subset (n_i)$ such that for each

$$k \geq 1, \ \mu_{k,\rho_i} \to \nu_k, \ \nu_k \nu = \nu_k, \text{ and } \nu_{\rho_i} \to \nu = \nu^{(2)}.$$

When G is abelian, the support of the idempotent probability measure ν is a compact subgroup and ν is the Haar measure.

(ii) Suppose that G is abelian. Then the measure ν is the same for <u>all</u> subsequences (n_i). Also the sequence $\mu_{k,n}$, as $n \to \infty$, converges for each $k \geq 1$ iff the sequence $\nu_k(\nu_k$'s are as in (i)) converges as $k \to \infty$.

(iii) Suppose that G is abelian. Then for each $k \geq 1$, $\mu_{k,n}$ converges as $n \to \infty$ iff there is a compact <u>subgroup</u> H with the following two properties:

(a) for every $\epsilon > 0$ and for any open set $U \supset H$, there exists a positive integer k_0 and a sequence of positive integers $n(k)$ such that for each $k \geq k_0$ and $n \geq n(k)$, we have $\mu_{k,n}(U) > 1 - \epsilon$

(b) if H_0 is a proper closed subgroup of H, then there exists $\epsilon > 0$, an open set $V \supset H_0$ and a subsequence (n_i) of positive integers such that for $j \geq i + 1$,

$\mu_{n_i, n_j}(V) < 1 - \epsilon.$

(iv) Suppose that G is abelian and for each $k \geq 1$, $\mu_{k,n}$ converges to ν_k as $n \to \infty$. Then there is a compact subgroup H of G such that $\nu_k \to \omega_H$, where ω_H is the Haar measure on H, and for any open set $U \supset H$,

$$\sum_{n=1}^{\infty} [1 - \mu_n(U U^{-1})] < \infty, \tag{1.1}$$

where the set AB^{-1} (in a semigroup) is defined by

$$AB^{-1} = \left\{y \in G \,|\, yx \in A \text{ for some } x \in B\right\}. \tag{1.2}$$

Also, the semigroup g(F) generated by the set F of essential points (defined earlier) is contained in the compact subsemigroup

$$HH^{-1} = \left\{y \in G \,|\, y\, x \in H \text{ for some } x \in H\right\}.$$

(v) If G is a finite semigroup (not necessarily abelian), then $g(F) \supset S_\nu \equiv$ the support of ν, where ν is as in (i). ∎

We prove this theorem in section 3. To show an application of our theorem, we consider a second example, namely the compact abelian semigroup (under usual topology and matrix multiplication) of matrices given by

$$(*) \quad S_0 = \left\{ \begin{bmatrix} a_1 & a_2 & \cdot & \cdot\cdot\cdot & a_n \\ 0 & a_1 & a_2 & \cdot\cdot\cdot & a_{n-1} \\ & & & & \\ 0 & 0 & 0 & \cdots\cdots & a_1 \end{bmatrix} : 0 \leq a_i \leq 1,\ 1 \leq i \leq n, \text{ and } \sum_{i=1}^{n} a_i \leq 1 \right\}.$$

Here we need to use our theorem to show that for any sequence (μ_n) of probability measures on these matrices, for each $k > 1$, the sequence $\mu_{k,m}$, as m tends to infinity, converges weakly to some ν_k, and the sequence ν_k, as k tends to infinity, converges weakly to either the unit mass at the zero matrix or the unit mass at the identity matrix.

2. <u>The Examples</u>: Let $X_1, X_2, \ldots.$ be independent random variables with values in [0,1] such that $\Pr(X_i \in B) = \mu_i(B)$, for Borel sets $B \subset [0,1]$. Write: $X_{k,n} = X_{k+1} \cdots X_n$. Notice that $Y_k = \lim X_{k,n}$ exists pointwise. Write: $A_k = \{Y_k > 0\}$. Then $A_k \subset A_{k+1}$. It is easily seen that on the set $E = \bigcap_{n=1}^{\infty} \bigcup_{k=n}^{\infty} A_k$, $\lim Y_m = 1$. By Kolmogorov's zero-one law, $\Pr(E) = 0$ or 1. Also, on E^c, $Y_m = 0$ eventually in m so that $Y = \lim Y_m$ exists almost surely and either $Y = 0$ a.s. or $y = 1$ a.s. Notice that on E, the series $\sum_{i=1}^{\infty} [1 - X_i]$ is convergent. Thus, $Y = 1$ almost surely iff this series converges almost surely.

This is now clear that given any sequence of probability measures μ_n on [0,1], for each $k \geq 1$, the sequence $\mu_{k,n}$ converges weakly to ν_k, and as k tends to infinity, ν_k converges weakly to either the unit mass at 1 or the unit mass at 0. (We could prove it directly also via our theorem.) Noting that $UU^{-1} = U$ for $U = (1 - \epsilon, 1]$, it follows routinely from our theorem that the unit mass at 1 is the limit of the measures ν_k above iff the following condition holds:

For any given sequence (s_n) of positive integers and given any positive number ϵ,

$$\sum_{n=1}^{\infty} [1 - \mu_{s_n, s_{n+1}}([1 - \epsilon, 1]] < \infty.$$

A simple sufficient condition for ν_k's to converge to $\delta_{\{0\}}$ is the following: There exists a subsequence of positive integers (n_i), a sequence ϵ_i of positive numbers converging to zero and a $\delta > 0$ such that $\mu_{n_i}([0,\epsilon_i]) \geq \delta \ \forall\ i \geq 1$. (A more general sufficient condition has been given at the end of this paper.)

A simple sufficient condition for ν_k's to converge to $\delta_{\{1\}}$ is the following: There exist (α_n), (β_n) such that

$$0 < \alpha_n < 1,\ 0 < \beta_n < 1,$$

$$\sum_{n=1}^{\infty}(1 - \alpha_n) < \infty,\ \sum_{n=1}^{\infty}(1 - \beta_n) < \infty \text{ and } \mu_n([\alpha_n,1]) \geq \beta_n. \ \blacksquare$$

Under the first condition, notice that for any $k \geq 1$ and given $\epsilon > 0$, if we choose $n_{i_0} > k$, $\epsilon_{i_0} < \epsilon$, then for any $m > n_{i_0}$,

$$\mu_{k,m}([0,\epsilon])$$

$$\geq \prod_{\substack{s=k+1 \\ s \neq n_{i_0}}}^{m} \mu_s([0,1]) \cdot \mu_{n_{i_0}}([0,\epsilon_{i_0}])$$

$$\geq \delta > 0$$

so that $\forall\ k \geq 1$, $\nu_k([0,\epsilon]) \geq \delta > 0 \ \forall\ \epsilon > 0$.

Under the second condition, notice that given $\epsilon > 0$, then exists k_o such that for $k > k_0$, $\prod_{i=k}^{\infty} \alpha_i > 1 - \epsilon$, $\prod_{i=k}^{\infty} \beta_i > 1 - \epsilon$ so that

for $m > k \geq k_0$, $\mu_{k,m}([1-\epsilon,1]) \geq \mu_{k,m}(\prod_{i=k+1}^{m}[\alpha_i,1])$

$$\geq \prod_{i=k+1}^{\infty} \mu_i([\alpha_i,1]) \geq \prod_{i=k+1}^{\infty} \beta_i > 1 - \epsilon.$$

This means that for $k > k_0$, $\nu_k([1-\epsilon,1]) \geq 1-\epsilon$, so that $\nu_k \to \delta_{\{1\}}$.

Now we consider our second example - the semigroup S_0 of matrices defined in (*). Actually the discussion that follows is also valid in any compact abelian semigroup where the only compact subgroups are the singleton elements the "zero" and the "one". Let (μ_m) be any sequence of probability measures and (m_i) be any subsequence of positive integers. Then there is a further subsequence (p_i) of (m_i) (see part (i) of our theorem) such that for each $k \geq 1$,

$$\mu_{k,p_i} \longrightarrow \nu_k \text{ and}$$

$$\nu_{p_i} \longrightarrow \omega_H,$$ the Haar probability measure on a compact subgroup, so that $H = \{0\}$ or $\{1\}$.

Suppose that $H = \{0\}$. Let (s_i) be another subsequence of positive integers such that ν_{s_i} converges weakly to a probability measure λ. Then, there exists $(t_i) \subset (s_i)$ such that for each $k \geq 1$,

$$\mu_{k,t_i} \longrightarrow \nu_{ko}$$ as i tends to infinity, and ν_{t_io} converges weakly, by part (ii) of our theorem, to the unit mass at 0. Now we fix j. Then for each i such that $p_i > t_j$, $\mu_{k,t_j} \cdot \mu_{t_j,p_i} = \mu_{k,p_i}$. Letting i to infinity, $\mu_{k,t_j} \cdot \nu_{t_j} = \nu_k$ for each $k \geq 1$ and for $t_j > k$, $\nu_{ko} \cdot \lambda = \nu_k$.

Thus, $\nu_{t_io} \cdot \lambda = \nu_{t_i}$ for each i so that $\lambda = \delta_{\{0\}} \cdot \lambda = \delta_{\{0\}}$. Hence, the sequence (ν_k) converges to the unit mass at 0. By part (ii) of our theorem, for each $k \geq 1$, $\mu_{k,m}$ converges weakly to ν_k. We can repeat the same argument in case $H = \{1\}$.

3. Proof of the Theorem.

Recall that the support of an idempotent probability measure on a locally compact Hausdorff (2nd countable) topological semigroup is a completely simple subsemigroup. In the case of an abelian semigroup, such a measure is always the Haar measure on its support, which must be a compact abelian group. (For these and other informations, see [6].) Let G be a compact (2nd countable) (not necessarily abelian) semi group and $\mu_n \in P(G)$ such that for each $k \geq 1$,

$$\mu_{k,n} \to \lambda_k \quad \text{as } n \to \infty.$$

Then for $k < m < n$,

$$\mu_{k,m}\mu_{m,n} = \mu_{k,n}$$

so that taking n to ∞, for $k < m$,

$$\mu_{k,m}\lambda_m = \lambda_k$$

If λ is a (weak) limit point (P(G) is compact in the weak topology) of (λ_n), then it is clear that for each $k \geq 1$,

$$\lambda_k\lambda = \lambda_k .$$

Thus, for any two limit points λ, λ' of (λ_m),

$$\lambda'\lambda = \lambda' , \quad \lambda^2 = \lambda. \tag{3.1}$$

Let G_0 be the closure of the union of the supports of the limit points of the (λ_m). Then it follows easily using (3.1) and the information given earlier that G_0 is a compact completely simple subsemigroup of the form X x S x Y, where S is a compact subgroup $\subset G_0$, X and Y are respectively a left-zero and a right-zero subsemigroup of G_0, Y X $\subset$ S, and

$$(x, s, y)(x', s', y') = (x, s(yx')s', y').$$

Also, then the support of a limit point λ of (λ_m) is of the form X_0 x S x Y, $X_0 \subset X$. (Proofs of these facts are not difficult and omitted.) Thus, when G is abelian, the semigroup G_0 becomes the compact group S and each limit point λ of (λ_m), being idempotent, becomes the Haar measure on S or G_{0_1} and the sequence λ_m then converges to the Haar measure on G_0.

It is well-known (see [8], pg. 45) that P(G), the Borel probability measures on G, when G is compact and second countable, is also compact and second countable with respect to weak topology. Then part (i) of our Theorem follows immediately from Theorem 2.1 in [2].

<u>Proof of</u> (ii). We now assume that G is abelian. It follows from (i) that,

$$\mu_{k,p_i} \to \nu_k \text{ as } i \to \infty, \; \nu_k \, \nu = \nu_k,$$

and $$\nu_{p_i} \to \nu \equiv \omega_H \text{ as } k \to \infty$$

where ω_H is the Haar measure on H, a compact subgroup.

Let q_i be another subsequence of positive integers such that for each $k \geq 1$,

$$\mu_{k,q_i} \to \nu'_k, \quad \nu'_k \omega_{H'} = \nu'_k$$

and $$\nu'_{q_i} \to \omega_{H'} \quad \text{as } i \to \infty,$$

where H' is another compact subgroup. Since P(G) is second countable, we can find subsequences $(p'_i) \subset (pi)$ and $(q'_i) \subset (q_i)$ such that for each $i \geq 1$,

$$p'_i < q'_i < p'_{i+1} < q'_{i+1},$$

$$\mu_{p'_i, p'_j} \to \omega_H \text{ as } i \to \infty \qquad (\text{for each } j \geq i + 1),$$

$$\mu_{q'_i, q'_j} \to \omega_H \text{ as } i \to \infty \qquad (\text{for each } j \geq i + 1).$$

Now we choose subsequences $(p'_{i_m}) \subset (p'_i)$ and $(q'_{i_m}) \subset (q'_i)$ such that

$\mu_{p'_{i_m}, q'_{i_m}} \to \lambda$ (some measure in P(G)) and $\mu_{q'_{i_m}, p'_{i_{m+1}}} \to \lambda'$ (some measure in P(G)).

Since

$$\mu_{p'_{i_m}, p'_{i_{m+1}}} = \mu_{p'_{i_m}, q'_{i_m}} \mu_{q'_{i_m}, p'_{i_{m+1}}}$$

and

$$\mu_{q'_{i_m}, q'_{i_{m+1}}} = \mu_{q'_{i_m}, p'_{i_{m+1}}} \mu_{p'_{i_{m+1}}, q'_{i_{m+1}}},$$

we have as $j \to \infty$,

$$\omega_H = \lambda.\lambda' = \lambda'\lambda = \omega_{H'}.$$

Hence, H = H'. This proves the first part of (ii).

Now we suppose that $\nu_k \to \omega_H$ as $k \to \infty$. Then since

$$\mu_{k, q'_{i_m}} = \mu_{k, p'_{i_m}} \mu_{p'_{i_m}, q'_{i_m}},$$

it follows as $m \to \infty$,

$$\nu'_k = \nu_k \cdot \lambda$$

This means that ν'_k also converges as $k \to \infty$ and

$$\omega_H = \omega_{H'} = \omega_H \cdot \lambda ,$$

so that

$$\nu'_k = \nu'_k \cdot \omega_{H'} = \nu'_k \cdot \omega_H$$

$$= (\nu_k \lambda)\omega_h$$

$$= \nu_k \, (\omega_H \lambda)$$

$$= \nu_k \omega_H = \nu_k \, .$$

Hence, for each $k \geq 1$, $\mu_{k,n}$ converges as $n \to \infty$. The "only if" part of (ii) has been established earlier in this section.

Proof of (iii). Suppose that G is abelian and properties (a) and (b) hold. Then, as in (i), given any sequence (n_i) of positive integers there exists a subsequence $(p_i) \subset (n_i)$ such that for each $k \geq 1$,

$$\mu_{k,p_i} \to \nu_k \text{ as } i \to \infty ,$$

$$\nu_{p_i} \to \omega_{H_0} \text{ as } i \to \infty, \ \nu_k \, \omega_{H_0} = \nu_k$$

and $$\mu_{p_i, p_{i+1}} \to \omega_{H_0} ,$$

where H_0 is a compact subgroup and ω_{H_0} is its Haar measure. By property (a) in (iii), $H_0 \subset H$. If H_0 is a proper subgroup of H, then by property (b) in (iii), there exist $\epsilon > 0$, an open set $V \supset H_0$ and a subsequence (q_i) of positive integers such that

$$\mu_{q_i, q_j}(V) < 1 - \epsilon \quad \forall \ j \geq i + 1. \tag{3.2}$$

Then using (i) and the first part of (ii), we see that there exists a subsequence $(r_i) \subset (q_i)$ such that

$$\mu_{r_i, \, r_{i+1}} \to \omega_{H_0} \text{ as } i \to \infty .$$

This contradicts (3.2) and thus, $H_0 = H$. Now let (p_i') be another subsequence of positive integers such that

$$\mu_{k, \, p_i'} \longrightarrow \nu_k' \text{ as } i \to \infty$$

and

$$\nu_{p_i'} \to \omega_H, \ \nu_k' = \nu_k' \, \omega_H .$$

Write using a subsequence of (p_i) and still calling it (p_i), so that for some $\lambda_0 \in P(G)$,

$$\nu_k = \nu_k' \, \lambda_0$$

Then, $\nu'_k = \nu'_k \omega_H = \nu'_k \cdot (\lambda_0\omega_H)$

$$= \nu_k \omega_H = \nu_k,$$

since we can assume, with no loss of generality, that $S_{\lambda_0} \subset H$ (because of property (a)) so that $\lambda_0\omega_H = \omega_H$. This establishes the "if" part of (iii). For the "only if" part, suppose that for each $k \geq 1$,

$$\mu_{k,n} \to \nu_k \text{ as } n \to \infty.$$

Then we know that there is a compact subgroup H such that

$$\nu_k \to \omega_H \text{ as } k \to \infty.$$

This H satisfies property (a) clearly. Now let H_0 be a proper closed subset of H. Then there exists $x \in H - H_0$, an open set N(x) containing x, an open set $U(H_0) \supset H_0$ such that $N(x) \cap U(H_0) = \phi$. Since $\omega_H(N(x)) > 0$, there exists k_0 and a positive δ such that for $k \geq k_0$,

$$\nu_k(N(x)) > \delta.$$

Property (b) now follows easily for the open set $U(H_0)$.

Proof of (iv). The first part of (iv) has been established earlier. We will prove only (1.1). We separate the proof in several steps. We assume that G is abelian.

Step I. Let S_0 be a closed subsemi-group of G. Let U be an open set $\supset S_0$. Then there is an open set V such that $S_0 \subset V$, $\bar{V} \cdot \bar{V} \subset U$ and an element $z \in U$ such that

$$VS_0^{-1} \subset Uz^{-1}. \tag{3.3}$$

(Here, VS_0^{-1} and Uz^{-1} are defined as in (1.2).)

To see this, notice that since $S_0 \subset U$, $S_0 \cdot S_0 \subset U$ and S_0 is compact, there exists an open V such that $S_0 \subset V$, $\bar{V} \cdot \bar{V} \subset U$. Now the class of sets

$$\{y \cdot \bar{V} \mid y \in S_0\}$$

has a finite intersection property since $y_1 y_2 \ldots y_n \in \bigcap_{i=1}^{n} y_i \bar{V}$ if each y_i is in S_0. Hence, there exists

$$z \in \bigcap_{y \in S_0} y . \bar{V} .$$

Let $\omega \in VS_0^{-1}$. Then there exists $\omega' \in S_0$ such that $\omega.\omega' \in V$. Now $z \in \omega' \bar{V}$ so that $z = \omega'\omega''$ for some $\omega'' \in \bar{V}$. Hence,

$$\omega z = (\omega\ \omega')\ \omega'' \in V \cdot \bar{V} \subset U$$

or $\omega \in Uz^{-1}$.

<u>Step II</u>. For this step, we do not need G to be abelian. For x, y in G, and an open set N(x) containing x, there exists open sets N'(x) and N'(y) containing x and y respectively such that

$$N'(x)\ N'(y)^{-1} \subset N(x)y^{-1}. \tag{3.4}$$

To see this, notice that if (3.4) does not hold for all open sets containing x and y respectively, then we can find $z_n \notin N(x)y^{-1}$, such that

$$z_n \in N_n(x)\ N_n(y)^{-1},\ \text{where}$$

$\{N_n(x) : n \geq 1\}$ and $\{N_n(y) : n \geq 1\}$ are countable local bases at x and y respectively. This means that there exist

$$y_n \to y, \; z_n y_n \to x,$$

where $y_n \in N_n(y)$, $z_n y_n \in N_n(x)$.

Since G is compact, there is a subsequence (n_i) such that

$$y_{n_i} \to y, \; z_{n_i} \to z, \; z_{n_i} y_{n_i} \to zy = x$$

so that $z \in N(x)y^{-1}$. This contradicts that $[N(x)y^{-1}]^c$ is closed. This establishes (3.4).

Step III. Let S_0 be a compact subset of G and V be an open subset such that $V \supset S_0$. Then there exists an open subset $W \supset S_0$ so that

$$WW^{-1} \subset VS_0^{-1} .$$

(In this step also, G need not be abelian.)

The proof of this step is routine. It follows routinely from Step II, using the compactness of S_0.

Step IV. Let $1 \leq i \leq m$. Let $\{A_1, A_2, \ldots, A_m\}$ and $\{B_1, B_2, \ldots, B_m\}$ be two families of events such that for each i, B_i is independent of $A_1^c, A_2^c, \ldots, A_{i-1}^c, A_i$. Then we have:

$$P\left(\bigcup_{i=1}^{m} (A_i \cap B_i)\right) \geq \left[\inf_{1 \leq i \leq m} P(B_i)\right]. \; P\left(\bigcup_{i=1}^{m} A_i\right).$$

To see this, notice that $P\left(\bigcup_{i=1}^{m} (A_i \cap B_i)\right)$

$$\geq P((A_1 \cap B_1) \cup (A_2 \cap B_2 \cap A_1^c) \cup \dots$$

$$\cup (A_m \cap B_m \cap A_1^c \cap \dots \cap A_{m-1}^c))$$

$$= \left[\sum_{i=2}^{m} P(B_i)\ P(A_i \cap A_{i-1}^c \cap \dots \cap A_1^c)\right] + P(B_1)\ P(A_1).$$

The proof of this step is now clear.

Step V. Here we complete the proof of (1.1). Let us then assume that for each $k \geq 1$, $\mu_{k,n} \to \upsilon_k$ as $n \to \infty$ and $\upsilon_k \to \omega_H$ as $k \to \infty$ where H is a compact subgroup of G. Given $\epsilon > 0$. Let U be an open set containing H. Then there exists k such that for $k \geq k_0$,

$$\upsilon_k\ (U) > 1 - \epsilon. \tag{3.5}$$

Since for $k < m$, $\upsilon_k = \mu_{k,m}\ \upsilon_m$, it follows that for $m > k \geq k_0$,

$$\mu_{k,m}\ (UU^{-1}) > 1 - 2\ \epsilon. \tag{3.6}$$

Let X_1, X_2, ... be a sequence of independent random variables with values in G such that $P(X_n \in B) = \mu_n(B)$. By steps I, II and III, there exists $z_0 \in U$ and an open set $V \supset H$ such that

$$VV^{-1} \subset Uz_0^{-1}. \tag{3.7}$$

By the same reason, there exists $z \in V$ and an open set $W \supset H$ such that

$$WW^{-1} \subset Vz^{-1}. \tag{3.8}$$

Notice that

$$\bigcup_{i=0}^{m-1} \Big\{X_k\, X_{k+1} \cdots X_{k+i} \notin VV^{-1}$$

$$\text{and } X_{k+i+1} \cdots X_{k+m} \in Vz^{-1}\Big\} \qquad (3.9)$$

$$\subset \Big\{X_k\, X_{k+1} \cdots X_{k+m} \notin Vz^{-1}\Big\}$$

since

$$X_k\, X_{k+1} \cdots X_{k+m} \in Vz^{-1},\ X_{k+i+1} \cdots X_{k+m} \in Vz^{-1}$$

$$\Rightarrow X_k \cdots X_{k+i} \in (Vz^{-1})(Vz^{-1})^{-1} \subset VV^{-1}.$$

Using (3.8) and the same idea as used in (3.6), we see that there exists k_1 such that $r > k \geq k_1$ implies

$$\mu_{k,r}(Vz^{-1}) > 1 - \epsilon. \qquad (3.10)$$

Using step IV, (3.9) and (3.10), we have for $k \geq k_1$,

$$P\left[\bigcup_{i=0}^{m-1} \Big\{X_k\, X_{k+1} \cdots X_{k+i} \notin VV^{-1}\Big\}\right]$$

$$\leq \frac{\epsilon}{1-\epsilon} < 2\epsilon, \text{ assuming } 0 < \epsilon < \tfrac{1}{2}.$$

This means that for $k \geq k_1$ and all $m \geq 1$,

$$P(X_k \in VV^{-1},\ X_k\, X_{k+1} \in VV^{-1},$$

$$\ldots,\ X_k\, X_{k+1} \cdots X_{k+m} \in VV^{-1})$$

$$> 1 - 2\epsilon,$$

so that

$$P(X_k \in Uz_0^{-1}, X_k X_{k+1} \in Uz_0^{-1}, \tag{3.11}$$

$$\ldots\ldots, X_k X_{k+1} \ldots. X_{k+m} \in Uz_0^{-1}) > 1 - 2\epsilon.$$

This last inequality is similar to the one obtained by Maximov in [4, p.61]. We not write:

$$P(X_k \in Uz_o^{-1}, \ldots\ldots, X_k \ldots X_{k+m+1} \in Uz_o^{-1}) \tag{3.12}$$

$$\leq \prod_{i=0}^{m+1} \mu_{k+1}(UU^{-1})$$

[Notice that $(Uz_o^{-1})(Uz_o^{-1})^{-1} \subset UU^{-1}$; also,

$$P(X_k \in Uz_o^{-1}, X_k X_{k+1} \in Uz_o^{-1})$$

$$\leq P(X_k \in Uz_o^{-1}, X_{k+1} \in (Uz_o^{-1})^{-1} (Uz_o^{-1}))$$

$$\leq P(X_k \in Uz_o^{-1})\, P\,(X_{k+1} \in (Uz_o^{-1})^{-1}(Uz_o^{-1}))$$

and then, we can use induction.]

The proof of (1.1) is now complete from (3.11) and (3.12). For the last part of (iv) in the Theorem, notice that if g(F) is not contained in HH^{-1}, then the set F is also not a subset of HH^{-1} since HH^{-1} is a semigroup. If possible, then let $x \in F$ and $x \notin HH^{-1}$. Then we have:

$$x\,H \cap H = \phi$$

Hence, there exists an open set $K \supset H$ and open N(x) containing x such that

$$(N(x).K) \cap K = \phi$$

or $$N(x) \cap KK^{-1} = \phi.$$

But this means that

$$\sum_{n=1}^{\infty} [1 - \mu_n(KK^{-1})]$$

$$\geq \sum_{n=1}^{\infty} \mu_n(N(x)) = \infty,$$

since x is an essential point. This is a contradiction to (1.1). Hence, $g(F) \subset HH^{-1}$.

Proof of (v). We now prove part (v) of the Theorem. Let (p_i) be a subsequence that for $k \geq 1$,

$$\mu_{k,p_i} \to \nu_k, \quad \nu_{p_i} \to \nu = \nu^{(2)}.$$

Choose $(p_i') \subset (p_i)$ such that

$$\| \nu_{p_i'} - \nu \| < \frac{1}{2^i}, \quad i \geq 1.$$

(This is possible since G is assumed finite.) Now for each $j \geq 1$,

$$\mu_{p_j', p_i'} \to \nu_{p_j'} \quad \text{as } i \to \infty.$$

Now notice that g(F) is a semigroup and therefore, for any s,t,

$$\mu_{s,s+t}(g(F)) \geq \prod_{i=s+1}^{s+t} \mu_i(g(F)).$$

Because of the definition of F and the finiteness of G, we have given $\epsilon > 0$, there exists s_0 such that $s \geq s_0$ implies that

$\forall\ t \geq 1$,

$$\prod_{i=s+1}^{s+t} \mu_i(g(F)) > 1 - \epsilon.$$

Thus, there exists a subsequence $(m_i) \subset (p_i')$ such that

$$\mu_{m_i, m_{i+1}} \to \nu \text{ as } i \to \infty$$

and $\mu_{m_i, m_{i+1}}(g(F)) \to 1$ as $i \to \infty$. This proves that $S_\nu \subset g(F)$. ■

Notice that in the semigroup {0,1} (under multiplication), for any probability measure μ with $\mu(0) > 0$, $\mu(1) > 0$, $\mu^{(n)} \to \delta_{\{0\}} \equiv \nu$ so that in this case S_ν is a proper subset of $g(F)$. This is completely different from what happens in groups. It is also worthwhile to mention very briefly how this zero-like behavior takes place in the context of a general compact (not necessarily abelian) semigroup S. Such a semigroup has always a smallest two-sided ideal I (which acts very much like the "0" in [0,1]). Suppose that $\mu_n \in P(S)$ are such that for any open set $U \supset I$,

$$\sum_{i=1}^{\infty} \mu_{k_i, k_i + r_i}(U) = \infty,$$

for two subsequences (k_i) and (r_i) of positive integers such that

$$1 \leq r_1 \leq r_2 \leq r_3 \leq \ldots \text{ and } \ k_i + r_i < k_{i+1}.$$

Then with one application of the Borel-Cantelli Lemma it can be verified that for each $k \geq 1$ and any open set $U \supset I$,

$$\lim_{n\to\infty} \mu_{k,n}(U) = 1.$$

This means that if I is a singleton, then calling I the zero of S, we have: for each $k \geq 1$, $\mu_{k,n} \to \delta_{\{0\}}$ as $n \to \infty$.

Now we make our last non-trivial observation. Notice that G/H, the family of cosets $\{gH: g \in G\}$ is again a compact abelian Hausdorff second countable topological semigroup with quotient topology and usual coset multiplication. Here, of course, H is a compact subgroup of the given compact abelian semigroup G. The assertions concerning G/H follow from the following observations:

(1) $g_1H \cap g_2H \neq \phi$ implies $g_1H = g_2H$.

(2) $g_1H = g_3H$, $g_2H = g_4H$ imply $g_1g_2H = g_3g_4H$.

(3) Suppose that $g_1H \neq g_2H$. Then, $g_1H \cap g_2H = \phi$. Therefore, there exist disjoint open subsets V_1 and V_2 in G such that $g_1H \subset V_1$ and $g_2H \subset V_2$. Define the sets U_1 and U_2 by

$$U_1 = \{y \in G: yH \subset V_1\}, \; U_2 = \{y \in G: yH \subset V_2\}.$$

Then, U_1 and U_2 are both open in G, $U_1H \subset V_1$, $U_2 \subset V_2$, and

$$p^{-1}(U_1H) = U_1, \quad p^{-1}(U_2H) = U_2,$$

where p is the natural projection from G to G/H. This shows that G/H is Hausdorff.

(4) Let $g_1g_2H \subset W$, where W is an open set in G/H. Then there exist sets V_1 and V_2 (both open in G) such that $g_1H \subset V_1$, $g_2H \subset V_2$ and $V_1V_2 \subset p^{-1}(W)$. The sets U_1 and U_2, as defined in (3) above, are then both open in G/H and $(U_1H)(U_2H) \subset W$. This proves that the coset multiplication is jointly continyous in G/H.

(5) Let B be a countable base for the topology of G such

that B is closed under finite unions. Let F be the countable family of sets open in G given by,

$$V_F = \{y \in G: yH \subset V\}, \ V \in B.$$

Let W be an open set in G/H and gH ∈ W. Then there exists V' in B such that $gH \subset V' \subset p^{-1}(W)$. Notice that $g \in V'_F$ and $p^{-1}(V'_F H) = V'_F$. Hence, the family $\{V_F H: V \in B\}$ generates a countable base for the topology of G/H.

We now state our final result.

Proposition. With the same hypotheses as in (iv) of our theorem, the following holds:

The sequence $p(X_1 X_2 \ldots X_n)$ converges almost surely as n tends to infinity, where $p: G \longrightarrow G/H$ is the natural projection and (X_i) is a sequence of G-valued independent random variables such that the distribution of X_i is μ_i. ■

Proof. We give the proof in several steps.

First, notice that $H = \bigcap_{i=1}^{\infty} V_i$, $\overline{V_{i+1}} \subset V_i$, where V_i's are open sets in G.

Step I. Given any subsequence (n_i) of positive integers and given u > 0, there exists a subsequence $(m_{ii}) \subset (n_i)$ such that

$$P\left(\bigcap_{t=1}^{\infty} \bigcap_{i=t}^{\infty} \{X_{m_{ii}} \ldots X_{m_{ii}+k} \in V_t V_t^{-1} \text{ for every } k \geq 0\}\right) > 1-u.$$

To see this, we already know from our theorem that there exists $(m_{1i}) \subset (n_i)$ such that for each i,

$$P(X_{m_{1i}} X_{m_{1i}+1} \ldots X_{m_{1i}+k} \in V_1 V_1^{-1} \text{ for } k \geq 0) > 1-(u/2^{i+2})$$

so that

$$P\Big(\bigcap_{i=1}^{\infty} \{X_{m_{1i}} \cdot\ldots\cdot X_{m_{1i}+k} \in V_1V_1^{-1} \text{ for } k \geq 0\}\Big) > 1-(u/2).$$

Similarly, we can find $(m_{2i}) \subset (m_{1i})$ such that

$$P\Big(\bigcap_{i=1}^{\infty} \{X_{m_{2i}} \cdot\ldots\cdot X_{m_{2i}+k} \in V_2V_2^{-1} \text{ for } k \geq 0\}\Big) > 1-(u/2^2),$$

and so on. It is now clear that the diagonal subsequence (m_{ii}) will meet the requirements of this step.

Step II. Given any two subsequences (n_i) and (n_i') of positive integers, there exist subsequences $(p_i) \subset (n_i)$ and $(p_i') \subset (n_i')$ such that $p_i < p_i'$ and almost surely, $X_{p_i+1}X_{p_i+2}\cdot\ldots\cdot X_{p_i'}.H$ converges to H, in the topology of G/H.

To see this, we use Step I. Given $u > 0$, there exists B_u such that $P(B_u) > 1-u$ and

$$B_u = \bigcap_{t=1}^{\infty} \bigcap_{i=t}^{\infty} \{X_{m_i} \cdot\ldots\cdot X_{m_i+k} \in V_tV_t^{-1} \quad \text{for } k \geq 0\}.$$

where (m_i) is a subsequence, depending on u, of (n_i). Let $B = \bigcup_{n=1}^{\infty} B_{(1/n)}$. Then $P(B) = 1$.

Let $\omega \in B$. Then, $\omega \in B_u$ for some $u > 0$. Choose a subsequence $(m_i') \subset (n_i')$ such that $m_i < m_i'$ for each i. Suppose that $X_{m_i}(\omega)\cdot\ldots\cdot X_{m_i'}(\omega).H$ does not converge to H in G/H. Then there exists an open set U(H) containing H in G/H such that for some subsequences $(m_{i_j}) \subset (m_i)$ and $(m_{i_j}') \subset (m_i')$, for each j we have:

$$X_{m_{i_j}}(\omega)\cdot\ldots\cdot X_{m_{i_j}'}(\omega).H \quad \notin U(H).$$

Using compactness and choosing further subsequences and still calling them (m_{i_j}) and (m'_{i_j}), we have:

$$X_{m_{i_j}}(\omega)\ldots.X_{m'_{i_j}}(\omega) \text{ converges to } x(\omega)$$

as $j \longrightarrow \infty$, in G. This implies that $x(\omega)H \cap U(H) = \phi$. Hence, there exists an open set N containing $x(\omega)$ and some V_t such that $NV_t \cap V_t = \phi$ or $N \cap (V_t V_t^{-1}) = \phi$. This means that for some j_o, for all $j \geq j_o$,

$$X_{m_{i_j}}(\omega)\ldots..X_{m'_{i_j}}(\omega) \notin V_t V_t^{-1}.$$

This contradicts that $\omega \in B_u$. Thus, for each $B_{(1/k)}$, we can find subsequences $(m_{ki}) \subset (n_i)$ and $(m'_{ki}) \subset (n'_i)$ such that

$$X_{m_{ki}+1}\ldots..X_{m'_{ki}}.H \text{ converges to H in G/H, on } B_{(1/k)}, \text{ where}$$

$$(m_{ki}) \supset (m_{k+1,i}),\ (m'_{ki}) \supset (m'_{k+1,i}).$$

Take $p_i = m_{ii}$ and $p'_i = m'_{ii}$. This step is complete.

Step III. In this step, we complete the proof of the proposition. Let ω be fixed. Let (n_i) and (n'_i) be two subsequences such that $n_i < n'_i$, $p(X_1(\omega)X_2(\omega)\ldots..X_{n_i}(\omega)) \longrightarrow gH$ and $p(X_1(\omega)X_2(\omega)\ldots..X_{n'_i}(\omega)) \longrightarrow g'H$, in G/H. [Note that G/H is compact.] Since $X_1X_2\ldots.X_{n'_i}.H = (X_1\ldots.X_{n_i}H)(X_{n_i+1}\ldots..X_{n'_i}H)$, it follows that $g'H = gH.H = gH$, by Step II above, and this completes the proof.

REFERENCE

1. B. Center and A. Mukherjea: More on limit theorems for iterates of probability measures on semigroups and groups, Z Wahrscheinlichkeitstheorie und verw. Gebiete 46 (1979), 259-275.

2. I. Csiszar: On infinite products of random elements and infinite convolutions of probability distributions on locally compact groups, Z. Wahrscheinlichkeitstheorie und verw. Gebiete 5 (1966), 279-295.

3. H. Heyer: Probabilistic characterization of certain classes of locally compact groups, Symp. Math. 16 (1975), 315-355.

4. V.M. Maximov: Composition convergent sequences of measures on compact groups, Theory of Probability and its Applications 16, No. 1 (1971), 55-73.

5. V.M. Maximov: Necessary and sufficient conditions for the convolution of non-identical distributions given on a finite group, Theory of Probability and its Applications 13 (1968), 287-298.

6. A. Mukherjea and N.A. Tserpes: Measures on topological semigroups: Convolution products and random walks, Lecture Notes in Math. 547, Springer-Verlag, Berlin and New york, 1976.

7. A. Tortrat: Lois de probabilité sur un espace topologique complètement régulier et produit infinis à termes indépendants dans un groupe topologique, Ann.lust. H. Poincaré Sect. B, 1 (1965), 217-237.

8. K.R. Parthasarathy: Probability Measures on Metric Spaces, Academic Press (1967).

THE LEVY LAPLACIAN AND MEAN VALUE THEOREM

NOBUAKI OBATA

Department of Mathematics, Faculty of Science,
Nagoya University, Nagoya, 464, Japan

Introduction

In his celebrated book [L] P.Lévy introduced an infinite dimensional Laplacian (called the Lévy Laplacian):

$$\Delta = \lim_{N \longrightarrow \infty} \frac{1}{N} \sum_{n=1}^{N} \frac{\partial^2}{\partial \xi_n^2}$$

and discussed many related problems. Keeping deep contact with Lévy's original ideas, T.Hida [H1] initiated the study of the Lévy Laplacian in terms of white noise calculus and, during the last decade the Lévy Laplacian has become important in the analysis of generalized Brownian functionals ([H2-5],[HS],[K]).

The paper aims at the study of harmonic functions with respect to the Lévy Laplacian. It is therefore interesting for us to consider the mean value property of functions on a Hilbert space. Motivated by Lévy's regular functionals, we introduce a notion of regularly analytic functions. These functions form a subclass of analytic functions in the usual sense and, remarkably, possess the mean value property. As a result regularly analytic functions are harmonic with respect to the Lévy Laplacian and, in particular, so are ordinary Brownian functionals.

Our setup is of great advantage to a group-theoretical approach to the Lévy Laplacian as is seen in the forthcoming paper [O4]. Further related topics, for example the Lévy group, are found in [O1-3].

§1. The Lévy Laplacian

Let H be a real separable Hilbert space with inner product $\langle\cdot,\cdot\rangle$ and norm $\|\cdot\|$. If $U \subset H$ is an open subset (with respect to the norm topology) we denote by $\mathscr{C}^2(U)$ the space of all twice continuously Fréchet differentiable functions on U. Let $\{e_n\}_{n=1}^{\infty}$ be a fixed complete orthonormal sequence (=CONS) in H. For $F \in \mathscr{C}^2(U)$ we put

$$(1\text{-}1)\qquad \Delta F(\xi) = \lim_{N\to\infty} \frac{1}{N} \sum_{n=1}^{N} < F''(\xi)e_n, e_n >, \qquad \xi \in U,$$

if the limit exists. The operator Δ is called the *Lévy Laplacian*. We denote by $\mathscr{D}(\Delta;U)$ the space of functions $F \in \mathscr{C}^2(U)$ which admits the limit (1-1) at every point $\xi \in U$. A function $F \in \mathscr{D}(\Delta;U)$ is called *harmonic* on U if ΔF vanishes on it.

For each $n \geqslant 1$ the unit sphere $S^{n-1} \subset \mathbf{R}^n$ is regarded as a subset of H by means of the mapping:

$$h = (h_1,\cdots,h_n) \longmapsto \sum_{k=1}^{n} h_k e_k \in H, \qquad h \in S^{n-1}.$$

We denote by $dS_{n-1}(h)$ the normalized uniform measure on S^{n-1}. The *(asymptotic) spherical mean* of a function F over the sphere of radius $\rho \in \mathbf{R}$ with center at $\xi \in H$ is defined by

$$(1\text{-}2)\qquad \mathfrak{M}F(\xi,\rho) = \lim_{n\to\infty} \int_{S^{n-1}} F(\xi+\rho h)\, dS_{n-1}(h),$$

provided the limit exists. We say that a function F possesses the *mean value property* on U if for each $\xi \in U$ there exists $R = R(\xi) > 0$ such that $\mathfrak{M}F(\xi,\rho) = F(\xi)$ whenever $|\rho| < R$.

We now give a basic identity which links the Lévy Laplacian and the spherical mean (cf. [L, Part 3,§69],[F]).

Proposition 1.1. Let $F \in \mathscr{C}^2(U)$ and $\xi \in U$. If F admits the spherical mean $\mathfrak{M}F(\xi,\rho)$ for $|\rho| < R$ with some $R > 0$, then

$$(1\text{-}3)\qquad \Delta F(\xi) = 2 \lim_{\rho\to 0} \frac{\mathfrak{M}F(\xi,\rho) - F(\xi)}{\rho^2},$$

in the sense that if either side exists, then so does the other and the two are equal.

Proof. We start with the identity:

$$(1\text{-}4)\qquad F(\xi+\rho h) - F(\xi) = \rho < F'(\xi), h > + \frac{\rho^2}{2} < F''(\xi)h, h > + R(\rho h),$$

where $h \in S^{n-1}$ and $|\rho| < R$. Integrating both sides of (1-4) over the sphere S^{n-1} we obtain

$$\int_{S^{n-1}} F(\xi+\rho h)dS_{n-1}(h) - F(\xi) =$$

$$= \frac{\rho^2}{2} \int_{S^{n-1}} < F''(\xi)h,h > dS_{n-1}(h) + \int_{S^{n-1}} R(\rho h)dS_{n-1}(h)$$

$$= \frac{\rho^2}{2n} \sum_{k=1}^{n} < F''(\xi)e_k, e_k > + \int_{S^{n-1}} R(\rho h)\, dS_{n-1}(h).$$

Since $R(\rho h) = o(\rho^2)$ uniformly in $h \in H$ with $\|h\| = 1$,

$$\lim_{\rho \to 0} \rho^{-2} \int_{S^{n-1}} R(\rho h)\, dS_{n-1}(h) = 0$$

uniformly in n. Hence the assertion follows immediately. Q.E.D.

The identity (1-3) illustrates that the Lévy Laplacian has a typical property of the finite dimensional Laplacian. Namely,

Corollary 1.2. A function $F \in \mathscr{E}^2(U)$ is harmonic on U if it possesses the mean value property on U.

The converse is false in general but we have

Proposition 1.3. A quadratic function $F(\xi) = < A\xi, \xi >$, A being a bounded symmetric operator on H, possesses the mean value property on H if and only if it is harmonic on H.

Proof. Taking Corollary 1.2 into account, we have only to prove the 'if' part. Suppose that F is harmonic on H. Then

$$(1\text{-}5) \qquad \Delta F(\xi) = 2 \lim_{N \to \infty} \frac{1}{N} \sum_{n=1}^{N} < Ae_n, e_n > = 0, \quad \xi \in H.$$

On the other hand, a direct calculation implies that

$$\mathfrak{M}F(\xi, \rho) = \lim_{n \to \infty} \int_{S^{n-1}} < A(\xi + \rho h), \xi + \rho h > dS_{n-1}(h)$$

$$= < A\xi, \xi > + \rho^2 \lim_{n \to \infty} \int_{S^{n-1}} < Ah, h > dS_{n-1}(h).$$

Hence

$$(1\text{-}6) \qquad \mathfrak{M}F(\xi, \rho) = F(\xi) + \rho^2 \lim_{n \to \infty} \frac{1}{n} \sum_{i=1}^{n} < Ae_i, e_i >.$$

Viewing (1-5) we have $\mathfrak{M}F(\xi, \rho) = F(\xi)$ for any $\xi \in H$ and $\rho \in \mathbf{R}$. Q.E.D.

§2. Regularly analytic functions

We first recall the standard notation of Fock space. The n-fold tensor product of H is denoted by $H^{\otimes n}$, $n \geqslant 1$. For $\xi_1, \cdots, \xi_n \in H$ put

$$\xi_1 \hat{\otimes} \cdots \hat{\otimes} \xi_n = \frac{1}{n!} \sum_{\sigma \in \mathfrak{S}_n} \xi_{\sigma(1)} \otimes \cdots \otimes \xi_{\sigma(n)} ,$$

where $\mathfrak{S}_n$ is the group of permutations of n symbols. Let $S^n H \subset H^{\otimes n}$ be the closed subspace spanned by such elements. We set $S^0 H = H^{\otimes 0} = \mathbf{R}$. The orthogonal direct sum

$$SH = \sum_{n=0}^{\infty} \oplus \, S^n H$$

equipped with the norm

$$\|a\|^2 = \sum_{n=0}^{\infty} \|a_n\|^2 , \quad a = \sum_{n=0}^{\infty} a_n , \quad a_n \in S^n H,$$

is called the *symmetric Hilbert space* or the *Fock space*.

It is convenient to introduce an operation $*$ in SH. For $a \in S^p H$ and $b \in S^q H$ with $p \geqslant q$ there exists an element $a*b$ uniquely determined by the condition that

$$\langle a , b \hat{\otimes} c \rangle = \langle a*b, c \rangle \quad \text{for all } c \in S^{p-q} H.$$

Obviously, $\|a*b\| \leqslant \|a\| \|b\|$.

Motivated by Lévy's regular functionals ([L, Part 1,§21]), we give the following

Definition. Let F be an $\mathbf{R}$-valued function defined in a neighborhood of $\xi_0 \in H$. Assume that F admits an expression

$$\text{(2-1)} \qquad F(\xi) = \sum_{n=0}^{\infty} \langle a_n , (\xi - \xi_0)^{\otimes n} \rangle , \quad a_n \in S^n H ,$$

in some neighborhood of ξ_0. We say that F is *regularly analytic at* ξ_0 if the power series $\sum_{n=0}^{\infty} \|a_n\| t^n$ has a non-zero radius of convergence.

If F is regularly analytic at ξ_0 the coefficients $\{a_n\}_{n=0}^{\infty}$ in the expression (2-1) are uniquely determined. We call (2-1) the *power series expansion* of F at ξ_0.

Definition. A function defined on an open subset $U \subset H$ is called *regularly analytic on* U if it is regularly analytic at every point of U. The space of all regularly analytic functions on U is denoted by $\mathcal{RA}(U)$.

We then have an obvious inclusion relation: $\mathcal{RA}(U) \subset \mathcal{A}(U) \subset \mathcal{C}^{\infty}(U)$, where $\mathcal{A}(U)$ denotes the space of analytic functions on U in the usual sense (e.g.,[N]). The followings are instructive examples.

Example 2.1. Consider a quadratic function $F(\xi) = \langle A\xi, \xi\rangle$, where A is a bounded symmetric operator on H. Obviously, $F \in \mathcal{A}(H)$. Moreover the following three conditions are equivalent:

(i) F is regularly analytic at some point $\xi_0 \in H$;

(ii) F is regularly analytic on H, i.e. $F \in \mathcal{RA}(H)$;

(iii) A is of Hilbert-Schmidt type.

Example 2.2. A finite linear combination of functions of the form $F_n(\xi) = \langle a_n, \xi^{\otimes n}\rangle$, $a_n \in S^nH$, is called a *Hilbert-Schmidt polynomial* in a literature (see [B]). It is clear that every Hilbert-Schmidt polynomial is regularly analytic on H.

Example 2.3. Given $a_n \in S^nH$, $n = 0,1,2,\cdots$, we put

$$F(\xi) = \sum_{n=0}^{\infty} \langle a_n, \xi^{\otimes n}\rangle, \quad \|\xi\| < R, \tag{2-2}$$

where R is the radius of convergence of the power series $\sum_{n=0}^{\infty} \|a_n\| t^n$. Then F is regularly analytic on $\mathcal{O} = \{ \xi \in H \ ; \ \|\xi\| < R \}$ and its power series expansion at $\xi_0 \in \mathcal{O}$ is given by

$$F(\xi) = \sum_{n=0}^{\infty} \langle a_n(\xi_0), (\xi-\xi_0)^{\otimes n}\rangle, \tag{2-3}$$

where

$$a_n(\xi_0) = \sum_{m=n}^{\infty} \binom{m}{n} a_m * \xi_0^{\otimes(m-n)} \in S^nH. \tag{2-4}$$

In fact, a formal calculation from which (2-3) follows is justified by the fact that the series (2-4) is convergent in S^nH and that the power series $\sum_{n=0}^{\infty} \|a_n(\xi_0)\| t^n$ admits a non-zero radius of convergence.

The above result immediately leads us to the following

Proposition 2.4. If F is regularly analytic at $\xi_0 \in H$, it is regularly analytic on some neighborhood of ξ_0.

§3. The mean value theorem

We shall prove the following main result.

Mean value theorem. Assume that a function F is regularly analytic at $\xi \in H$. Then there exists $R > 0$ such that $\mathfrak{M}F(\xi,\rho) = F(\xi)$ whenever $|\rho| < R$.

A function F on H is called a *homogeneous polynomial* of degree p if there is a symmetric p-form F_p on H such that $F(\xi) = F_p(\xi,\cdots,\xi)$ for any $\xi \in H$.

Lemma 3.1. Let F be a homogeneous polynomial of degree p.
(1) If p is odd, $\mathfrak{M}F(0,\rho) = F(0) = 0$ for any $\rho \in \mathbf{R}$.
(2) If $p = 0$, $\mathfrak{M}F(0,\rho) = F(0)$ for any $\rho \in \mathbf{R}$.
(3) If $p = 2q$ with an integer $q \geq 1$,

$$\mathfrak{M}F(0,\rho) = \rho^p (p-1)!! \lim_{n \to \infty} n^{-q} \sum_{i_1,\cdots,i_q=1}^{n} F_p(e_{i_1}, e_{i_1}, \cdots, e_{i_q}, e_{i_q}).$$

Proof. (1) is easily verified with the invariane of the measure $dS_{n-1}(h)$ under reflections. (2) is obvious. We shall prove (3). By definition we have

$$\mathfrak{M}F(0,\rho) = \lim_{n \to \infty} \int_{S^{n-1}} F(\rho h)\, dS_{n-1}(h) = \rho^p \lim_{n \to \infty} Q_n , \tag{3-1}$$

where

$$Q_n = \int_{S^{n-1}} F(h)\, dS_{n-1}(h) = \int_{S^{n-1}} F\left(\sum_{k=1}^{n} h_k e_k \right) dS_{n-1}(h) \tag{3-2}$$

$$= \sum_{k_1,\cdots,k_p=1}^{n} F_p(e_{k_1}, \cdots, e_{k_p}) \int_{S^{n-1}} h_{k_1} \cdots h_{k_p}\, dS_{n-1}(h).$$

We shall devote ourselves to computing the integral Q_n for $n > q$. Some notation is needed. For non-negative integers $\ell_1,\cdots,\ell_n$ we put

$$J_n(\ell_1,\cdots,\ell_n) = \frac{1}{\ell_1! \cdots \ell_n!} \int_{S^{n-1}} h_1^{\ell_1} \cdots h_n^{\ell_n}\, dS_{n-1}(h) , \tag{3-3}$$

where $h = (h_1,\cdots,h_n) \in S^{n-1}$. Obviously $J_n(\ell_1,\cdots,\ell_n) = 0$ unless every ℓ_j is even. In case where every ℓ_j is even, the integral (3-3) may be computed explicitly with the polar coordinate and we obtain

(3-4) $$J_n(\ell_1,\cdots,\ell_n) = \left\{(n+\ell-2)(n+\ell-4)\cdots(n+2)n \prod_{j=1}^{n} \ell_j!!\right\}^{-1},$$

where $\ell = \Sigma_{j=1}^{n} \ell_j$. With these notations we continue to compute (3-2).

$$Q_n = p! \sum_{\substack{\ell_1,\cdots,\ell_n \geq 0 \\ \ell_1+\cdots+\ell_n=p}} F_p(\underbrace{e_1,\cdots,e_1}_{\ell_1},\cdots,\underbrace{e_n,\cdots,e_n}_{\ell_n})J_n(\ell_1,\cdots,\ell_n)$$

$$= p! \sum_{\substack{k_1,\cdots,k_n \geq 0 \\ k_1+\cdots+k_n=q}} F_p(\underbrace{e_1,\cdots,e_1}_{2k_1},\cdots,\underbrace{e_n,\cdots,e_n}_{2k_n})J_n(2k_1,\cdots,2k_n).$$

We devide the above summation into two parts. Put

(3-5) $$S_n = p! \sum_{\substack{1 \geq k_1,\cdots,k_n \geq 0 \\ k_1+\cdots+k_n=q}} F_p(\underbrace{e_1,\cdots,e_1}_{2k_1},\cdots,\underbrace{e_n,\cdots,e_n}_{2k_n})J_n(2k_1,\cdots,2k_n)$$

and

(3-6) $$Q_n = S_n + T_n .$$

Since the number of terms appearing in T_n is $\binom{n+q-1}{q} - \binom{n}{q}$, in view of (3-4) we have

$$|T_n| \leq p!\left\{\binom{n+q-1}{q} - \binom{n}{q}\right\} \|F_p\| \{(n+p-2)(n+p-4)\cdots(n+2)n\}^{-1}$$

$$= \frac{p!}{q!} \|F_p\| \left\{\frac{(n+q-1)(n+q-2)\cdots(n+1)}{(n-1)(n-2)\cdots(n-q+1)} - 1\right\} \frac{n(n-1)\cdots(n-q+1)}{(n+p-2)(n+p-4)\cdots n}$$

$$\longrightarrow 0 \quad \text{as } n \longrightarrow \infty .$$

We therefore see from (3-4),(3-5) and (3-6) that

$$\lim_{n\to\infty} Q_n = \lim_{n\to\infty} S_n$$

$$= \lim_{n\to\infty} p! \sum_{\substack{1 \geq k_1,\cdots,k_n \geq 0 \\ k_1+\cdots+k_n=q}} F_p(\underbrace{e_1,\cdots,e_1}_{2k_1},\cdots,\underbrace{e_n,\cdots,e_n}_{2k_n})$$

$$\times 2^{-q} \{(n+p-2)(n+p-4)\cdots(n+2)n\}^{-1}$$

$$= \lim_{n\to\infty} p!\, 2^{-q}\{(n+p-2)(n+p-4)\cdots(n+2)n\}^{-1}$$

$$\times \frac{1}{q!} \sum_{\substack{i_1,\cdots,i_q=1 \\ \text{distinct}}}^{n} F_p(e_{i_1},e_{i_1},\cdots,e_{i_q},e_{i_q})$$

$$= (p-1)!! \lim_{n\to\infty} n^{-q} \sum_{i_1,\cdots,i_q=1}^{n} F_p(e_{i_1}, e_{i_1}, \cdots, e_{i_q}, e_{i_q}).$$

Inserting the last expression into (3-1) we get the desired result.

Q.E.D.

Lemma 3.2. If F is a Hilbert-Schmidt polynomial, $\mathfrak{M}F(0,\rho) = F(0)$ for any $\rho \in \mathbf{R}$.

Proof. Taking Lemma 3.1 into account, we have only to show the assertion for a homogeneous polynomial F of the form $F(\xi) = \langle a, \xi^{\otimes p} \rangle$, $a \in S^pH$, $p = 2q$ with $q \geqslant 1$. It follows from Lemma 3.1(3) that

$$\mathfrak{M}F(0,\rho) = \rho^p (p-1)!! \lim_{n\to\infty} n^{-q} \sum_{i_1,\cdots,i_q=1}^{n} \langle a, e_{i_1}^{\otimes 2} \otimes \cdots \otimes e_{i_q}^{\otimes 2} \rangle$$

$$= \rho^p (p-1)!! \lim_{n\to\infty} n^{-q} \sum_{i_1,\cdots,i_q=1}^{n} \langle a, e_{i_1}^{\otimes 2} \otimes \cdots \otimes e_{i_q}^{\otimes 2} \rangle .$$

Since $\{ e_{i_1}^{\otimes 2} \otimes \cdots \otimes e_{i_q}^{\otimes 2} \}_{i_1,\cdots,i_q=1}^{\infty}$ is a subset of a CONS of $H^{\otimes p}$,

$$\lim_{i_1,\cdots,i_q\to\infty} \langle a, e_{i_1}^{\otimes 2} \otimes \cdots \otimes e_{i_q}^{\otimes 2} \rangle = 0.$$

Hence $\mathfrak{M}F(0,\rho) = 0 = F(0)$. Q.E.D.

Proof of mean value theorem. Suppose that F is regularly analytic at $\xi_0 \in H$. Then there exists some $R > 0$ such that F is expressed by the power series:

$$F(\xi) = \sum_{n=0}^{\infty} \langle a_n, (\xi-\xi_0)^{\otimes n} \rangle, \quad \|\xi-\xi_0\| < R$$

and that the power series $\sum_{n=0}^{\infty} \|a_n\| t^n$ is convergent whenever $|t| < R$. We put

$$F_m(\xi) = \sum_{n=0}^{m} \langle a_n, (\xi-\xi_0)^{\otimes n} \rangle .$$

It then follows from Lemma 3.2 that

$$\mathfrak{M}F_m(\xi_0,\rho) = a_0 = F(\xi_0) .$$

Since $F_m(\xi)$ converges to $F(\xi)$ uniformly on the sphere of radius ρ ($|\rho| < R$) with center at ξ_0, F also admits the mean $\mathfrak{M}F(\xi_0,\rho)$ and

$$\mathfrak{M}F(\xi_0,\rho) = \lim_{m\to\infty} \mathfrak{M}F_m(\xi,\rho) = F(\xi_0) ,$$

whenever $|\rho| < R$. Q.E.D.

As an immediate consequence of the mean value theorem we obtain

Corollary 3.3. Every regularly analytic function on U is harmonic on it.

§4. Application to Brownian functionals

We start with a Gelfand triple: $E \subset H \subset E^*$. It follows from the famous Bochner-Minlos theorem that there exists a unique probability measure μ on E^* such that

$$e^{-\|\xi\|^2/2} = \int_{E^*} e^{i<x,\xi>} \, d\mu(x) \quad , \quad \xi \in E,$$

where $<\cdot,\cdot>$ stands for the canonical bilinear form on $E^* \times E$. The measure μ is called the *standard Gaussian measure* on E^*. Let us denote by $(L^2) = L^2(E^*,\mu)$ the Hilbert space of all square-integrable functions on E^* with values in $\mathbb{R}$. An *ordinary Brownian functional* is by definition a member of (L^2).

Following [KT] we introduce the *S-transform* which carries functions of $L^p(E^*,\mu)$, $1 < p \leqslant \infty$, into functions on E:

$$Sf(\xi) = \int_{E^*} f(x+\xi) \, d\mu(x)$$

$$= e^{-\|\xi\|^2/2} \int_{E^*} f(x) e^{<x,\xi>} \, d\mu(x) \quad , \quad \xi \in E,$$

where $f \in L^p(E^*,\mu)$, $1 < p \leqslant \infty$. Then Sf is continuously extended to a function of $\mathscr{E}^\infty(H)$ which will be denoted by the same symbol.

Proposition 4.1. For any $f \in (L^2)$ there exists a unique $a \in SH$ such that

$$Sf(\xi) = < a \; , \; \exp \xi > \; , \; \xi \in H,$$

where

$$\exp \xi = \sum_{n=0}^{\infty} (n!)^{-1/2} \, \xi^{\otimes n} \quad , \; \xi \in H.$$

Moreover, the correspondence $f \longmapsto a$ gives an isometric isomorphism from (L^2) onto SH.

Proof. We first recall a CONS in (L^2) consisting of Fourier-Hermite polynomials based on a previously fixed CONS $\{e_j\}_{j=1}^{\infty}$ in H. Let N_0^{∞} be the set of all sequences $\mathbf{n} = (n_j)_{j=1}^{\infty}$ of non-negative integers such that $|\mathbf{n}| = \sum_{j=1}^{\infty} n_j < \infty$. With each $\mathbf{n} = (n_j)_{j=1}^{\infty} \in N_0^{\infty}$ we associate a function on E^* :

$$h_{\mathbf{n}}(x) = \left(2^{|\mathbf{n}|}\ \mathbf{n}!\right)^{-1/2} \prod_{j=1}^{\infty} H_{n_j}(\langle x, e_j\rangle/\sqrt{2}\ ), \quad x \in E^* ,$$

where $\mathbf{n}! = \prod_{j=1}^{\infty} n_j!$ and H_n $(n \geqslant 0)$ is the Hermite polynomial defined by the generating function

$$e^{-t^2+2ts} = \sum_{n=0}^{\infty} \frac{t^n}{n!} H_n(s) .$$

It is known that $\{ h_{\mathbf{n}} ; \mathbf{n} \in N_0^{\infty} \}$ forms a CONS in (L^2). By a direct calculation one may obtain

$$Sh_{\mathbf{n}}(\xi) = (|\mathbf{n}|!)^{-1/2} < e_{\mathbf{n}} , \xi^{\otimes|\mathbf{n}|}> = < e_{\mathbf{n}} , \exp \xi > ,$$

where

$$e_{\mathbf{n}} = \left(\frac{|\mathbf{n}|!}{\mathbf{n}!}\right)^{1/2} e_1^{\otimes n_1} \otimes e_2^{\otimes n_2} \otimes \cdots .$$

With the help of the Fourier series expansion we see that, for any $f \in (L^2)$ there is some $a \in SH$ such that $Sf(\xi) = < a , \exp \xi >$. Since $\{ e_{\mathbf{n}} ; \mathbf{n} \in N_0^{\infty} \}$ is a CONS in SH, the correspondence $f \longmapsto a$ becomes an isometric isomorphism from (L^2) onto SH. Q.E.D.

Proposition 4.2. The S-transform of any $f \in (L^2)$ possesses the mean value property on H.

Proof. It follows from Proposition 4.1 that Sf is regularly analytic on H (cf. Example 2.3). Hence Sf possesses the mean value property on H by the mean value theorem. Q.E.D.

The following result, which was investigated in somewhat weaker forms in [H2,§8.5] and [K], is now immediate from Proposition 4.2 and Corollary 1.2.

Corollary 4.3. The Lévy Laplacian annihilates (L^2) in the sense that $\Delta(Sf) = 0$ for any $f \in (L^2)$.

Remark. The Lévy Laplacian is sometimes denoted by Δ_L to avoid confusing it with the number operator (or the Ornstein-Uhlenbeck operator) Δ_∞ defined uniquely by the condition:

$$\Delta_\infty f = -nf, \ f \in \mathcal{H}_n \ , \ n \geqslant 0.$$

Through the S-transform the number operator acts on $\mathcal{E}^\infty(H)$ as

$$\Delta_\infty F(\xi) = - < F'(\xi), \ \xi > \ , \ F \in \mathcal{E}^\infty(H), \ \xi \in H.$$

This illustrates an essential difference between the number operator and the Lévy Laplacian.

Acknowledgement. I am very grateful to the organizers of the conference for the kind invitation. Financial support by the Japan Association for Mathematical Sciences is also acknowledged.

References

[B] Balakrishnan,A.V.: A white noise version of the Girsanov formula. In: Itô, K. (ed.) Proceedings of the international symposium on stochastic differential equations, Kyoto, 1976, pp.1-19. Kinokuniya, Tokyo, 1978.

[F] Feller,M.N.: Infinite dimensional elliptic equations and operators of Lévy type. Russian Math.Surveys, **41**-4 (1986), 119-170.

[H1] Hida,T.: Analysis of Brownian functionals (2nd.ed.). Carleton Math.Lect.Notes No.13, 1978.

[H2] ———: Brownian motion. Springer-Verlag, 1980.

[H3] ———: Brownian motion and its functionals. Ricerche Mat. **34** (1985), 183-222.

[H4] ———: Analysis of Brownian functionals. Lecture Notes, IMA, University of Minnesota, 1986.

[H5] ———: in these proceedings.

[HS] Hida,T., Saitô,K.: White noise analysis and the Lévy Laplacian. In: Albeverio,S. et al. (eds.) Stochastic processes in physics and engineering. pp.177-184. D.Reidel Pub.Co., Dordrecht/Boston/Lancaster/Tokyo, 1988.

[KT] Kubo,I., Takenaka,S.: Calculus on Gaussian white noise I - IV. Proc.Japan Acad. **56A** (1980), 376-380; 411-416; **57A** (1981), 433-437; **58A** (1982), 186-189.

[K] Kuo,H.-H.: On Laplacian operators of generalized Brownian functionals. In: Itô,K., Hida,T. (eds.) Stochastic processes and their applications. Proceedings, Nagoya, July 1985 (Lect.Notes Math. vol.1203), pp.119-128. Springer-Verlag, 1986.

[L] Lévy,P.: Problèmes concrets d'analyse fonctionnelle, Gauthier-Villars, Paris, 1951.

[N] Nachbin,L.: Topology on spaces of holomorphic mappings. Springer Verlag, 1969.

[O1] Obata,N.: A note on certain permutation groups in the infinite dimensional rotation group. Nagoya Math.J. **109** (1988), 91-107.

[O2] ———: Analysis of the Lévy Laplacian. Soochow J.Math. **14** (1988), 115-119.

[O3] ———: Density of natural numbers and the Lévy group. To appear in J. Number Theory, **30** (1988).

[O4] ———: The Lévy Laplacian and infinite dimensional rotation groups. Submitted to Nagoya Math.J.

BIMEASURES AND HARMONIZABLE PROCESSES

(Analysis, Classification, and Representation)

M. M. Rao

Introduction. Bimeasures arise in studies such as multiparameter martingales, second order random processes, representation theory of linear operators, harmonic analysis, and are related to (often induced by) bilinear or sesquilinear forms on suitable function spaces. These play a key role in the theory of harmonizable random fields in their structural analyses as well as classifications. Although one may regard bimeasures often as extensions of noncartesian products of pairs of scalar measures, their integration, in the general case, departs significantly from the standard product integration and new techniques are needed for their employment in applications. If the underlying measure space has a group structure, then it is also possible to study extensions of the classical measure algebra theory for bimeasures. Thus the purpose of this article is to present a somewhat detailed description of some of these results, leading to a classification of harmonizable functions including certain new developments. For instance, the material in Section 2.2.3 below did not appear in print before. A brief account of the work, given in three parts, will be described here since then one gets a better perspective of the theory covered. It largely complements the recent detailed account presented in Chang and Rao (1986).

The first part is on bimeasure theory. Starting with a general concept, the bimeasure integrals in the sense of Morse and Transue (1955), termed MT-integrals hereafter, are introduced. The Lebesgue type limit theorems are not valid for them. So a subclass, termed strict MT-integrals, is isolated for use in stochastic theory. Specializing to groups, bimeasure algebras with a suitable convolution product are described. If these bimeasures are also positive definite, then the structure of function spaces on them are treated. A few extensions, if some of these objects are vector valued, are also included since such results are useful for multidimensional harmonizable random fields to be discussed later on in the paper.

The preceding work is applied, in Part II, for the stochastic theory. The primary class is the harmonizable random fields. It is used in classifying harmonizability into weak, strong, ultraweak, strict, and other types. Integral representations of harmonizable fields on LCA (= locally compact abelian) groups are given. For the nonabelian case, several new problems arise. Here a novel treatment is given for a class of the so-called type I groups. For them an integral representation is obtained, basing it on the structural analysis of these groups due to Mautner, Segal and others. This extends Yaglom's (1960) fundamental work on stationary random fields in many ways. Then stationary dilations and linear filtering problems are discussed. Further, some applications here lead to a study of harmonizability on hypergroups. This is briefly sketched. Also extensions to strictly harmonizable functions as a subclass of stable processes are included. Open problems suggested by this analysis abound, and several are pointed out at many places.

The final part discusses multidimensional extensions of the foregoing, and are motivated by applications. However, to keep the article in bounds, this part is unfortunately curtailed. Here weak and strong V-boundedness of Bochner's concept have to be separated, since one of them admits a dilation and the other does not. To see these distinctions clearly, a discussion of vector (and module) harmonizability is sketched. Also, the multiplicity problem of these processes and some related ones such as almost harmonizability are touched on. Let us now turn to the details. (Notation is given at the end of the paper.)

Part I: Bimeasure Theory

1.1 *The general concept.* A systematic study of bimeasures originated with the work of Fréchet (1915) and was continued, after a long lapse, in several papers by Morse and Transue (1949-56), and also Lévy (1946). The concept may be introduced as follows. Let $(\Omega_i, \Sigma_i), i = 1,2$ be a pair of measurable spaces and $\beta : \Sigma_1 \times \Sigma_2 \to \mathbb{C}$ be a mapping such that $\beta(A, \cdot)$ and $\beta(\cdot, B)$ are (complex) measures for each $A \in \Sigma_1$ and $B \in \Sigma_2$. Then β is called a (complex) *bimeasure.* It should be noted that this definition does not assume (or imply) that β has an extension to be a (complex) measure on the generated product σ-algebra $\Sigma_1 \otimes \Sigma_2$.

The preceding remark is better understood if the concept is given an alternative form through tensor products and bilinear mappings on them. Since these products will come up again later on, let us recall them. Thus, if $\mathcal{X}_1, \mathcal{X}_2$ are Banach spaces, $\mathcal{X}_1 \otimes \mathcal{X}_2$ denotes the vector space of all formal sums of the form $x = \sum_{i=1}^{n} x_{1i} \otimes x_{2i}, x_{ji} \in \mathcal{X}_j, j = 1,2$. If there is a norm α on this space such that $\|x_1 \otimes x_2\|_\alpha = \|x_1\| \|x_2\|$, termed a cross-norm α, let $\mathcal{X}_1 \otimes_\alpha \mathcal{X}_2$ be the completion for $\|\cdot\|_\alpha$. There exist several such norms, but those of interest below are the greatest and least cross-norms denoted $\|\cdot\|_\gamma$ and $\|\cdot\|_\lambda$, and are given by

$$\|x\|_\gamma = \inf\{\sum_{i=1}^{n} \|x_{1i}\|_{\mathcal{X}}\|x_{2i}\|_{\mathcal{X}} : x = \sum_{i=1}^{n} x_{1i} \otimes x_{2i}, n \geq 1\}, \tag{1}$$

and $\mathcal{X}_i^*$ denoting the adjoint space of $\mathcal{X}_i$,

$$\|x\|_\lambda = \sup\{|\sum_{i=1}^{n} \ell_1(x_{1i})\ell_2(x_{2i})| : \ell_i \in \mathcal{X}_i^*, \|\ell_i\| \leq 1, i = 1, 2, n \geq 1\}. \tag{2}$$

The corresponding completed spaces are denoted by $\mathcal{X}_1 \otimes_\gamma \mathcal{X}_2$ or $\mathcal{X}_1 \widehat{\otimes} \mathcal{X}_2$, and $\mathcal{X}_1 \otimes_\lambda \mathcal{X}_2$ or $\mathcal{X}_1 \widehat{\widehat{\otimes}} \mathcal{X}_2$. These are termed the tensor product spaces. If $L(\mathcal{X}_1, \mathcal{X}_2)$ stands for the space of continuous linear operators on $\mathcal{X}_1$ into $\mathcal{X}_2$, the following relations are true and classical:

Proposition 1. *Let $\mathcal{X}_1, \mathcal{X}_2$ be Banach spaces. Then the following identifications hold:*

(a) $(\mathcal{X}_1 \widehat{\otimes} \mathcal{X}_2)^* \cong L(\mathcal{X}_1, \mathcal{X}_2^*), (\cong L(\mathcal{X}_2, \mathcal{X}_1^*))$,

(b) $(\mathcal{X}_1 \widehat{\widehat{\otimes}} \mathcal{X}_2)^* \hookrightarrow L(\mathcal{X}_1^*, \mathcal{X}_2)$

where "$\cong$" is an isometric isomorphism in (a) *and "$\hookrightarrow$" in* (b) *is an isometric imbedding into the second space.*

A specialization of (a) gives an alternative definition of bimeasures. Thus let Ω_i be locally compact and $C_0(\Omega_i)$ be the continuous scalar functions on Ω_i vanishing at "∞", $i = 1, 2$. If $V(\Omega_1, \Omega_2) = C_0(\Omega_1) \widehat{\otimes} C_0(\Omega_2)$ where $\mathcal{X}_i = C_0(\Omega_i)$ in the above result, with uniform ($= \|\cdot\|_\infty$) norms, thus $V(\Omega_1, \Omega_2)^* \cong L(C_0(\Omega_1), C_0(\Omega_2)^*)$, then the correspondence is given by (the topology is always Hausdorff for the work here)

$$B(f, g) = F(f \otimes g) = T(f)g,\ f \in C_0(\Omega_1), g \in C_0(\Omega_2), \tag{2'}$$

with $F \in V(\Omega_1, \Omega_2)^*$ and $T : C_0(\Omega_1) \to C_0(\Omega_1)^*$, a bounded linear operator. Here $B : C_0(\Omega_1) \times C_0(\Omega_2) \to \mathbb{C}$ is the bounded bilinear form corresponding to F. Since $C_0(\Omega_2)^* = M(\Omega_2)$, the space of regular bounded scalar (= Radon) measures with total variation norm, so that $Tf \in M(\Omega_2)$, by the general Riesz-Markov theorem one has

$$(Tf)(\cdot) = \int_{\Omega_1} f(\omega_1)\mu(d\omega_1, \cdot) \in M(\Omega_2). \tag{3}$$

This holds since $M(\Omega_2)$ is weakly sequentially complete so that T is a weakly compact operator (cf. Dunford-Schwartz (1958), Theorems VI.7.3 and IV.9.9). Letting $\mu^f(\cdot) = (Tf)(\cdot)$ of

$(3), (2')$ becomes

$$
\begin{aligned}
F(f \otimes g) = B(f,g) &= (Tf)(g) \\
&= \int_{\Omega_2} g(\omega_2)\mu^f(d\omega_2) \\
&= \int_{\Omega_2} g(\omega_2)[\int_{\Omega_1} f(\omega_1)\mu(d\omega_1,\cdot)](d\omega_2), \\
&= \int_{\Omega_2}\int_{\Omega_1} (f,g)(\omega_1,\omega_2)\mu(d\omega_1,d\omega_2),
\end{aligned}
\tag{4}
$$

by definition of this last symbol. Thus, $B(\cdot,\cdot)$ may be identified with $\mu(\cdot,\cdot)$, a bimeasure. For this reason Voropoulos (1967) calls each member of $V(\Omega_1,\Omega_2)^*$, also denoted $BM(\Omega_1,\Omega_2)$, a bimeasure. (Although he defined this for Ω_i compact, the local compactness presents no difficulty here.) Also

$$
\begin{aligned}
\|F\| = \|B\| &= \sup\{|B(f,g)| : \|f\|_\infty \le 1, \|g\|_\infty \le 1\} \\
&= \sup\{|\int_{\Omega_2}\int_{\Omega_1} (f,g)(\omega_1,\omega_2)\mu(d\omega_1,d\omega_2)| : \|f\|_\infty \le 1, \|g\|_\infty \le 1\} \\
&= \|\mu\|, \text{ the ``semi-variation'' of } \mu.
\end{aligned}
\tag{5}
$$

For a bimeasure μ, the symbol $\|\mu\|$ should be termed Fréchet variation, following the work by Fréchet (1915). Since $V(\Omega_1,\Omega_2)^* \underset{\neq}{\supset} M(\Omega_1 \times \Omega_2)[= C_0(\Omega_1 \times \Omega_2)^*]$, in general μ is not a restriction of some scalar measure $\tilde{\mu}$ on the Borel σ-algebra $\mathcal{B}(\Omega_1 \times \Omega_2)$ to $\mathcal{B}(\Omega_1) \times \mathcal{B}(\Omega_2)$. This makes the last integration symbol of (4) different from the Lebesgue definition, so that a generalized concept of integration, retaining the property (5), is needed. This is provided in the next section and is used throughout the article.

1.2 *Bimeasure integrals.* Let (Ω_i, Σ_i) be a measurable space and $f_i : \Omega_i \to \mathbb{C}$ be measurable for $\Sigma_i, i = 1, 2$. If $\beta : \Sigma_1 \times \Sigma_2 \to \mathbb{C}$ is a bimeasure, suppose that f_1 is $\beta(\cdot, B)$ and f_2 is $\beta(A,\cdot)$ - integrable for each $A \in \Sigma_1, B \in \Sigma_2$. Then

$$
\beta^{f_1}(B) = \int_{\Omega_1} f(\omega_1)\beta(d\omega_1, B);\, \beta^{f_2}(A) = \int_{\Omega_2} f_2(\omega_2)\beta(A, d\omega_2), \tag{6}
$$

exist and the β^{f_i} are σ-additive. Suppose moreover that f_2 is β^{f_1}-integrable, f_1 is β^{f_2}-integrable (Lebesgue's sense). Then (f_1, f_2) is said to be β*-integrable* (in the MT-sense) if the following equality holds:

$$
\int_{\Omega_1} f_1(\omega_1)\beta^{f_2}(d\omega_1) = \int_{\Omega_2} f_2(\omega_2)\beta^{f_1}(d\omega_2)(= \int_{\Omega_1}\int_{\Omega_2} (f_1, f_2)\beta(d\omega_1, d\omega_2)), \tag{7}
$$

where the double integral on the right of (7) (and in (4)) is, by definition, the common value of the first two terms. [The original definitions of MT-integral is slightly different but is equivalent to this one for the work below.] If $\Omega_1 = \Omega_2$, then f is termed *β-integrable* if (f, f) is such.

Regarding this definition, it must be noted that the left side integrals of (7) can exist and be not equal. In comparing (4) and (7) one can show that all bounded Borel functions (f, g) are β-integrable, although β is only a bimeasure. It was remarked that (5) is Fréchet variation. This can be written abstractly as follows, i.e., the *Fréchet variation* of μ is:

$$\begin{aligned}\|\mu\| &= \|\mu\|(\Omega_1, \Omega_2)\\ &= \sup\{|\sum_{i=1}^{n}\sum_{j=1}^{n} a_i\bar{b}_j\mu(A_i, B_j)| : \{A_i\}_1^n, \{B_j\}_1^n \text{ disjoint sets from } \Sigma_1, \Sigma_2, |a_i| \le 1,\\ &\quad |b_j| \le 1, n \ge 1\}. \end{aligned} \tag{8}$$

A more restricted concept is *Vitali's variation* which is given by

$$\begin{aligned}|\mu| &= |\mu|(\Omega_1, \Omega_2)\\ &= \sup\{\sum_{i=1}^{n}\sum_{j=1}^{n} |\mu(A_i, B_j)| : \{A_i\}_1^n, \{B_j\}_1^n \text{ are as in (8)}, n \ge 1\}. \end{aligned} \tag{9}$$

It is clear that $\|\mu\| \le |\mu|$; and there is strict inequality when $|\mu| = +\infty$. Some properties of these variations and of the MT-integral of (7) will be included to distinguish it from the Lebesgue integral.

If a bimeasure μ has finite Vitali variation then the integral in (7) coincides with the standard Lebesgue concept, and μ can also be extended to a scalar measure $\tilde{\mu}$ onto $\Sigma_1 \otimes \Sigma_2$ uniquely. However, this statement fails if $|\mu|(\Omega_1, \Omega_2) = +\infty$ without further restrictions. It can be shown that $\|\mu\|(\Omega_1, \Omega_2) < \infty$ always, and the MT-integral of (7), in general, is not absolutely continuous in that (f, g) is μ-integrable does not imply the same of $(|f|, |g|)$. Also the dominated convergence theorem does not hold for the MT-integral. The following approximation and a sufficient condition for the existence of the MT-integral can be stated from the work of Morse and Transue, and it is useful in applications.

Theorem 1. *Let Ω_i be locally compact, $C_c(\Omega_i)$ the space of scalar continuous functions with compact supports and Σ_i be Borel σ -algebras of $\Omega_i, i = 1, 2$. If $\beta : \Sigma_1 \times \Sigma_2 \to \mathbb{C}$ is a* *bimeasure and* $B : C_c(\Omega_1) \times C_c(\Omega_2) \to \mathbb{C}$ *is the bilinear form defined by* (4) *for this* β, *let*

$$B_*(p,q) = \sup\{|B(f_1,f_2)| : |f_1| \le p, |f_2| \le q, f_i \in C_c(\Omega_i)\}$$

$$B^*(u,v) = \inf\{B_*(p,q) : p \ge u \ge 0, q \ge v \ge 0\},$$

where p,q *are lower semicontinuous and* u,v *are Borel functions on* Ω_1,Ω_2. *Suppose that*

$$(i) B^*(u,f_2) < \infty \text{ and} (ii) B^*(f_1,v) < \infty, 0 \le f_i \in C_c(\Omega_i).$$

Then $\beta^u(\cdot)$ *and* $\beta^v(\cdot)$ *of* (6) *are* σ *-additive. If moreover* $B^*(|u|,|v|) < \infty$, *then the equality* (7) *holds so that* (u,v) *is* β *-integrable. In particular each bounded Borel pair* (u,v) *is* β*-integrable.*

Although the dominated convergence statement is not available, the following type of special approximation is nevertheless true.

Theorem 2. *Let* Ω_i, β *and* B *be as in the above theorem and* (f_1,f_2) *be* β*-integrable. If* $|f_i| \le p_i$ *with* p_i *lower semi-continuous (in particular if* $|f_i|$ *itself is lower semi-continuous,* $|f_i| = p_i$ *can be taken) and* $\epsilon > 0$ *is given, then there exist* $u_i \in C_c(\Omega_i)$ *such that* $|u_i| \le p_i, i =$ $1,2$, *and one has*

$$|B(f_1,f_2) - B(u_1,u_2)| < \epsilon. \tag{10}$$

This type of approximation from below by compactly based continuous functions, given by (10), is quite useful. Both these results, simple in form, are not easy. The details are given in Morse and Transue (1956). However, this approximation is not strong enough for applications in stochastic theory and elsewhere. Note that the MT-integral is weaker than the Lebesgue concept since the β-integrability of (f,g) does not imply that of $(|f|,|g|)$. It is of Riemann type and an additional uniformity condition can and should be added to obtain a dominated convergence statement. Also in the MT-integral it is significant that $(f,g)(\omega_1,\omega_2) = f(\omega_1)g(\omega_2)$ is a product of an ω_1 and an ω_2-function and not one of the form: $h : \Omega_1 \times \Omega_2 \to \mathbb{C}$, measurable relative to $\Sigma_1 \otimes \Sigma_2$. An extension of the integral to the latter type relative to a bimeasure has been given by Kluvanek (1981). A brief discussion of this will be included to illuminate the problem at hand, since the definition is patterned after that of Dunford and Schwartz [(1958), Sec. IV.10], the latter playing a pivotal role in our theory.

Let $f_n : \Omega = \Omega_1 \times \Omega_2 \to \mathbb{C}$ be a simple function so that $f_n = \sum\limits_{i=1}^{k_n}\sum\limits_{j=1}^{k_n} a_{ij}^n \chi_{A_i^n}\chi_{B_i^n}, A_i^n \in$ $\Sigma_1, B_j^n \in \Sigma_2$, disjoint. If $\beta : \Sigma_1 \times \Sigma_2 \to \mathbb{C}$ is a bimeasure, define

$$\int_\Omega f_n d\beta = \sum_{i=1}^{k_n}\sum_{j=1}^{k_n} a_{ij}^n \beta(A_i^n, B_j^n). \tag{11}$$

If $f_n \to h$ pointwise and $\{\int_\Omega f_n d\beta, n \geq 1\} \subset \mathbb{C}$ is Canchy with a limit α_o, one can define $\alpha_o = \int_\Omega h d\beta$. It can be verified that $h \longmapsto \int_\Omega h d\beta$ is well-defined, and is linear. An equivalent form of this is given by Kluvanek (1981) even when the range $\mathbb{C}$ is replaced by a Banach space. He showed the usefulness of this extension with examples including the disintegration problem. However, the dominated or bounded convergence theorem does not hold for it. Thus a restriction of the MT-integral is needed for these limit operations to hold. Such a result will now be given since one needs it in the stochastic analysis below.

1.3 *The strict integral as a subclass.* If (Ω_i, Σ_i) is a measurable space, $f_i : \Omega_i \to \mathbb{C}$ is measurable for $\Sigma_i, i = 1, 2$, and $\beta : \Sigma_1 \times \Sigma_2 \to \mathbb{C}$ is a bimeasure, suppose that f_1 is $\beta(d\omega_1, \cdot)$ and f_2 is $\beta(\cdot, d\omega_2)$ -integrable in the D-S sense, treating these as vector measures. Then define

$$\tilde{\beta}_1^F : A \longmapsto \int_F f_2(\omega_2)[= (\int_F f_2(\omega_2)\beta(\cdot, d\omega_2))(A)], A \in \Sigma_1,$$

$$\tilde{\beta}_2^E : B \longmapsto \int_E f_1(\omega_1)\beta(d\omega_1, B)[= (\int_E f_1(\omega_1)\beta(d\omega_1, \cdot))(B)], B \in \Sigma_2,$$

for each $E \in \Sigma_1, F \in \Sigma_2$. $\tilde{\beta}_1^F$ and $\tilde{\beta}_2^E$ are complex measures. The pair (f_1, f_2) is termed *strictly β-integrable* if $(i) f_1$ is $\tilde{\beta}_1^F$ and f_2 is $\tilde{\beta}_2^E$-integrable for each pair E and F, and (ii) one has

$$\int_E f_1(\omega_1)\tilde{\beta}_1^F(d\omega_1) = \int_F f_2(\omega_2)\tilde{\beta}_2^E(d\omega_2). \tag{12}$$

The common value is denoted $\int_E \int_F^* f_1(\omega_1)f_2(\omega_2)\beta(d\omega_1, d\omega_2)$. Note that the difference between this and the original concept is that (12) must hold for each pair E, F now, whereas in (7) it should only hold for $E = \Omega_1$ and $F = \Omega_2$. This stregthening of the definition (motivated by Theorem 1.2.1 above) renders the validity of a useful dominated convergence theorem and restores the absolute continuity property, although it is still weaker than the Lebesgue concept. One should also observe that only the class of β-integrable functions is restricted here, but not the class of bimeasures. For instance, β need not admit a Jordan decomposition and thus the corresponding Lebesgue theory does not also hold for strict β-integrals.

The following two results will illuminate the structure:

Theorem 1. *Let $(\Omega_i, \Sigma_i), i = 1, 2$ and β be as above. If $f_{in} : \Omega_i \to \mathbb{C}$, is Σ_i-measurable, $n \geq 1$, and $f_{in} \to f_i$ pointwise as $n \to \infty$, suppose that $|f_{in}| \leq g_i, i = 1, 2$ where (g_1, g_2) is strictly β-integrable. Then both (f_{1n}, f_{2n}) and (f_1, f_2) are β-integrable and*

$$\int_A \int_B^* f_1(\omega_1)f_2(\omega_2)\beta(d\omega_1, d\omega_2) = \lim_{n\to\infty} \int_A \int_B^* f_{1n}(\omega_1)f_{2n}(\omega_2)\beta(d\omega_1, d\omega_2). \tag{13}$$

In particular, if $\{f_{in}, n \geq 1\}$ is bounded, then (13) holds so that the bounded convergence criterion is valid.

A proof of this result is given in Chang and Rao [(1986), p.45]. It is of interest to note that when β is also positive definite with $\Omega_1 = \Omega_2$ and $\Sigma_1 = \Sigma_2 = \Sigma$, so that

$$\sum_{i=1}^{n}\sum_{j=1}^{n} a_i \bar{a}_j \beta(A_i, A_j) \geq 0, A_i \in \Sigma, a_i \in \mathbb{C}, \tag{14}$$

one can obtain a (simpler) characterization of the strict β-integral through its relation to the D-S definition. This is given in Part II below. On the other hand, if $\beta : \Sigma_1 \times \Sigma_2 \to \overline{\mathbb{R}}^+$, then the variations of Vitali and Fréchet coincide, and the theory simplifies; and β essentially extends to a measure on $\Sigma_1 \otimes \Sigma_2$. More precisely one has:

Theorem 2. *Let $(\Omega_i, \Sigma_i), i = 1,2$ be Borelian spaces where Ω_i is a Hausdorff space. If $\beta : \Sigma_1 \times \Sigma_2 \to \overline{\mathbb{R}}^+$ is a Radon bimeasure,* i.e. *$\beta(K_1, K_2) < \infty$ for each compact set $K_i \subset \Omega_i$ and is inner regular in the sense that*

$$\beta(A, B) = \sup\{\beta(K_1, K_2) : K_1 \subset A, K_2 \subset B, K_i \text{ compact}\}, \tag{15}$$

$A \in \Sigma_1, B \in \Sigma_2$, then β admits an extension to a Radon measure μ on $\Sigma_1 \otimes \Sigma_2$ so that $\beta(A, B) = \mu(A \times B)$.

The method of proof is standard but several details have to be filled in. These are available from Berg, Christensen, and Ressel [(1985), p.24]. This result allows an immediate extension of the Fubini-Tonelli type theorems for noncartesian product measures. In particular, this result gives a characterization of the *positive* elements of $V(\Omega_1, \Omega_2)^*$ of Section 1.1 above. The general set $V(\Omega_1, \Omega_2)^*$ will now be examined if the Ω_i have a group structure in addition.

1.4 *Bimeasure algebras on locally compact groups.* A refinement and specialization of the preceding work for locally compact groups Ω_i is of interest not only for applications of harmonizable random fields, but also because it generalizes the study of the classical group algebra $M(\Omega_1 \times \Omega_2)$ and unifies other results. Thus if β is a bimeasure on $\Sigma_1 \times \Sigma_2$ with Ω_i as LCA groups, one can define the Fourier transform $\widehat{\beta}$ of β by the formula (with the strict integrals on which "$*$" is dropped):

$$\widehat{\beta}(\gamma, \delta) = \int_{\widehat{\Omega}_1}\int_{\widehat{\Omega}_2} < \gamma, \gamma' >< \overline{\delta, \delta'} > \beta(d\gamma', d\delta'), \tag{16}$$

for $(\gamma, \gamma',) \in \Omega_1 \times \Omega_2, <\gamma, \gamma\prime>$ being the duality pairing ($\widehat{\Omega}_i$ is the dual of Ω_i). It follows from (5) and (16) that

$$\|\widehat{\beta}\|_\infty = \sup\{|\widehat{\beta}(\gamma,\delta)| : \gamma \in \Omega_1, \delta \in \Omega_2\} \leq \|\beta\|, \beta \in BM(\Omega_1, \Omega_2). \tag{17}$$

In a similar way one can define the *convolution* operation in the space of bimeasures $BM(\Omega_1,\Omega_2)$, denoted $\beta_1 * \beta_2$, by

$$(\beta_1 * \beta_2)(A, B) = \int_{\Omega_1}\int_{\Omega_2} \beta_1(A-\gamma, B-\delta)\beta_2(d\gamma, d\delta), \tag{18}$$

for each pair β_1, β_2 in $BM(\Omega_1, \Omega_2)$, and $A \in \Sigma_1, B \in \Sigma_2$. It is not hard to see that $\beta_1 * \beta_2 \in BM(\Omega_1, \Omega_2)$, and $\|\beta_1 * \beta_2\| \leq \|\beta_1\|\|\beta_2|$. However, a more refined analysis is possible only after proving a uniqueness theorem for the bimeasure Fourier transform and an employment of Grothendieck's inequality. One form of the latter states that for each $\beta \in BM(\Omega_1, \Omega_2)(\Omega_i$ need not be groups for this) there exists a pair of Radon probability measures μ_1, μ_2 on Ω_1, Ω_2 such that

$$\left|\int_{\Omega_1}\int_{\Omega_2} f(\omega_1)g(\omega_2)\beta(d\omega_1, d\omega_2)\right|^2 \leq C\int_{\Omega_1} |f(\omega_1)|^2\mu_1(d\omega_1)\int_{\Omega_2} |g(\omega_2)|^2\mu_2(d\omega_2), \tag{19}$$

where C is an absolute constant. If $\Omega_1 = \Omega_2$, one may choose $\mu_1 = \mu_2$. Using (19), Graham and Schreiber (1984) have made a detailed study of BM (Ω_1, Ω_2) for the LCA groups Ω_i, where $\widehat{\beta}$ and the convolution are defined differently. With the MT-integral one can show that both these definitions agree. Moreover, the work of these authors shows that $\|\beta_1 * \beta_2\| \leq C^2\|\beta_1\|\|\beta_2\|$, so that $BM(\Omega_1, \Omega_2)$ is a Banach algebra with this norm constant. Using still different techniques Ylinen (1987) recently showed that, with an *equivalent* norm, one can take $C = 1$ in the last inequality so that $BM(\Omega_1, \Omega_2)$ is a (standard) Banach algebra. It may be noted that (18) extends to noncommutative groups without any change.

If $VM(\Omega_1, \Omega_2)$ is the subspace of $BM(\Omega_1, \Omega_2)$ consisting of those bimeasures of finite Vitali variation, then it is known (and easily verified) that $(BM(\Omega_1, \Omega_2), \|\cdot\|)$ and $(VM(\Omega_1, \Omega_2), |\cdot|)$ are Banach spaces. Since by (8) and (9), $\|\cdot\| \leq |\cdot|$, a question of interest here is about the density of $VM(\Omega_1, \Omega_2)$ in the topology of the latter. This was raised in Chang and Rao [(1986), p. 33], but a negative solution is obtained from Graham and Schreiber [(1984), corollary 5.10], when Ω_1, Ω_2 are groups. This involves a delicate analysis.

To appreciate the structure of $BM(\Omega_1, \Omega_2)$, which is of interest even if $\Omega_1 = \Omega_2 = \mathbb{R}$ in stochastic theory, the above solution and a related result will be presented.

Let Ω_i be an LCA group with Γ_i as its dual group, $i = 1, 2$. Let $S(\Gamma_1, \Gamma_2) = BM(\Omega_1, \Omega_2)\hat{} = \{\hat{\beta} : \beta \in BM(\Omega_1, \Omega_2)\}$, and $\Delta(\Omega) = \{(\omega, \omega) : \omega \in \Omega\}$ the diagonal set of $\Omega \times \Omega$. Then one has:

Theorem 1. *If $f : \hat{G} \to \mathbb{C}$ is a uniformly continuous bounded function on the dual of an LCA group G, then there is a $\hat{\beta} \in S(\hat{G}, \hat{G})$ such that $\hat{\beta}|\Delta(\hat{G}) = f$, i.e.,*

$$\hat{\beta}(x, x) = f(x), x \in \hat{G}. \tag{20}$$

If further G is nondiscrete, then $VM(G, G)$ is not dense in $BM(G, G)$ in the (norm) topology of the latter.

A proof of this result is based on several other propositions, and is given in the above authors' paper. In passing one should note another fact about $BM(\Omega_1, \Omega_2)$. An element $\beta \in BM(\Omega_1, \Omega_2)$ is termed *continuous* or *diffuse* if $\beta(A_1, A_2) = 0$ for all finite sets $A_i \subset \Omega_i$, and *discrete* if there are increasing sequences of finite sets A_{in} such that, letting $\beta_n = \beta|\Sigma_1(A_{1n}) \times \Sigma_2(A_{2n})$, then $\|\beta - \beta_n\| \to 0$ as $n \to \infty$. With these concepts the following result, from Graham and Schreiber (1984-88) and Gilbert, Ito, and Schreiber (1985), clarifies the structure of $BM(\Omega_1, \Omega_2)$ further.

Theorem 2. *Let Ω_i be a locally compact space, $i = 1, 2$, and $BM(\Omega_1, \Omega_2)$ be the Banach space of bimeasures on (Σ_1, Σ_2) as before. Then one has:*

(i)

$$BM(\Omega_1, \Omega_2) = BM_c(\Omega_1, \Omega_2) \oplus BM_d(\Omega_1, \Omega_2) \tag{21}$$

where $BM_c(\Omega_1, \Omega_2)(BM_d(\Omega_1, \Omega_2))$ is the set of diffuse (discrete) bimeasures of $BM(\Omega_1, \Omega_2)$ which is a closed subspace. Further, the mapping $Q : \beta \longmapsto \beta_c, \beta \in BM(\Omega_1, \Omega_2)$ is a norm-decreasing projection whose kernel is $BM_d(\Omega_1, \Omega_2)$.

(ii) *If Ω_1, Ω_2 are also groups, then $BM_c(\Omega_1, \Omega_2)$ is a closed ideal, and if $BM_a(\Omega_1, \Omega_2) \subset BM_c(\Omega_1, \Omega_2)$ is the set of bimeasures of finite Vitali variation whose extensions are absolutely continuous relative to a (left) Haar measure on $\Omega_1 \times \Omega_2$ then $BM_a(\Omega_1, \Omega_2)$ is also a closed ideal in $BM(\Omega_1, \Omega_2)$ onto which there is no bounded projection.*

It may be observed that although $BM_a(\Omega_1, \Omega_2)$ is an $(AL-)$ space $BM(\Omega_1, \Omega_2)$ is not, and the decomposition (21) does not imply norm additivity. A simple counter example is given to this effect in the first of the above papers.

Using these ideas one may study algebras of multimeasures. In fact, Voropoulos (1968) considered the space $V(\Omega_1, \ldots, \Omega_n) = \widehat{\otimes}_{1\leq i\leq n} C_0(\Omega_i)$, with $C_0(\Omega_i)$ replaced by $L^\infty(\Omega_i)$, under the name tensor algebras, and calling the members of $V(\Omega_1, \ldots, \Omega_n)^*$ the n-linear forms and multimeasures. Using a procedure that is similar to that of the bimeasure case discussed earlier, one can consider Fourier transforms of $u \in V(\Omega_1, \ldots, \Omega_n)^*$ where Ω_i is an LCA group. Thus $\widehat{u}$ is given by

$$\widehat{u}(\gamma_1, \ldots, \gamma_n) = < u, \gamma_1 \otimes \cdots \otimes \gamma_n >, \gamma_i \in C_b(\widehat{\Omega}_i), \tag{22}$$

$\widehat{\Omega}_i$ being the dual group of Ω_i. However, this extension brings in some unpleasant phenomenon. For instance, if $n = 3$, it was observed by Graham and Schreiber (1987) that $\widehat{u}$ need not be uniformly continuous, and that for $u, v \in V(\Omega_1, \Omega_2, \Omega_3)^*$, $\widehat{uv}$ need not be a Fourier transform, in contrast to the bimeasure case. Thus several new problems arise in this extension. These authors also considered questions of sets of interpolation, and many other problems of the classical harmonic analysis can be studied on these spaces.

1.5 *Positive definite bimeasures.* Although there are many applications, such as those in martingale theory and operator representations, of general bimeasures one finds the positive definite class playing a key role in second order stochastic processes. Such a bimeasure β satisfies the inequality (14). This restriction allows an introduction of inner product structure into the class of β-integrable functions, and the resulting space plays a useful role in the spectral analysis of these processes to be detailed in Part II below. Hence these measures are discussed here.

The following result of Grothendieck's is needed (cf., Pisier (1986), p. 55, which can be stated in an alternative form for locally compact spaces):

Theorem 1. *Let (Ω_i, Σ_i) be Borelian spaces with Ω_i locally compact, $i = 1, 2$, and $\beta : \Sigma_1 \times \Sigma_2 \to \mathbb{C}$ be a bimeasure. Then there exists a pair of Radon probability measures $\mu_i : \Sigma_i \to \mathbb{R}^+$ such that*

$$|B(f_1, f_2)| = \left| \int_{\Omega_1} \int_{\Omega_2} f_1(\omega_1) f_2(\omega_2) \beta(d\omega_1, d\omega_2) \right|$$

$$\leq K_G \|B\| \, \|f_1\|_{2,\mu_1} \|f_2\|_{2,\mu_2}, \tag{23}$$

for all $f_i \in C_0(\Omega_i), i = 1, 2$, *with* K_G *as an absolute constant. Here* $\|B\| = \|\beta\|$, *where* $B(\cdot, \cdot)$ *is the bilinear form determined by* β *through the* MT*-integral.*

As a consequence of (23), one can extend B to a bounded bilinear form on $L^2(\Omega_1, \Sigma_1, \mu_1) \times L^2(\Omega_2, \Sigma_2, \mu_2)$. Hence this extended B denoted by the same symbol, can be expressed as:

$$B(f, g) = (Tf, g) = \int_{\Omega_2} (Tf) g d\mu_2 \tag{24}$$

for a bounded linear operator $T : L^2(\mu_1) \to L^2(\mu_2)$, and then

$$\begin{aligned} \|T\| &= \sup\{|B(f, g)| : \|f\|_{2,\mu_1} \leq 1, \|g\|_{2,\mu_2} \leq 1\} \\ &\leq K_G \|B\|, \text{ by (23).} \end{aligned} \tag{25}$$

In the following, let $\Omega = \Omega_1 = \Omega_2$ and $\mu = \mu_1 = \mu_2$ so that μ is a finite Radon measure ($\mu(\Omega) = C$), and $C_0(\Omega)$ is dense in $L^2(\Omega, \Sigma, \mu)$. So T is still a bounded linear operator on $L^2(\mu)$ with bound CK_G in (25). If β is moreover positive definite, then so is (B and) T. Hence T has a positive square root $T^{1/2}$ on $L^2(\mu)$. (The best known estimate of K_G, due to J.L. Krivine, is: $K_G \leq 1 \cdot 782$, to three places.) These facts are used in our applications.

Let $\mathcal{L}^2 = \{f \in \mathbb{C}^\Omega : f$ is μ-measurable, $\|f\|^2 = B(f, \bar{f}) < \infty\}$ where B and β are positive definite. Then $\|f\|^2 = (T^{1/2}f, T^{1/2}f) = \|T^{1/2}f\|_2^2$ and $\|\cdot\|$ is a norm on $\mathcal{L}^2$. In fact, $\|f\| = 0$ implies $T^{1/2}f = 0$ a.e., and since T (hence $T^{1/2}$) is a bounded positive (hence invertible) operator, $f = 0$ a.e. $[\mu]$. Also $\|f_1 + f_2\| = \|T^{1/2}(f_1 + f_2)\|_2 \leq \|T^{1/2}f_1\|_2 + \|T^{1/2}f_2\|_2 = \|f_1\| + \|f_2\|$, and $\|af\| = \|aT^{1/2}f\|_2 = |a| \, \|T^{1/2}f\|_2 = |a| \, \|f\|$. Let $\{f_n, n \geq 1\} \subset \mathcal{L}^2$ be a Canchy sequence so that $\|f_m - f_n\| = \|T^{1/2}f_m - T^{1/2}f_n\|_2 \to 0$. By the completeness of $L^2(\mu)$, there is a $g \in L^2(\mu)$ such that $\|T^{1/2}f_n - g\|_2 \to 0$. Let $f = T^{-1/2}g$. Then f is in $\mathcal{L}^2$ since $\|f\| = \|T^{1/2}f\|_2 = \|g\|_2 < \infty$, the range of T being $L^2(\mu)$. Also

$$\|f_n - f\| = \|T^{1/2}f_n - T^{1/2}f\|_2 = \|T^{1/2}f_n - g\|_2 \to 0 \text{ as } n \to \infty. \tag{26}$$

Hence $f_n \to f$ in $\mathcal{L}^2$. This proves the following result:

Theorem 2. *Let* $\{\mathcal{L}^2, \|\cdot\|\}$ *be the space of scaler Borel functions on* Ω *with* β *as a positive definite bimeasure, introduced above. Then it is a complete inner product space, i.e., a Hilbert space in the usual sense.*

This fact will be of interest in connection with the covariance analysis of harmonizable processes, and $\mathcal{L}^2$ will be termed the spectral domain space there. It is somewhat curious that the result should depend on Grothendieck's theorem so intimately.

In the preceding argument the positive definiteness of β is used. If it is semi-definite, then one has to extend it using, for instance, the concept of a generalized inverse of operators to achieve a similar conclusion or by considering a suitable quotient space. If β is merely a bimeasure, then also (24) and (25) are available, but the argument for (26) does not work. Here one may perhaps use the polar decomposition of T in (24) and manipulate the rest of the analysis. This will not be considered here, since it is not necessary for the stochastic applications below.

1.6 *Vector bimeasures.* In some applications, including the analysis of noncommutative harmonizable fields in the next part, one needs to employ vector valued bimeasures also. As a straightforward extension of the scalar case, which however has some new problems to deal with, the concept may be introduced as follows. Thus let $(\Omega_i, \Sigma_i), i = 1, 2$, be measurable spaces, $\mathcal{X}$ a Banach space, and $\beta : \Sigma_1 \times \Sigma_2 \to \mathcal{X}$ be a mapping. Then β is termed a *vector* (or $\mathcal{X}$-valued) *bimeasure* if $\beta(\cdot, B)$ and $\beta(A, \cdot)$ are vector measures for each $A \in \Sigma_1, B \in \Sigma_2$. The β-integral can be given, generalizing Sections 1.2 and 1.3, as follows.

Definition 1. Let $(\Omega_i, \Sigma_i), i = 1, 2, \mathcal{X}$ and β be as above. If $f_i : \Omega_i \to \mathbb{C}$ is Σ_i-measurable, then (f_1, f_2) is said to be *strictly β-integrable* provided the following four conditions hold:

(i) for each $A \in \Sigma_1, B \in \Sigma_2, f_1$ is $\beta(\cdot, B)$-and f_2 is $\beta(A, \cdot)$-integrable (D-S sense for vector measures),

(ii) $\tilde{\beta}_1^F : A \longmapsto \int_F f_2(\omega_2)\beta(A, d\omega_2), \tilde{\beta}_2^E : B \longmapsto \int_E f_1(\omega_1)\beta(d\omega_1, B)$, exist as vector measures for each $E \in \Sigma_1, F \in \Sigma_2$,

(iii) f_1 is $\tilde{\beta}_1^F$-and f_2 is $\tilde{\beta}_2^E$-integrable in the D-S sense for each pair $E \in \Sigma_1, F \in \Sigma_2$; and

(iv) $\int_E f_1(\omega_1)\tilde{\beta}_1^F(d\omega_1) = \int_F f_2(\omega_2)\tilde{\beta}_2^E(d\omega_2)\Big[= \int_E \int_F^* f_1 f_2 \beta(d\omega_1, d\omega_2)\Big]$
where the common value is symbolically denoted by the double integral.

If (iv) is required to hold only for $E = \Omega_1, F = \Omega_2$, then (f_1, f_2) is termed simply β-*integrable.* This is the vector MT-integral. In case $\Omega_1 = \Omega_2, f$ is (strictly) β- *integrable* if (f, f) is such.

It should be noted that these concepts are of interest even in the integral (measure kernel) representation of bilinear operators with analogs in the multilinear theory. Some of these aspects are actively studied by abstract analysts (cf. e.g., Dobrakov (1987) with references given there to his and others' works). For instance, a vector bimeasure β has finite Fréchet variation in that

$$\|\beta\|(\Omega_1, \Omega_2) = \sup\{\| \sum_{i,j=1}^{n} a_j \bar{b}_j \beta(A_i, B_j)\|_{\mathcal{X}} : A_i \in \Sigma_1, B_j \in \Sigma_2, \text{ disjoint}, |a_i| \leq 1, |b_j| \leq 1, \text{ scalers }, n \geq 1\} < \infty. \tag{27}$$

This follows from the earlier case since $\ell o \beta$ is a scalar bimeasure with finite Fréchet variation for each $\ell \in \mathcal{X}^*$, and then the uniform boundedness principle applies. It follows that all bounded measurable functions on $(\Omega_i, \Sigma_i), i = 1, 2$, are β-integrable (in either sense). Some extensions of the scalar results for vector bimeasures have been given by Ylinen (1978), culminating in the following analog of the Riesz-Markov representation.

Theorem 2. *Let Ω_i be locally compact, $(\Omega_i, \mathcal{B}_i)$ be Borelian, $C_0(\Omega_i)$ be continuous scalar functions vanishing at "∞", and $\mathcal{X}$ be a reflexive Banach space. If $B : C_0(\Omega_1) \times C_0(\Omega_2) \to \mathcal{X}$ is a bounded bilinear form then there exists a unique vector bimeasure $\beta : \mathcal{B}_1 \times \mathcal{B}_2 \to \mathcal{X}$, regular in each component, such that*

$$B(f_1, f_2) = \int_{\Omega_1} \int_{\Omega_2} f_1(\omega_1) f_2(\omega_2) \beta(d\omega_1, d\omega_2), f_i \in C_0(\Omega_i), \tag{28}$$

and

$$\|B\| = \sup\{\|B(f_1, f_2)\|_{\mathcal{X}} : \|f_i\|_\infty \leq 1, i = 1, 2\} = \|\beta\|(\Omega_1, \Omega_2). \tag{29}$$

If $\mathcal{X}$ is not reflexive, then the same conclusion holds provided B is assumed to map bounded sets of $V(\Omega_1, \Omega_2) = C_0(\Omega_1)\widehat{\otimes} C_0(\Omega_2)$ into relatively weakly compact sets.

This is an extension of a classical theorem [cf. Dunford and Schwartz (1958), VI.7.3], and the details are given by Ylinen (1978). The result admits a generalization to the case that the $\mathcal{X}, \mathcal{Y}, \mathcal{Z}$ are Banach spaces with a bilinear mapping on $\mathcal{Y} \times \mathcal{Z} \to \mathcal{X}$. Then $B : C_0(\Omega_1, \mathcal{Y}) \times C_0(\Omega_2, \mathcal{Z}) \to \mathcal{X}$ can be studied for a similar representation as in (28). In case $\Omega_1 = \Omega_2 = \Omega$ is a group one can also extend the theory of bimeasure algebras of Section 1.4. These are potential areas of research and some of these ideas will appear in Section 2.3 below.

1.7 *Bimeasures originating from classical problems.* Applications of non-negative bimeasures appear frequently in classical analysis. Some of these are included in the book by Berg, Christensen and Ressel (1984). A natural place is the theory of (regular) conditional probability functions. For instance, if $b : \Omega \times \Sigma \to \mathbb{R}^+$ is such that $b(\omega, \cdot)$ is a probability and $b(\cdot, A)$ is measurable relative to a σ-algebra $\mathcal{B}$ of Ω (need not be related to Σ), then

$$\tilde{\beta}^{\bullet}(A, B) = \int_A b(\omega, B)\mu(d\omega), A \in \mathcal{B}, B \in \Sigma, \tag{30}$$

where μ is a probability on $\mathcal{B}$, defines $\tilde{\beta}^{\bullet} : \mathcal{B} \times \Sigma \to \mathbb{R}^+$ as a positive bimeasure. It may be identified as the restriction of a measure on $\mathcal{B} \otimes \Sigma$. If $\mathcal{B} \subset \Sigma$, then $\tilde{\beta}^{\bullet}(A, B) = \mu(A \cap B)$, and such a $\tilde{\beta}^{\bullet}$ is familiar. From another point of view (30) is of interest also. If ν denotes a measure on $\mathcal{B} \otimes \Sigma$ extending $\tilde{\beta}^{\bullet}$, of the form (30), it is said to be disintegrated into a family $\{b(\omega, \cdot), \omega \in \Omega\}$. An application of bimeasure theory to the disintegration problem has recently been considered by Calbrix (1981), and much further work remains to be done. Other applications of bimeasures, of finite Vitali variation, to planar semi-martingales have been considered by Merzbach and Zakai (1986), and earlier in a special case by Horowitz (1977). By specializing $(\Omega_i, \Sigma_i), i = 1, 2$, and restricting the bimeasures suitably, a detailed analysis was presented by M. Cotlar and his associates (cf. (1982), and references there). It has a close relation to Toeplitz matrices. An indication will be given here for comparison, and to show its potential.

Suppose $\Omega = \mathbb{Z}$ and $\mathbb{Z}_1 = \{n \in \mathbb{Z} : n \geq 0\} = \mathbb{Z}^+, \mathbb{Z}_2 = \mathbb{Z} - \mathbb{Z}^+$. Let $\beta : \mathbb{Z} \times \mathbb{Z} \to \mathbb{C}$ be defined by a self-adjoint matrix such that in each quadrant $\mathbb{Z}_i \times \mathbb{Z}_j, \beta$ can be expressed as

$$\beta(m, n) = \rho_{ij}(m - n), (m, n) \in \mathbb{Z}_i \times \mathbb{Z}_j, 1 \leq i, j \leq 2. \tag{31}$$

Then $\beta = (\beta(m, n) : (m, n) \in \mathbb{Z} \times \mathbb{Z})$ is called a *generalized Toeplitz kernel* (and it is the Toeplitz kernel if $\rho_{ij} = \rho$ all i, j). The studies based on stationary random sequences whose covariance functions turn out to be positive definite Toeplitz kernels, is a motivation for the present investigation. Thus a generalized Toeplitz kernel is said to be positive definite if for each vector $\underline{a}_i = (a_i(1), a_i(2), \ldots, a_i(k)), i = 1, 2$ with $a_i(k) \in \mathbb{C}$, the following quadratic form is positive for each $k \geq 1$:

$$[\underline{a}_1 \ \underline{a}_2] \begin{bmatrix} \rho_{11} & \rho_{12} \\ \rho_{21} & \rho_{22} \end{bmatrix} \begin{bmatrix} \underline{a}_1^* \\ \underline{a}_2^* \end{bmatrix} \geq 0, \tag{32}$$

where ρ_{ij} is given by (31). If M is the 2-by-2 block matrix of ρ_{ij}'s, then one can associate (non-uniquely) a matrix of complex measures $\mu = \begin{pmatrix} \mu_{11} & \mu_{12} \\ \mu_{21} & \mu_{22} \end{pmatrix}$ on the torus. This representation leads to a semi-group theory and the corresponding convolution, with the Lebesgue-Stieltjes integration, has a parallel analysis developed by Professor M. Cotlar and his associates, who also have obtained an interesting stochastic application, generalizing the stationary sequences.

Thus the bimeasure theory is sufficiently flexible, and general enough, that a variety of applications are possible. Since the interest here is in probability, let us concentrate on that aspect and utilize the bimeasure theory which is developed thus far.

Part II: Harmonizability and Integral Representations

2.1 *Concepts and classification.* Let $L_0^2(P)$ be the usual Hilbert space of square integrable centered random variables on a probability space (Ω, Σ, P). A mapping $X : \mathbb{R} \to L_0^2(P)$ is called a second order process with covariance function $r : (s,t) \longmapsto E(X_s\overline{X}_t) = \int_\Omega X_s\overline{X}_t dP$. The process can be classified according to the structure of r. If r is continuous on $\mathbb{R}^2$ and is of the form $r(s,t) = \tilde{r}(s-t)$, then the process is called *weakly* (or *Khintchine*) *stationary.* But by a classical theorem of Bochner such $\tilde{r}$ is representable as

$$\tilde{r}(s-t) = \int_{\mathbb{R}} e^{i(s-t)\lambda}\mu(d\lambda), \tag{1}$$

for a unique bounded Borel measure μ on $\mathbb{R}$. Less restrictively, suppose only that the continuous covariance r can be expressed as:

$$r(s,t) = \int_{\mathbb{R}}\int_{\mathbb{R}} e^{is\lambda - it\lambda'}\beta(d\lambda, d\lambda'), \tag{2}$$

for a (unique) positive definite bimeasure β on $\mathbb{R}^2$ which is of finite Vitali variation. A process whose covariance admits the representation (2) is called *strongly harmonizable.* This class was introduced by Loève (1947). If β in (2) concentrates on the diagonal $\lambda = \lambda'$ so that $\beta(A,B) = \mu(A \cap B)$, then it reduces to (1) so that each weakly stationary process is strongly harmonizable. However, there are simple strongly harmonizable processes that are not stationary. For instance, if $\hat{f}$ is the Fourier transform of a Lebesgue integrable function f on $\mathbb{R}$, then $r(s,t) = \hat{f}(s)\overline{\hat{f}}(t)$ defines the covariance of such a process.

Suppose now that β of (2) is a bimeasure which is thus only of Fréchet variation finite (but its Vitali variation is infinite). Then the integral in (2) cannot be given in the Lebesgue sense. It can be defined as the (even strict-) MT-integral. When this is adapted, then the corresponding process is termed *weakly harmonizable.* This concept was introduced by Bochner

(1956) in a more general form under the name "V-bounded" process, and by Rozanov (1959) under the name "harmonizable", a term which was used by Loève in the more restricted (the strong) sense. Thus weak stationarity is stronger than strong harmonizability which in turn is stronger than weak harmonizability.

The preceding concepts extend immediately if $\mathbb{R}$ is replaced by any LCA group G. Thus if $\widehat{G}$ is the dual group of G, then $X : G \to L_0^2(P)$ is *weakly (or strongly) harmonizable* accordingly as its covariance $r(g_1, g_2) = E(X_{g_1}\overline{X}_{g_2})$ is representable as ($< g, \cdot > \in \widehat{G}$ being a character of G):

$$r(g_1, g_2) = \int_{\widehat{G}}\int_{\widehat{G}} < g_1, \gamma_1 >< \overline{g_2, \gamma_2} > \beta(d\gamma_1, d\gamma_2), \tag{3}$$

where β is a positive definite bimeasure, on $\mathcal{B}(\widehat{G}) \times \mathcal{B}(\widehat{G})$, of Fréchet (or finite Vitali) variation. If $G = \mathbb{R}^n$ or $\mathbb{Z}^n, n > 1$. Then $\{X_g, g \in G\}$ is usually called a random field. The latter term is used here for any locally compact space G, indexing the family.

It is of interest to note a few related second order random functions before proceding to the integral representation of harmonizable fields, since this helps in a better understanding of the subject. Thus let T be an index set and $\{X_t, t \in T\} \subset L_0^2(P)$ be a family with covariance $r(s,t) = E(X_s\overline{X}_t)$. If $\mathcal{B}$ is a σ-algebra of subsets of T, then the family is said to be of *class* (C) (Cramér class), if there is a measurable space $(S, \mathcal{S})$ and a positive definite bimeasure $\beta : \mathcal{S} \times \mathcal{S} \to \mathbb{C}$, of finite Vitali variation such that

$$r(s,t) = \int_S\int_S g_s(\lambda)\overline{g_t(\lambda')}\beta(d\lambda, d\lambda'), s, t \in T, \tag{4}$$

relative to a collection $\{g_s, s \in T\}$ of (Lebesgue) β-integrable scalar functions so that $r(s,s) < \infty$ for each $s \in T$. If β has only a finite Fréchet variation and the integral is the strict β-integral, then the corresponding family is of *weak class* (C). Clearly this reduces to the strong or weak harmonizability if $S = T = \mathbb{R}, g_s(\lambda) = e^{is\lambda}$. The class (4) (with $S = T = \mathbb{R}$) was introduced by Cramér (1951). If β in (4) concentrates on the diagonal of $S \times S$, so that $\beta(A, B) = \mu(A \cap B)$ for some positive finite (or σ-finite) measure μ on $\mathcal{S}$. Then it becomes

$$r(s,t) = \int_S g_s(\lambda)\overline{g_t(\lambda)}\mu(d\lambda), s, t \in T, \tag{5}$$

and the corresponding family is of *Karhunen class* introduced by him in 1947. Again if $S = T = \mathbb{R}, g_s(\lambda) = e^{is\lambda}$, then it reduces to the stationary class. Similarly, if $T = G$, an LCA group, $S = \widehat{G}$ and $g_s(\lambda) =< s, \lambda >$, then (3) is recovered. Beyond these identifications one has the following nontrivial result.

Proposition 1. *Every harmonizable random field $X : G \to L_0^2(P), G$ is an LCA group, belongs to a Karhunen class. More explicitly, if the given family is weakly harmonizable, then there is a finite Borel measure μ on $\widehat{G}$ and a suitable family $\{g_s, s \in G\} \subset L^2(\widehat{G}, \mu)$, such that* (5) *holds with $T = G$ and $S = \widehat{G}$ there.*

A proof of this result and certain other related extensions of strong harmonizability may be found in, e.g., Rao (1985).

2.2 *Integral representation of harmonizable fields on LCA groups.* For the integral representations it will be helpful to restate precisely the D-S integral of a scalar function relative to a vector measure. Thus if (Ω, Σ) is a measurable space, $f : \Omega \to \mathbb{C}$ is measurable for $\Sigma, Z : \Sigma \to \mathcal{X}$ (a Banach space) is a vector measure then f is D-S integrable relative to Z whenever the following two conditions hold:

(i) there is a sequence $f_n : \Omega \to \mathbb{C}$of simple (measurable for Σ) functions such that $f_n \to f$ pointwise, and

(ii) if $f_n = \sum_{i=1}^{k_n} a_i^n \chi_{A_i^n}, \int_E f_n dZ = \sum_{i=1}^{k_n} a_i^n Z(E \cap A_i^n) \in \mathcal{X}$, then $\{\int_E f_n dZ, n \geq 1\}$ is a Canchy sequence in $\mathcal{X}, E \in \Sigma$.

Then the unique limit of this sequence in $\mathcal{X}$ is denoted $\int_E f dZ, E \in \Sigma$. It is standard (but not trivial) to show that the D-S integral is a uniquely defined element of $\mathcal{X}$, is linear, and the dominated convergence theorem is valid for it. However, if $\mathcal{X}$ is infinite dimensional, then the D-S integral should *not* be confused with the Lebesgue-Stieltjes integral, and the evaluation of $\int_E f dZ$ as a Stieltjes integral is generally *false*. Also the convergence in (i) is pointwise, and strengthening it to uniformity restricts the generality of the D-S integral. These points should be kept in mind in its applications.

Now let $L^1(Z)$ denote the space of scalar functions on $(S, \mathcal{S})$, D-S integrable relative to Z, and $\mathcal{L}_*^2(\beta)$ be the collection of strictly β-integrable (MT-integration) $f : S \to \mathbb{C}$ where $\beta : (A, B) \longmapsto E(Z(A)\overline{Z(B)})$ is the bimeasure associated with Z, when $\mathcal{X} = L_0^2(P)$ on a probability space (Ω, Σ, P). In this case $Z(\cdot)$ is called a *stochastic measure* and β its *spectral bimeasure* of a second order process related by the following result.

Theorem 1. *Let $(S, \mathcal{S})$ be a measurable space and $\beta : \mathcal{S} \times \mathcal{S} \to \mathbb{C}$ be a positive definite bimeasure. Then there exists a probability space (Ω, Σ, P) and a stochastic measure $Z : \Sigma \to L_0^2(P)$ such that*

(i) $E(Z(A)\overline{Z(B)}) = \beta(A, B)$ *for all* $A, B, \in \mathcal{S}$, *and*

(ii) $L^1(Z) = \mathcal{L}_*^2(\beta)$, *equality as sets of functions.*

This result can be established quickly by using the Aronszajn theory of reproducing kernels. Then it is used in representing second order random fields. A general form of the latter is obtained as follows. If $(S, \mathcal{S})$ is a Borelian space, S being a topological space, a bimeasure β on $\mathcal{S} \times \mathcal{S}$ is said to have *locally finite Fréchet* (or Vitali) variation if $\beta : \mathcal{S}(E) \times \mathcal{S}(E) \to \mathbb{C}$ has finite Fréchet (or respectively Vitali) variation for each bounded Borel set $E \subset S$ (i.e., E is included in a compact set). [Regarding the clear distinction of these concepts, see also Edwards (1955).] Then the following general representation, to be specialized later, holds:

Theorem 2. *Let $(S, \mathcal{S})$ be a Borelian space with S locally compact. Suppose that $\{X_t, t \in T\} \subset L_0^2(P)$, on a probability space (Ω, Σ, P), is a (locally) weakly class (C) process relative to a positive definite bimeasure $\beta : \mathcal{S}_0 \times \mathcal{S}_0 \to \mathbb{C}$ of (locally) finite Fréchet variation and a family $g_t : S \to \mathbb{C}, t \in T$, of functions each of which is (locally) strictly β-integrable, where $\mathcal{S}_0$ is the δ-ring of bounded (Borel) sets of $\mathcal{S}$. Then there exists a σ-additive $Z : \mathcal{S}_0 \to L_0^2(P)$ such that (T being an index set)*

(i) $X_t = \int_S g_t(\lambda) Z(d\lambda), t \in T$, (D-S *integral*)

(ii) $E(Z(A)\overline{Z(B)}) = \beta(A, B), A, B \in \mathcal{S}_0$.

Conversely, if $\{X_t, t \in T\}$ is defined by (i) *for a stochastic measure Z, then the process is of (local) weak class* (C) *relative to a bimeasure β given by* (ii) *and the g_t of* (i) *being (locally) strictly β-integrable. The process is of (local) Karhunen class iff* (i) *and* (ii) *hold with $\beta(A, B) = \mu(A \cap B)$ for a σ-finite measure μ on $\mathcal{S}$.*

In fact if $K \subset S$ is a compact set, consider the trace $\mathcal{S}(K)$, of $\mathcal{S}$, on K which is a σ-algebra and $\beta : \mathcal{S}(K) \times \mathcal{S}(K) \to \mathbb{C}$ is a positive definite bimeasure for which the preceding theorem applies. If $\tilde{Z} : \mathcal{S}(K) \to L_0^2(P)$ is the representing stochastic measure then one has

$$\int_K g_t(\lambda)\tilde{Z}(d\lambda) = j\Big(\int_K g_t(\lambda)\beta(d\lambda, \cdot)\Big) \in L_0^2(P), \tag{6}$$

where j is the isometric isomorphism between $\tilde{Z}$ and β guaranteed by that result. By the local compactness of S, one can define a vector measure $Z : \mathcal{S} \to L_0^2(P)$ and extend (6) uniquely using a familiar procedure (cf. Hewitt and Ross (1963), pp. 133-134). Without local compactness of S this method of piecing together does not work. From here on the details are as in Chang and Rao [(1986), p. 53].

Since the functions $\{g_t, t \in T\}$ are not explicitly given in the above case, and are somewhat arbitrary, it will be interesting to specialize the result for harmonizable and stationary fields and show how these functions are naturally obtained in their representations.

Theorem 3. *Let G be an LCA group and $\{X_t, t \in G\} \subset L_0^2(P)$ be given. Then this family is weakly (resp. strongly) harmonizable relative to a positive definite bimeasure $\beta : \mathcal{B}(\widehat{G}) \times \mathcal{B}(\widehat{G}) \to \mathbb{C}$ (also of finite Vitali variation) iff there is a stochastic measure $Z : \mathcal{B}(\widehat{G}) \to L_0^2(P)$ such that*

$$\text{(i) } X_t = \int_{\widehat{G}} < t, \lambda > Z(d\lambda), t \in G, \text{ (D-S } \textit{integral}) \tag{7}$$

where $< t, \cdot >$ is a character of G, and

$$\text{(ii) } E(Z(A)\overline{Z(B)}) = \beta(A, B), A, B \in \mathcal{B}(\widehat{G}). \tag{7'}$$

When these conditions are met the mapping $t \longmapsto X_t$ is strongly uniformly continuous in $L_0^2(P)$. Further the random field $\{X_t, t \in G\}$ is weakly stationary iff (i) *and* (ii) *hold with $\beta(A, B) = \mu(A \cap B)$ for a bounded Borel measure μ so that Z also has orthogonal increments.*

An obvious question is to extend this result when G is not necessarily abelian. However, this needs several new concepts. Let us start with a vector analog of the above theorem which will be useful in the desired extension. Thus if $\mathcal{X}$ is a reflexive Banach space and $X : T \to L_0^2(P; \mathcal{X})$ is an $\mathcal{X}$-valued strongly measurable process or field on (Ω, Σ, P) with $E(\|X_t\|^2) < \infty$, then it is termed of weakly class (C), Karhunen class, harmonizable or stationary accordingly as the scalar process or field $\ell(X) : T \to L_0^2(P)$ is of the corresponding class as defined before, for each $\ell \in \mathcal{X}^*$ and $\sup_t E(\|X_t\|^2) < \infty$. [If $\mathcal{X}$ is not reflexive, the last condition should be replaced by the relative weak compactness of $\{X_t, t \in T\}$ in $L_0^2(P; \mathcal{X})$, and the work extends. For simplicity the reflexive case is considered.]

To see how this is accomplished, let us discuss the harmonizable case, so that $\ell(X) : G \to$

$L^2_0(P), \ell \in \mathcal{X}^*$, admits a representation as in (7):

$$\ell(X_t) = \int_{\widehat{G}} < t, \lambda > Z_\ell(d\lambda), t \in G, \tag{8}$$

where Z_ℓ is a stochastic measure. The mapping $\ell \longmapsto Z_\ell$ is linear and Z_ℓ is a regular vector measure with semi-variation $\|Z_\ell\|(\widehat{G}) < \infty$. Moreover,

$$\|Z_\ell\|(\widehat{G}) = \sup_t \|\ell(X_t)\|_2 \leq \|\ell\| \sup_t \|X_t\|_2 < \infty,$$

since $X(G)$ is bounded. By the uniform boundedness, $\sup_{\|\ell\|\leq 1} \|Z_\ell\|(\widehat{G}) < \infty$ and there is a $\tilde{Z}$ such that $Z_\ell = \ell(\tilde{Z})$, where $E(\tilde{Z}(A)) \in \mathcal{X}^{**} \cong \mathcal{X}$ by reflexivity, $A \in \Sigma$. Hence (8) can be expressed as

$$\ell(X_t) = \int_{\widehat{G}} < t, \lambda > \ell(\tilde{Z})(d\lambda) = \ell\Big(\int_{\widehat{G}} < t, \lambda > \tilde{Z}(d\lambda)\Big). \tag{9}$$

Since $\ell \in \mathcal{X}^*$ is arbitary, one gets

$$X_t = \int_{\widehat{G}} < t, \lambda > \tilde{Z}(d\lambda), t \in G, \text{ (D-S integral) }. \tag{10}$$

Thus one has

Theorem 4. *Let G be an LCA group, $\mathcal{X}$ a reflexive Banach space and $X : G \to L^2_0(P; \mathcal{X})$, a second order random function such that $X(G)$ is norm bounded (or $X(G)$ is relatively weakly compact if $\mathcal{X}$ is not reflexive). Then $\mathcal{X}$ is weakly harmonizable iff there is a stochastic measure $\tilde{Z}$ such that the representation* (10) *holds.*

This suggests that one may characterize weakly harmonizable random fields differently without using bimeasure integration. Such a procedure was given by Bochner (1956) with $\mathcal{X} = \mathbb{C}$. This will be employed when G is not necessarily abelian. The weakly harmonizable case when $G = \mathbb{R}$ and $\mathcal{X} = \mathbb{C}$ was first considered by Niemi (1975) who analyzed this class for certain other properties (cf. e.g., (1975-76)); and some special representations are given in Chang and Rao (1988).

2.3 *Noncommutative harmonizable random fields.* For a definition and integral representation of harmonizable functions in this case, one should define a suitable Fourier transform extending the LCA case above. A general form of the latter can be obtained through a use of C^*-algebras when G is any locally compact group. But an integral representation which usually depends on a Plancherel measure is then not possible since there is no dual group of G, and the analysis loses any resemblance with the previous theory. (See Ylinen (1975), (1984) and (1987) who has investigated the general case through C^*-algebra theory employing the techniques developed by Eymard (1964).) However, if we restrict G to be separable and (for simplicity here) unimodular, then the desired result can be derived, as shown below. Thus, in this section, G will be a *separable locally compact unimodular group.*

To proceed further, it is necessary to recall some results from the representation theory of such groups. Thus a locally compact group G is of type I if each unitary representation u of G into a Hilbert space $\mathcal{H}$ has the property that the weakly closed self-adjoint algebra $\mathcal{A}$ generated by $\{u_g, g \in G\}$ is isomorphic to some weakly closed self-adjoint subalgebra $\tilde{\mathcal{A}}$ of $L(\mathcal{H})$ such that $\tilde{\mathcal{A}}'$ is abelian. Here $\tilde{\mathcal{A}}'$ is the set of elements, of the algebra of bounded linear operations $L(\mathcal{H})$, that commute with $\tilde{\mathcal{A}}$. The group G is of type II if there is a normal semi-finite trace functional τ on $\mathcal{A}$ so that τ is linear, and for each $A \in \mathcal{A}$ there is a $B \leq A$ such that $|\tau(B)| < \infty$ and a monotone convergence theorem holds for it. One knows that each separable unimodular group is of type I or type II, and the following important facts are available (cf., Segal (1950), Mautner (1955) and especially Tatsuuma (1967); also Naĭmark (1964), Ch. 8):

(i) If $\widehat{G}$ denotes the set of all irreducible (strongly) continuous unitary representations of G into a Hilbert space, then one can endow $\widehat{G}$ a topology relative to which it becomes a locally compact Hausdorff space. And if μ is a Haar measure on G, then there is a unique Radon measure ν on $\widehat{G}$ such that $(\widehat{G}, \nu)$ becomes a dual object (or dual gauge) of (G, μ), and a Plancherel formula holds.

(ii) The representation Hilbert space $\mathcal{H}$ may be taken as $L^2(G, \mu) = L^2(G)$, and $\mathcal{H}$ can be expressed as a direct sum $\mathcal{H} = \bigoplus_{y \in \widehat{G}} \mathcal{H}_y$, with $\mathcal{H}_y$ as the representation space for each y in $\widehat{G}$. If $\mathcal{A}_y$ is the weakly closed self-adjoint subalgebra of $L(\mathcal{H}_y)$ generated by the strongly continuous unitary operators $\{u_y(g), g \in G\}$, then $\mathcal{A}_y$ is of type I or type II, and

$$L^2(G) = \int_{\widehat{G}}^{\oplus} \mathcal{H}_y \nu(dy), \text{ (direct integral).} \tag{11}$$

Moreover, if $(L_a f)(x) = f(a^{-1}x), x \in G$, then the weakly closed self-adjoint algebra $\mathcal{A}$ generated by $\{L_a, a \in G\}$ of $L(\mathcal{H})$, admits a direct sum decomposition of $\mathcal{A}_y, y \in \widehat{G}$, and for each

$f \in L^1(G) \cap L^2(G)$, the following (Bochner) integral exists

$$\hat{f}(y) = \int_G u_y(g) f(g) \mu(dg), u_y(g) \in \mathcal{A}_y, y \in \widehat{G}, \tag{12}$$

and defines a bounded linear mapping on $\mathcal{H}$. Also $\hat{f}(y)$ may be extended uniquely to a dense subspace of $\mathcal{H}$ containing $L^1(G) \cap L^2(G)$ so that it is closed and self-adjoint. This extended function $y \longmapsto \hat{f}(y)$, denoted by the same symbol, is the (generalized) *Fourier transform* of f.

(iii) There is a trace functional $\tau_y : \mathcal{A}_y \to \mathbb{C}$ which is positive, normal, semi-finite, and faithful, in terms of which one has the Plancherel formula for $f_i \in L^2(G), i = 1, 2$ ($\hat{f}_i^*$ denoting the adjoint of $\hat{f}_i$):

$$\int_G f_1(g)\overline{f_2(g)}\mu(dg) = \int_{\widehat{G}} \tau_y(\hat{f}_1(g)\hat{f}_2^*(y))\nu(dy). \tag{13}$$

The measurability of $\hat{f}_i$ as well as that of $y \longmapsto \tau_y(\hat{f}_1(y)\hat{f}_2^*(y))$ relative to ν are nontrivial facts and are established in the theory. An important result here is that there is a one-to-one correspondence between f and $\hat{f}$, and there is an inversion formula as well, (cf. Mautner, 1955). This is given next.

(iv) If $A(y) \in \mathcal{A}_y, y \longmapsto A(y)$ measurable, $y \longmapsto \|A(y)\|$ bounded and $\int_{\widehat{G}} \tau_y(A(y) A^*(y))\nu(dy) < \infty$, then there exists $f \in L^2(G)$ such that $\hat{f}(y) = A(y), y \in \widehat{G}$. On the other hand, if $h \in L^2(G)$ such that $h = f \star f$ for some $f \in L^2(G)$,then one has (the inversion formula):

$$h(g) = \int_{\widehat{G}} \tau_y(u_y(g)^* \hat{h}(y))\nu(dy), g \in G. \tag{14}$$

With these results, especially the (generalized) Fourier transform, the concept and a characterization of weak harmonizability for noncommutative groups can be given. The general concept is motivated by Bochner's classical notion of V-boundedness.

Definition 1. Let G be a separable locally compact unimodular group, and $X : G \to X_g \in L_0^2(P), g \in G$, be a random field. Then X is *weakly harmonizable* if it is weakly continuous and the set

$$\{\int_G X_g \varphi(g)\mu(dg) : \|\widehat{\varphi}\|_\infty = \sup_{y \in \widehat{G}} \|\widehat{\varphi}(y)\| \leq 1, \varphi \in L^1(G) \cap L^2(G)\},$$

is bounded in the Hilbert space $L_0^2(P), \widehat{\varphi}$ being the generalized Fourier transform of φ defined above.

With this concept at hand, the main integral representation of X is in:

Theorem 2. *Let $X : g \longmapsto X_g \in L_0^2(P), g \in G$, be a weakly harmonizable random field. Then there is a weakly σ-additive regular operator measure $\mathbf{m}(dy)$ on $\widehat{G}$, operating on $\mathcal{H}_y \to L_0^2(P)$, vanishing on ν-null sets and a trace functional $\tau_y : \mathcal{A}_y \to \mathbb{C}$, such that one has:*

$$X_g = \int_{\widehat{G}} \tau_y(u_g(y)\mathbf{m}(dy)), g \in G \text{ (Bartle integral)}, \tag{15}$$

and $X_{(\cdot)}$ is uniformly continuous in the strong topology of $L_0^2(P)$. On the other hand, a weakly continuous $X : g \to X_g$ defined by (15) is weakly harmonizable. Further, the covariance function r of the weakly harmonizable X, satisfying (15), is given by (a corresponding MT-integral for vector functions):

$$r(g_1, g_2) = \int_{\widehat{G}} \int_{\widehat{G}} \tau_{y_1} \otimes \tau_{y_2}\{(u_{g_1}(y_1) \otimes u_{g_2}(y_2))\beta(dy_1, dy_2)\}, \tag{16}$$

where β is an operator valued bimeasure (cf. Section 1.6) on $\mathcal{B}(\widehat{G}) \times \mathcal{B}(\widehat{G})$, with $\mathcal{B}(\widehat{G})$ as the Borel σ-algebra of $\widehat{G}$.

Proof. If $f \in L^1(G) \cap L^2(G)$, let $\hat{f}$ be defined by (12), which is a measurable operator function. To see that it is bounded, considering $\mathcal{H} = \int_{\widehat{G}}^{\oplus} \mathcal{H}_y \nu(dy)$, embed $\mathcal{H}_y$ in $\mathcal{H}$ and treat it as a closed subspace. Then $u_y(g) = u(g, y)$ in $L(\mathcal{H}_y)$ may be extended as $\tilde{u}(g, y) = u(g, y)$ on $\mathcal{H}_y$,= identity on $\mathcal{H}_y^{\perp}$ so that $\{\tilde{u}(g, y), g \in G\}$ is a family of unitaries in $L(\mathcal{H})$, and $\tilde{u}(g, \cdot) \in L(\mathcal{H}), g \in G$. If the corresponding operator of (12), obtained by replacing u by $\tilde{u}$, is again denoted by $\hat{f}$, then it is measurable. Let $\mathcal{A}(\mathcal{H}) = \{\hat{f} : \hat{f}(y) \in L(\mathcal{H}_y), y \in \widehat{G}\}$ which is identifiable with a subalgebra of $L(\mathcal{H})$. If $T : f \longmapsto \hat{f}$, then T is one-to-one and is a contraction. The former is a consequence of the general theory and the latter follows from the computation: ($\|\cdot\|_{op}$ denotes the operator norm)

$$\begin{aligned} \|\hat{f}(y)\|_{op} &= \| \int_G f(g)\tilde{u}(g, y)\mu(dg)\|_{op} \\ &\leq \int_G |f(g)| \, \|\tilde{u}(g, y)\|_{op} \, \mu(dg), \text{ by a property of the vector integral,} \\ &\leq \int_G |f(g)|\mu(dg) = \|f\|_1. \end{aligned} \tag{17}$$

Hence

$$\sup_{y \in \widehat{G}} \|\hat{f}(y)\|_{op} \leq \|f\|_1 < \infty. \tag{18}$$

Thus $T : L^1(G) \cap L^2(G) \to \mathcal{A}(\mathcal{H})$ is a contraction, and since X is weakly harmonizable, one has for each $f \in L^1(G) \cap L^2(G)$,

$$T_1(f) = \int_G f(g) X_g \, \mu(dg) \in L_0^2(P), \tag{19}$$

and T_1 is bounded. Let $\tilde{T} = T_1 \circ T^{-1}$ so that

$$\tilde{T}(\hat{f}) = T_1(T^{-1}(\hat{f})) = T_1(f), f \in L^1(G) \cap L^2(G). \tag{20}$$

Then $\tilde{T}$ is unambiguous and by Definition 1,

$$\|\tilde{T}(\hat{f})\|_2 \leq C\|f\|_2. \text{ (C is a constant.)} \tag{21}$$

Thus $\tilde{T}$ can be expressed as a direct sum of bounded operators from $\mathcal{A}(\mathcal{H}_y)$ into $L_0^2(P), y \in \widehat{G}$, by the general theory. Since the range of $\tilde{T}$ is reflexive, $\tilde{T}$ is weakly compact. Applying a suitable form of the Riesz-Markov theorem (cf. Dinculeanu (1967), p. 398, Thn. 9), and using the theory of direct integral for which the hypothesis on G and the separability of G are needed crucially at this point (cf. Naǐmark (1964), Ch. 8, Sec. 4), one gets a regular weakly σ-additive operator measure $\mathbf{m}$ on $\mathcal{B}(\widehat{G})$ into $L(\mathcal{A}(\mathcal{H}), L_0^2(P))$ such that

$$\tilde{T}(\hat{f}) = \int_{\widehat{G}} \tau_y(\hat{f}(y)\mathbf{m}(dy)), \hat{f} \in \mathcal{A}(\mathcal{H}), \tag{22}$$

where τ_y is a trace on $\mathcal{A}(\mathcal{H}_y)$ and where the integral is a suitable D-S (or Bartle (1956)) extension. Here $\mathbf{m}(\cdot)x : \mathcal{B}(\widehat{G}) \to L_0^2(P)$ is σ-additive and regular for each $x \in \mathcal{A}(\mathcal{H})$, and that property does not necessarily hold for $\mathbf{m}(\cdot)$ itself. It follows from (12), (19) and (22) that

$$\begin{aligned}\int_G f(g) X_g \, \mu(dg) &= \int_{\widehat{G}} \tau_y(\hat{f}(y)\mathbf{m}(dy)) \\ &= \int_{\widehat{G}} \tau_y\Big[\int_G f(g)\tilde{u}(g,y)\mu(dg)\mathbf{m}(dy)\Big] \\ &= \int_G f(g) \int_{\widehat{G}} \tau_y[\tilde{u}(g,y)\mathbf{m}(dy)]\mu(dg),\end{aligned}$$

since f is scalar and τ_y is linear and commutes with the integral over G, and a Fubini type argument applies. The above can be rearranged;

$$\int_G f(g)(X_g - \int_{\widehat{G}} \tau_y[\tilde{u}(g,y)\mathbf{m}(dy)])\mu(dg) = 0. \tag{23}$$

Since $f \in L^1(G) \cap L^2(G)$ is arbitrary and the latter is a dense set in $L^1(G)$ (as well as $L^2(G)$), the (continuous) function inside the parenthesis must vanish. Now replacing $\tilde{u}$ by u which is legitimate, (23) gives (15).

The reverse direction is obtained similarly. In fact, if (15) holds and $\varphi \in L^1(G) \cap L^2(G)$, then by the familiar reasoning one has

$$\begin{aligned}
\int_G X_g \varphi(g)\mu(dg) &= \int_G \varphi(g)\Big[\int_{\widehat{G}} \tau_y(u(g,y)\mathbf{m}(dy))\Big]\mu(dg) \\
&= \int_G \int_{\widehat{G}} \tau_y[\varphi(g)u(g,y)\mathbf{m}(dy)]\mu(dg) \\
&= \int_{\widehat{G}} \tau_y\Big[\Big(\int_G \varphi(g)u(g,y)\mu(dg)\Big)\mathbf{m}(dy)\Big] \\
&= \int_{\widehat{G}} \tau_y[\widehat{\varphi}(y)\mathbf{m}(dy)].
\end{aligned}$$

From this it follows that

$$\Big\| \int_G X_g \varphi(g)\mu(dg)\Big\| \le \|\widehat{\varphi}\|_{op}\|\mathbf{m}\|(\widehat{G}), \tag{24}$$

where $\|\mathbf{m}\|(\cdot)$ is the semi-variation of $\mathbf{m}$ (cf. Dinculeanu (1967), Sec. 19). Letting $C = \|\mathbf{m}\|(\widehat{G}) < \infty$ in (24) it follows that Definition 1 holds, and X is weakly harmonizable since it is clearly weakly continuous.

Finally to establish (16), one may calculate the covariance r of the $X(g)$'s using some properties of the extended MT-integral:

$$\begin{aligned}
r(g_1,g_2) &= E(X_{g_1}\overline{X}_{g_2}) \\
&= E\Big(\int_{\widehat{G}} \tau_{y_1}(\tilde{u}(g_1,y_1)\mathbf{m}(dy_1)\cdot \int_{\widehat{G}} \tau_{y_2}(\tilde{u}(g_2,y_2)\mathbf{m}(dy_2))^*\Big) \\
&= E\Big[\int_{\widehat{G}}\int_{\widehat{G}} \tau_{y_1}\otimes\tau_{y_2}\{(\tilde{u}(g_1,y_2)\otimes\tilde{u}(g_2,y_2)^*)\mathbf{m}(dy_1)\otimes\mathbf{m}^*(dy_2)\}\Big]
\end{aligned}$$

where one uses the properties of tensor products of

trace functionals (cf. Hewitt and Ross (1970), Appendix D),

$$\begin{aligned}
&= \int_{\widehat{G}}\int_{\widehat{G}} \tau_{y_1}\otimes\tau_{y_2}\{(\tilde{u}(g_1,y_1)\otimes\tilde{u}(g_2,y_2)^*)E(\mathbf{m}(dy_1)\otimes\mathbf{m}(dy_2))\} \\
&= \int_{\widehat{G}}\int_{\widehat{G}} \tau_{y_1}\otimes\tau_{y_2}[(\tilde{u}(g_1,y_1)\otimes\tilde{u}(g_2,y_2)^*)\beta(dy_1,dy_2)],
\end{aligned}$$

where $\beta(\cdot\,,\cdot)$ is an operator valued positive

definite bimeasure,

$$= \int_{\widehat{G}}\int_{\widehat{G}} \tau_{y_1}\otimes\tau_{y_2}[\tilde{u}(g_1,y_1)\beta(dy_1,dy_2)\tilde{u}(g_2,y_2)],$$

which is (16) written in a different form, and the result follows.

Remarks. Some complements to the above theorem will be included here in the form of remarks:

1. The integral representation (15) may be used for solving filter equations and in other applications. If G is abelian, then $\widehat{G}$ is a group, and each $\mathcal{H}_y = \mathbb{C}$ so that the result reduces to a previously known case (cf. Rao (1982)).

2. The corresponding representation for stationary random fields was first obtained by Yaglom (1960, 1961). For this class, several other interesting results for homogeneous spaces as well as multidimensional fields were also given there. I plan to consider the corresponding theory for weakly harmonizable fields later.

3. The measure $\mathbf{m}(\cdot)$ in (15) need not be σ-additive in the uniform operator topology, as known counter examples show. In the LCA case this difficulty disappears since $\mathcal{H}$ is $\mathbb{C}$, and by the classical Pettis's theorem weak and strong σ-additivities coincide.

4. If G is not a separable group, the decomposition theory runs into difficulties, and one may have to settle with the C^*-algebra approach, as was done by Ylinen (1975). Here the representation (15) is not available.

5. Using the Vitali variation of $\beta(\cdot,\cdot)$ in (16), one can present a result on *strongly harmonizable random* fields by a similar (and simpler) argument. A class intermediate to this and weak harmonizability is isolated by Ylinen (1988), who termed it "completely bounded". It coincides with weak harmonizability in case G is an LCA group.

6. If G is a compact group, then no separability assumptions are needed, and the Fourier transform can be derived through the Peter-Weyl theory. Thus a representation corresponding to (15) can be obtained using some classical computations as done, for instance in another context in Rao (1968), (cf. also the general theory on compact groups in Hewitt and Ross (1970)).

2.4 *The linear filter equation* $\Lambda X = Y$. Let G be an LCA group, and $X, Y : G \to L_0^2(P; \mathbb{C}^k)$ be k-dimensional random fields. If Λ is a linear operator on the class of such second order functions satisfying $T_h(\Lambda X)(g) = \Lambda(T_h X)(g)$ for all $g, h \in G$, where $(T_h X)(g) = X(g+h)$, then Λ is called a linear *filter* and $(\Lambda X)(g) = Y(g), g \in G$, is a *filter equation.* Thus Λ

commutes with translations. For instance if $G = \mathbb{R}^n$, then Λ can be an integro-difference-differential operator (with constant coefficients). In G the group operation is denoted "+". The problem now is that, given a random field Y (e.g., harmonizable or stationary) called an "output", find conditions on Λ in order that there is a solution X, called an "input", of the filter equation belonging to the same class. Here a solution is described if Λ is of the form:

$$Y(g) = (\Lambda X)(g) = \int_G A(s)X(g-s)ds \tag{25}$$

where 'ds' is an invariant measure on G, and A is a k-by-k matrix of scalar integrable Borel functions on G.

In the context of harmonizable functions one has:

Theorem 1. *Let the output Y be a k-dimensional weakly harmonizable random field with β_y as its k-by-k matrix spectral bimeasure. For the filter equation* (25), *there is a weakly harmonizable solution X iff*

$$\text{(i)} \quad \int_D \int_D^* (I - FF^{-1})(\lambda)\beta_y(d\lambda, d\lambda')(I - FF^{-1})^*(\lambda') = 0$$

for all Borel sets D of $\widehat{G}$, where $F = \widehat{A}$, the Fourier transform of A, F^{-1} is the generalized inverse of F and '' denotes the adjoint operation of the matrix, the integral being in the strict* MT-*sense, and where*

$$\text{(ii)} \quad \int_{\widehat{G}} \int_{\widehat{G}}^* F(\lambda)^{-1}\beta_y(d\lambda, d\lambda')(F(\lambda')^{-1})^* \textit{ exists.}$$

When these conditions hold, the solution X can be given by:

$$X_t = \int_{\widehat{G}} < t, \lambda > F^{-1}(\lambda)Z_y(d\lambda), t \in G, \tag{26}$$

where Z_y is the stochastic measure representing Y. The solution is unique iff $F(\lambda)$ is nonsingular for each $\lambda \in \widehat{G}$.

Here $F(\lambda)$ is often called the *spectral characteristic* of the filter Λ. Under further restrictions on $A(\cdot)$ one can obtain a simpler condition, such as that given by the following:

Proposition 2. *Let F be the spectral characteristics of the filter Λ of* (25). *If conditions* (i) *and* (ii) *of Theorem* 1 *hold, and if there is an integrable k-by-k matrix function f whose*

Fourier transform $\hat{f}$ *satisfies* $\|F^{-1} - \hat{f}^*\|_{2,\beta_y} = 0$, *with the norm used in Thm.* 1.5.2 *before, then the solution can be given by*

$$X_t = \int_G f(s)Y(t-s)ds, t \in G. \tag{27}$$

When $G = \mathbb{R}^n$, but Λ is more general, similar problems were considered by Chang and Rao (1986), and their methods yield the last two results for LCA groups G. Since a stationary random field is also harmonizable, the preceding work implies that for stationary Y, (25) has a weakly harmonizable solution X under the given hypothesis. What else is needed to assert that X is also stationary? This was studied by Bochner (1956) who gave conditions for a positive solution. Those considerations have been analyzed in more detail and the corresponding results are given in Rao (1984). So further discussion of the problem will be omitted here.

2.5 *Harmonizability over hypergroups.* Some statistical applications such as sample means of stationary or harmonizable sequences can lead to classes of second order processes which are not of the same type but are closely related to the original family. Many of these can be described as second order processes not on topological groups but on objects which are a generalization of these, called hypergroups. The latter have an algebraic group structure, but the topology they are endowed with does not always make the group operation continuous. Since it has a potential for future developments in this area, harmonizability on such spaces will be defined and a result on its integral representation given here.

One of the origins of hypergroups K may be traced to a study of the double coset spaces $H \setminus G/H$, (also denoted $G//H$) of a locally compact group G with H as a compact subgroup. It is clear that such K are locally compact spaces which are not groups in general. However, a group operation through convolution can often be introduced in such a space and the corresponding representation theory developed. Thus the hypergroups may be considered as objects between topological groups and the homogeneous spaces G/H, with interesting structure, and hence they have applicational potential. Abstraction of this remark will now be stated, following Jewett (1975) and others, for further development:

Definition 1. A locally compact space K is called a *hypergroup* if the following conditions are met:

(i) There exists an operation $\star : K \times K \to M_1(K)$, called convolution, such that $(x,y) \to \delta_x \star \delta_y, (x,y \in K)$ where δ_x is the Dirac measure at $x, M_1(K)$ is the set of Radon probability measures on K endowed with the vague (or weak*-) topology when $M(K)$ is regarded as the dual space of $C_0(K)$, and $\delta_x \star (\delta_y \star \delta_z) = (\delta_x \star \delta_y) \star \delta_z$;

(ii) $\delta_x \star \delta_y$ has compact support;

(iii) There is an involution, denoted by "$\sim$", on K such that $x^{\approx} = x$ and $(\delta_x \star \delta_y)^{\sim} = \delta_{\tilde{y}} \star \delta_{\tilde{x}}, x, y \in K$, where for a measure $\mu \in M_1(K), \tilde{\mu}(A) = \mu(\tilde{A})$ with $\tilde{A} = \{\tilde{x} : x \in A\}$, and there is a unit e in K satisfying $\delta_e \star \delta_x = \delta_x \star \delta_e = \delta_x$; and

(iv) $e \in \text{supp}(\delta_x \star \delta_{\tilde{y}})$ iff $x = y$, and that $(x,y) \longmapsto \text{supp}(\delta_x \star \delta_y)$ is continuous when 2^K is given the Kuratowski topology.

If (iv) is not assumed, then the object K for which (i)-(iii) hold is called a *weak hypergroup*. A number of concrete examples of these objects are given by Lasser (1983). For instance, several classical orthogonal polynomials on $K = \mathbb{Z}^+$, such as the Jocobi, Čebyšev, q-ultraspherical, Pollaczek, and certain Legendre polynomials are hypergroups. Also if K is a *commutative* hypergroup, i.e., $\delta_x \star \delta_y = \delta_y \star \delta_x$ holds in addition (the above examples are commutative hypergroups), its dual $\widehat{K}$ is defined as:

$$\widehat{K} = \{\alpha \in C_b(K) : (\delta_x \star \delta_y)(\alpha) = \int_K \alpha(t)(\delta_x \star \delta_y)(dt) = \alpha(x)\alpha(y), \\ x, y \in K \text{ and } \alpha(\tilde{x}) = \overline{\alpha}(x)\}. \tag{28}$$

Here $C_b(K)$ is the space of bounded continuous complex functions on K, with the topology of uniform convergence on compact sets. Then $\widehat{K}$ becomes a locally compact space which however need not be a hypergroup in general, the binary operation in $\widehat{K}$ being pointwise multiplication.

If K is a commutative hypergroup, then it admits an invariant (or Haar) measure, as shown by Spector (1978), and if $\widehat{K}$ its dual, also happens to be a hypergroup then $K \subset \widehat{\widehat{K}}$; and is termed a *strong hypergroup* provided $K = \widehat{\widehat{K}}$. A great deal of classical harmonic analysis is being extended to hypergroups (cf. e.g. Vren (1979), Lasser (1987), and references there). Our interest here is in the following stochastic application. For other developments of probability theory on these structures, one should refer to a detailed account in Heyer (1984).

If $X : K \to L_0^2(P)$ is a mapping such that its covariance function ρ, $\rho(a,b) = E(X_a\overline{X}_b)$, is bounded, continuous and representable as

$$\rho(a,b) = \int_K \rho(x,o)(\delta_a \star \delta_b)(dx), a, b \in K, \tag{29}$$

then X is termed a stationary random field on the commutative hypergroup K, or simply a *hyper-weakly stationary random field.* This concept is due to Lasser and Leitner (1988), except that they termed it "K-stationary". Since Bochner (1956) already used this term for Khintchine stationary, to avoid confusion the above term with the prefix "hyper" will be used here and below. It includes the sequences of symmetric Cesàro averages of ordinary stationary sequences, with $K = \mathbb{Z}^+$. For this concept the authors infer, via an analog of Bochner's theorem on positive definite functions, that X is hyper-weakly stationary on a commutative hypergroup K, iff

$$\rho(a,b) = \int_{\widehat{K}} \alpha(a)\overline{\alpha(b)} d\nu(\alpha), \tag{30}$$

for a unique bounded Borel measure ν on $\widehat{K}$. This allows an integral representation of X itself from the classical Karhunen-Cramér theorem.

The corresponding concept for harmonizability can be given as:

Definition 2. Let $X : K \to L_0^2(P)$ be a second order random field on a commutative hypergroup K whose dual object is denoted by $\widehat{K}$. If $\rho : (a,b) \longmapsto E(X_a\overline{X}_b), a, b \in K$, is its covariance function then X is called a *hyper-weakly (strongly) harmonizable* random field if ρ admits a representation

$$\rho(a,b) = \int_{\widehat{K}} \int_{\widehat{K}}^{*} \alpha_1(a)\overline{\alpha_2(b)} \beta(d\alpha_1, d\alpha_2), \tag{31}$$

where $\beta : \mathcal{B}(\widehat{K}) \times \mathcal{B}(\widehat{K}) \to \mathbb{C}$ is a positive definite bimeasure (of finite Vitali variation), and the integral is a strict MT-integral (a Lebesgue-Stieltjes integral).

It is well-known (cf. e.g., Chang and Rao (1986), p. 21) that β has always a finite Fréchet variation on the Borel σ-algebra $\mathcal{B}(\widehat{K})$. This definition reduces to the hyper-weakly stationary case if β concentrates on the diagonal of $\widehat{K} \times \widehat{K}$. The Fourier transform is well-defined, one-to-one and contractive, as in the LCA group case (cf. e.g., Heyer (1984), p. 491). Using these properties and the arguments of Sections 2.2 and 2.3, the following representation can be established.

Theorem 3. *Let $X : K \to L_0^2(P)$ be a hyper-weakly harmonizable random field in the sense of Definition 2. Then there is a stochastic measure $Z : \mathcal{B}(\widehat{K}) \to L_0^2(P)$ such that*

$$X_a = \int_{\widehat{K}} \alpha(a) Z(d\alpha), a \in K, \tag{32}$$

with $E(Z(A_1)\overline{Z}(A_2)) = \beta(A_1, A_2)$ defining the bimeasure β in (31). In fact, a second order weakly continuous random field on a commutative hypergroup K admits the representation (32), hence hyper-weakly harmonizable, iff the following set is norm bounded:

$$\{\int_K \varphi(a) X(a) d\mu(a) : \|\widehat{\varphi}\|_\infty \leq 1, \varphi \in L^1(K,\mu) \cap L^2(K,\mu)\} \subset L_0^2(P),$$

where μ is a Haar measure on K and $\widehat{\varphi}$ is the Fourier transform of φ so that $\widehat{\varphi}(\alpha) = \int_K \alpha(a)\varphi(a)\mu(a), \alpha \in \widehat{K}$, holds.

In the original talk at Oberwolfach this result was not included. It is a natural outgrowth of a study of some material on these objects. I should like to thank Drs. R. Lasser and R.C. Vrem for sending me their interesting published and some unpublished work on hypergroups which inspired it.

2.6 *Strict harmonizability and V-boundedness.* All the preceding study is based on the covariance properties of second order random functions. However, it is possible to analyze processes based on their distributional structure without asking for the existence of two moments, motivated by the strict stationarity concept. Since the study is linked to the Fourier transform of a stochastic measure eventually, the classical probability theory indicates that these stochastic measures be related to stability, so that one can define the corresponding integrals, if the values of the measures are independently scattered. Thus the desired concept can be presented as follows:

Definition 1. Let $(S, \mathcal{S})$ be a measurable space and $Z : \mathcal{S} \to L^p(P), 0 < p \leq 2$, be a mapping. Then $Z(\cdot)$ is an *independently scattered random stable measure of exponent* p, if the following conditions are met:

(i) $A_k \in \mathcal{S}, k = 1, \ldots, n, n \geq 1$, disjoint, implies $\{Z(A_k), 1 \leq k \leq n, n \geq 1\}$ is a mutually independent collection,

(ii) for each $A \in \mathcal{S}, Z(A)$ is a stable random variable of exponent p, and

(iii) $A_n \in \mathcal{S}$, disjoint, $A = \bigcup_n A_n$ implies $X(A) = \sum_{n=1}^{\infty} X(A_n)$, the series converging in probability (hence also with prob. 1 here).

For such a measure Z, a Wiener type stochastic integral can be defined, and if S is compact, ($\mathcal{S}$ Borel) then each continuous scalar function on S will be integrable relative to Z. Taking S as the torus, identified with $(-\pi, \pi)$ and $\mathcal{S}$ as its Borel σ-algebra, one says that a stochastic measure $Z : \mathcal{S} \to L^p(P)$ is *isotropic* if for each $A \in \mathcal{S}$ and each $\omega \in S$ (i.e., $-\pi < \omega < \pi$), $Z(A)$ and $e^{i\omega} Z(A)$ are identically distributed.

Recalling that a strictly stationary process is one whose (joint) finite dimensional distributions are invariant under translations of the index set, assumed an LCA group, one can present the following result when that index is the integers (considered as a group under addition and the torus as its dual):

Theorem 2. *Let $X : Z \to L^\alpha(P), 1 \le \alpha \le 2$ be a process. Then there is an independently scattered random stable measure Z of exponent α, which is isotropic on $\mathcal{S}$ (the Borel σ-algebra of $(-\pi, \pi)$) such that*

$$X_n = \int_{-\pi}^{\pi} e^{in\,t} Z(dt), n \in Z, \tag{33}$$

iff $\{X_n, n \in Z\}$ is strictly stationary and V*-bounded,* i.e. *for $1 < \alpha \le 2$*

$$\|\sum_{k=1}^{n} a_k X_k\|_\alpha \le C \sup\{|\sum_{k=1}^{n} a_k\, e^{-2\pi i t n_k}| : -\pi < t \le \pi\}, \tag{34}$$

and for $\alpha = 1, \{\sum_{k=1}^{n} a_k X_k, n \ge 1, a_k \in \mathbb{C}\}$ is relatively weakly compact in $L^1(P)$ in addition. When these conditions hold, the finite dimensional distributions, or equivalently their characteristic functions, are given by

$$\begin{aligned}\varphi_{n_1,\dots,n_k}(u_1, \dots, u_k) &= E\Big(\exp\Big[\sum_{j=1}^{k} i u_j X_{n_j}\Big]\Big) \\ &= \exp\Big\{-\int_{-\pi}^{\pi}\Big|\sum_{j=1}^{k} u_j\, e^{i\lambda n_j}\Big|^\alpha dG(\lambda)\Big\},\end{aligned} \tag{35}$$

for a non-negative, bounded, and non-decreasing function G.

The representation (33) is obtained from (34) through Bochner's criterion of V-boundedness for processes in a Banach space. The rest of the calculation through (35) is due to Hosoya (1982) who gives it for $0 < \alpha < 1$ also, when (33) is assumed. There are several extensions and related studies of processes given by the integral representation (33). Some of the references are Cambanis (1983), Weron (1985), Urbanik (1968), Kuelbs (1973), Okazaki (1979), Rosinski (1986) and Marcus (1987) among others.

Thus $X : Z \to L^{\alpha}(P), 0 < \alpha \leq 2$, admitting the representation (33) relative to a stable random measure of exponent α, and which is infinitely divisible, will be called *strictly harmonizable*. It follows from the above work and references, that there exist strictly stationary processes which are not strictly harmonizable and in fact an example is readily obtained from one in Sinaǐ (1963). Thus the hierarchy of the classes available for the second order processes is no longer valid for the strict sense concepts. Also a characterization of (33) for $0 < \alpha < 1$ is not available since the $L^{\alpha}(P)$ is not a Banach space and the V-boundedness theory has not been extended for these Fréchet (or invariant metric) spaces.

The next result gives an indication of how the Karhunen class undergoes a change for the strict sense case, with a Wiener type integral. It is due to Kuelbs (1973) and is an extension of an earlier "finite process" case by Schilder (1970):

Proposition 3. *Let $\{X_t, t \in T\}$ be a random field, with T as a second countable Hausdorff space, whose finite dimensional distributions belong to a symmetric stable class of index $\alpha, 0 < \alpha \leq 2$. Then there is an independently scattered stable random measure of index α on $[-\frac{1}{2}, \frac{1}{2}]$ and a family $\{f_t, t \in T\} \subset L^{\alpha}(-\frac{1}{2}, \frac{1}{2})$, the Lebesgue space, such that*

$$W_t = \int_{-\frac{1}{2}}^{\frac{1}{2}} f_t(\lambda) Z(d\lambda), t \in T, \tag{36}$$

and the X_t and W_t- processes have the same finite dimensional distributions.

The integral (36) is a kind of Wiener integral. The replacement of $f_k(\lambda)$ by $e^{ik\lambda}$ is not generally possible without further hypothesis. It appears that some analog of the V-boundedness is needed in addition. Probabilistic studies here concentrate on the sample path behavior of the process. Detailed analysis of such families have been given by Okazaki (1979) and Marcus (1987), (see also references to earlier results on these problems). For some other applications and extensions see Pourahmadi (1984) and Rajput and Rama-Murthy (1987).

Part III: Vector Harmonizable Random Fields

3.1 *Multidimensional harmonizability.* If a second order process is vector valued, new questions and classifications arise even when the group $G = \mathbb{R}$. In this part the vector (and operator) valued casses will be discussed briefly to focus on the new problems. Suppose then $X : G \to L_0^p(P;\mathcal{X}), 1 \leq p < \infty$, is an $\mathcal{X}$-valued process where $\mathcal{X}$ is a separable Banach space and G is an LCA group. First it is useful to present V-boundedness at two levels, since it is needed to classify vector harmonizability.

Definition 1. A mapping $X : G \to L_0^p(P;\mathcal{X}), 1 \leq p < \infty$, is *weakly* V-*bounded* if for each $\ell \in \mathcal{X}^*$, the dual of $\mathcal{X}$, the scalar process $\ell \circ X : G \to L_0^p(P)$, is V-bounded in that the function $t \longmapsto (\ell \circ X)(t), T \in G$, is continuous in $L_0^p(P)$ and the set (with the sample path [or Lebesgue] integral)

$$\{\int_G (\ell \circ X)(t)\varphi(t)dt : \|\widehat{\varphi}\|_\infty \leq 1, \varphi \in L^1(G)\} \subset L_0^p(P) \tag{1}$$

is relatively weakly compact and $\ell \circ X(G)$ is bounded in $L_0^p(P)$, where $\widehat{\varphi}$ is the Fourier transform of $\varphi \in L^1(G)$, and 'dt' is a Haar measure on G. The mapping X is *strongly* V-*bounded* if $t \longmapsto X(t)$ is continuous in $L_0^p(P;\mathcal{X})$ and the set (with the Bochner integral)

$$\{\int_G X(t)\varphi(t)dt : \|\widehat{\varphi}\|_\infty \leq 1, \varphi \in L^1(G)\} \subset L_0^p(P;\mathcal{X}), \tag{2}$$

is relatively weakly compact.

It is seen that strong V-boundedness implies weak V-boundedness but not necessarily conversely. Taking $p = 2$ and $\mathcal{X} = \mathcal{H}$, a Hilbert space, the corresponding concept of interest here is given by:

Definition 2. A mapping $X : G \to L_0^2(P;\mathcal{H})$ is *weakly harmonizable* if it is strongly V-bounded (and continuous), and it is *ultra weakly harmonizable*, if it is weakly V-bounded or equivalently $\ell \circ X : G \to L_0^2(P)$ is a weakly harmonizable (scalar) random field for each $\ell \in \mathcal{H}^*$.

Unless $\mathcal{H}$ is finite dimensional, the ultra weak class properly contains the weak class. This fact can be demonstrated by means of an example due to O.E. Lanford, discussed for a related purpose in the paper of R.I. Loebl (1976). This example has again been detailed for a similar purpose in a recent survey of dilation problems in Makagon and Salehi (1987). The distinction can be anticipated as the following remarks indicate.

In both cases of V-boundedness, standard arguments of abstract analysis imply that X is representable as a suitable D-S integral:

$$X(t) = \int_{\widehat{G}} < t, \lambda > Z(d\lambda), t \in G \tag{3}$$

where $Z : \mathcal{B}(\widehat{G}) \to (L_0^2(P;\mathcal{H}))^* \supseteq L_0^2(P)\widehat{\widehat{\otimes}}\mathcal{H}$, the tensor product space with the least cross-norm and $L_0^2(P)\widehat{\otimes}\mathcal{H} \cong (L_0^2(P)\widehat{\widehat{\otimes}}\mathcal{H})^*$. If $\mathcal{H}$ or $L_0^2(P)$ is not finite dimensional, then the above containment is proper (cf. e.g., Rao (1975), p. 1178 with $L^\rho = L_0^2$ and $\mathcal{X} = \mathcal{H}$, and the general exposition of Gilbert and Leih (1980)). In any case the semi-variation of Z in (3) can be defined in different senses:

$$\|Z\|(A) = \sup\{\|\ell \circ Z\|(A) : \|\ell\| \leq 1\}, \tag{4}$$

where $\|\ell \circ Z\|(\cdot)$ is the ordinary semi-variation of $\ell \circ Z : \mathcal{B}(\widehat{G}) \to L_0^2(P)$. On the other hand, let

$$|||Z|||(A) = \sup\{|\sum_{i=1}^{n}(\ell_i \circ Z)(A_i)| : A_i \in \mathcal{B}(A),\ \text{disjoint},\ \|\ell_i\| \leq 1, n \geq 1,$$
$$\ell_i \in (L_0^2(P;\mathcal{H})^*\}. \tag{5}$$

Restricting ℓ_i, one gets (4) from (5), so that $\|Z\|(A) \leq |||Z|||(A)$ always, with strict inequality sometimes. [In the above mentioned Lanford example, $|||Z|||(\widehat{G}) = +\infty$, with strict inequality here.] Since by hypothesis $\sup_t \|X_t\|_2 < \infty$, (4) is always finite even if $A = \widehat{G}$. Thus the distinction is necessary. Viewing $L_0^2(P;\mathcal{H})$ differently, a general form of an integral representation of weak vector harmonizability will be presented, and then the dilation problem will be discussed.

3.2 *More on vector integral representations.* The range space of the random field X considered above has an alternative description: if f, g are in $L_0^2(P;\mathcal{H}), A \in L(\mathcal{H})$, then $Af \in L_0^2(P;\mathcal{H})$ and if one defines

$$[f,g] = \int_\Omega f \otimes \overline{g}\, dP, \text{ then } [f,g] \in L(\mathcal{H}), |\text{ trace } [f,g]\,| < \infty.$$

Hence letting

$$\tau([f,g]) = \int_\Omega < f, g > dP, \|f\|_{2,\tau}^2 = \tau([f,f]) \tag{6}$$

one has $(L_0^2(P;\mathcal{H}), \|\cdot\|_{2,\tau})$ to be a Hilbert space where $f \otimes \overline{g}$ is the formal tensor product of the vectors f, g and $< f, g > (\omega) \in \mathcal{H}$. These properties motivate the next abstraction.

If $\mathcal{H}$ is a Hilbert space and $\mathcal{T}(\mathcal{H})$ is the set of trace class operators in $L(\mathcal{H})$, let $\mathcal{X}_0$ be the vector space on which is given a mapping: $[\cdot,\cdot] : \mathcal{X}_0 \times \mathcal{X}_0 \to \mathcal{T}(\mathcal{H})$ with the following properties:

(i) $[x,x] \geq 0, = 0$ iff $x = 0$, (ii) $[x+y,z] = [x,z] + [y,z]$, (iii) $[Ax,y] = A[x,y]$, for each $A \in L(\mathcal{H})$, and (iv) $[x,y]^* = [y,x]$, where '*' is the adjoint operation in $L(\mathcal{H})$.

The mapping $[\cdot,\cdot]$ is called a Gramian on $\mathcal{X}_0$. If $\|x\|_\tau^2 = \tau([x,x]) = \text{trace}\,([x,x])$, let $\mathcal{X}$ be the completion of $\mathcal{X}_0$ under the norm $\|\cdot\|_\tau$. The space $(\mathcal{X}, \|\cdot\|_\tau)$ is termed the *normal Hilbert $L(\mathcal{H})$-module*, and $L_0^2(P;\mathcal{H})$ is an example. If $\mathcal{H},\mathcal{K}$ are two Hilbert spaces over the same scalar field, then Kakihara (1985), who studied harmonizability on these spaces, shows that the space $HS(\mathcal{K},\mathcal{H})$ of Hilbert-Schmidt operators from $\mathcal{K}$ to $\mathcal{H}$ is a normal $L(\mathcal{H})$-module if $[x,y] = xy^*$ and $A \cdot x = Ax$, for $x, y \in HS(\mathcal{K},\mathcal{H})$ and $A \in L(\mathcal{H})$. Another example is $\mathcal{K}^q, 1 \leq q \leq \infty$, the cartesian product of Hilbert spaces $\mathcal{K}$, which can be made a normal $L(\mathcal{H}_q)$-module if $\mathcal{H}_q = \ell_q^2$, the coordinate Hilbert space.

Thus let $X : G \to \mathcal{X}$ be a mapping on an LCA group G into a normal $L(\mathcal{H})$-module $\mathcal{X}$. It is *weakly harmonizable* if one has

$$X(t) = \int_{\widehat{G}} < t, \lambda > Z(d\lambda), t \in G, \tag{7}$$

where the integral is in the D-S sense and $Z(\cdot)$ is a bounded regular σ-additive function on $\mathcal{B}(\widehat{G})$ into $\mathcal{X}$. The regularity here is in the strong sense, i.e., for each $A \in \mathcal{B}(\widehat{G})$ and $\varepsilon > 0$, there exist compact F and open O of $\widehat{G}$ such that $F \subset A \subset O$ and $\|Z\|(O-F) < \varepsilon$, $\|Z\|$ being the semi-variation of Z. Considering other variations one gets other harmonizabilities. The V-boundedness is defined similarly. Then one has the following:

Theorem 1. *A random field $X : G \to \mathcal{X}$, a normal $L(\mathcal{H})$-module, is weakly harmonizable iff it is* V*-bounded and continuous in the norm topology of $\mathcal{X}$.*

Although the statement is familiar in view of the earlier work, there is considerable technical machinary to be developed for its proof. Kakihara (1985, 1986) has done this and obtained other extensions.

3.3 *Dilation of harmonizable processes.* The dilation problem in the present context is the statement that (under minimal conditions) a given harmonizable process in $L_0^2(P)$ is the orthogonal projection of some stationary process from a super Hilbert space containing $L_0^2(P)$. That every such projection defines a weakly harmonizable process is the easy part. The reverse direction, depending on a suitable construction is hard and depends generally on the Grothendieck inequality, given as Theorem 1.5.1. [The details of this construction can be found, e.g., in Rao (1982), p. 326.] The corresponding result can be continued for a normal $L(\mathcal{H})$-module valued harmonizable process under some restrictions resulting in the finiteness of the Fréchet variation of the bimeasure of the representing stochastic measure. But such a construction fails for ultra weakly harmonizable processes since the corresponding bimeasure has infinite Fréchet variation. This is also verifiable with Lanford's example.

It is a surprizing fact that the Grothendieck inequality should play a vital role in the dilation problem. [For the strongly harmonizable case, one does not need this inequality, cf. Abreu (1970).] On the other hand, given the existence of a stationary dilation, one can prove Grothendieck's inequality for positive definite bimeasures by considering its Fourier transform, through the MT-integration, which qualifies to be a covariance function. Then one can construct a centered Gaussian harmonizable process with this covariance function via the Kolmogorov existence theorem and dilate it. The desired inequality follows from this. It is also observed by Chatterji (1982) and others. However, a general form of Grothendieck's inequality for not necessarily positive definite bimeasures does not seem possible in this way. The simplest known proof of the general inequality, due to Blei (1987), uses a probabilistic argument in its key parts. On the other hand, the Lanford example shows that there can be no infinite dimensional analog of Grothendieck's inequality. Some special types of dilations weaker than the above are possible. To understand this situation, a problem with the noncommutative harmonizable random field will be indicated here (cf., Rosenberg (1982), for a related study).

If $X : G \to L_0^2(P)$ is a random field with covariance r given by $r(g_1, g_2) = E(X_{g_1}\overline{X}_{g_2})$, then there is a right, a left, and a two sided stationary concepts available, and so one has to discuss the dilation problem for each class. Thus X is *left* [*right*] *stationary* iff

$$r(gg_1, gg_2) = \tilde{r}(g_2^{-1}g_1), [r(g_1g, g_2g) = \tilde{r}(g_1g_2^{-1})], \tag{8}$$

and it is *two sided stationary* if it is both right and left stationary. Thus for a reasonable dilation problem one restricts the class of dilations admitted. A weaker condition is obtained from a combination of the left-right properties. Thus X is termed *hemihomogeneous*, by Ylinen (1986), if its covariance r can be expressed as:

$$r(g_1, g_2) = \rho_1(g_2^{-1}g_1) + \rho_2(g_1g_2^{-1}), g_1, g_2 \in G, \tag{9}$$

where ρ_1, ρ_2 are positive definite covariances on G. Then Ylinen's result implies the following statement wherein the weak harmonizability of Definition 2.3.1 is used.

Theorem 1. *Let G be a separable unimodular group and $X : G \to L_0^2(P)$ be a continuous random field in $L_0^2(P)$. Then X is weakly harmonizable iff it has a hemihomogeneous dilation $Y : G \to L_0^2(\tilde{P}) \supset L_0^2(P)$, so that $X(g) = (QY)(g), g \in G$, where Q is an orthogonal projection of $L_0^2(\tilde{P})$ onto $L_0^2(P)$.*

Actually the result was given by Ylinen (1987) for all locally compact groups using his treatment of Fourier transforms through Eymard's (1964) approach and C^*algebras. It reduces to the present case, and the treatment simplifies slightly for G as given here. Thus the dilation problem has additional difficulties to consider for vector valued random fields.

3.4 *Multiplicity and least squares prediction.* The problem is usually considered in two stages. First, one wants to predict a future value of the process or field based on the past and present, and this assumes that the indexing group G must have a partial order (or a cone) in its structure. The most natural examples are $G = \mathbb{R}$ or $\mathbb{Z}$, and in this case one proceeds as follows.

Let $X : \mathbb{R} \to L_0^2(P)$ be a process, with a continuous covariance, and be nondeterministic in that $\bigcap_{t \in \mathbb{R}} \overline{\text{sp}}\{X(s) : s \leq t\} = \{0\}$. This is not a serious restriction in view of Wold's decomposition. Then there is a minimal integer $N \geq 1$, called the *multiplicity* of the process, jointly Borel measurable functions $F_n : \mathbb{R} \times \mathbb{R} \to \mathbb{C}$, and orthogonally scattered measures Z_n such that

$$X(t) = \sum_{n=1}^{N} \int_{-\infty}^{t} F_n(t, \lambda) Z_n(d\lambda), t \in \mathbb{R}, \tag{10}$$

with $F_n(t, \lambda) = 0$ for $t < \lambda$ and $\sum_{n=1}^{N} \int_{-\infty}^{\infty} |F_n(t, \lambda)|^2 \mu_n(d\lambda) < \infty$ where $\mu_n(A) = E(|Z_n(A)|^2)$, $A \in \mathcal{B}(\mathbb{R})$. Even when X is strongly harmonizable it is possible that $1 \leq N \leq \infty$. If $N = 1$, one has a simple Karhunen process. For stationary processes $N = 1$ always. Here harmonizability and bimeasure theory play a secondary role. Also in any given problem, the F_n's are not unique. They arise from the Hellinger-Hahn theory and are not easily obtained.

The second approach is to study the (simpler) strongly harmonizable case when its bimeasure has also a (spectral) density that is rational. Then one may extend the classsical theory of multivariate prediction in analogy with various results in, e.g., Rozanov's (1967) monograph. For a subclass of these processes having "factorizable spectral measures" the corresponding analysis, worked out in a preliminary study, is promising and nontrivial. The sample path behavior of the harmonizable process is another avenue, and most of it is a potentially interesting area.

3.5 *A final remark.* The main idea underlying the analysis of harmonizable processes and fields is a use of the powerful Fourier analytic methods. It is thus true that on an LCA group G, one has the representation of a harmonizable function X as:

$$X(g) = \int_{\widehat{G}} < g, \lambda > Z(d\lambda), g \in G. \tag{11}$$

Now the function $< \cdot, \cdot >: G \times \widehat{G} \to \mathbb{C}$ is jointly continuous, bounded and $< \cdot, \lambda >$ is periodic uniformly relative to λ in relatively compact open sets. But this fact motivates a study of $X(\cdot)$ in (11) in which $< \cdot, \cdot >$ is replaced by $f : G \times \widehat{G} \to \mathbb{C}$, such that $f(\cdot, \lambda)$ is almost periodic uniformly (in λ) relative to $D \subset \widehat{G}$, bounded open sets. Thus the resulting random field X becomes

$$X(g) = \int_{\widehat{G}} f(g, \lambda) Z(d\lambda), g \in G, \tag{12}$$

where $Z : \mathcal{B}(\widehat{G}) \to L_0^2(P)$ is a stochastic measure. Such a random field may be termed *almost weakly harmonizable*; and if the bimeasure induced by Z has finite Vitali variation then one has the case of an *almost strongly harmonizable* family. These form a subfamily of class (C) of Section 2.1, but have a better structure than the general members. A few properties of the latter class when $G = \mathbb{R}$, have been discussed in Rao (1978). It has a good potential for further study because there is a considerable amount of available results on almost periodic functions with important applications both when $G = \mathbb{R}$ and general locally compact groups. These and many of the (vector) extensions, having interesting structure, present a rich source of problems for research.

Notation: Throughout the paper, definitions, propositions, theorems are serially numbered. Thus m.n.p. denotes that object in part m, section n, and name p. In a given part, m is omitted and in a section n is also dropped. All unexplained symbols, if any, are as in Dunford and Schwartz (1958). Also $\mathbb{R}$ denotes reals, $\mathbb{C}$ - complex numbers, and $\mathbb{Z}$ for the integers. Almost all the notation used is standard.

Acknowledgment. This work was supported, in part, by the ONR Contract No. N00014-84-K-0356 (mod P00004).

REFERENCES

1. J. L. Abreu, "A note on harmonizable and stationary sequences," *Bol. Soc. Mat. Mexicana* **15**(1970), 48–51.

2. R. G. Bartle, "A general bilinear vector integral," *Studia Math.* **15**(1956), 337–352.

3. R. C. Blei, "An elementary proof of the Grothendieck inequality," *Proc. Amer. Math. Soc.* **100**(1987), 58–60.

4. S. Bochner, "Stationarity, boundedness, almost periodicity of random valued functions," *Proc. Third Berkeley Symp. Math. Statist. & Prob.* **2**(1956), 7–27.

5. J. Calbrix, "Mesures non σ-finies: désintegration et quelques autres proprietés," *Ann. Inst. H. Poincaré* **17**(1981), 75–95.

6. S. Cambanis, "Complex symmetric stable variables and processes," *in: Contributions to Statistics*, North-Holland, Amsterdam, 1983, 63–79.

7. D. K. Chang and Rao, M. M., "Bimeasures and nonstationary processes," *in: Real and Stochastic Analysis*, Wiley, New York, 1986, 7–118.

8. ________, "Special representations of weakly harmonizable processes," *Stochastic Anal. Appl.* **6**(1988), 169–189.

9. S. D. Chatterji, "Orthogonally scattered dilation of Hilbert space valued set functions," *in: Measure Theory Proceedings, Lect. Notes in Math.* No. 945, Springer-Verlag, 1982, 269–281.

10. M. Cotlar and Sadosky, C., "Majorized Toeplitz forms and weighted inequalities with general norms," *in: Lect. Notes in Math.* No. 908, Springer-Verlag, 1982, 139–168.

11. H. Cramér, "A contribution to the theory of stochastic processes," *Proc. Second Berkeley Symp. Math. Statist. & Prob.* 1951, 329–339.

12. N. Dinculeanu, *Vector Measures*, Pargamon Press, London, 1967.

13. I. Dobrakov, "On integration in Banach spaces-VIII," *Čzech Math. J.* **37**(1987), 487–506.

14. N. Dunford and Schwartz, J. T., *Linear Operators, Part I: General Theory*, Wiley-Interscience, New York, 1958.

15. D. A. Edwards, "Vector-valued measure and bounded variation in Hilbert space," *Math. Scand.* **3**(1955), 90–96.

16. P. Eymard, "L'algèbre de Fourier d'un groupe localement compact," *Bull. Soc. Math. France* **92**(1964), 181–236.

17. M. Fréchet, "Sur les fonctionnelles bilineares," *Trans. Amer. Math. Soc.* **16**(1915), 215–254.

18. J. E. Gilbert and Leih, T. J., "Factorization, tensor products, and bilinear forms in Banach space theory," *in: Notes in Banach Spaces*, Univ. of Texas Press, Austin, 1980, 182–305.

19. J. E. Gilbert, Ito, T., and Schreiber, B. M., "Bimeasure algebras on locally compact groups," *J. Functional Anal.* **64**(1985), 134–162.

20. C. C. Graham and Schreiber, B. M., "Bimeasure algebras on LCA groups,," *Pacific J. Math.* **115**(1984), 91–127.

21. _______, "Sets of interpolation for Fourier transforms of bimeasures," *Colloq. Math.* **51**(1987), 149–154.

22. _______, "Projections in spaces of bimeasures," *Canad. Math. Bull.*, to appear.

23. E. Hewitt and Ross, K. A., *Abstract Harmonic Analysis - I*, Springer-Verlag, New York, 1963.

24. _______, *Abstract Harmonic Analysis - II*, Springer-Verlag, New York, 1970.

25. H. Heyer, "Probability theory on hypergroups: a survey," *in: Probability Measures on Groups VII, Lect. Notes in Math.* No. 1064, Springer-Verlag, New York, 1984, 481–550.

26. J. Horowitz, "Une remarque sur le bimesures," *in: Sem. Prob. XI, Lect. Notes in Math.* No. 581, Springer-Verlag, New York, 1977, 59–64.

27. Y. Hosoya, "Harmonizable stable processes," *Z. Wahrs.* **60**(1982), 517–533.

28. R. I. Jewett, "Spaces with an abstract convolution of measures," *Adv. Math.* **18**(1975), 1–101.

29. Y. Kakihara, "A note on harmonizable and V-bounded processes," *J. Multivar. Anal.* **16**(1985), 140–156.

30. _______, "Strongly and weakly harmonizable stochastic processes of $\mathcal{H}$-valued random variables," *J. Multivar. Anal.* **18**(1986), 127–137.

31. K. Karhunen, "Über lineare Methoden in der Wahrscheinlichkeitsrechnung," *Ann. Acad. Sci. Fenn, Ser AI Math.* **37**(1947), 3–79.

32. I. Kluvanek, "Remarks on bimeasures," *Proc. Amer. Math. Soc.* **81**(1981), 233–239.

33. J. Kuelbs, "A representation theorem for symmetric stable processes and stable measures on $\mathcal{H}$," *Z. Wahrs.* **26**(1973), 259–271.

34. R. Lasser, "Orthogonal polynomials and hypergroups," *Rend. Math., Ser. 7* **3**(1983), 185–208.

35. _______, "Convolution semigroups on hypergroups," *Pacific J. Math.* **127**(1987), 353–371.

36. R. Lasser and Leitner, M., "Stochastic processes indexed by hypergroups," *J. Theoretical Prob.*, to appear.

37. P. Lévy, "Sur les fonctionnelles bilineares," *C. R. Acad. Sci., Paris, Ser I, Math.* **222** (1946), 125–127.

38. R. I. Loebl, "A Hahn decomposition for linear maps," *Pacific J. Math.* **65**(1976), 119–133.

39. Loève, M., "Fonctions aléatories du second ordres," Note in P. Lévy's *Processes Stochastiques et Movement Brownien*, Gauthier-Villars, Paris, 1947, 228–352.

40. A. Makagon and Salehi, H., "Spectral dilation of operator-valued measures and its application to infinite dimensional harmonizable processes," *Studia Math.* **85**(1987), 257–297.

41. M. B. Marcus, "ξ-radial processes and random Fourier series," *Mem. Amer. Math. Soc.* **368**(1987), 181pp.

42. F. I. Mautner, "Unitary representations of locally compact groups - II," *Ann. Math.* **52**(1950), 528–556.

43. _______, "Note on Fourier inversion formula on groups," *Trans. Amer. Math. Soc.* **78**(1955), 371–384.

44. E. Merzbach and Zakai, M., "Bimeasures and measures induced by planar stochastic integrators," *J. Multivar. Anal.* **19**(1986), 67–87.

45. M. Morse and Transue, W., "Functions of bounded Fréchet variation," *Canad. J. Math.* **1** (1949), 153–165.

46. _______, "$\mathbb{C}$-bimeasures Λ and their superior integrals Λ^*," *Rend. Circolo Mat. Palermo* **4**(1955), 270–300.

47. _______, "$\mathbb{C}$-bimeasures Λ and their integral extensions," *Ann. Math.* **64**(1956), 480–504.

48. _______, "The representation of a $\mathbb{C}$-bimeasure on a general rectangle," *Proc. Nat. Acad. Sci., USA* **42**(1956), 89–95.

49. M. A. Naǐmark, *Normed Rings*, Noordhoff, Groningen, 2d edition (translation), 1964.

50. H. Niemi, "Stochastic processes as Fourier transforms of stochastic measures," *Ann. Acad. Sci. Fenn, Ser AI Math.* **591**(1975), 1–47.

51. _______, "On orthogonally scattered dilations of bounded vector measures," *Ann. Acad. Sci. Fenn, Ser AI Math.* **3**(1977), 43–52.

52. Y. Okazaki, "Wiener integral by stable random measure," *Mem. Fac. Sci. Kyushu Univ., Ser A Math.* **33**(1979), 1–70.

53. G. Pisier, *Factorization of Linear Operators and Geometry of Banach Spaces*, CBMS Monograph No. 60, Amer. Math. Soc., Providence, 1986.

54. M. Pourahmadi, "On minimality and interpolation of harmonizable stable processes," *SIAM J. Appl. Math.* **44**(1984), 1023–1030.

55. B. S. Rajput and Rama-Murthy, K., "Spectral representation of semistable processes and semistable laws on Banach spaces," *J. Multivar. Anal.* **21**(1987), 139–157.

56. M. M. Rao, "Extensions of the Hausdorff-Young theorem," *Israel J. Math.* **6**(1968), 133–149.

57. _______, "Compact operators and tensor products," *Bull. Acad. Pol. Sci.* **23**(1975), 1175–1179.

58. ________, "Covariance analysis of nonstationary times series," *in: Developments in Statistics - I*, Academic Press, New York, 1978, 171–225.

59. ________, "Harmonizable processes: structure theory," *L'Enseign. Math.* **28**(1982), 295–351.

60. ________, "Harmonizable signal extraction, filtering, and sampling," *in: Topics in Non-Gaussian Signal Processes*, Springer-Verlag, New York, 1989, 98-117.

61. ________, "Harmonizable Cramér, and Karhunen classes of processes," *Handbook of Statistics, Vol 5*, North-Holland, Amsterdam, 1985, 279–310.

62. M. Rosenberg, "Quasi-isometric dilation of operator valued measures and Grothendieck's inequality," *Pacific J. Math.* **103**(1982), 135–161.

63. J. Rosinski, "On stochastic integral representations of stable processes with sample paths in Banach spaces," *J. Multivar. Anal.* **20**(1986), 277–302.

64. Yu. A. Rozanov, "Spectral analysis of abstract functions," *Theor. Prob. Appl.* **4**(1959), 271–287.

65. ________, *Stationary Random Processes*, Holden-Day, Inc., San Francisco, 1967 (translation).

66. M. Schilder, "Some structure theorems for the symmetric stable laws," *Ann. Math. Statist.* **41**(1970), 412–421.

67. I. E. Segal, "An extension of Plancherel's formula to separable unimodular groups," *Ann. Math.* **52**(1950), 272–292.

68. ________, "Decomposition of operator algebras - I," *Mem. Amer. Math. Soc.* **9**(1951), 67 pp.

69. Ya. G. Sinaǐ, "On properties of spectra of ergodic dynamical systems," *Dokl. Acad. Nauk USSR* **150**(1963), 1235–1237.

70. R. Spector, "Mesures invariantes sur les hypergroupes," *Trans. Amer. Math. Soc.* **239** (1978), 147–165.

71. N. Tatsuuma, "A duality theorem for locally compact groups," *J. Math. Kyoto Univ.* **6**(1967), 187–293.

72. K. Urbanik, "Random measures and harmonizable sequences," *Studia Math.* **31**(1968), 61–88.

73. N. Th. Voropoulos, "Tensor algebras and harmonic analysis," *Acta Math.* **119**(1967), 51–112.

74. R. C. Vrem, "Harmonic analysis on compact hypergroups," *Pacific J. Math.* **85**(1979), 239–251.

75. ________, "Idempotent sets and lacunarity for hypergroups," preprint.

76. A. Weron, "Harmonizable stable processes on groups: spectral, ergodic, and interpolation properties," *Z. Wahrs.* **68**(1985), 473–491.

77. A. M. Yaglom, "Positive definite functions and homogeneous random fields on groups and homogeneous spaces," *Soviet Math.* **1**(1960), 1402–1405.

78. ________, "Second order homogeneous random fields," *Proc. Fourth Berkeley Symp. Math. Statist. & Prob.* **2**(1961), 593–622.

79. K. Ylinen, "Fourier transforms of noncommutative analogs of vector measures and bimeasures with applications to stochastic processes," *Ann. Acad. Sci. Fenn, Ser AI Math.***1** (1975), 355–385.

80. ________, "On vector bimeasures," *Ann. Mat. Pura Appl.* **117**(1978), 115–138.

81. ________, "Dilations of V-bounded stochastic processes indexed by a locally compact group," *Proc. Amer. Math. Soc.* **90**(1984), 378–380.

82. ________, "Random fields on noncommutative locally compact groups," *in: Probability Measures On Groups - VIII*, Lect. Notes in Math. No. 1210, Springer-Verlag, New York, 1986, 365–386.

83. ________, "Noncommutative Fourier transformations of bounded bilinear forms and completely bounded multilinear operators," (UCR Tech. Report No. 12), 1987, *J. Functional Anal.*, to appear.

Department of Mathematics
University of California
Riverside, CA 92521

A CONJECTURE CONCERNING MIXTURES OF CHARACTERS FROM A GIVEN CLOSED SUBSEMIGROUP IN THE DUAL

Paul Ressel

Let $S = (S,+,0,*)$ denote an abelian $*$-semigroup with neutral element 0 , and let $K \subseteq \hat{S}$ denote a closed $*$-subsemigroup of bounded characters, containing 1, the neutral element of $\hat{S}$.

Let $\mu \in M_+(\hat{S})$ be a (positive) Radon measure on $\hat{S}$ and denote by $\hat{\mu}(s) := \int \rho(s)\, d\mu(\rho)$ its (generalized Laplace-) transform, $s \in S$. Then $\hat{\mu}$ is a bounded positive definite function on S , abbreviated $\hat{\mu} \in \mathcal{P}^b(S)$, and we have the fundamental result that the mapping $\mu \longmapsto \hat{\mu}$ is an affine homeomorphism, with the further property $\widehat{\mu * \nu} = \hat{\mu}\,\hat{\nu}$, cf. [1], 4.2.11.

If $\operatorname{supp}(\mu) \subseteq K$ then $\varphi = \hat{\mu}$ has obviously the following property:

(*) For any square matrix (s_{jk}) of elements of S such that $(\rho(s_{jk}))$ is positive (semi)definite for each $\rho \in K$ (let us call (s_{jk}) K-*definite* in this case), the matrix $(\varphi(s_{jk}))$ is positive definite, too.

Indeed, if (s_{jk}) is of size $n \times n$, then

$$\sum_{j,k=1}^{n} c_j \bar{c}_k\, \varphi(s_{jk}) = \sum_{j,k=1}^{n} c_j \bar{c}_k \int_K \rho(s_{jk})d\mu(\rho) \int_K \Sigma c_j \overline{c_k}\, \rho(s_{jk})d\mu(\rho) \geq 0,$$

the integrand being nonnegative by assumption.

Let us call any function $\varphi : S \longrightarrow \mathbb{C}$ fulfilling (*) a K-*positive definite* function; the set of all these K-positive definite functions will be denoted $\mathcal{P}(S,K)$ and is obviously a convex cone in $\mathbb{C}^S$, closed under pointwise convergence, stable under multiplication and conjugation and containing the nonnegative constants. Since for any $s \in S$ the 2×2-matrix $\begin{pmatrix} 0 & s \\ s* & 0 \end{pmatrix}$ is K-definite (even $\hat{S}$-definite) a function $\varphi \in P(S,K)$ is automatically bounded (by $\varphi(0) \geq 0$). If $P^b(S|K) := \{\hat{\mu} | \mu \in M_+(S),\ \mathrm{supp}(\mu) \subseteq K\}$ denotes those bounded positive definite functions whose representing measure is concentrated on K, then

$$P^b(S|K) \subseteq P(S,K)$$

as we saw before.

We are now able to state our

CONJECTURE: $P^b(S|K) = P(S,K)$ *for each closed* $*$*-subsemigroup* $K \subseteq \hat{S}$.

Note that $P(S,K) \subseteq P^b(S)$, since any matrix (s_{jk}) of the form $s_{jk} = s_j + s_k^*$ is obviously (even) $\hat{S}$-definite. Therefore a function $\varphi \in P(S,K)$ has a unique disintegration $\varphi(s) = \int_{\hat{S}} \rho(s)\, d\mu(\rho)$; the problem is to see that necessarily $\mathrm{supp}(\mu) \subseteq K$.

As long as this conjecture remains unsettled, let us agree to call K *regular* iff $P^b(S|K) = P(S,K)$, and to call S *normal* if this holds for each $K \in K(S)$, the family of all closed $*$-subsemigroups of $\hat{S}$. The subfamily consisting of all regular $K \in K(S)$ will be denoted $R(S)$. It is easy to see that $R(S)$ ist stable under arbitrary intersections, for if $K_\lambda \in R(S)$, $\lambda \in \Lambda$, and $K = \bigcap_{\lambda\in\Lambda} K$, then

$$P(S,K) \subseteq \bigcap_{\lambda\in\Lambda} P(S,K_\lambda) = \bigcap_{\lambda\in\Lambda} P^b(S|K_\lambda) = P^b(S|K) \subseteq P(S,K) .$$

Since $P(S,S) \subseteq P^b(S) = P^b(S|\hat{S})$ we have $\hat{S} \in R(S)$; for any $K \in K(S)$ therefore there exists a smallest $\tilde{K} \in R(S)$ containing K , namely

$$\tilde{K} = \bigcap_{K\subseteq R\in R(S)} R .$$

We shall call $\tilde{K}$ the *regular hull* of K .

The already mentioned isomorphie between $P^b(S)$ and $M_+(\hat{S})$ is proved essentially by showing $P_1^b(S) := \{\varphi \in P^b(S) | \varphi(0) = 1\}$ to be a Bauer simplex with $\hat{S}$ as extreme boundary. Of course $P_1^b(S|K)$ is a Bauer simplex, too, and $\mathrm{ex}(P_1^b(S|K)) = K$. Every $\varphi \in P(S,K)$ being bounded by $\varphi(0)$, it is clear that $P_1(S,K) := \{\varphi \in P(S,K) | \varphi(0) = 1\}$

is a compact convex base of $P(S,K)$. But we can say much more:

Theorem 1. *The compact convex set* $P_1(S,K)$ *of all normalized K-positive definite functions is a Bauer simplex whose extreme points coincide with* $\tilde{K}$ *, the regular hull of* K *, i.e.* $P_1(S,K) = P_1^b(S|\tilde{K})$ *and* $P(S,K) = P^b(S|\tilde{K})$.

Proof. We shall first show that the extreme points of $P_1(S,K)$ are characters. Let $(s_{jk})_{j,k\leq n}$ be a given K-definite matrix, and let $\{c_1,\ldots,c_n\} \subseteq \mathbb{C}$ also be fixed. Given $\varphi \in P(S,K)$, consider $F : S \longrightarrow \mathbb{C}$, defined by

$$F(s) := \sum_{j,k=1}^{n} c_j \overline{c_k} \varphi(s + s_{jk}) .$$

Then $F \in P(S,K)$ because if (t_{pq}) is another K-definite matrix, so is the "tensor sum" $((j,p),(k,q)) \longmapsto s_{jk} + t_{pq}$, implying

$$\Sigma\, d_p \overline{d_q} F(t_{pq}) = \Sigma\, \Sigma\, c_j d_p \overline{c_k} \overline{d_q} \varphi(s_{jk} + t_{pq}) \geq 0 .$$

In particular $|F(s)| \leq F(0)$ for all $s \in S$.

For $a \in S$, $z \in \mathbb{C}$, $|z| \leq 1$ we put

$$(T_{a,z}\varphi)(s) := \varphi(s) + \tfrac{1}{2}[z\varphi(s + a) + \bar{z}\, \varphi(s + a^*)] , \quad s \in S.$$

Then

$$\sum_{j,k=1}^{n} c_j \overline{c_k} (T_{a,z}\varphi)(s_{jk})$$

$$= \sum_{j,k=1}^{n} c_j \overline{c_k} \{\varphi(s_{jk}) + \tfrac{1}{2} [z\, \varphi(s_{jk} + a) + \bar{z}\, \varphi(s_{jk} + a^*)]\}$$

$$= F(0) + \mathrm{Re}\, [z\, F(a)] \geq 0 ,$$

so that $\{T_{a,z}\varphi | a \in S,\ |z| \leq 1\} \subseteq P(S,K)$.

Let now $\varphi \in \text{ex}\, P_1(S,K)$. Using $T_{a,z}\,\varphi + T_{a,-z}\,\varphi = 2\varphi$ for $z \in \{1,i\}$ we get two nonnegative constants λ_1, λ_2 such that

$$\varphi(s) + \frac{1}{2}\,[\varphi(s+a) + \varphi(s+a^*)] = \lambda_1\,\varphi(s)$$

$$\varphi(s) + \frac{i}{2}\,[\varphi(s+a) - \varphi(s+a^*)] = \lambda_2\,\varphi(s)$$

for all $s \in S$. It follows $\lambda_1 = 1 + \frac{1}{2}\,[\varphi(a) + \varphi(a^*)]$, $\lambda_2 = 1 + \frac{i}{2}\,[\varphi(a) - \varphi(a^*)]$, from there

$$\varphi(s+a) \pm \varphi(s+a^*) = \varphi(s)\,[\varphi(a) \pm \varphi(a^*)]$$

and finally $\varphi(s+a) = \varphi(s)\,\varphi(a)$, i.e. $\varphi \in \hat{S}$.

From $\text{ex}\, P_1(S,K) \subseteq \hat{S} \cap P_1(S,K) = (\text{ex}\, P_1^b(S)) \cap P_1(S,K) \subseteq \text{ex}\, P_1(S,K)$ we get $\text{ex}\, P_1(S,K) = \hat{S} \cap P_1(S,K)$. Different measures on $\hat{S}$ having different Laplace transforms we see that $P_1(S,K)$ is indeed a Bauer simplex. It remains to identify the set $K' := \hat{S} \cap P_1(S,K)$ of all K-positive definite characters with the regular hull $\tilde{K}$ of K.

By definition K is regular iff $P(S,K) = P^b(S|K)$ iff $\text{ex}\, P_1(S,K) = \text{ex}\, P_1^b(S|K) (= K)$ iff any K-positive definite character belongs already to K.

For a given square matrix $M = (s_{jk})$ put $R_M := \{\rho \in \hat{S} \mid \rho(M)$ is positive definite$\}$, $\rho(M) := (\rho(s_{jk}))$. Then R_M is a closed $*$-subsemigroup of $\hat{S}$; M being R_M-definite, R_M is even regular. Therefore $K' = \bigcap \{R_m \mid M \text{ K-definite}\}$ is regular, too, and contains K. If there was a regular L such that $K \subseteq L \subsetneq K'$, we would arrive at the contradiction

$$P(S,K) \subseteq P(S,L) = P^b(S|L) \subsetneq P^b(S|K') = P(S,K)\ .$$

This finishes the proof of Theorem 1. □

Our conjecture, that $P^b(S|K) = P(S,K)$, is now reduced to the question, if any K-positive definite <u>character</u> already belongs to K. We will give many examples where the answer is yes.

Example 1. S arbitrary, $K = \hat{S}_+$. If $\rho \in \hat{S} \cap \mathcal{P}(S,\hat{S}_+)$ and $a \in S$, then the 1×1-matrix (a) is K-definite, hence so is $(\rho(a))$, i.e. $\rho(a) \geq 0$; hence $\rho \in \hat{S}_+$.

Example 2. S arbitrary, $K = \{1\}$. If $\rho \in \hat{S} \cap \mathcal{P}(S,\{1\})$ and $a \in S$ then $\begin{pmatrix} a & 0 \\ 0 & a \end{pmatrix}$ is K-definite implying $\begin{pmatrix} \rho(a) & 1 \\ 1 & \rho(a) \end{pmatrix}$ to be positive (semi-)definite; we get $\rho(a) = 1$, i.e. $\rho \equiv 1$.

Example 3. $S = \mathbb{N}_0$ (with usual addition); $\hat{S}$ may canonically be identified with the multiplicative semigroup $[-1,1]$. Consider $K = \{-1,0,1\}$. If $\rho(\simeq \alpha) \in \hat{S} \cap \mathcal{P}(\mathbb{N}_0,\{-1,0,1\})$ we use the K-definite matrix $\begin{pmatrix} 4 & 2 \\ 2 & 4 \end{pmatrix}$ to get $\alpha^8 \geq \alpha^4$ or $\alpha^2 \in \{0,1\}$, i.e. $\alpha \in K$. In a similar way regularity of $K' = \{-1,1\}$ may be seen with help of the matrix $\begin{pmatrix} 2 & 0 \\ 0 & 2 \end{pmatrix}$, and the matrix $\begin{pmatrix} 2 & 1 \\ 1 & 1 \end{pmatrix}$ shows $K'' := \{0,1\}$ to be regular.

Example 4. $S = \mathbb{R}_+$ (with usual addition); $\hat{S}$ may canonically be identified with the additive semigroup $[0,\infty]$. Consider $K = \{0,1,2,\ldots,\infty\}$. If $\rho(\simeq \alpha) \in \hat{S} \setminus K \simeq \mathbb{R}_+ \setminus \mathbb{N}_0$, there is a positive definite matrix (x_{jk}) with entries from $]0,1]$ such that (x_{jk}^{α}) is <u>not</u> positive definite (see f. ex. [1], Lemma 5.3.1). Put $s_{jk} = -\log x_{jk}$. Then for all $n \in \overline{\mathbb{N}}_0 = K$ the matrix $(e^{-n\, s_{jk}}) = (x_{jk}^n)$ is positive definite (by Schur's theorem), however $(e^{-\alpha\, s_{jk}}) = (x_{jk}^{\alpha})$ has not this property, therefore $\rho \notin \mathcal{P}(S,K)$.

Example 5. $S = [-1,1]$ (with multiplication); consider K = closure of $\{1,t,t^2,\ldots\}$, i.e. the closure of the set of monomials (as functions on S) in $\hat{S}$, being a two-point compactification. Here a matrix (s_{jk}) is K-definite iff it is positive (semi-) definite in the usual sense. That every K-positive definite character on S belongs already to K has been shown in [2].

Example 6. Let S be the complex (closed) unit disk, viewed as a $*$-semigroup w.r.t. to ordinary multiplication and complex conjugation as involution. As an extension of the previous example we consider the closure K of the functions $z \longmapsto z^m \bar{z}^n$ on S, within $\hat{S}$. The regularity of K was proved in [3].

Example 7. Let S be an abelian group; S is then a $*$-semigroup w.r. to the group involution $s^* = -s$, and the characters in this case are precisely the ordinary group characters. Let K be any closed subgroup (= $*$-subsemigroup in this case). If $\rho \in \hat{S}$ is K-positive definite we will see that $\rho \in (K^{\perp})^{\perp}$. Indeed, for $s \in K^{\perp}$ the 1×1-matrix (s) is K-definite, hence $\rho(s) \geq 0$ which means of course $\rho(s) = 1$. By Pontryagin's duality theorem $(K^{\perp})^{\perp}$ equals K. We see that every K is regular, i.e. S is (what we called) normal.

Example 8. $S = \mathbb{N}_0^2$ (with usual addition and the involution $(m,n)^* = (n,m)$); $\hat{S}$ may canonically be identified with the closed unit disc $\mathbb{D} \subseteq \mathbb{C}$. Consider $K := \{z \in \mathbb{D} \mid |z| = 1\}$ and suppose $\rho(\simeq \alpha) \in \mathbb{D}$ is K-positive definite. The 2×2-matrix $(s_{jk}) = \begin{pmatrix} (1,1) & (0,0) \\ (0,0) & (1,1) \end{pmatrix}$ is K-definite, hence so is $(\rho(s_{jk})) = \begin{pmatrix} |\alpha|^2 & 1 \\ 1 & |\alpha|^2 \end{pmatrix}$, i.e. $|\alpha| = 1$.

Example 9. $S = \mathbb{N}_0^2$ as before, $K = [-1,1]$. We shall use the 3×3-matrix

$$(s_{jk}) = \begin{pmatrix} (1,1) & (2,0) & (2,0) \\ (0,2) & (1,1) & (2,0) \\ (0,2) & (0,2) & (1,1) \end{pmatrix} .$$

The identification of $\hat{S}$ with $\mathbb{D}$ being given by $\rho_z(m,n) = z^m \bar{z}^n$, $z \in \mathbb{D}$, we have $\rho_t(s_{jk}) = t^2$ for $t \in K$, i.e. (s_{jk}) is K-definite. Hence if $\rho = \rho_z \in \mathcal{P}(S,K)$, $z \neq 0$ without restriction, then $(\rho_z(s_{jk})) = |z|^2 . M_\zeta$ is positive (semi-) definite, where $\zeta := (z/|z|)^2$ and

$$M_\zeta := \begin{pmatrix} 1 & \zeta & \zeta \\ \bar{\zeta} & 1 & \zeta \\ \bar{\zeta} & \bar{\zeta} & 1 \end{pmatrix} .$$

By [1], Lemma 5.4.1 this implies $\zeta = 1$ or $z = \pm |z|$, i.e. $z \in K$.

Example 10. $S = \mathbb{N}_0$, $K = [-\gamma, 1]$, where $0 < \alpha < 1$. Suppose $\rho(\simeq \alpha) \in \hat{S} \setminus K$, i.e. $-1 \leq \alpha < \gamma$. Choose $k,n \in \mathbb{N}$ such that

$$\alpha < -\frac{1}{\sqrt[2k+1]{n}} < -\gamma$$

and consider the $(n+1)\times(n+1)$-matrix (s_{jk}) defined by $s_{jj}=0$ and $s_{jk}=2k+1$ for $j\neq k$. Then, with $a := x^{2k+1}$, we have

$$(x^{s_{jk}}) = \begin{pmatrix} 1 & a & \cdots & a \\ a & 1 & \cdots & \vdots \\ \vdots & \vdots & \ddots & \vdots \\ \vdots & \vdots & 1 & a \\ a & \cdots & a & 1 \end{pmatrix} =: M(a) .$$

Now $\det M(a) = (1+na)(1-a)^n$; so for $a \in [-1,1]$ the matrix $M(a)$ is positive (semi-) definite iff $a \geq -1/n$ iff $x \geq -n^{1/(2k+1)}$. Therefore (s_{jk}) is K-definite, however $(\rho(s_{jk})) = (\alpha^{s_{jk}})$ is not positive (semi-) definite, i.e. $\rho \notin \mathcal{P}(S,K)$. Hence K is regular.

<u>Example 11.</u> $S = \mathbb{N}_0$, $K = [-\gamma,\gamma] \cup \{1\}$, where $0 < \gamma < 1$. Consider for $x \in [-1,1]$ the $(n+1)\times(n+1)$-matrix

$$M(x) := \begin{pmatrix} 1 & x^k & x^k & \cdots & x^k \\ x^k & 1 & 0 & \cdots & 0 \\ x^k & 0 & 1 & \cdots & 0 \\ \vdots & \vdots & \vdots & \ddots & \vdots \\ x^k & 0 & 0 & \cdots & 1 \end{pmatrix}$$

An easy calculation gives $D(x) := \det M(x) = 1 - n\,x^{2k}$, and (strict) positive definiteness of $M(x)$ turns out to be equivalent with $D(x) \geq 0$ (> 0), i.e. with $|x| \leq \frac{1}{\sqrt[2k]{n}}$ (resp. $<$). Note that $\{n^{-1/2k} \mid k,n \geq 1\}$ is dense in $[0,1]$.

Let now $\rho_0 \simeq \alpha \in [-1,1]\setminus K$.

<u>1. case:</u> $\alpha = -1$: then $\rho_0 \notin \mathcal{P}(S,K)$ by the previous example (even $\rho_0 \notin \mathcal{P}(S,[-\gamma,1])$).

<u>2. case:</u> $|\alpha| \in]\gamma,1[$: choose $k,n \geq 1$ such that $\gamma < \frac{1}{\sqrt[2k]{n}} < |\alpha|$, and consider the $(n+1)\times(n+1)$-matrix

$$(s_{ij}) := \begin{pmatrix} 0 & k & k & \dots & k \\ k & 0 & N & \dots & N \\ k & N & 0 & \dots & N \\ \vdots & \vdots & \vdots & \ddots & \vdots \\ k & N & N & \dots & 0 \end{pmatrix}$$

with elements from S . If $\rho \simeq x \in [-\gamma,\gamma]$,

$$(\rho(s_{ij})) = (x^{s_{ij}}) = \begin{pmatrix} 1 & x^k & x^k & \dots & x^k \\ x^k & 1 & x^N & \dots & x^N \\ x^k & x^N & 1 & \dots & x^N \\ \vdots & \vdots & \vdots & \ddots & \vdots \\ x^k & x^N & x^N & \dots & 1 \end{pmatrix} .$$

Since $\det(\rho(s_{ij}))_{i,j\leq p}$ approaches the value $1-(p-1)x^{2k}$ as N tends to infinity, $p = 1,2,\dots,n+1$, this limit being strictly positive on $[-\gamma,\gamma]$, and since x^N tends to zero uniformly on $[-\gamma,\gamma]$, we find an $N \in \mathbb{N}$ such that $(x^{s_{ij}})$ is (strictly) positive definite for all $x \in [-\gamma,\gamma]$, and of course N may be chosen in such a way that $\det(\alpha^{s_{ij}}) < 0$, implying $(\rho_o(s_{ij}))$ to be not positive definite, whence $\rho_o \notin P(S,K)$.

Example 12. $S = \mathbb{N}_0$, $K = [0,\gamma] \cup \{1\}$, $0 < \gamma < 1$. Here K is regular since $K = ([-\gamma,\gamma] \cup \{1\}) \cap \hat{S}_+$.

Example 13. $S = \mathbb{N}_0$, $K = [-\gamma,\delta] \cup \{1\}$, $0 < \gamma \leq \delta \leq 1$. We have $K = ([-\delta,\delta] \cup \{1\}) \cap [-\gamma,1]$, so K is regular.

Example 14. $S = ([0,1],\vee)$; a description of $\hat{S}$ can be found in [1], 4.4.19. Let $K \subsetneq \hat{S}$ be an arbitrary closed subsemigroup and let $\rho_o \in \hat{S} \setminus K$.

1. case: $\exists\, a \in [0,1]$ such that $\rho_o = 1_{[0,a]}$. Since K is closed, $\exists\, \delta \in\,]0,1-a[\; : \{1_{[0,b]} | a \leq b \leq a+\delta\} \cap K = \emptyset$ and $\{1_{[0,b[} | a < b \leq a+\delta\} \cap K = \emptyset$. Consider the 2×2-matrix

$(s_{jk}) := \begin{pmatrix} a+\delta, & a \\ a & ,a+\delta \end{pmatrix}$, then $(\rho(s_{jk})) \in \left\{ \begin{pmatrix} 0 & 0 \\ 0 & 0 \end{pmatrix} , \begin{pmatrix} 1 & 1 \\ 1 & 1 \end{pmatrix} \right\}$ for all $\rho \in K$, so that (s_{jk}) is K-definite, but $(\rho_o(s_{jk})) = (1_{[0,a]}(s_{jk}))$ $= \begin{pmatrix} 0 & 1 \\ 1 & 0 \end{pmatrix}$ is not positive definite.

2. case: $\exists\, a \in]0,1]$ such that $\rho_o = 1_{[0,a[}$. Again $\exists\, \delta \in]0,a[$ with $\{1_{[0,b]} | a - \delta \leq b \leq a\} \cap K = \emptyset$ and $\{1_{[0,b[} | a - \delta \leq b \leq a\} \cap K = \emptyset$. Put $(s_{jk}) := \begin{pmatrix} a, & a-\delta \\ a-\delta, & a \end{pmatrix}$, then (s_{jk}) is K-definite but $(\rho_o(s_{jk})) = (1_{[0,a[}(s_{jk})) = \begin{pmatrix} 0 & 1 \\ 1 & 0 \end{pmatrix}$.

Hence S is a normal semigroup.

Example 15. One more class of closed $*$-subsemigroups of $\hat{S}$ turns out to be regular: let $a, b \in S$ and $\alpha, \beta \in]0,\infty[$ be fixed and consider

$$K := \{\rho \in S \,|\, |\rho(b)|^{\beta} \leq |\rho(a)|^{\alpha}\} \quad .$$

Obviously it suffices to take $\beta = 1$. If $\alpha = \frac{m}{n}$ is rational, $m,n \in \mathbb{N}$, then $K = \{\rho \,|\, |\rho(b)|^n \leq |\rho(a)|^m\}$, and for $\rho_o \in \hat{S} \setminus K$ the 2×2-matrix $(s_{jk}) := \begin{pmatrix} m(a + a^*) & n(b + b^*) \\ n(b + b^*) & m(a + a^*) \end{pmatrix}$ is K-definite, $(\rho_o(s_{jk}))$ being not positive definite. For other values of α let α_j be rationals decreasing to α ; since $\mathcal{R}(S)$ is stable under intersections, the result follows.

Let f. ex. $S = \mathbb{N}_o^2$ with <u>identical</u> involution, then $\hat{S} \simeq [-1,1]^2$ and with $a = (1,0)$, $b = (0,1)$, $\alpha = \beta = 1$ we see that $\{(x,y) \in \hat{S} \,|\, |y| \leq |x|\}$ is regular. Similarly, for $0 < \gamma < \delta < \infty$ the subsemigroup $\{(x,y) \in \hat{S} \,|\, |x|^{\delta} \leq |y| \leq |x|^{\gamma}\}$ is regular, too. For $S = \mathbb{N}_o^p$ (and $\hat{S} \simeq [-1,1]^p$) we get regularity of $\{x \,|\, |x_1| \leq |x_2| \leq \ldots \leq |x_p|\}$, f. ex., and likewise for the ordered simplex $\{x \,|\, 0 \leq x_1 \leq \ldots \leq x_p\}$.

The reader having looked through all the examples above will perhaps be inclined to find our conjecture not totally unrealistic. The semigroup $\mathbb{N}_o^2$ with $(m,n)^* = (n,m)$ (cf. Examples 8 and 9), being the free abelian $*$-semigroup with one generator, plays a particular role, as the following result shows.

Theorem 2. *If* $\mathbb{N}_o^2$ *and all its finite powers are normal then our conjecture holds.*

Proof. Let S be any semigroup and $K \subsetneqq \hat{S}$ a closed $*$-subsemigroup. If $\rho_o \in \hat{S} \setminus K$, then also $\rho_o \notin \mathcal{P}^b(S|K)$, so that by Hahn-Banach's theorem there exist finitely many $s_1, \ldots, s_p \in S$ and $c_1, \ldots, c_p \in \mathbb{C}$ with

$$(*) \qquad \mathrm{Re}[\sum_{j=1}^{p} c_j \, \rho_o(s_j)] < 0 \leq \mathrm{Re}[\sum_{j=1}^{p} c_j \, \varphi(s_j)]$$

for all $\varphi \in \mathcal{P}^b(S|K)$, in particular for all $\rho \in K$. Consider the $*$-subsemigroup $T := \langle s_1, \ldots, s_p \rangle$ generated by $s_1, \ldots, s_p$, as well as the $*$-homomorphism $h : (\mathbb{N}_o^2)^p \longrightarrow T$, defined by $h((1,0),(0,0),\ldots,(0,0)) := s_1$, $h((0,0),(1,0),(0,0),\ldots) := s_2$, etc., which of course is onto. Then $L := K \circ h := \{\rho \circ h | \rho \in K\}$ is a closed $*$-subsemigroup of $[(\mathbb{N}_o^2)^p]\hat{}$, not containing $\xi_o := \rho_o \circ h$ because of $(*)$. By assumption there is an L-definite matrix (u_{jk}) in $(\mathbb{N}_o^2)^p$ s.th. $(\xi_o(u_{jk}))$ is not positive definite, therefore $(s_{jk}) := (h(u_{jk}))$ is K-definite, $(\rho_o(s_{jk}))$ being not positive definite, showing that indeed $\rho_o \notin \mathcal{P}(S,K)$. □

Similarly, if the conjecture is true for $\mathbb{N}_o^p$, $p = 1,2,\ldots$, then it holds for every semigroup with identical involution. However, already for $p = 1$ the answer is not yet known and perhaps difficult to find, as some of our examples indicate.

As long as the general conjecture remains unsettled, one might try to get some partial results. Here are some open questions:

(a) What about $S = \mathbb{N}_o$ and $K := [0,\gamma^2] \cup \{\gamma,1\}$, $0 < \gamma < 1$?

(b) What about $S = \mathbb{N}_o$ and $K := [-\gamma,\gamma^2] \cup \{1\}$, $0 < \gamma < 1$?

(c) If K and L are regular $(\subseteq \hat{S})$ what about $\overline{\langle K \cup L \rangle}$?

(d) For two closed $*$-subsemigroups $K, L \subseteq \hat{S}$ one has obviously $\widetilde{K \cap L} \subseteq \tilde{K} \cap \tilde{L}$. Is there equality?

(e) Is every idempotent semigroup normal (cf. Example 14)?

(f) It is easy to see that a homomorphic image of a normal semigroup is again normal. In order that the product of two semigroups be normal it is therefore necessary that both factors are normal. Is this condition also sufficient? - In view of Theorem 2 a positive answer to this question would reduce our conjecture to the question if $\mathbb{N}_0^2$ (with $(m,n)^* = (n,m)$) is normal. (Since regularity is preserved under intersections it is clear that the product of two regular *-subsemigroups $K \subseteq \hat{S}$ and $L \subseteq \hat{T}$ is regular in $\widehat{S \times T} = \hat{S} \times \hat{T}$. With the K from Example 5 and $L = K$ this gives a characterisation of bivariate generating functions, cf. Exercise 5.3.20 in [1].)

REFERENCES

[1] C. BERG, J.P.R. CHRISTENSEN and P. RESSEL, *Harmonic Analysis on Semigroups. Theory of positive definite and related functions* (Graduate texts in Math. 100). Springer-Verlag, Berlin - Heidelberg - New York, 1984.

[2] J.P.R. CHRISTENSEN and P. RESSEL (1978). *Functions operating on positive definite matrices and a theorem of Schoenberg.* Trans. Amer. Math. Soc. 243, 89 - 95.

[3] J.P.R. CHRISTENSEN and P. RESSEL (1982). *Positive definite kernels on the complex Hilbert sphere.* Math. Z. 180, 193 - 201.

Paul Ressel
Kath. Universität Eichstätt
D-8078 Eichstätt

Infinitely Divisible States on Cocommutative Bialgebras*

Michael Schürmann**

Institut für Angewandte Mathematik, Universität Heidelberg
Im Neuenheimer Feld 294, D-6900 Heidelberg 1
Federal Republic of Germany

ABSTRACT

We prove an embedding theorem for the Gelfand-Naimark-Segal (GNS) representation given by the convolution exponential of a hermitian, conditionally positive linear functional on a cocommutative *-bialgebra. Our theory generalises a well-known construction [3, 7, 12, 18] for infinitely divisible positive definite functions on a group. As an application of our result we prove that the GNS representation given by an infinitely divisible state on a tensor algebra can be embedded into a representation by polynomials in annihilation, creation and second quantisation operators on a Fock space.

1. Introduction

Let G be a group and denote by $\mathbb{C}G$ the *-algebra of formal finite complex linear combinations of elements of G where the multiplication is given by the group multiplication and the involution is given by the forming of inverses. We define the linear mappings $\Delta: \mathbb{C}G \to \mathbb{C}G \otimes \mathbb{C}G$ and $\delta: \mathbb{C}G \to \mathbb{C}$ by setting $\Delta x = x \otimes x$ and $\delta x = 1$ for $x \in G$. The mappings Δ and δ are *-algebra homomorphisms and satisfy

$$(\Delta \otimes \mathrm{id}) \circ \Delta = (\mathrm{id} \otimes \Delta) \circ \Delta \quad \text{(coassociativity)} \tag{1.1}$$

and

$$(\delta \otimes \mathrm{id}) \circ \Delta = (\mathrm{id} \otimes \delta) \circ \Delta = \mathrm{id} \quad \text{(counit property).} \tag{1.2}$$

The space $\mathbb{C}G$ is an example of a **-bialgebra* with *comultiplication* Δ and *counit* δ. The elements Δb, $b \in \mathbb{C}G$, of $\mathbb{C}G \otimes \mathbb{C}G$ are elements of the *symmetric* tensor product of $\mathbb{C}G$ with itself which means that $\mathbb{C}G$ is a *cocommutative* *-bialgebra. Linear functionals on $\mathbb{C}G$ and complex-valued functions on G are in one-to-one correspondence and positive definite (hermitian) functions on G are identified with positive (hermitian) linear functionals on $\mathbb{C}G$. A conditionally positive definite function on G becomes a conditionally positive linear functional on the *-bialgebra $\mathbb{C}G$, that is a linear functional α satisfying

$$\alpha(b^*b) \geqslant 0 \text{ for all } b \in \mathbb{C}G \text{ with } \delta(b) = 0. \tag{1.3}$$

The pointwise multiplication of functions on G can be written as the commutative 'convolution' product of linear functionals μ and λ on $\mathbb{C}G$ given by

$$\mu * \lambda = (\mu \otimes \lambda) \circ \Delta. \tag{1.4}$$

The notion (1.3) of conditional positivity and the commutative convolution product (1.4) of linear functionals are introduced for arbitrary cocommutative *-bialgebras [20] which consist of a *-algebra $\mathcal{B}$ and *-algebra homomorphisms $\Delta: \mathcal{B} \to \mathcal{B} \otimes \mathcal{B}$ and $\delta: \mathcal{B} \to \mathbb{C}$ satisfying the coassociativity, counit and

---

*This work was supported by an SERC visiting fellowship.

**Present address: Mathematics Department, University of Nottingham, University Park, Nottingham, NG7 2RD, England

cocommutativity conditions. A state (that is a normalised, positive linear functional) μ on $\mathcal{B}$ is called *infinitely divisible* if for each $n \in \mathbb{N}$ there is a state μ_n on $\mathcal{B}$ such that μ is the n-th convolution power of μ_n. (In the case of $\mathbb{C}G$ this is the notion of an infinitely divisible, normalised positive definite function.) Especially, if μ is the convolution exponential $\exp_*\alpha$ of a hermitian, conditionally positive linear functional α on $\mathcal{B}$ and $\alpha(1) = 0$, then μ is an infinitely divisible state on $\mathcal{B}$. We are interested in the GNS representation given by μ. In the case of $\mathbb{C}G$ the Araki-Woods embedding theorem says that this representation can be embedded into a representation of G mapping the elements of G to operators on a Fock space which are characterised by the property that they map exponential vectors to multiples of exponential vectors; see [3,7,12,18].

It turns out that the situation is very similar in the general case of a cocommutative *-bialgebra. There is a concept generalising the concept of exponential vectors, and the GNS representation given by $\mu = \exp_*\alpha$ can be embedded into a representation on a Fock space. In this paper we are especially interested in the case of a tensor algebra, that is the free associative, unital algebra generated by a set of indeterminates. For the Hopf algebra structure on a free algebra, or more generally on the universal enveloping algebra of a Lie algebra, see [11]. Infinitely divisible states on a tensor algebra were also considered in [5,8,17]. As an application of our general theory we prove that the GNS representation of an infinitely divisible state on a tensor algebra can be embedded into a representation mapping a free generator of the tensor algebra to a sum of annihilation, creation and second quantisation operators on a Fock space. This result was obtained in [15] using a different method, the method presented in this paper having the advantage of giving an explicit construction of the representation. The theory developed in this paper has applications in the field of noncommutative stochastic processes with independent and stationary increments [3]; see also [14,15].

2. Cocommutative coalgebras

Vector spaces will be over the complex numbers. For two vector spaces $\mathcal{V}$ and $\mathcal{W}$ we denote by $\mathbf{L}(\mathcal{V},\mathcal{W})$ the vector space of linear mappings from $\mathcal{V}$ to $\mathcal{W}$. An *involutive* vector space is a vector space $\mathcal{V}$ together with a map $v \mapsto v^*$ on $\mathcal{V}$ such that

$$(v+zw)^* = v^* + \bar{z}w^* \text{ and } (v^*)^* = v$$

for $v,w \in \mathcal{V}$ and $z \in \mathbb{C}$. The complex numbers with complex conjugation form an example of an involutive vector space. An element v of an involutive vector space is called hermitian if $v^* = v$. For two involutive vector spaces $\mathcal{V}$ and $\mathcal{W}$ we turn $\mathbf{L}(\mathcal{V},\mathcal{W})$ into an involutive vector space by setting

$$R^*(v) = R(v^*)^*$$

for $R \in \mathbf{L}(\mathcal{V},\mathcal{W})$ and $v \in \mathcal{V}$. In this sense, a linear map between involutive vector spaces is hermitian if and only if it maps hermitian elements to hermitian elements. An algebra is understood to be associative and unital, and algebra homomorphisms are assumed to preserve the unit elements. A subalgebra of an algebra is a subalgebra in the usual sense containing the unit element. A *-algebra is an algebra $\mathcal{A}$ which is also an involutive vector space such that $(ab)^* = b^*a^*$ for $a,b \in \mathcal{A}$.

We sometimes regard an algebra as a triplet $(\mathcal{A},\mathrm{M},m)$ consisting of a vector space $\mathcal{A}$ and linear mappings $\mathrm{M}\colon \mathcal{A}\otimes\mathcal{A} \to \mathcal{A}$ and $m\colon \mathbb{C} \to \mathcal{A}$ satisfying the associativity and unit element conditions. In the usual notation $\mathrm{M}(a\otimes b) = ab$ and $m(z) = z\mathbf{1}$, $a,b \in \mathcal{A}$, $z \in \mathbb{C}$. A coalgebra is a triplet $(\mathcal{C},\Delta,\delta)$ consisting of a vector space $\mathcal{C}$ and linear mappings $\Delta\colon \mathcal{C} \to \mathcal{C}\otimes\mathcal{C}$ and $\delta\colon \mathcal{C} \to \mathbb{C}$ satisfying the coassociativity and counit conditions (1.1) and (1.2); see [1,19] for an introduction to the theory of coalgebras. A subcoalgebra $\mathcal{D}$ of a coalgebra $(\mathcal{C},\Delta,\delta)$ is a linear subspace of $\mathcal{C}$ such that $\Delta\mathcal{D} \subset \mathcal{D}\otimes\mathcal{D}$. If $\mathcal{D}$ is a subcoalgebra $(\mathcal{D},\Delta\lceil\mathcal{D},\delta\lceil\mathcal{D})$ is a coalgebra. The intersection and the sum $\mathcal{D}_1+\mathcal{D}_2 = \{d_1+d_2 : d_1 \in \mathcal{D}_1, d_2 \in \mathcal{D}_2\}$ of two subcoalgebras $\mathcal{D}_1$ and $\mathcal{D}_2$ are again subcoalgebras. For a finite-dimensional subspace $\mathcal{V}$ of a coalgebra $\mathcal{C}$ there is a finite-dimensional subcoalgebra of $\mathcal{C}$ containing $\mathcal{V}$ by the Fundamental Theorem on Coalgebras [19]. We denote by $\mathrm{D}(\mathcal{V})$ the smallest subcoalgebra containing $\mathcal{V}$. It follows that every coalgebra is the inductive limit of the system $\Gamma(\mathcal{C})$ of its finite-dimensional subcoalgebras. We frequently make use of the following construction. Let $\mathcal{C}$ be a coalgebra and let $\mathcal{W}$ be a vector space. Let $(R_{\mathcal{D}})_{\mathcal{D}\in\Gamma(\mathcal{C})}$ be a family of linear mapppings $R_{\mathcal{D}}\colon \mathcal{D} \to \mathcal{W}$ satisfying the consistency condition

$$R_{\mathcal{D}}\lceil\mathcal{D}' = R_{\mathcal{D}'}$$

for $\mathcal{D},\mathcal{D}'\in\Gamma(\mathcal{C}),\mathcal{D}'\subset\mathcal{D}$. Then there is a unique linear map $R:\mathcal{C}\to\mathcal{W}$, called the inductive limit of $(R_{\mathcal{D}})_{\mathcal{D}\in\Gamma(\mathcal{C})}$, such that

$$R\restriction\mathcal{D}=R_{\mathcal{D}}$$

for $\mathcal{D}\in\Gamma(\mathcal{C})$. Especially, if S is a linear operator on $\mathcal{C}$ leaving invariant all subcoalgebras of $\mathcal{C}$ we define the linear operator $\exp S$ on $\mathcal{C}$ as the inductive limit of the system $(\exp(S\restriction\mathcal{D}))_{\mathcal{D}\in\Gamma(\mathcal{C})}$. Moreover, for such an operator S we can define $\exp(\operatorname{ad}S)$ as an operator on $\mathbf{L}(\mathcal{C})=\mathbf{L}(\mathcal{C},\mathcal{C})$ as follows. For $\mathcal{D},\mathcal{D}'\in\Gamma(\mathcal{C})$ we have that $\operatorname{ad}S$ leaves $\mathbf{L}(\mathcal{D},\mathcal{D}')$ invariant. Thus we can define $\exp(\operatorname{ad}S)$ as an operator on $\mathbf{L}(\mathcal{D},\mathcal{D}')$. For $T\in\mathbf{L}(\mathcal{C})$ and $\mathcal{D}\in\Gamma(\mathcal{C})$ let $R_{\mathcal{D}}$ be the linear map $\exp(\operatorname{ad}S)(T)$ from $\mathcal{D}$ to $\mathrm{D}(T(\mathcal{D}))$. Define $\exp(\operatorname{ad}S)(T)$ as the inductive limit of $(R_{\mathcal{D}})_{\mathcal{D}\in\Gamma(\mathcal{C})}$. We have the relation

$$(\exp S)\circ T\circ(\exp(-S))=\exp(\operatorname{ad}S)(T). \tag{2.1}$$

For an algebra $(\mathcal{A},\mathrm{M},m)$ and a coalgebra $(\mathcal{C},\Delta,\delta)$ the vector space $\mathbf{L}(\mathcal{C},\mathcal{A})$ becomes an algebra with multiplication

$$R*S=\mathrm{M}\circ(R\otimes S)\circ\Delta,$$

and unit element

$$m\circ\delta.$$

Especially, the algebraic dual space $\mathbf{L}(\mathcal{C},\mathbb{C})$ of a coalgebra becomes an algebra with multiplication (1.4) and unit element δ. For a coalgebra $\mathcal{C}$ we define the linear map

$$\mathcal{R}:\mathbf{L}(\mathcal{C},\mathbb{C})\to\mathbf{L}(\mathcal{C},\mathcal{C})$$

by

$$\mathcal{R}(\lambda)=(\mathrm{id}\otimes\lambda)\circ\Delta,$$

$\lambda\in\mathbf{L}(\mathcal{C},\mathbb{C})$. Using the coassociativity and counit conditions, one can show that $\mathcal{R}$ is an algebra homomorphism, and

$$\mu*\lambda=\mu\circ\mathcal{R}(\lambda),$$

$\mu,\lambda\in\mathbf{L}(\mathcal{C},\mathbb{C})$. Moreover, the operator $\mathcal{R}(\lambda)$ leaves invariant all subcoalgebras of $\mathcal{C}$ and we can define $\exp_*\lambda$ to be the linear functional $\delta\circ\exp\mathcal{R}(\lambda)$. We have

$$(\exp_*\lambda)(b)=\sum_{n=0}^{\infty}\frac{1}{n!}\lambda^{*n}(b)$$

for $b\in\mathcal{C}$, where λ^{*n} denotes the n-th convolution power of λ.

For two vector spaces $\mathcal{V}$ and $\mathcal{W}$ denote by $\tau_{\mathcal{V},\mathcal{W}}$ the linear map

$$\tau_{\mathcal{V},\mathcal{W}}:\mathcal{V}\otimes\mathcal{W}\to\mathcal{W}\otimes\mathcal{V}$$

given by

$$\tau_{\mathcal{V},\mathcal{W}}(v\otimes w)=w\otimes v,$$

$v\in\mathcal{V},w\in\mathcal{W}$. A coalgebra $\mathcal{C}$ is called *cocommutative* if

$$\tau_{\mathcal{C},\mathcal{C}}\circ\Delta=\Delta$$

holds. $\mathcal{C}$ is cocommutative if and only if $\mathbf{L}(\mathcal{C},\mathbb{C})$ is commutative. The *tensor product* of two coalgebras $(\mathcal{C}_i,\Delta_i,\delta_i)$, $i=1,2$, is defined to be the coalgebra

$$(\mathcal{C}_1\otimes\mathcal{C}_2,(\mathrm{id}\otimes\tau_{\mathcal{C}_1,\mathcal{C}_2}\otimes\mathrm{id})\circ(\Delta_1\otimes\Delta_2),\delta_1\otimes\delta_2).$$

The conjugate $\overline{\mathcal{V}}$ of a vector space $\mathcal{V}$ is the vector space $\{\overline{v}:v\in\mathcal{V}\}$ with the linear structure given by

$$z\overline{v_1}+\overline{v_2}=\overline{\bar{z}v_1+v_2},$$

$z \in \mathbb{C}$, $v_1, v_2 \in \mathcal{V}$. For two vector spaces $\mathcal{V}$ and $\mathcal{W}$ an isomorphism Φ from $\overline{\mathbf{L}(\mathcal{V},\mathcal{W})}$ to $\mathbf{L}(\overline{\mathcal{V}},\overline{\mathcal{W}})$ is given by

$$\Phi(\overline{R})(\overline{v}) = \overline{R(v)}.$$

We identify the elements of $\overline{\mathbf{L}(\mathcal{V},\mathcal{W})}$ and $\mathbf{L}(\overline{\mathcal{V}},\overline{\mathcal{W}})$. Similarly, $\overline{\mathcal{V}\otimes\mathcal{W}}$ is identified with $\overline{\mathcal{V}}\otimes\overline{\mathcal{W}}$. The vector space $\overline{\mathbb{C}}$ is identified with $\mathbb{C}$ in the obvious way. The conjugate of a coalgebra $(\mathcal{C},\Delta,\delta)$ is the coalgebra $(\overline{\mathcal{C}},\overline{\Delta},\overline{\delta})$. If $\mathcal{C}$, $\mathcal{C}_1$ and $\mathcal{C}_2$ are cocommutative, so are $\mathcal{C}_1\otimes\mathcal{C}_2$ and $\overline{\mathcal{C}}$.

By regarding sesquilinear forms on a coalgebra $\mathcal{C}$ as elements of $\mathbf{L}(\overline{\mathcal{C}}\otimes\mathcal{C},\mathbb{C})$, we obtain the notion of convolution product of sesquilinear forms on a coalgebra. For a vector space $\mathcal{V}$ the vector space $\overline{\mathcal{V}}\otimes\mathcal{V}$ is turned into an involutive vector space by setting $(\overline{v}\otimes w)^* = \overline{w}\otimes v$ for $v,w\in\mathcal{V}$. A sesquilinear form L on a vector space $\mathcal{V}$ is called positive if $L(\overline{v}\otimes v) \geqslant 0$ for all $v\in\mathcal{V}$. For a coalgebra $\mathcal{C}$ and $b\in\mathcal{C}$ we denote by L^b the sesquilinear form on $\mathcal{C}$ with

$$L^b(\overline{c}\otimes d) = L(\overline{(c-\delta(c)b)}\otimes(d-\delta(d)b)).$$

A sesquilinear form A on a coalgebra $\mathcal{C}$ is called *conditionally positive* if

$$A(\overline{b}\otimes b) \geqslant 0 \text{ for all } b\in\mathcal{C} \text{ with } \delta(b) = 0.$$

The following result was proved in [20] for the cocommutative case and in [13] for arbitrary coalgebras.

Theorem 2.1. Let A be a sesquilinear form on a coalgebra $\mathcal{C}$. We consider the following statements

(i) A is conditionally positive

(ii) there is a $b\in\mathcal{C}$ with $\delta(b) = 1$ such that A^b is positive

(iii) A^b is positive for all $b\in\mathcal{C}$ with $\delta(b) = 1$

(iv) A is hermitian and conditionally positive

(v) $\exp_*(tA)$ is positive for all $t \geqslant 0$.

Then the implications

$$\begin{array}{ccccc} \text{(i)} & \Leftrightarrow & \text{(ii)} & \Leftrightarrow & \text{(iii)} \\ \Uparrow & & & & \\ \text{(iv)} & \Leftrightarrow & \text{(v)} & & \end{array}$$

hold. □

The following lemma is the starting point of our considerations.

Lemma 2.1. Let A be a hermitian sesquilinear form on a cocommutative coalgebra $\mathcal{C}$. Then we have for $e\in\mathcal{C}$ with $\delta(e) = 1$

$$e^{-A(\overline{e}\otimes e)}\exp_* A^e = (\exp_* A)\circ\left(\exp(-\overline{\mathcal{R}(\alpha)})\otimes\exp(-\mathcal{R}(\alpha))\right)$$

where α denotes the linear functional on $\mathcal{C}$ given by

$$\alpha(b) = A(\overline{e}\otimes b).$$

Proof: We have

$$A^e = A + A'$$

where

$$A' = -\overline{\delta}\otimes\alpha - \overline{\alpha}\otimes\delta + A(\overline{e}\otimes e)\overline{\delta}\otimes\delta.$$

As A commutes with A'

$$\exp_* A^e = \exp_*(A+A') = (\exp_* A) * (\exp_* A') = (\exp_* A) \circ \exp \mathcal{R}(A').$$

But as

$$\mathcal{R}(A') = -\overline{\mathcal{R}(\alpha)} \otimes \mathrm{id}_{\mathcal{C}} - \mathrm{id}_{\overline{\mathcal{C}}} \otimes \mathcal{R}(\alpha) + A(\bar{e} \otimes e)\, \mathrm{id}_{\overline{\mathcal{C}}} \otimes \mathrm{id}_{\mathcal{C}}$$

we have

$$\exp \mathcal{R}(A') = e^{A(\bar{e} \otimes e)} \exp(-\overline{\mathcal{R}(\alpha)} \otimes \mathrm{id}_{\mathcal{C}} - \mathrm{id}_{\overline{\mathcal{C}}} \otimes \mathcal{R}(\alpha))$$
$$= e^{A(\bar{e} \otimes e)} \exp(-\overline{\mathcal{R}(\alpha)}) \otimes \exp(-\mathcal{R}(\alpha)). \square$$

Let $\mathcal{V}$ be a vector space. The tensor algebra $\mathrm{T}(\mathcal{V})$ of $\mathcal{V}$ is defined to be the algebra

$$\mathrm{T}(\mathcal{V}) = \mathbb{C}\mathbf{1} \oplus \mathcal{V} \oplus (\mathcal{V} \otimes \mathcal{V}) \oplus (\mathcal{V} \otimes \mathcal{V} \otimes \mathcal{V}) \oplus \ldots$$

with unit element **1** and multiplication given by

$$vb = v \otimes b,$$

$v \in \mathcal{V}$, $b \in \mathrm{T}(\mathcal{V})$. Denote by s_n, $n \in \mathbb{N}$, the group of permutations of the set $\{1,\ldots,n\}$. We let s_n act on $\mathcal{V}^{\otimes n}$ by setting for $p \in \mathrm{s}_n$ and $v_1,\ldots,v_n \in \mathcal{V}$

$$p(v_1 \otimes \ldots \otimes v_n) = v_{p(1)} \otimes \ldots \otimes v_{p(n)}.$$

Define the linear operator P on $\mathrm{T}(\mathcal{V})$ by

$$P(v_1 \otimes \ldots \otimes v_n) = \frac{1}{n!} \sum_{p \in \mathrm{s}_n} p(v_1 \otimes \ldots \otimes v_n).$$

The range of P is denoted by $\mathrm{F}(\mathcal{V})$. For a Hilbert space H the linear operator P on $\mathrm{T}(H)$ extends to an orthogonal projection on the completion $\mathcal{T}(H)$ of the pre-Hilbert space $\mathrm{T}(H)$. The range of this orthogonal projection is the Bose Fock space over H which we denote by $\mathcal{F}(H)$; see [4].

Let $\mathcal{C}$ be a coalgebra. We define mappings

$$\Delta_n : \mathcal{C} \to \mathcal{C}^{\otimes n}$$

inductively by

$$\Delta_0 = \delta$$
$$\Delta_{n+1} = (\Delta_n \otimes \mathrm{id}) \circ \Delta.$$

We have $\Delta_1 = \mathrm{id}$ and $\Delta_2 = \Delta$, and

$$\lambda_1 * \ldots * \lambda_n = (\lambda_1 \otimes \ldots \otimes \lambda_n) \circ \Delta_n$$

for $n \in \mathbb{N}$, $\lambda_1,\ldots,\lambda_n \in \mathrm{L}(\mathcal{C},\mathbb{C})$. If $\mathcal{C}$ is cocommutative we have

$$p \circ \Delta_n = \Delta_n \tag{2.2}$$

for $n \in \mathbb{N}$ and $p \in \mathrm{s}_n$.

Let L be a positive sesquilinear form on a vector space $\mathcal{V}$. We form the quotient space $D_L = \mathcal{V}/N_L$ where $N_L = \{v \in \mathcal{V} : L(\bar{v} \otimes v) = 0\}$. Let $\eta_L : \mathcal{V} \to D_L$ be the canonical map. D_L is a pre-Hilbert space with inner product

$$\langle \eta_L(v), \eta_L(w) \rangle = L(\bar{v} \otimes w),$$

$v, w \in \mathcal{V}$. Denote by $\tilde{D}_L$ the completion of D_L. For $n \in \mathbb{N}$, $V \in \mathcal{V}^{\otimes n}$ we write $[V]$ for $\eta_L^{\otimes n}(V)$ if no confusion can arise.

Lemma 2.2. Let L be a positive sesquilinear form on a coalgebra $\mathcal{C}$. Then

(i) for $b \in \mathscr{C}$ there exist positive constants D and C such that

$$\| [\Delta_n b] \|^2 \leqslant \mathrm{D}\,\mathrm{C}^n$$

for all $n \in \mathbb{N}$.

(ii) the series

$$\sum_{n=0}^{\infty} \frac{[\Delta_n b]}{\sqrt{n!}}$$

converges in $\mathcal{I}(\tilde{D}_L)$ for all $b \in \mathscr{C}$ (we denote the limit by $\mathrm{E}_L(b)$). The map

$$\mathrm{E}_L : \mathscr{C} \to \mathcal{I}(\tilde{D}_L)$$

is linear.

(iii) $\langle \mathrm{E}_L(b), \mathrm{E}_L(c) \rangle = (\exp_* L)(\bar{b} \otimes c)$

for all $b, c \in \mathscr{C}$.

(iv) if $\mathscr{C}$ is cocommutative

$$\mathrm{E}_L(\mathscr{C}) \subset \mathcal{F}(\tilde{D}_L).$$

Proof: For $b, c \in \mathscr{C}$ we have

$$\langle ([\Delta_n b], [\Delta_n c] \rangle = L^{*n}(\bar{b} \otimes c), \tag{2.3}$$

so

$$\| [\Delta_n b] \|^2 = L^{*n}(\bar{b} \otimes b).$$

Moreover,

$$L^{*n} = \delta \circ \mathcal{R}(L)^n.$$

Let $\mathscr{D}_b$ be a finite-dimensional subcoalgebra of $\bar{\mathscr{C}} \otimes \mathscr{C}$ containing $\bar{b} \otimes b$ equipped with a norm $\| \;\; \|$. Setting

$$\mathrm{D} = \| \bar{\delta} \otimes \delta \lceil \mathscr{D}_b \| \; \| \bar{b} \otimes b \| \text{ and } \mathrm{C} = \| \mathcal{R}(L) \lceil \mathscr{D}_b \|$$

(i) follows. Now (ii) and (iii) follow from (i) and (2.3). (iv) is a direct consequence of (2.2). □

Theorem 2.2. Let A be a hermitian, conditionally positive sesquilinear form on a cocommutative coalgebra $\mathscr{C}$ and let e be an element of $\mathscr{C}$ such that $\delta(e) = 1$. We set $M = \exp_* A$ and $L = A^e$, and define α as in Lemma 2.1. Then the equation

$$S_{A,e} \circ \eta_M = \mathrm{e}^{-\frac{1}{2}A(\bar{e} \otimes e)} \mathrm{E}_L \circ \exp \mathcal{R}(\alpha)$$

defines an isometrie

$$S_{A,e} : D_M \to \mathcal{F}(\tilde{D}_L)$$

and

$$S_{A,e}(D_M) = \mathrm{E}_L(\mathscr{C}).$$

Thus $S_{A,e}$ can be extended to a unitary map from $\tilde{D}_M$ to the closure of $\mathrm{E}_L(\mathscr{C})$.

Proof: For $b, c \in \mathscr{C}$ we have

$$\langle (\mathrm{E}_L \circ \exp \mathcal{R}(\alpha))(b), (\mathrm{E}_L \circ \exp \mathcal{R}(\alpha))(c) \rangle = (\exp_* L) \circ (\overline{\exp \mathcal{R}(\alpha)(b)} \otimes \exp \mathcal{R}(\alpha)(c))$$

by Lemma 2.2.(iii). By Lemma 2.1. this is equal to

$$\mathrm{e}^{A(\bar{e} \otimes e)} (\exp_* A)(\bar{b} \otimes c) = \mathrm{e}^{A(\bar{e} \otimes e)} \langle \eta_M(b), \eta_M(c) \rangle. \; \square$$

3. Cocommutative *-bialgebras

If $\mathcal{A}$ is an algebra we turn $\mathcal{A} \otimes \mathcal{A}$ into an algebra by defining the multiplication

$$(a \otimes b)(a' \otimes b') = aa' \otimes bb',$$

$a, b, a', b' \in \mathcal{A}$. If $\mathcal{A}$ is a *-algebra $\mathcal{A} \otimes \mathcal{A}$ becomes a *-algebra with the involution

$$(a \otimes b)^* = a^* \otimes b^*,$$

$a, b \in \mathcal{A}$. A bialgebra $(\mathcal{B}, \mathrm{M}, m, \Delta, \delta)$ is an algebra $(\mathcal{B}, \mathrm{M}, m)$ and a coalgebra $(\mathcal{B}, \Delta, \delta)$ such that Δ and δ are algebra homomorphisms. A *-bialgebra is a bialgebra $\mathcal{B}$ together with a map $b \mapsto b^*$ on $\mathcal{B}$ such that $(\mathcal{B}, \mathrm{M}, m)$ is turned into a *-algebra and Δ and δ are hermitian. A linear functional λ on a *-algebra $\mathcal{A}$ is called positive if $\lambda(a^*a) \geq 0$ for all $a \in \mathcal{A}$. A linear functional α on a *-bialgebra $\mathcal{B}$ is called *conditionally positive* if $\alpha(b^*b) \geq 0$ for all $b \in \mathcal{B}$ with $\delta(b) = 0$. If λ is a linear functional on a *-algebra $\mathcal{A}$ we define the sesquilinear form $M(\lambda)$ on $\mathcal{A}$ by

$$M(\lambda)(\bar{a} \otimes b) = \lambda(a^*b),$$

$a, b \in \mathcal{A}$, and if λ is a linear functional on a *-bialgebra $\mathcal{B}$ we define the sesquilinear form $L(\lambda)$ on $\mathcal{B}$ by

$$L(\lambda)(\bar{b} \otimes c) = M(\lambda)^1(\bar{b} \otimes c) = \lambda((b - \delta(b)\mathbf{1})^*(c - \delta(c)\mathbf{1})),$$

$b, c \in \mathcal{B}$. For a *-bialgebra $\mathcal{B}$ the map

$$M: \mathrm{L}(\mathcal{B}, \mathbb{C}) \to \mathrm{L}(\overline{\mathcal{B}} \otimes \mathcal{B}, \mathbb{C})$$

is an algebra monomorphism and it maps hermitian, positive and conditionally positive linear functionals on $\mathcal{B}$ to hermitian, positive and conditionally positive sesquilinear forms on $\mathcal{B}$ resp. Theorem 2.1. yields the following result for *-bialgebras; see [13].

Theorem 3.1. Let α be a linear functional on a *-bialgebra $\mathcal{B}$. We consider the following statements

(i) α is conditionally positive

(ii) $M(\alpha)$ is conditionally positive

(iii) $L(\alpha)$ is positive

(iv) α is hermitian and conditionally positive

(v) $\exp_*(t\alpha)$ is positive for all $t \in \mathbb{R}_+$.

(vi) $M(\alpha)$ is hermitian and conditionally positive

(vii) $\exp_*(tM(\alpha))$ is positive for all $t \geq 0$.

Then the implications

$$\begin{array}{ccccccc} \text{(i)} & \Leftrightarrow & \text{(ii)} & \Leftrightarrow & \text{(iii)} & & \\ \Uparrow & & & & & & \\ \text{(iv)} & \Leftrightarrow & \text{(v)} & \Leftrightarrow & \text{(vi)} & \Leftrightarrow & \text{(vii)} \end{array}$$

hold. □

For a pre-Hilbert space D a linear operator R on D is called hermitian if

$$\langle R\xi, \zeta \rangle = \langle \xi, R\zeta \rangle$$

for all $\xi, \zeta \in D$. The linear span $\mathrm{H}(D)$ of all hermitian operators on D is a subalgebra of $\mathrm{L}(D)$ and it is a *-algebra with $R^* = R^\dagger \lceil D$ where $R^\dagger$ denotes the adjoint of R considered as an unbounded linear operator on D. A *-representation of a *-algebra $\mathcal{A}$ on a pre-Hilbert space D is a *-algebra homomorphism from $\mathcal{A}$ to $\mathrm{H}(D)$.

Let $\mathcal{A}$ be a *-algebra. For a positive $\lambda \in \mathrm{L}(\mathcal{A}, \mathbb{C})$ the GNS representation given by λ is the *-representation π_λ of $\mathcal{A}$ on the pre-Hilbert space $D_{M(\lambda)}$ given by

$$\pi_\lambda(a)\eta_{M(\lambda)}(b) = (\eta_{M(\lambda)} \circ \mathrm{L}_a)(b),$$

$a, b \in \mathcal{A}$, where L_a denotes the left multiplication $\mathrm{L}_a(b) = ab$ on $\mathcal{A}$.

Theorem 3.2. Let α be a hermitian, conditionally positive linear functional on a cocommutative *-bialgebra $\mathcal{B}$. Let $\mu = \exp_* \alpha$. We write S, π, E and $\mathcal{R}$ for $S_{M(\alpha),1}, \pi_\mu, \mathrm{E}_{L(\alpha)}$ and $\mathcal{R}(\alpha)$ resp. Then for $b \in \mathcal{B}$

$$S \circ \pi(b) \circ S^{-1} \circ \mathrm{E} = \mathrm{E} \circ \exp \mathcal{R} \circ \mathrm{L}_b \circ \exp(-\mathcal{R}) = \mathrm{E} \circ \exp(\operatorname{ad} \mathcal{R})(\mathrm{L}_b),$$

and

$$\mu(b) = \langle \Omega, (S \circ \pi(b) \circ S^{-1})\Omega \rangle$$

where Ω denotes the vacuum vector $(1, 0, 0, \ldots)$ in $\mathcal{F}(\tilde{D}_{L(\alpha)})$.

Proof: We write η for $\eta_{M(\mu)}$. From

$$S \circ \eta = \mathrm{E} \circ \exp \mathcal{R} \tag{3.1}$$

we have

$$S^{-1} \circ \mathrm{E} = \eta \circ \exp(-\mathcal{R}).$$

Thus

$$S \circ \pi(b) \circ S^{-1} \circ \mathrm{E} = S \circ \pi(b) \circ \eta \circ \exp(-\mathcal{R}). \tag{3.2}$$

Using

$$\pi(b) \circ \eta = \eta \circ \mathrm{L}_b,$$

it follows that (3.2) equals

$$\mathrm{E} \circ \exp \mathcal{R} \circ \mathrm{L}_b \circ \exp(-\mathcal{R})$$

which is equal to

$$\mathrm{E} \circ \exp(\operatorname{ad} \mathcal{R})(\mathrm{L}_b)$$

by (2.1).- As $\Delta_0(1) = \delta(1) = 1$ and $\Delta_n(1) = \mathbf{1}^{\otimes n}$ for $n \in \mathbb{N}$, it follows from $L(\alpha)(\bar{1} \otimes 1) = 0$ that $\mathrm{E}(1) = \Omega$. But then

$$\langle \Omega, (S \circ \pi(b) \circ S^{-1})\Omega \rangle = \langle \eta(1), \pi(b)\eta(1) \rangle = \langle \eta(1), \eta(b) \rangle = M(\mu)(\bar{1} \otimes b) = \mu(b). \square$$

Let G be a group. We treat the example of the *-bialgebra $\mathbb{C}G$ of the introduction. Let α be a hermitian, conditionally positive linear functional on $\mathbb{C}G$. The map

$$\mathrm{E}: \mathbb{C}G \to \mathcal{F}(\tilde{D}_{L(\alpha)})$$

is the linear extension of the map from G to $\mathcal{F}(\tilde{D}_{L(\alpha)})$ which sends a group element x to the coherent state (or exponential vector)

$$\mathrm{EXP}\,[x] = 1 + [x] + \frac{1}{\sqrt{2!}}[x \otimes x] + \frac{1}{\sqrt{3!}}[x \otimes x \otimes x] + \ldots$$

As $D_{L(\alpha)}$ is dense in $\tilde{D}_{L(\alpha)}$, it follows that $\{\mathrm{EXP}\,\xi : \xi \in D_{L(\alpha)}\}$ is total in $\mathcal{F}(\tilde{D}_{L(\alpha)})$. But it is not true in general that $\mathrm{E}(\mathbb{C}G)$ is dense in $\mathcal{F}(\tilde{D}_{L(\alpha)})$. The closure of $\mathrm{E}(\mathbb{C}G)$ is equal to the closure of the linear span of the set $\{\mathrm{EXP}\,[x] : x \in G\}$. We have for $x \in G$

$$\mathcal{R}(x) = \alpha(x)x$$

and

$$S\eta(x) = \mathrm{E}((\exp \mathcal{R})(x)) = \mathrm{e}^{\alpha(x)}\mathrm{E}(x) = \mathrm{e}^{\alpha(x)}\mathrm{EXP}\,[x].$$

Moreover, for $x, y \in G$

$$(S \circ \pi(x) \circ S^{-1})\mathrm{EXP}\,[y] = (S \circ \pi(x) \circ S^{-1} \circ \mathrm{E})(y)$$

$$= \mathrm{E}\big((\exp \mathcal{R} \circ \mathrm{L}_x \circ \exp(-\mathcal{R}))(y)\big)$$

$$= \mathrm{e}^{\alpha(xy)-\alpha(y)}\,\mathrm{EXP}\,[xy];$$

cf. [7, 12].

The Araki-Woods embedding theorem says more. There is a representation of G associated to α in a natural way (see also our Theorem 4.1.), and the GNS representation given by e^{α} can be written in terms of this representation. This result can be generalised to arbitrary cocommutative *-bialgebras. In the present paper we restrict ourselves to the case of special elements of a cocommutative *-bialgebra.

4. Group-like elements and Lie elements

An element b of a coalgebra $\mathscr{C}$ is *group-like* if

$$\Delta(b) = b \otimes b \text{ and } \delta(b) = 1.$$

An element b of a bialgebra $\mathscr{B}$ is a *Lie element* if

$$\Delta(b) = b \otimes \mathbf{1} + \mathbf{1} \otimes b \text{ and } \delta(b) = 0.$$

Proposition 4.1. Let $\mathscr{B}$ be a bialgebra and let λ be a linear functional on $\mathscr{B}$.

(i) For a group-like element $b \in \mathscr{B}$

$$\mathrm{ad}\,\mathcal{R}(\lambda)(\mathrm{L}_b) = \mathrm{L}_b \circ \mathcal{R}(\lambda \circ (\mathrm{L}_b - \mathrm{id})),$$

and if $\mathscr{B}$ is cocommutative

$$(\mathrm{ad}\,\mathcal{R}(\lambda))^n(\mathrm{L}_b) = \mathrm{L}_b \circ \mathcal{R}(\lambda \circ (\mathrm{L}_b - \mathrm{id}))^n$$

for all $n \in \mathbb{N}$, and

$$\exp(\mathrm{ad}\,\mathcal{R}(\lambda))(\mathrm{L}_b) = \mathrm{L}_b \circ \exp \mathcal{R}(\lambda \circ (\mathrm{L}_b - \mathrm{id})).$$

(ii) For a Lie element $b \in \mathscr{B}$

$$\mathrm{ad}\,\mathcal{R}(\lambda)(\mathrm{L}_b) = \mathcal{R}(\lambda \circ \mathrm{L}_b),$$

and if $\mathscr{B}$ is cocommutative

$$(\mathrm{ad}\,\mathcal{R}(\lambda))^n(\mathrm{L}_b) = 0$$

for all $n \in \mathbb{N}$, $n \geq 2$, and

$$\exp(\mathrm{ad}\,\mathcal{R}(\lambda))(\mathrm{L}_b) = \mathrm{L}_b + \mathcal{R}(\lambda \circ \mathrm{L}_b).$$

Proof: Let $c \in \mathscr{B}$ and

$$\Delta c = \sum_{i=1}^{d} c_{1i} \otimes c_{2i}$$

with $d \in \mathbb{N}$ and $c_{1i}, c_{2i} \in \mathscr{B}$, $i = 1, \ldots, d$. If b is a group-like element

$$(\mathcal{R}(\lambda) \circ \mathrm{L}_b)(c) = (\mathrm{id} \otimes \lambda)(\Delta b \Delta c)$$

$$= (\mathrm{id} \otimes \lambda)(b \otimes b) \sum_{i=1}^{d} (c_{1i} \otimes c_{2i})$$

$$= \sum_{i=1}^{d} bc_{1i}\,\lambda(bc_{2i})$$

$$= (\mathrm{L}_b \circ \mathcal{R}(\lambda \circ \mathrm{L}_b))(c)$$

and

$$[\mathcal{R}(\lambda), \mathrm{L}_b] = \mathrm{L}_b \circ \mathcal{R}(\lambda \circ \mathrm{L}_b - \lambda).$$

If b is a Lie element we have

$$(\mathcal{R}(\lambda) \circ \mathrm{L}_b)(c) = (\mathrm{id} \otimes \lambda)(b \otimes \mathbf{1} + \mathbf{1} \otimes b) \sum_{i=1}^{d} (c_{1i} \otimes c_{2i})$$

$$= \sum_{i=1}^{d} (bc_{1i}\,\lambda(c_{2i}) + c_{1i}\,\lambda(bc_{2i}))$$

$$= (\mathrm{L}_b \circ \mathcal{R}(\lambda) + \mathcal{R}(\lambda \circ \mathrm{L}_b))(c).$$

Now let $\mathcal{B}$ be cocommutative. For a group-like element b, suppose that we have for $n \in \mathbb{N}$

$$(\mathrm{ad}\,\mathcal{R}(\lambda))^n(\mathrm{L}_b) = \mathrm{L}_b \circ \mathcal{R}(\lambda \circ (\mathrm{L}_b - \mathrm{id}))^n.$$

Then

$$(\mathrm{ad}\,\mathcal{R}(\lambda))^{n+1}(\mathrm{L}_b) = [\mathcal{R}(\lambda), \mathrm{L}_b \circ \mathcal{R}(\lambda \circ (\mathrm{L}_b - \mathrm{id}))^n]$$

$$= \mathrm{L}_b \circ \mathcal{R}(\lambda \circ \mathrm{L}_b) \circ \mathcal{R}(\lambda \circ (\mathrm{L}_b - \mathrm{id}))^n - \mathrm{L}_b \circ \mathcal{R}(\lambda \circ (\mathrm{L}_b - \mathrm{id}))^n \circ \mathcal{R}(\lambda)$$

$$= \mathrm{L}_b \circ (\mathcal{R}(\lambda \circ \mathrm{L}_b) - \mathcal{R}(\lambda)) \circ \mathcal{R}(\lambda \circ (\mathrm{L}_b - \mathrm{id}))^n$$

$$= \mathrm{L}_b \circ \mathcal{R}(\lambda \circ (\mathrm{L}_b - \mathrm{id}))^{n+1}.$$

For a Lie element b

$$(\mathrm{ad}\,\mathcal{R}(\lambda))^2(\mathrm{L}_b) = [\mathcal{R}(\lambda), \mathcal{R}(\lambda \circ \mathrm{L}_b)] = \mathcal{R}(\lambda * (\lambda \circ \mathrm{L}_b) - (\lambda \circ \mathrm{L}_b) * \lambda) = 0. \square$$

As a corollary of the preceeding proposition we have

Theorem 4.1. Let α be a hermitian and conditionally positive linear functional on a cocommutative *-bialgebra $\mathcal{B}$.

(i) For a group-like element $b \in \mathcal{B}$

$$S \circ \pi(b) \circ S^{-1} \circ \mathrm{E} = \mathrm{E} \circ \mathrm{L}_b \circ \exp \mathcal{R}(\alpha \circ (\mathrm{L}_b - \mathrm{id})).$$

(ii) For a Lie element $b \in \mathcal{B}$

$$S \circ \pi(b) \circ S^{-1} \circ \mathrm{E} = \mathrm{E} \circ (\mathrm{L}_b + \mathcal{R}(\alpha \circ \mathrm{L}_b)). \square$$

We now concentrate on Lie elements. Before we go on to compute the representation of a Lie element we proof the following general result for *-bialgebras [14] which establishes the generalisation of the representation associated to the logarithm of the infinitely divisible positive definite function in the Araki-Woods embedding theorem.

Theorem 4.2. Let α be a conditionally positive linear functional on a *-bialgebra $\mathcal{B}$. Then the equation

$$\rho_\alpha(b)[c] = [b(c - \delta(c)\mathbf{1})] = [bc] - [b]\delta(c),$$

$b, c \in \mathcal{B}$, defines a *-representation ρ_α of $\mathcal{B}$ on the pre-Hilbert space $D_{L(\alpha)}$.

Proof: We have for $b, c \in \mathcal{B}$

$$\| [b(c-\delta(c)\mathbf{1})] \|^2 = \alpha((c-\delta(c)\mathbf{1})^* b^* b(c-\delta(c)\mathbf{1}))$$

$$= \langle [c], [b^*b](c-\delta(c)\mathbf{1})] \rangle$$

$$\leq \| [c] \| \, \| [b^*b(c-\delta(c)\mathbf{1})] \|$$

which shows that ρ_α is well-defined. The rest follows from the fact that δ is a *-algebra homomorphism. □

For a Hilbert space H and $\xi \in H$ we denote by $a(\xi)$ and $a^\dagger(\xi)$ the annihilation and creation operators resp. on $\mathcal{F}(H)$ with domain of definition

$$\{ \sum_{n=0}^{\infty} \xi^{(n)} : \xi^{(n)} \text{ is in the closure of } P(H^{\otimes n}) \text{ and } \sum_{n=0}^{\infty} n \| \xi^{(n)} \|^2 < \infty \};$$

see [4]. For a pre-Hilbert space D and a linear operator R on D we define the 'differential second quantisation' of R to be the operator $\Lambda(R)$ on $\mathcal{F}(\bar{D})$, with domain of definition $\mathrm{F}(D)$, given by

$$\Lambda(R) \lceil P(D^{\otimes n}) = R \otimes \mathrm{id} \otimes \ldots \otimes \mathrm{id} + \ldots + \mathrm{id} \otimes \ldots \otimes \mathrm{id} \otimes R.$$

Notice that $\Lambda(R)$ leaves $\mathrm{F}(D)$ invariant, and if $R \in \mathbf{H}(D)$ then $\Lambda(R) \in \mathbf{H}(\mathrm{F}(D))$ and $\Lambda(R)^* = \Lambda(R^*)$.

Theorem 4.3 Let α be a hermitian, conditionally positive linear functional on a cocommutative *-bialgebra $\mathcal{B}$. We set $\rho = \rho_\alpha$. Then for $b, c \in \mathcal{B}$, b a Lie element, and $n \in \mathbb{N} \cup \{0\}$

(i) $[(\Delta_{n+1} \circ \mathrm{L}_b) c] = \sqrt{n+1}\, a^\dagger([b])[\Delta_n c] + \Lambda(\rho(b))[\Delta_{n+1} c]$

(ii) $[(\Delta_n \circ \mathcal{R}(\alpha \circ \mathrm{L}_b)) c] = \alpha(b)\, [\Delta_n c] + \sqrt{n+1}\, a([b^*])[\Delta_{n+1} c]$.

Proof: We use the notation

$$\Delta_n c = \sum_{i=1}^{d_n} c_{1i}^{(n)} \otimes \ldots \otimes c_{ni}^{(n)},$$

$n \in \mathbb{N}$, $d_n \in \mathbb{N}$, $c_{ki}^{(n)} \in \mathcal{B}$, $k = 1, \ldots, n$, $i = 1, \ldots, d_n$. We have

$$[\Delta_{n+1}(bc)] = \sum_{i=1}^{d_{n+1}} ([b\, c_{1i}^{(n+1)}] \otimes [c_{2i}^{(n+1)}] \otimes \ldots \otimes [c_{n+1,i}^{(n+1)}] + \ldots + [c_{1i}^{(n+1)}] \otimes \ldots \otimes [c_{ni}^{(n+1)}] \otimes [b\, c_{n+1,i}^{(n+1)}])$$

$$= \sum_{i=1}^{d_{n+1}} (\delta(c_{1i}^{(n+1)})\, [b] \otimes [c_{2i}^{(n+1)}] \otimes \ldots \otimes [c_{n+1,i}^{(n+1)}] + \ldots + \delta(c_{n+1,i}^{(n+1)})\, [c_{1i}^{(n+1)}] \otimes \ldots \otimes [c_{ni}^{(n+1)}] \otimes [b])$$

$$+ \sum_{i=1}^{d_{n+1}} (\rho(b)[c_{1i}^{(n+1)}] \otimes [c_{2i}^{(n+1)}] \otimes \ldots \otimes [c_{n+1,i}^{(n+1)}] + \ldots + [c_{1i}^{(n+1)}] \otimes \ldots \otimes [c_{ni}^{(n+1)}] \otimes \rho(b)[c_{n+1,i}^{(n+1)}])$$

$$= \sum_{i=1}^{d_n} ([b] \otimes [c_{1i}^{(n)}] \otimes \ldots \otimes [c_{ni}^{(n)}] + \ldots + [c_{1i}^{(n)}] \otimes \ldots \otimes [c_{ni}^{(n)}] \otimes [b])$$

$$+ \Lambda(\rho(b))[\Delta_{n+1} c]$$

$$= \sqrt{n+1}\, a^\dagger([b])[\Delta_n c] + \Lambda(\rho(b))[\Delta_{n+1} c]$$

and

$$[\Delta_n(\mathcal{R}(\alpha \circ \mathrm{L}_b)(c))] = \sum_{i=1}^{d_{n+1}} \alpha(b\, c_{1i}^{(n+1)})\, [c_{2i}^{(n+1)}] \otimes \ldots \otimes [c_{n+1,i}^{(n+1)}]$$

$$= \sum_{i=1}^{d_{n+1}} \alpha(b)\delta(c_{1i}^{(n+1)})\,[c_{2i}^{(n+1)}]\otimes\ldots\otimes[c_{n+1,i}^{(n+1)}]$$

$$+ \sum_{i=1}^{d_{n+1}} \langle[b^*],[c_{1i}^{(n+1)}]\rangle[c_{2i}^{(n+1)}]\otimes\ldots\otimes[c_{n+1,i}^{(n+1)}]$$

$$= \alpha(b)\,[\Delta_n c] + \sqrt{n+1}\,a([b^*])[\Delta_{n+1}c]. \square$$

For a positive sesquilinear form L on a cocommutative coalgebra $\mathscr{C}$ the linear subspace $\mathrm{E}_L(\mathscr{C})$ of $\mathscr{F}(\tilde{D}_L)$ consists of analytic vectors for $a(\xi)$ and $a^\dagger(\xi)$, $\xi\in\tilde{D}_L$, by Lemma 2.2.(i). In the situation of Theorem 4.3. the unbounded operator $\Lambda(\rho(b))$ on $\mathscr{F}(\tilde{D}_{L(\alpha)})$ is closable because

$$\Lambda(\rho(b))^\dagger\lceil \mathrm{F}(D_{L(\alpha)}) = \Lambda(\rho(b^*)).$$

Denote the closure again by $\Lambda(\rho(b))$. Then by Theorem 4.3.(i) we have that $\mathrm{E}_{L(\alpha)}(\mathscr{B})$ lies in the domain of $\Lambda(\rho(b))$ and

$$\Lambda(\rho(b))\mathrm{E}_{L(\alpha)}(c) = \mathrm{E}_{(L(\alpha)}(bc) - a^\dagger([b])\mathrm{E}_{L(\alpha)}(c).$$

We have as a corollary of Theorem 4.1. (ii) and Theorem 4.3.

Theorem 4.4. Let α be a hermitian, conditionally positive linear functional on a cocommutative *-bialgebra $\mathscr{B}$. Then for a Lie element b

$$S\circ\pi(b)\circ S^{-1} = a([b^*]) + a^\dagger([b]) + \Lambda(\rho(b)) + \alpha(b)\,\mathrm{id}. \square$$

Let $\mathscr{V}$ be an involutive vector space. The tensor algebra $\mathrm{T}(\mathscr{V})$ over the vector space $\mathscr{V}$ becomes a *-algebra by extending $v\mapsto v^*$ to an involution on $\mathrm{T}(\mathscr{V})$. We define a *-bialgebra structure un $\mathrm{T}(\mathscr{V})$ by setting

$$\Delta(v) = v\otimes\mathbf{1} + \mathbf{1}\otimes v \text{ and } \delta(v) = 0$$

for $v\in\mathscr{V}$ and extending Δ and δ to algebra homomorphisms $\Delta\colon \mathrm{T}(\mathscr{V})\to\mathrm{T}(\mathscr{V})\otimes\mathrm{T}(\mathscr{V})$ and $\delta\colon\mathrm{T}(\mathscr{V})\to\mathbb{C}$; see [11]. The *-bialgebra $\mathrm{T}(\mathscr{V})$ is cocommutative.

Proposition 4.2. Let $\mathscr{V}$ be an involutive vector space and let L be a positive sesquilinear form on $\mathrm{T}(\mathscr{V})$ such that $L(\bar{\mathbf{1}}\otimes\mathbf{1}) = 0$. Then $\mathrm{E}_L(\mathrm{T}(\mathscr{V})$ is contained in $\mathrm{F}(D_L)$ and for $k\in\mathbb{N}$, $v_1,\ldots,v_k\in\mathscr{V}$

$$\mathrm{E}_L(v_1....v_k) = \sum_{n\in\mathbb{N}} \frac{1}{\sqrt{n!}} \sum_{\substack{X_1,\ldots,X_n\subset\{1,\ldots,k\}\\ X_i\cap X_j=\varnothing \text{ for } i\neq j\\ X_i\neq\varnothing}} [v_{X_1}]\otimes\ldots\otimes[v_{X_n}] \tag{4.1}$$

where for $X = \{r_1,\ldots,r_l\}\subset\{1,\ldots,k\}$, $r_1<\ldots<r_l$, we set $v_X = v_{r_1}\ldots v_{r_l}$.

Proof: One only must prove (4.1). Then $\mathrm{E}_L(\mathrm{T}(\mathscr{V}))\subset\mathrm{F}(D_L)$ follows, because the sums on the right hand side of (4.1) are finite. One easily computes that

$$\Delta_n(v_1\ldots v_k) = \sum_{\substack{X_1,\ldots,X_n\subset\{1,\ldots,k\}\\ X_i\cap X_j=\varnothing \text{ for } i\neq j}} v_{X_1}\otimes\ldots\otimes v_{X_n}$$

where we set $v_\varnothing = \mathbf{1}$. It follows from $L(\bar{\mathbf{1}}\otimes\mathbf{1}) = 0$ that $[\mathbf{1}] = 0$, and (4.1) holds. $\square$

For an involutive vector space $\mathscr{V}$ a *generator* on $\mathscr{V}$ is a quadruple $(D,\rho_0,\theta_0,\alpha_0)$ consisting of

a pre-Hilbert space D

a hermitian linear map $\rho_0\colon\mathscr{V}\to\mathbf{H}(D)$

a linear map $\theta_0: \mathcal{V} \to D$ such that $\{R\xi : R \in \mathcal{A}, \xi \in \theta_0(\mathcal{V})\} = D$ where $\mathcal{A}$ is the subalgebra of $\mathbf{H}(D)$ generated by $\rho_0(\mathcal{V})$

a hermitian linear functional α_0 on $\mathcal{V}$.

For a generator $(D, \rho_0, \theta_0, \alpha_0)$ we define the linear mappings $\rho: \mathrm{T}(\mathcal{V}) \to \mathbf{H}(D)$, $\theta: \mathrm{T}(\mathcal{V}) \to D$ and $\alpha: \mathrm{T}(\mathcal{V}) \to \mathbb{C}$ to be the extensions of ρ_0, θ_0 and α_0 resp. given by

$$\rho(1) = \mathrm{id},\ \rho(bv) = \rho(b)\rho_0(v),$$

$$\theta(1) = 0,\ \theta(bv) = \rho(b)\theta_0(v),$$

and

$$\alpha(1) = 0,\ \alpha(v) = \alpha_0(v),\ \alpha(vbw) = \langle\theta(v^*), \theta(bw)\rangle, \tag{4.2}$$

for $v, w \in \mathcal{V}$, $b \in \mathrm{T}(\mathcal{V})$.

The linear functional α defined by (4.2) is hermitian and conditionally positive. Conversely, if $\alpha \in \mathbf{L}(\mathrm{T}(\mathcal{V}), \mathbb{C})$ is hermitian and conditionally positive the pre-Hilbert space $D_{L(\alpha)}$ and the restrictions of $\rho_\alpha, \eta_{L(\alpha)}$ and α to $\mathcal{V}$ form a generator on $\mathcal{V}$. One gets a one-to-one correspondence between hermitian, conditionally positive linear functionals on $\mathrm{T}(\mathcal{V})$ and equivalence classes of generators on $\mathcal{V}$; see [15].

For a vector space $\mathcal{V}$ let $\mathrm{T}(\mathcal{V})_\mathrm{L}$ be the Lie algebra associated to $\mathrm{T}(\mathcal{V})$, that is $\mathrm{T}(\mathcal{V})_\mathrm{L} = \mathrm{T}(\mathcal{V})$ as a vector space and $[b,c] = bc - cb$, $b, c \in \mathrm{T}(\mathcal{V})$. We denote by $\mathrm{L}(\mathcal{V})$ the smallest Lie subalgebra of $\mathrm{T}(\mathcal{V})_\mathrm{L}$ containing $\mathcal{V}$. We are now ready to state our result on the GNS representation given by an infinitely divisible state on a tensor algebra.

Theorem 4.5. Let $\mathcal{V}$ be an involutive vector space and let μ be an infinitely divisible state on $\mathrm{T}(\mathcal{V})$. Then there exist a generator $(D, \rho_0, \theta_0, \alpha_0)$ on $\mathcal{V}$ and an isometry

$$S: D_{M(\mu)} \to \mathrm{F}(D)$$

such that for $b \in \mathrm{L}(\mathcal{V})$

$$S \circ \pi_\mu(b) \circ S^{-1} = a(\theta(b^*)) + a^\dagger(\theta(b)) + \Lambda(\rho(b)) + \alpha(b)\,\mathrm{id}.$$

Proof: It is well-known that b lies in $\mathrm{L}(\mathcal{V})$ if and only if b is a Lie element; see e.g. [10]. There is a hermitian, conditionally positive linear functional α' on $\mathrm{T}(\mathcal{V})$ such that $\mu = \exp_* \alpha'$; see [5,8,17]. We set $D = D_{L(\alpha')}$, and for $v \in \mathcal{V}$ we set $\rho_0(v) = \rho_{\alpha'}(v)$, $\theta_0(v) = [v] = \eta_{L(\alpha')}(v)$ and $\alpha_0(v) = \alpha'(v)$. Then $\alpha = \alpha'$, $\rho = \rho_\alpha$ and $\theta = \eta_{L(\alpha)}$, and the theorem follows from Proposition 4.2. and Theorem 4.4. □

We treat some examples. First let $\mathcal{V} = \mathbb{C}$. Then $\mathrm{T}(\mathcal{V})$ is the (commutative) *-algebra $\mathbb{C}[\mathrm{x}]$ of polynomials in one indeterminate x, and $\mathrm{x}^* = \mathrm{x}$. We define a generator $(D, \rho_0, \theta_0, \alpha_0)$ on $\mathcal{V}$ by setting $D = \mathbb{C}^2$,

$$\rho_0(\mathrm{x}) = \begin{pmatrix} 0 & 1 \\ 1 & 0 \end{pmatrix},\ \theta_0(\mathrm{x}) = \begin{pmatrix} 1 \\ 0 \end{pmatrix} \text{ and } \alpha_0(\mathrm{x}) = 0.$$

Then for $n \in \mathbb{N}$, $n \geqslant 2$,

$$\alpha(\mathrm{x}^n) = \begin{cases} 1 & \text{if } n \text{ is even} \\ 0 & \text{if } n \text{ is odd} \end{cases}$$

and

$$S \circ \pi_\mu(\mathrm{x}) \circ S^{-1} = (a + a^\dagger)\left(\begin{pmatrix} 1 \\ 0 \end{pmatrix}\right) + \Lambda\left(\begin{pmatrix} 0 & 1 \\ 1 & 0 \end{pmatrix}\right).$$

One easily shows that $\begin{pmatrix} 0 \\ 1 \end{pmatrix}$ is not in the linear span of $\{S \circ \pi_\mu(\mathrm{x}^n) \circ S^{-1}(\Omega) : n \in \mathbb{N} \cup \{0\}\}$. Thus we have an example for α such that $S(D_{M(\mu)}) = \mathrm{E}_{L(\alpha)}(\mathrm{T}(\mathcal{V}))$ is not equal to $\mathrm{F}(D_{L(\alpha)})$.

Next let α_0 be a hermitian linear functional on $\mathcal{V}$, let Q be a positive sesquilinear form on $\mathcal{V}$ and let σ be a positive linear functional on $\mathrm{T}(\mathcal{V})$. We define a linear functional α on $\mathrm{T}(\mathcal{V})$ by

$$\alpha(1) = 0,\ \alpha(v) = \alpha_0(v),\ \alpha(vbw) = Q(\overline{v^*} \otimes w)\,\sigma(b),$$

for $v, w \in \mathcal{V}$, $b \in \mathrm{T}(\mathcal{V})$. Then α is hermitian and conditionally positive [16]. (Our first example actually is a special case with $\alpha_0 = 0$, $Q = 1$ and $\sigma = \alpha + \delta$.) The equation

$$\mathcal{U}([b] \otimes [v]) = [bv],$$

$b \in \mathrm{T}(\mathcal{V})$, $v \in \mathcal{V}$, defines a bijective isometry

$$\mathcal{U}\colon D_{M(\sigma)} \otimes D_Q \to D_{L(\alpha)}.$$

Moreover, identifying $D_{M(\sigma)} \otimes D_Q$ with $D_{L(\alpha)}$, we have for $b \in \mathrm{T}(\mathcal{V})$

$$\rho_\alpha(b) = \pi_\sigma(b) \otimes \mathrm{id}$$

and for $v \in \mathcal{V}$

$$S \circ \pi_\mu(v) \circ S^{-1} = a([1] \otimes [v^*]) + a^\dagger([1] \otimes [v]) + \Lambda(\pi_\sigma(v) \otimes \mathrm{id}) + \alpha_0(v)\,\mathrm{id}.$$

Especially, for $\sigma = \delta$ the state $\exp_* \alpha$ is a noncommutative gaussian state with covariance matrix Q in the sense of [6], and if $\mathcal{V} = \mathbb{C}^2$,

$$Q = \begin{pmatrix} |z_1|^2 & \bar{z}_1 z_2 \\ z_1 \bar{z}_2 & |z_2|^2 \end{pmatrix} \text{ with } z_1, z_2 \in \mathbb{C}$$

and σ the homomorphism such that $\sigma(v) = 1$, then $\exp_* \alpha$ is a noncommutative Poisson state [9, 16] with 'intensities' z_1 and z_2.

Acknowledgement

The author thanks J. Quaegebeur for useful discussions.

References

[1] Abe, E., Hopf Algebras, Cambridge University Press, 1980

[2] Accardi, L., Schürmann, M. and von Waldenfels, W., Quantum independent increment processes on superalgebras, to appear in Math. Zeitschrift

[3] Araki, H., Factorizable representation of current algebra, Publ. RIMS, Kyoto Univ. 5, 361-422 (1970)

[4] Bratelli, O. and Robinson, D. W., Operator Algebras and Quantum Statistical Mechanics II, Texts and Monographs in Physics, Springer, Berlin Heidelberg New York, 1981

[5] Canisius, J., Algebraische Grenzwertsätze und unbegrenzt teilbare Funktionale, Diplomarbeit, Heidelberg, 1978

[6] Giri, N. and von Waldenfels, W., An algebraic version of the central limit theorem, Z. Wahrscheinlichkeitstheorie verw. Gebiete 42, 129-134 (1978)

[7] Guichardet, A., Symmetric Hilbert spaces and related topics, Lect. Notes Math. 261, Springer, Berlin Heidelberg New York, 1972

[8] Hegerfeld, G. C., Noncommutative analogs of probabilistic notions and results, J. Funct. Anal. 64, 436-456 (1985)

[9] Hudson, R. L. and Parthasarathy, K. R., Quantum Ito's formula and stochastic evolutions, Comm. Math. Phys. 93, 301-323 (1984)

[10] Jacobson, N., Lie Algebras, Wiley, New York London, 1962

[11] Milnor, J. W. and Moore, J. C., On the structure of Hopf algebras, Ann. of Math. 81, 211-264 (1965)

[12] Parthasarathy, K. R. and Schmidt, K., Positive definite kernels, continuous tensor products, and central limit theorems of probability theory, Lect. Notes Math. 272, Springer, Berlin Heidelberg New York, 1972

[13] Schürmann, M., Positive and conditionally positive linear functionals on coalgebras, in: Accardi, L. and von Waldenfels, W. (Eds), Quantum probability and applications II, Proceedings, Heidelberg

1984, Lect. Notes Math. 1136, Springer, Berlin Heidelberg New York, 1985

[14] Schürmann, M., Noncommutative stochastic processes with independent and stationary increments satisfy quantum stochastic differential equations, submitted for publication

[15] Schürmann, M., Noncommutative stochastic processes with independent and stationary additive increments, submitted for publication

[16] Schürmann, M. and von Waldenfels, W., A central limit theorem on the free Lie group, in: Accardi, L. and von Waldenfels, W. (Eds), Quantum probability and applications III, Proceedings, Oberwolfach 1987, Lect. Notes Math. 1303, Springer, Berlin Heidelberg New York, 1988

[17] Streater, R., Infinitely divisible representations of Lie algebras, Z. Wahrscheinlichkeitstheorie verw. Geb. 19, 67-80 (1971)

[18] Streater, R., Current commutation relations, continuous tensor products and infinitely divisible group representations, in: Jost, R. (Ed), Local quantum theory, Academic Press, London New York, 1969

[19] Sweedler, M. E., Hopf Algebras, Benjamin, New York, 1969

[20] von Waldenfels, W., Positive and conditionally positive sesquilinear forms on anticocommutative coalgebras, in: Heyer, H. (Ed), Probability measures on groups VII, Proceedings, Oberwolfach 1983, Lect. Notes Math. 1064, Springer, Berlin Heidelberg New York, 1984

SEMISTABLE CONVOLUTION SEMIGROUPS AND THE TOPOLOGY OF CONTRACTION GROUPS

Eberhard Siebert

Introduction. Let G be a topological group and τ a topological automorphism of G. It is well known that every τ-semistable convolution semigroup of probability measures on G is supported by the contraction subgroup $C(\tau)$ of τ (cf. [8]). In general $C(\tau)$ is not closed in G; for example this is always the case if G is compact. Thus G being a locally compact group or a Banach space, the subgroup $C(\tau)$ does not necessarily belong to the same category.

But if G is a Lie group then $C(\tau)$ is an analytic subgroup [8]. Hence $C(\tau)$ may be retopologized to yield a Lie group $\tilde{C}(\tau)$ again, and the investigation of τ-semistable convolution semigroups may be performed on $\tilde{C}(\tau)$. On the other hand, by adapting an idea of Z.J.Jurek [6], for every Banach space G one can renorm $C(\tau)$ in such a way that it becomes a Banach space $\tilde{C}(\tau)$ by itself.

These observations led to the question wether $C(\tau)$ may be always retopologized to yield a topological group $\tilde{C}(\tau)$ that has the same properties as G and that supports the τ-semistable convolution semigroups of G. In Section 1 we show that this is possible in the category of complete and metrizable groups; but not in the category of locally compact groups. But we give a sufficient condition for $\tilde{C}(\tau)$ to be locally compact. In Section 2 we discuss several examples of groups G and automorphisms τ for which the structure of $\tilde{C}(\tau)$ can be exhibited. In particular, the Lie group case and the Banach space case are considered. In Section 3 we illustrate the general results for the character group G of the discrete rationals $\mathbb{Q}_d$ and for its topological automorphisms $\tau_{1/p}$ and τ_p (where p is an integer ≥ 2). This group has already been considered in connection with semistability [3,8]. It turns out that the groups $\tilde{C}(\tau_{1/p})$ are locally compact and connected whereas the groups $\tilde{C}(\tau_p)$ are locally compact and totally disconnected.

In Appendix 1 we briefly discuss the consequences of our results for semistable convolution semigroups. In Appendix 2 we sketch an extension of our construction to continuous one-parameter groups $T = (\tau_t)_{t>0}$ of topological automorphisms of G.

Preliminaries. Let $\mathbb{Z}$, $\mathbb{Q}$, $\mathbb{R}$, $\mathbb{C}$ denote the sets of integers, rational numbers, real numbers, and complex numbers respectively. Moreover let $\mathbb{N} = \{n \in \mathbb{Z} : n \geq 0\}$, $\mathbb{N}_1 = \{n \in \mathbb{Z} : n \geq 1\}$, $\mathbb{Q}^* = \{r \in \mathbb{Q} : r \neq 0\}$, $\mathbb{R}_+^* = \{r \in \mathbb{R} : r > 0\}$, and $\mathbb{T} = \{z \in \mathbb{C} : |z| = 1\}$.

Let G always denote a non-trivial topological Hausdorff group with identity e. Then $\mathfrak{U}(G)$ denotes the system of all neighbourhoods of e. Let $\mathfrak{U}_o(G) = \{U \in \mathfrak{U}(G) : U \text{ open}\}$. If A is a subset of G then A^- denotes its closure. Finally Aut(G) denotes the (topological) group of topological automorphisms of G (cf. [5], (26.5)).

1. General results

Let G be a topological group and $\tau \in \operatorname{Aut}(G)$. The contraction group $C(\tau)$ of τ is defined by $C(\tau) = \{x \in G : \lim_{n \geq 1} \tau^n(x) = e\}$. The automorphism τ is said to be (pointwise) contractive if $C(\tau) = G$.

For every $U \in \mathfrak{U}(G)$ and for every $n \in \mathbb{Z}$ we put

$$U_{(n)} = \bigcap_{-\infty < k \leq n} \tau^k(U) \qquad \text{and} \qquad U_n = U_{(n)} \cap C(\tau) .$$

1. LEMMA. Let G be complete and metrizable, and let $\tau \in \operatorname{Aut}(G)$ be contractive. Then the following holds:

a) For every $U \in \mathfrak{U}(G)$ and $n \in \mathbb{Z}$ we have also $U_{(n)} \in \mathfrak{U}(G)$.

b) τ is compactly contractive i.e. for every compact subset C of G and for every $U \in \mathfrak{U}(G)$ there exists some $n_o \in \mathbb{N}$ such that $\tau^n(C) \subset U$ for all $n \geq n_o$.

Proof (cf. [8], Lemma 1.4 and Remark 1.5, 1.).

a) Let $U \in \mathfrak{U}(G)$ and choose some closed $V \in \mathfrak{U}(G)$ such that $VV^{-1} \subset U$. Then every $V_{(n)}$ is closed too, and in view of $C(\tau) = G$ we have $\bigcup_{n \in \mathbb{Z}} V_{(n)} = G$. By assumption G is a Baire space. Hence there exists some $n_o \in \mathbb{Z}$ such that $V_{(n_o)}$ has an interior point x. Then $e = xx^{-1}$ is an interior point of $U_{(n_o)}$. In view of $U_{(n)} = \tau^{n-n_o}(U_{(n_o)})$ we conclude $U_{(n)} \in \mathfrak{U}(G)$ for all $n \in \mathbb{Z}$.

b) Let C be a compact subset of G and let $U \in \mathfrak{U}(G)$. In view of a) there exists some $V \in \mathfrak{U}_o(G)$ such that $V \subset U_{(o)}$. Hence $\tau^{-n}(V) \subset U_{(-n)}$

for all $n \in \mathbb{N}$. In view of $C(\tau) = G$ we have $G = \bigcup_{n\geq 1} \tau^{-n}(V)$. Hence there exists some $n_o \in \mathbb{N}_1$ such that

$$C \subset \bigcup_{1\leq n\leq n_o} \tau^{-n}(V) \subset \bigcup_{1\leq n\leq n_o} U_{(-n)} = U_{(-n_o)} .$$

Thus $\tau^n(C) \subset U_{(n-n_o)} \subset U$ for all $n \geq n_o$. ┘

2. PROPOSITION. There exists a unique topology $\mathcal{O}_\tau$ turning $C(\tau)$ into a topological Hausdorff group $\tilde{C}(\tau)$ such that $\mathfrak{V}_\tau = \{U_n : n \in \mathbb{Z}, U \in \mathfrak{U}_o(G)\}$ is a basis for $\mathfrak{U}_o(\tilde{C}(\tau))$.

Proof. It suffices to prove that $\mathfrak{V}_\tau$ satisfies conditions (i)-(v) of Theorem 4.5 in [5]. For this reason we fix some $U_n \in \mathfrak{V}_\tau$.

(i) There exists some $V \in \mathfrak{U}_o(G)$ such that $V^2 \subset U$. Consequently, $(\tau^k(V))^2 \subset \tau^k(U)$ for all $k \in \mathbb{Z}$ and thus $(V_n)^2 \subset U_n$.

(ii) There exists some $V \in \mathfrak{U}_o(G)$ such that $V^{-1} \subset U$. Consequently, $(V_n)^{-1} \subset U_n$,

(iii) Let $x \in U_n$. Then $x \in C(\tau)$. Hence $F = \{\tau^{-k}(x) : k \leq n\} \cup \{e\}$ is a compact subset of the open set U. Consequently, there exists some $V \in \mathfrak{U}_o(G)$ such that $FV \subset U$ ([5], (4.10)). Thus for all $k \leq n$ we have

$$xV_n \subset x\tau^k(V) = \tau^k(\tau^{-k}(x)V) \subset \tau^k(FV) \subset \tau^k(U) ;$$

hence $xV_n \subset U_n$.

(iv) Let $x \in C(\tau)$. Since $F = \{\tau^{-k}(x) : k \leq n\} \cup \{e\}$ is a compact subset of G there exists some $V \in \mathfrak{U}_o(G)$ such that $yVy^{-1} \subset U$ for all $y \in F$ ([5], (4.9)). Thus for all $k \leq n$ we have

$$xV_nx^{-1} \subset x\tau^k(V)x^{-1} = \tau^k(\tau^{-k}(x)V(\tau^{-k}(x))^{-1}) \subset \tau^k(U) .$$

Moreover $xV_nx^{-1} \subset C(\tau)$. Hence $xV_nx^{-1} \subset U_n$.

(v) Let $V_m \in \mathfrak{V}_\tau$. We put $p = \max(m,n)$ and $W = U \cap V$. Then

$$W_p \subset (\bigcap_{k\leq n} \tau^k(W)) \cap C(\tau) \subset (\bigcap_{k\leq n} \tau^k(U)) \cap C(\tau) = U_n ;$$

and analogously $W_p \subset V_m$.

In view of $\bigcap\{U_o : U \in \mathfrak{U}_o(G)\} = \{e\}$ the topology $\mathcal{O}_\tau$ is also Hausdorff. ┘

REMARK. If G is complete and metrizable and if $\tau \in \mathrm{Aut}(G)$ is contractive then we have $\tilde{C}(\tau) = C(\tau) = G$ (in view of Lemma 1 a)).

3. COROLLARY. a) $\mathfrak{O}_\tau$ is stronger than the relative topology of $C(\tau)$ (as a subspace of G).

b) The restriction of τ to $C(\tau)$ yields a contractive topological automorphism $\tilde{\tau}$ of $\tilde{C}(\tau)$.

Proof. Assertion a) is obvious in view of $U_o \subset U$ for all $U \in \mathfrak{U}_o(G)$.
b) Let $\tilde{\tau}$ denote the restriction of τ to $C(\tau)$. Then $\tilde{\tau}(U_n) = \tau(U_n) = U_{n+1}$ for all $n \in \mathbb{Z}$ and $U \in \mathfrak{U}_o(G)$. Hence $\tilde{\tau} \in \mathrm{Aut}(\tilde{C}(\tau))$. The contractivity of $\tilde{\tau}$ follows from the definition of $\mathfrak{O}_\tau$. ┘

4. LEMMA. We endow $G^{\mathbb{N}}$ with the topology $\mathfrak{O}_u$ of uniform convergence on $\mathbb{N}$. For every $x \in G$ we define $\hat{x}(n) = \tau^n(x)$ (all $n \in \mathbb{N}$). Then by $\Phi(x) = \hat{x}$ for all $x \in G$ there is given a monomorphism Φ of G into $G^{\mathbb{N}}$. Then the following holds:

a) $\Phi(G)$ and $\Phi(C(\tau))$ are closed in $(G^{\mathbb{N}}, \mathfrak{O}_u)$.

b) Φ is a homeomorphism of $(C(\tau), \mathfrak{O}_\tau)$ into $(G^{\mathbb{N}}, \mathfrak{O}_u)$.

Proof. a) Let $(x_\alpha)_{\alpha \in I}$ be a net in G such that $(\hat{x}_\alpha)_{\alpha \in I}$ converges to some $f \in G^{\mathbb{N}}$. Then $(x_\alpha)_{\alpha \in I}$ converges to $f(0) =: x$; and hence $(\tau^n(x_\alpha))_{\alpha \in I}$ converges to $\tau^n(x) = f(n)$ (all $n \in \mathbb{N}$). Thus $f = \hat{x}$.

In addition let $x_\alpha \in C(\tau)$ for all $\alpha \in I$. Given $U \in \mathfrak{U}(G)$ there exists some $\alpha = \alpha(U) \in I$ such that $\tau^n(x) \in \tau^n(x_\alpha)U$ for all $n \in \mathbb{N}$. Moreover there exists some $n(U) \in \mathbb{N}$ such that $\tau^n(x_\alpha) \in U$ for all $n \geq n(U)$. Hence $\tau^n(x) \in U^2$ for all $n \geq n(U)$. Thus $x \in C(\tau)$.

b) The system $\{U_o: U \in \mathfrak{U}_o(G)\}$ is a basis for $\mathfrak{U}_o(\tilde{C}(\tau))$ (in view of $U_n = (\tau^n(U))_o$ for all $n \in \mathbb{Z}$). Let $(x_\alpha)_{\alpha \in I}$ be a net in $C(\tau)$. Then $(x_\alpha)_{\alpha \in I}$ converges to $x \in C(\tau)$ with respect to $\mathfrak{O}_\tau$ iff for every U in $\mathfrak{U}_o(G)$ there exists some $\alpha(U) \in I$ such that $x_\alpha^{-1}x \in U_o$ for all $\alpha > \alpha(U)$; hence iff $\tau^n(x_\alpha)^{-1}\tau^n(x) \in U$ for all $n \in \mathbb{N}$, $\alpha > \alpha(U)$, and $U \in \mathfrak{U}_o(G)$; and thus iff $(\hat{x}_\alpha)_{\alpha \in I}$ converges to $\hat{x}$ uniformly on $\mathbb{N}$. ┘

5. PROPOSITION. a) If G is metrizable then $\tilde{C}(\tau)$ is metrizable too.

b) If G has a countable basis then $\tilde{C}(\tau)$ has a countable basis too.

c) If G is complete then $\tilde{C}(\tau)$ is complete too.

Proof. Taking into account Lemma 4, assertion a) follows from [1], §3, Proposition 1; and assertion c) follows from [1], §1, Théorème 1. As for b) we endow $\mathbb{N}$ with the discrete topology. Then for every $x \in C(\tau)$ the function $\hat{x}$ has limit e at infinity. Thus assertion b) follows from [1], §3, Corollary of Théorème 1 (taking into account Lemma 4). ⌋

6. PROPOSITION. Let G be complete and metrizable. Then $\mathcal{O}_\tau$ is the weakest topology $\mathcal{O}$ on $C(\tau)$ with the following properties:

a) $(C(\tau), \mathcal{O})$ is a complete and metrizable group.

b) $\mathcal{O}$ is stronger than the relative topology of $C(\tau)$ (as a subspace of G).

c) The restriction of τ to $C(\tau)$ yields a contractive topological automorphism σ of $(C(\tau), \mathcal{O})$.

Proof. In view of Proposition 2, Corollary 3 and Proposition 5 the topology $\mathcal{O}_\tau$ has the desired properties.

Let $\mathcal{O}$ be another topology on $C(\tau)$ with these properties. We abbreviate $(C(\tau), \mathcal{O})$ by H. Let $U \in \mathcal{U}_o(G)$. In view of b) there exists some $V \in \mathcal{U}(H)$ such that $V \subset U$. In view of Lemma 1 a) we have $V_{(n)} := \bigcap_{k \leq n} \sigma^k(V) \in \mathcal{U}(H)$ for all $n \in \mathbb{Z}$. Moreover

$$V_{(n)} = \bigcap_{k \leq n} \tau^k(V) \subset (\bigcap_{k \leq n} \tau^k(U)) \cap C(\tau) = U_n .$$

Thus the identity mapping on $C(\tau)$ yields a continuous monomorphism χ of H onto $\tilde{C}(\tau)$. ⌋

7. COROLLARY. If there exists a topology $\mathcal{O}$ on $C(\tau)$ with properties a), b), c) of Proposition 6 and if $(C(\tau), \mathcal{O})$ is locally compact then we already have $\mathcal{O} = \mathcal{O}_\tau$.

Proof. Let us keep the notations of the proof of Proposition 6 and let $V \in \mathcal{U}(H)$ be compact. In view of $C(\tau) = H$ we have $H = \bigcup_{n \in \mathbb{Z}} V_{(n)}$ and consequently $\tilde{C}(\tau) = \bigcup_{n \in \mathbb{Z}} \chi(V_{(n)})$. But every $\chi(V_{(n)})$ is compact (since in view of Proposition 6 the mapping χ is continuous) and hence closed (in $\tilde{C}(\tau)$). But $\tilde{C}(\tau)$ is a Baire space (Propositon 5). Hence there exists some $n_o \in \mathbb{Z}$ such that $\chi(V_{(n_o)})$ has non-void interior. Thus $\tilde{C}(\tau)$ is locally compact too. Taking into account the open mapping theorem for locally compact groups ([5], (5.29)) thus χ is an open mapping. Hence the assertion. ⌟

8. <u>COROLLARY</u>. Let G be complete and metrizable. If $C(\tau)$ can be turned into a locally compact group $(C(\tau), \mathcal{O})$ with properties a), b), c) of Proposition 6 then the topology $\mathcal{O}$ is unique and coincides with $\mathcal{O}_\tau$.

9. <u>PROPOSITION</u>. Let there exist some $U \in \mathcal{U}(G)$ such that U_o is compact in G and such that $\bigcap_{n \in \mathbb{Z}} U_n = \{e\}$. Then the following holds:

a) $(U_n)_{n \in \mathbb{Z}}$ is a basis for $\mathcal{U}(\tilde{C}(\tau))$.

b) $\tilde{C}(\tau)$ is locally compact.

<u>Proof</u>. a) By assumption $(U_n)_{n \in \mathbb{Z}}$ is a family of compact subsets of G (observe $U_n = \tau^n(U_o)$) descending to e as n tends to $+\infty$. Hence given $V \in \mathcal{U}_o(G)$ there exists some $m \in \mathbb{Z}$ such that $U_m \subset V$. Then $U_{m+n} \subset V_n$ for all $n \in \mathbb{Z}$.

b) Let $V \in \mathcal{U}(G)$ be closed and such that $VV^{-1} \subset U$. Then V_o is a closed neighbourhood of e in $\tilde{C}(\tau)$ such that $V_oV_o^{-1} \subset U_o$. We are going to show that V_o is compact in $\tilde{C}(\tau)$. Thus let $(x_\alpha)_{\alpha \in I}$ be a net in V_o. In view of $V_o \subset U_o$ and the compactness of U_o we can assume without loss of generality that the net $(x_\alpha)_{\alpha \in I}$ converges in G to some $x \in V_o$. Hence there exists some $\alpha(n) \in I$ such that $x_\alpha x^{-1} \in \bigcap_{o \leq k \leq n} \tau^k(U)$ for all $\alpha > \alpha(n)$ (all $n \in \mathbb{N}$). Since $x_\alpha x^{-1} \in V_oV_o^{-1} \subset U_o$ for all $\alpha \in I$ we conclude $x_\alpha x^{-1} \in U_n$ for all $\alpha > \alpha(n)$ (and for all $n \in \mathbb{N}$). Hence in view of a) we have shown that $(x_\alpha)_{\alpha \in I}$ converges in $\tilde{C}(\tau)$ (to x). Thus every net in V_o contains a convergent subnet; whence the assertion. ⌟

<u>REMARK</u>. The assumptions of Proposition 9 are also necessary for $\tilde{C}(\tau)$ to be locally compact (cf. [8], Lemma 1.4).

10. CORROLLARY. Let there exist some compact $U \in \mathfrak{U}(G)$ such that $\bigcap_{k\in\mathbb{Z}} \tau^k(U) = \{e\}$. Then we have:

a) $U_{(n)} \subset C(\tau)$, whence $U_{(n)} = U_n$ for every $n \in \mathbb{Z}$.

b) $(U_n)_{n\in\mathbb{Z}}$ is a basis for $\mathfrak{U}(\tilde{C}(\tau))$.

c) $\tilde{C}(\tau)$ is locally compact.

Proof. We first prove assertion a). In view of $U_{(n)} = \tau^n(U_{(o)})$ it suffices to prove $U_{(o)} \subset C(\tau)$. Hence let $x \in U_{(o)}$ i.e. $\tau^n(x) \in U$ for all $n \in \mathbb{N}$. Thus $L(x) = \bigcap_{m\geq o} (\{\tau^n(x): n \geq m\}^-)$ is a non-void compact subset of U. Obviously, $\tau(L(x)) = L(x)$ and consequently L(x) is contained in $\bigcap_{k\in\mathbb{Z}} \tau^k(U)$. Hence $L(x) = \{e\}$ by assumption, and thus $x \in C(\tau)$.

Taking into account a), the assumptions of Proposition 9 are fulfilled. This proves assertions b) and c). ⌟

REMARK. The sufficient condition of Corollary 10 is not necessary for $\tilde{C}(\tau)$ to be locally compact (cf. assertion (iii) of the Theorem in Section 3).

2. Examples

1. EXAMPLE (cf. [8], Example 1.2). Let G be a Lie group and let τ be in Aut(G) such that $C(\tau)$ is dense in G. Then there exists some compact $U \in \mathfrak{U}(G)$ such that $\bigcap_{k\in\mathbb{Z}} \tau^k(U) = \{e\}$.

[Let $\mathfrak{g} = \mathfrak{g}^+ \oplus \mathfrak{g}^- \oplus \mathfrak{g}^o$ be the decomposition of the Lie algebra $\mathfrak{g}$ of G according to the spectrum of $d\tau$ (cf. [4], Proposition 1.10). We proceed as in the discussion of Example 1.2 in [8]: Let $N \in \mathfrak{U}(\mathfrak{g})$ be compact such that exp is a homeomorphism of N onto $U := \exp N$ and such that exp is injective on $N \cup d\tau(N)$. Let $x \in G$ such that $\tau^k(x) \in U$ for all $k \in \mathbb{Z}$. Then there exist $X_k \in N$ such that $\tau^k(x) = \exp X_k$ (all $k \in \mathbb{Z}$). But $\exp X_{k+1} = \tau^{k+1}(x) = \tau(\tau^k(x)) = \tau(\exp X_k) = \exp d\tau(X_k)$ and $d\tau(X_k) \in d\tau(N)$ yield $X_{k+1} = d\tau(X_k)$. Consequently, $X_k = d\tau^k(X_o)$ for all $k \in \mathbb{Z}$. Thus $\{d\tau^k(X_o): k \in \mathbb{Z}\}$ is bounded; hence $X_o \in \mathfrak{g}^o$.

On the other hand, $C(\tau)$ being dense in G there exists a sequence $(x_n)_{n\geq 0}$ in $U \cap C(\tau)$ converging to x. Let $Y_n \in N$ such that $x_n = \exp Y_n$ (all $n \in \mathbb{N}$). Since exp is a homeomorphism of N onto U we conclude $\lim_{n\geq 0} Y_n = X_o$. But $Y_n \in \mathfrak{y}^-$ for all $n \in \mathbb{N}$ (cf. [8], Example 1.2) ; hence $X_o \in \mathfrak{y}^-$.

Consequently, $X_o \in \mathfrak{y}^o \cap \mathfrak{y}^- = \{0\}$ and thus $x = e$.]

Thus $\tilde{C}(\tau)$ is locally compact (in view of Corollary 1.10). On the other hand, $C(\tau)$ is an analytic subgroup of G ([8], 1.2); hence may be endowed with a unique topology $\mathcal{O}$ turning it into a connected Lie group. But $\mathcal{O}$ has properties a), b) c) of Proposition 1.6 (cf. [8], 1.2), and $(C(\tau), \mathcal{O})$ is locally compact. Consequently, in view of Corollary 1.8, we have $\mathcal{O} = \mathcal{O}_\tau$.

2. EXAMPLE. (cf. [8], Example 3.5 c)). Let F be a countable group endowed with the discrete topology. Then $G = F^{\mathbb{Z}} = \{(x_\nu)_{\nu\in\mathbb{Z}}: x_\nu \in F \text{ for all } \nu \in \mathbb{Z}\}$ endowed with the product topology becomes a totally disconnected, separable, complete, and metrizable topological group whose identity will be denoted by 1. Moreover G is (locally) compact iff F is finite.

By $\tau((x_\nu)_{\nu\in\mathbb{Z}}) = (x_{\nu-1})_{\nu\in\mathbb{Z}}$, $(x_\nu)_{\nu\in\mathbb{Z}} \in G$, there is defined an element $\tau \in \mathrm{Aut}(G)$. It is easily seen that

$$C(\tau) = \{(x_\nu)_{\nu\in\mathbb{Z}} \in G: \exists\, \nu_o \in \mathbb{Z} \cup \{\infty\} \text{ such that } x_\nu = e \quad \forall \nu < \nu_o\}.$$

Moreover, $U = \{(x_\nu)_{\nu\in\mathbb{Z}} \in G: x_o = e\}$ is an open subgroup of G. In view of $\tau^k(U) = \{(x_\nu)_{\nu\in\mathbb{Z}} \in G: x_k = e\}$ for all $k \in \mathbb{Z}$ we have $\bigcap_{k\in\mathbb{Z}} \tau^k(U) = \{1\}$.

Evidently, the sets $\bigcap_{|k|\leq n} \tau^k(U) = \{(x_\nu)_{\nu\in\mathbb{Z}} \in G: x_\nu = e \text{ if } |\nu| < n\}$, $n \in \mathbb{N}$, constitute a basis of $\mathfrak{U}(G)$, and hence the sets

$$\{(x_\nu)_{\nu\in\mathbb{Z}} \in G: x_\nu = e \text{ if } \nu \leq n\} = \bigcap_{k\leq n} \tau^k(U) = U_n \ , \ n \in \mathbb{Z},$$

constitute a basis of $\mathfrak{U}(\tilde{C}(\tau))$.

$\tilde{C}(\tau)$ is locally compact iff F is finite i.e. **iff G is (locally) compact.**

[If G is compact $\tilde{C}(\tau)$ is locally compact in view of Corollary 1.10. Conversely, if $\tilde{C}(\tau)$ is locally compact, some U_n is compact in $\tilde{C}(\tau)$ and hence in G too. Thus F must be finite.]

Consequently, if the topological group G is not locally compact, then the condition $\bigcap_{k\in\mathbb{Z}} \tau^k(U) = \{e\}$ (for some $U \in \mathcal{U}(G)$) is not sufficient for the local compactness of $\tilde{C}(\tau)$ (cf. Corollary 1.10).

3. EXAMPLE. Let K be a compact group and $\rho \in \mathrm{Aut}(K)$ such that $C(\rho)$ is dense in K (for an example see Section 3). For every $\nu \in \mathbb{N}$ let $K_\nu = K$ and $\rho_\nu = \rho$. Then $G = \prod_{\nu\in\mathbb{N}} K_\nu$ endowed with the product topology becomes a compact group. By $\tau((x_\nu)_{\nu\in\mathbb{N}}) = (\rho_\nu(x_\nu))_{\nu\in\mathbb{N}}$, $(x_\nu)_{\nu\in\mathbb{N}} \in G$, there is defined an element $\tau \in \mathrm{Aut}(G)$. Obviously $C(\tau) = \prod_{\nu\in\mathbb{N}} C(\rho_\nu)$; hence $C(\tau)$ is dense in G.

Let $m \in \mathbb{N}$ and let $U^{(\nu)} \in \mathcal{U}(K_\nu)$, $0 \leq \nu \leq m$, be closed. Then $U = U^{(o)} \times \ldots \times U^{(m)} \times \prod_{\nu>m} K_\nu$ is a closed neighbourhood of the identity of G. Since

$$\tau^k(U) = \rho_o^k(U^{(o)}) \times \ldots \times \rho_m^k(U^{(m)}) \times \prod_{\nu>m} K_\nu \quad \text{for all } k \in \mathbb{Z}$$

we conclude: $U_n = (U^{(o)})_n \times \ldots \times (U^{(m)})_n \times \prod_{\nu>m} C(\rho_\nu)$.

Since $C(\rho)$ is not closed in K (for $C(\rho)$ cannot be compact) the set U_n never can be closed in G. But sets of this type constitute a basis of closed neighbourhoods of $\mathcal{U}(\tilde{C}(\tau))$. Consequently, $\tilde{C}(\tau)$ cannot be locally compact. In fact $\tilde{C}(\tau)$ is topologically isomorphic with the topological direct product $\prod_{\nu\in\mathbb{N}} \tilde{C}(\rho_\nu)$.

4. REMARKS. a) In view of Example 3 (locally) compact groups G and automorphisms $\tau \in \mathrm{Aut}(G)$ such that $\tilde{C}(\tau)$ is not locally compact exist in abundance.

b) On the other hand there exist topological groups G that are not locally compact and automorphisms $\tau \in \mathrm{Aut}(G)$ such that $C(\tau) = G$ and such that $\tilde{C}(\tau)$ is locally compact: Let H be a topological group and let $\sigma \in \mathrm{Aut}(H)$ such that $C(\sigma) \neq C(\sigma)^- = H$ and such that $\tilde{C}(\sigma)$ is locally compact (for an example see Section 3). Let G denote the group $C(\sigma)$ endowed with its relative topology (as a subspace of H) and let τ denote the restriction of σ to G. Then G and τ have the desired properties (observe $\tilde{C}(\tau) = \tilde{C}(\sigma)$).

5. EXAMPLE (cf. [6]). Let $(E, |.|)$ be a Banach space and let $\tau \in \mathrm{Aut}(E)$. If $x \in C(\tau)$ then $|x|_1 := \sup\{|\tau^n(x)|: n \in \mathbb{N}\}$ is finite. Thus $(C(\tau), |.|_1)$ is a normed space. Obviously $|x| \leq |x|_1$ for all $x \in C(\tau)$. If $\tilde{\tau}$ denotes the restriction of τ to $C(\tau)$, then one easily sees that $|\tilde{\tau}|_1 \leq \min(1, |\tau|)$.

Let $U = \{x \in E: |x| \leq \varepsilon\}$ for some $\varepsilon \in \mathbb{R}_+^*$. Then $x \in U_o$ iff $x \in C(\tau)$ and $|\tau^n(x)| \leq \varepsilon$ for all $n \in \mathbb{N}$. Hence $U_o = \{x \in C(\tau): |x|_1 \leq \varepsilon\}$. Thus $\tilde{C}(\tau)$ is the topological group underlying $(C(\tau), |.|_1)$. In view of Proposition 1.5 then $(\tilde{C}(\tau), |.|_1)$ is a Banach space.

Let on $C(\tau)$ there exist another norm $|.|_2$ turning $C(\tau)$ into a Banach space such that $|x| \leq c|x|_2$ and $\lim_{n\geq 1} |\tau^n(x)|_2 = 0$ for all $x \in C(\tau)$ (with some $c \in \mathbb{R}_+^*$). Then $|.|_1$ and $|.|_2$ are equivalent. [Obviously the topology on $C(\tau)$ underlying $|.|_2$ enjoys properties a), b), c) of Proposition 1.6. Hence the identity mapping on $C(\tau)$ yields a continuous monomorphism of $(C(\tau), |.|_2)$ onto $(C(\tau), |.|_1)$. Taking into account the open mapping theorem for Banach spaces the assertion follows.]

3. A further example: the character group of the discrete rationals

By $\mathbb{Q}_d$ we denote the additive group of rational numbers endowed with the discrete topology. Then the character group G of $\mathbb{Q}_d$ is a compact solenoid. We are going to illustrate our general results in some detail for this group G. Our source for background information will be [5]. For convenience the elements of G will be denoted by x (instead of χ), and e denotes again the identity of G.

For every $r \in \mathbb{Q}^*$ there are given topological automorphisms σ_r

and τ_r of $\mathbb{Q}_d$ and G respectively, defined by

$$\sigma_r(s) = rs \quad \text{and} \quad \tau_r(x)(s) = x(\sigma_r(s))$$

for all $s \in \mathbb{Q}_d$ and $x \in G$. Moreover we define $\tau_o(x) = e$ for all $x \in G$.

At first we list some well-known properties of G:

a) G is connected and torsion-free.

b) Every proper closed subgroup of G is totally disconnected (but not discrete).

c) By $\varphi(t)(s) := \exp(its)$ for all $s \in \mathbb{Q}_d$ and $t \in \mathbb{R}$ there is given a continuous monomorphism φ of $\mathbb{R}$ into G. The image G_a of φ is the arc-component of e in G; this is a dense proper subgroup.

d) Aut(G) is discrete. By $r \longrightarrow \tau_r$ there is given an isomorphism of the multiplicative group $\mathbb{Q}^*$ onto Aut(G); in particular Aut(G) is abelian.

e) Let $x \in G$. Then $\langle x \rangle := \{\tau_r(x) : r \in \mathbb{Q}\}$ is a divisible subgroup of G. If $x \neq e$ then $r \longrightarrow \tau_r(x)$ is a monomorphism of $\mathbb{Q}_d$ onto $\langle x \rangle$. In view of $\tau_n(x) = x^n$ for all $n \in \mathbb{Z}$ we may interprete $\tau_r(x)$ as x^r (all $r \in \mathbb{Q}$).

f) Let $x \in G$. Then $L(x) := \bigcap_{n \in \mathbb{N}_1} \{\tau_r(x) : r \in \mathbb{Q} \cap [0,1/n[\}^-$ is a closed connected subgroup of G. It is $L(x) = \{e\}$ iff $x \in G_a$; and $L(x) = G$ iff $x \notin G_a$.

g) For every $m \in \mathbb{N}_1$ and $\varepsilon \in \mathbb{R}_+^*$ let $N(m,\varepsilon) := \{x \in G : |x(1/m) - 1| < \varepsilon\}$. These sets constitute a basis for $\mathcal{U}_o(G)$.

Furthermore we need some preparations:

a) For every $p \in \mathbb{N}_1$ let $Z(p^\infty) := \{z \in \mathbb{T} : z^{p^k} = 1$ for some $k \in \mathbb{N}\}$. Then $T := \bigcup_{p \in \mathbb{N}_1} Z(p^\infty)$ is the torsion subgroup of $\mathbb{T}$.

b) If P denotes the set of all primes, then T is isomorphic with the weak direct product $\prod^*_{q \in P} Z(q^\infty)$ ([5], (A.3)); hence $Z(p^\infty)$ is isomorphic with the direct product $\prod_{q \in P, q | p} Z(q^\infty)$ (all $p \in \mathbb{N}_1$).

The projection ψ_p of T onto $Z(p^\infty)$ is a homomorphism (all $p \in \mathbb{N}_1$).

c) T and all $Z(p^\infty)$ are divisible groups.

d) For all $m,p \in \mathbb{N}_1$ and $n \in \mathbb{Z}$ we define

$\mathbb{Q}(p^n/m) = \{kp^n/m : k \in \mathbb{Z}\}$ and $\mathbb{Q}(p^\infty/m) = \bigcup_{n\in\mathbb{Z}} \mathbb{Q}(p^n/m)$.

Then $\mathbb{Q}(p^n/m)$ and $\mathbb{Q}(p^\infty/m)$ are proper subgroups of $\mathbb{Q}_d$ such that $\mathbb{Q}(p^{n+1}/m) \subset \mathbb{Q}(p^n/m)$ and $\bigcap_{n\in\mathbb{Z}} \mathbb{Q}(p^n/m) = \{0\}$. Moreover, $\sigma_p(\mathbb{Q}(p^n/m)) = \mathbb{Q}(p^{n+1}/m)$ and hence $\sigma_p(\mathbb{Q}(p^\infty/m)) = \mathbb{Q}(p^\infty/m)$.

e) The annihilators $G(p^n/m)$ and $G(p^\infty/m)$ of $\mathbb{Q}(p^n/m)$ and $\mathbb{Q}(p^\infty/m)$ respectively are proper closed subgroups of G such that

$G(p^n/m) \subset G(p^{n+1}/m)$, $\bigcap_{n\in\mathbb{Z}} G(p^n/m) = G(p^\infty/m)$, $\bigcup_{n\in\mathbb{Z}} G(p^n/m) = G$.

Moreover, $\tau_p(G(p^{n+1}/m)) = G(p^n/m)$ and $\tau_p(G(p^\infty/m)) = G(p^\infty/m)$.

<u>THEOREM</u>. Let p be an integer ≥ 2. Then the following assertions hold:

(i) $C(\tau_{1/p}) = G_a$.

(ii) $\tilde{C}(\tau_{1/p})$ is locally compact; in fact it is topologically isomorphic with $\mathbb{R}$.

(iii) $\bigcap_{k\in\mathbb{Z}} \tau_{1/p}^k(U) = \bigcap_{k\in\mathbb{Z}} \tau_p^k(U) \neq \{e\}$ for every $U \in \mathcal{U}(G)$.

(iv) $C(\tau_p) = \{x \in G: x(\mathbb{Q}_d) \subset Z(p^\infty)\}$; and $C(\tau_p)$ is dense in G.

(v) $\tilde{C}(\tau_p)$ is locally compact and totally disconnected.

(vi) $(G(p^n) \cap C(\tau_p))_{n\in\mathbb{Z}}$ is a basis for $\mathcal{U}(\tilde{C}(\tau_p))$.

(vii) $\tilde{C}(\tau_p)$ is topologically isomorphic with the topological direct product $\prod_{q\in P, q|p} \tilde{C}(\tau_q)$. In particular, $\tilde{C}(\tau_{p^n}) = \tilde{C}(\tau_p)$ for all $p \in P$ and $n \in \mathbb{N}_1$.

The <u>proof</u> follows from a series of lemmatas.

<u>LEMMA 1</u>. $G(p^\infty) \cap C(\tau_p) = \{e\}$ and $G(p^\infty) \cap C(\tau_{1/p}) = \{e\}$.

<u>Proof</u>. For every $n \in \mathbb{N}_1$ there is given a subgroup $F_n := \{r \in \mathbb{Q}: n!r \in \mathbb{Q}(p^\infty)\}$ of $\mathbb{Q}_d$ that is a finite extension of $\mathbb{Q}(p^\infty)$. Moreover, in view of Preparation d) we have $\sigma_p(F_n) = F_n$ for all $n \in \mathbb{N}_1$.

Finally, $\bigcup_{n\in\mathbb{N}_1} F_n = \mathbb{Q}$.

Let H_n denote the annihilator of F_n. Then H_n is a closed subgroup of $G(p^\infty)$ that has finite index and hence is open in $G(p^\infty)$. Moreover, $\tau_p(H_n) = H_n$ for all $n \in \mathbb{N}_1$.

Let $x \in G(p^\infty) \cap C(\tau_p)$. Then for every $n \in \mathbb{N}_1$ there is some $k(n) \in \mathbb{N}$ such that $\tau_p^{k(n)}(x) \in H_n$ and hence $x \in \tau_p^{-k(n)}(H_n) = H_n$. Thus $x = e$ in view of $\bigcap_{n\in\mathbb{N}_1} H_n = \{e\}$. By observing $\tau_{1/p} = (\tau_p)^{-1}$ the second assertion follows analogously. ┘

<u>LEMMA 2</u>. Let $\delta := \pi/p$, $\varepsilon := \sin\delta$, and $U := N(1,\varepsilon)$. We put $U_{(n)} = \bigcap_{k\leq n} \tau_{1/p}^k(U)$ for all $n \in \mathbb{Z}$. Then $C(\tau_{1/p}) \cap U_{(n)} = G_a \cap U_{(n)}$ for all $n \in \mathbb{Z}$.

<u>Proof</u>. Obviously we have $G_a = \varphi(\mathbb{R}) \subset C(\tau_{1/p})$. Thus we only have to show $C(\tau_{1/p}) \cap U_{(n)} \subset G_a$. Now let $x \in C(\tau_{1/p}) \cap U_{(n)}$. In view of $x \in U_{(n)}$ we have $|x(p^k) - 1| < \varepsilon$ for all $k \leq n$. Hence for every $k \leq n$ there is some $\vartheta_k \in]-\delta,\delta[$ such that $x(1)^{p^k} = x(p^k) = \exp(i\vartheta_k)$. Thus for all $k \leq n$ we have $\exp(i\vartheta_k) = x(1)^{p^k} = (x(1)^{p^{k-1}})^p = \exp(ip\vartheta_{k-1})$ and $|p\vartheta_{k-1}| < p\delta = \pi$. Consequently, $\vartheta_k = p\vartheta_{k-1}$ and thus $\vartheta_k = p^{k-n}\vartheta_n$ for all $k \leq n$. This implies $x(p^k) = \exp(i\vartheta_k) = \exp(ip^{k-n}\vartheta_n) = \varphi(p^{-n}\vartheta_n)(p^k)$, hence $x^{-1}\varphi(p^{-n}\vartheta_n) \in G(p^k)$ for all $k \leq n$. This yields $x^{-1}\varphi(p^{-n}\vartheta_n) \in G(p^\infty)$.

On the other hand we have $x \in C(\tau_{1/p})$ and $\varphi(p^{-n}\vartheta_n) \in C(\tau_{1/p})$. Consequently, $x^{-1}\varphi(p^{-n}\vartheta_n) \in G(p^\infty) \cap C(\tau_{1/p})$. Taking into account Lemma 1 we conclude $x = \varphi(p^{-n}\vartheta_n) \in G_a$. ┘

<u>COROLLARY</u>. Assertions (i) and (ii) are true.

<u>Proof</u>. Given $x \in C(\tau_{1/p})$ there exists some $n \in \mathbb{N}$ such that $\tau_{1/p}^k(x) \in U$ for all $k \geq n$ (where U as in Lemma 2). Consequently, $x \in \tau_{1/p}^k(U)$ for all $k \leq -n$, i.e. $x \in U_{(-n)}$. Taking into account Lemma 2 we conclude $x \in G_a$. This proves (i).

We endow $C(\tau_{1/p})$ with the unique topology $\mathcal{O}$ for which φ becomes a homeomorphism. Then $(C(\tau_{1/p}), \mathcal{O})$ is a locally compact group. It is easy to check that $\mathfrak{O}$ has properties a), b), c) of Proposition 1.6. Thus assertion (ii) follows from Corollary 1.8. ⌋

LEMMA 3. Let $\delta \in]0,\pi/p]$, $\varepsilon := \sin\delta$, and $U := N(m,\varepsilon)$ for some $n \in \mathbb{N}_1$. We put $U_{(n)} = \bigcap_{k\leq n} \tau_p^k(U)$ for all $n \in \mathbb{Z}$ and $U_{(\infty)} = \bigcap_{n\in\mathbb{Z}} U_{(n)}$.

Then $U_{(-n)} = G(p^n/m)$ for all $n \in \mathbb{Z}$ and $U_{(\infty)} = G(p^\infty/m)$.

Proof. In view of $U_{(\infty)} = \bigcap_{n\in\mathbb{Z}} U_{(n)}$ and $G(p^\infty/m) = \bigcap_{n\in\mathbb{Z}} G(p^n/m)$ we only have to prove $U_{(-n)} = G(p^n/m)$ for every $n \in \mathbb{Z}$. In view of

$$\tau_p^k(U) = \tau_{p^k}(U) = \{x \in G: |x(p^{-k}/m) - 1| < \varepsilon\} \quad \text{we have}$$

$$U_{(-n)} = \{x \in G: |x(p^k/m) - 1| < \varepsilon \text{ for all } k \geq n\}.$$

This shows first of all $G(p^n/m) \subset U_{(-n)}$.

Conversely, let $x \in U_{(-n)}$. We put $z := x(p^n/m)$. Thus $|z^{p^k} - 1| < \varepsilon$ for all $k \in \mathbb{N}$. We will show $z = 1$, i.e. $x \in G(p^n/m)$:

At first for every $k \in \mathbb{N}$ there exists some $\vartheta_k \in]-\delta,\delta[$ such that $z^{p^k} = \exp(i\vartheta_k)$. We prove by induction (*) $|\vartheta_0| < \delta p^{-k}$ for all $k \in \mathbb{N}$. This yields $\vartheta_0 = 0$ and hence $z = 1$.

Obviously, (*) holds for $k = 0$. Let us assume that (*) holds for some $k \in \mathbb{N}$. Then $\exp(i\vartheta_{k+1}) = z^{p^{k+1}} = \exp(ip^{k+1}\vartheta_0)$ and $|p^{k+1}\vartheta_0| = p|p^k\vartheta_0| < p\delta \leq \pi$ imply $\vartheta_{k+1} = p^{k+1}\vartheta_0$ and thus $|\vartheta_0| = |\vartheta_{k+1}|/p^{k+1} < \delta/p^{k+1}$. Hence (*) also holds for k+1. This finishes the proof. ⌋

COROLLARY. Assertion (iii) is true.

PROOF. Without loss of generality let $U \in \mathfrak{U}(G)$ as in Lemma 3. Hence $\bigcap_{k\in\mathbb{Z}} \tau_p^k(U) = G(p^\infty/m)$ by this very Lemma. But $G(p^\infty/m)$ is the annihilator of $\mathbb{Q}(p^\infty/m) \neq \mathbb{Q}_d$ (cf. Preparations d) and e)), whence $G(p^\infty/m) \neq \{e\}$. This proves assertion (iii) (observe that $\tau_{1/p} = (\tau_p)^{-1}$). ⌋

LEMMA 4. For every $r \in \mathbb{Q}^*$ the contraction group $C(\tau_r)$ is either dense or trivial.

Proof. Since Aut(G) is abelian we conclude $\tau_s(C(\tau_r)) = C(\tau_r)$ and hence also $\tau_s(C(\tau_r)^-) = C(\tau_r)^-$ for all $s \in \mathbb{Q}^*$. Now Property e) of G implies that $C(\tau_r)^-$ is divisible and hence also connected ([5], (24.25)). In view of Property b) of G then either $C(\tau_r)^- = G$ or $C(\tau_r) = \{e\}$. ⌟

COROLLARY. Assertion (iv) is true.

Proof. Let $x \in C(\tau_p)$. Then for $U := N(m, \sin(\pi/p))$ there exists some $n \in \mathbb{N}$ such that $x \in U_{(-n)}$. Hence $x(1/m)^{p^n} = x(p^n/m) = 1$ in view of Lemma 3. Thus $x(1/m) \in Z(p^\infty)$ (for all $m \in \mathbb{N}_1$), i.e. $x(\mathbb{Q}_d) \subset Z(p^\infty)$.

Conversely, let $x \in G$ such that $x(\mathbb{Q}_d) \subset Z(p^\infty)$. Then for every $m \in \mathbb{N}_1$ there exists some $n \in \mathbb{N}$ such that $x(1/m)^{p^n} = 1$. Consequently, for all $j \in \mathbb{Z}$ and $k \geq n$, we have

$$\tau_p^k(x)(j/m) = x(jp^k/m) = (x(1/m)^{p^n})^{jp^{k-n}} = 1 .$$

Thus $x \in C(\tau_p)$. This proves the first part of assertion (iv).

Now let $a_1 \in Z(p^\infty) \setminus \{1\}$. Since $Z(p^\infty)$ is divisible (Preparation c)), there exists a sequence $(a_n)_{n \geq 2}$ in $Z(p^\infty)$ such that $a_n^n = a_{n-1}$ for all $n \geq 2$. Then there is exactly one $x \in G$ such that $x(1/n!) = a_n$ for all $n \in \mathbb{N}_1$ ([5], (25.5)). Obviously, we have $x(\mathbb{Q}_d) \subset Z(p^\infty)$, hence $x \in C(\tau_p)$. But $x(1) \neq 1$. Thus $C(\tau_p)$ is dense in G in view of Lemma 4. This completes the proof of (iv). ⌟

LEMMA 5. $G(1) \cap C(\tau_p)$ is closed.

Proof. Let $H := G(1) \cap C(\tau_p)$. Then H^- is a totally disconnected compact subgroup of G (by Property b) of G). Hence $H(r) := \{x(r) : x \in H\}$ is a finite subgroup of $\mathbb{T}$ for every $r \in \mathbb{Q}$ ([5], (24.26)). Moreover, $H(r) \subset Z(p^\infty)$ by assertion (iv).

Now let $(x_n)_{n \in \mathbb{N}}$ be a sequence in H converging to some $x \in G$. First of all $x \in G(1)$ (since G(1) is closed). Moreover, $\lim x_n(r) = x(r)$ and $x_n(r) \in H(r)$ for all $n \in \mathbb{N}$ yield $x(r) \in H(r)$ for all $r \in \mathbb{Q}$. Consequently $x(\mathbb{Q}_d) \subset Z(p^\infty)$ and thus $x \in C(\tau_p)$ by assertion (iv). Hence we have $x \in H$. Therefore H is closed. ⌟

COROLLARY. Assertions (v) and (vi) are true.

Proof. $U := N(1,\sin(\pi/p))$ is a neighbourhood of $e \in G$ such that $U_n = G(p^{-n}) \cap C(\tau_p)$ for all $n \in \mathbb{Z}$ (Lemma 3). But U_o is compact by Lemma 5. Thus $U_\infty = \bigcap_{n\in\mathbb{Z}} U_n = \bigcap_{n\in\mathbb{Z}} \tau_p^n(U_o)$ is a compact τ_p-invariant subgroup of $C(\tau_p)$. Consequently, $U_\infty = \{e\}$. Taking into account Proposition 1.9 this proves assertion (vi) and the first part of assertion (v). But every U_n is a subgroup of $\tilde{C}(\tau_p)$ hence is open in $\tilde{C}(\tau_p)$. This proves the second part of assertion (v). ⌋

COROLLARY. Assertion (vii) is true.

Proof. Let $Q = \{q_1,\ldots,q_m\}$ denote the set of prime divisors of p. Then $Z(p^\infty)$ is isomorphic with the direct product $\prod_{1\leq i\leq m} Z(q_i^\infty)$ (cf. Preparation b)).

First of all we prove $L := C(\tau_{q_1})\ldots C(\tau_{q_m}) = C(\tau_p)$: In view of $Z(q^\infty) \subset Z(p^\infty)$ and of assertion (iv) we have $C(\tau_q) \subset C(\tau_p)$ for all $q \in Q$, hence $L \subset C(\tau_p)$. Conversely, let $x \in C(\tau_p)$ and put $x_q := \psi_q \circ x$ for all $q \in Q$ (cf. assertion (iv)). Then $x_q \in G$ and $x_q(\mathbb{Q}_d) \subset Z(q^\infty)$ (cf. Preparation b)), hence $x_q \in C(\tau_q)$ in view of assertion (iv). Moreover, $x = x_{q_1}\ldots x_{q_m}$ (since $Z(p^\infty) = Z(q_1^\infty)\ldots Z(q_m^\infty)$). Hence $x \in L$, i.e. $C(\tau_p) \subset L$.

Now let $m > 1$ and $k \in \{1,\ldots,m-1\}$. Then we have

$$C(\tau_{q_1})\ldots C(\tau_{q_k}) \cap C(\tau_{q_{k+1}}) = \{e\}$$

(observe Preparation b) and assertion (iv)). Hence $C(\tau_p)$ is isomorphic with the direct product $\prod_{1\leq i\leq m} C(\tau_{q_i})$ (cf. [5], (2.4)).

Now we endow $C(\tau_p)$ with the unique topology $\mathfrak{O}$ for which the canonical mapping $(x_1,\ldots,x_m) \longrightarrow x_1\ldots x_m$ of $\prod_{1\leq i\leq m} \tilde{C}(\tau_{q_i})$ onto $C(\tau_p)$ becomes a homeomorphism. This turns $C(\tau_p)$ into a locally compact group $(C(\tau_p),\mathfrak{O})$ (cf. assertion (v)). It is easy to check that $\mathfrak{O}$ has properties a), b), c) of Proposition 1.6. Hence $\mathfrak{O} = \mathfrak{O}_{\tau_p}$ in view of Corollary 1.8. This finishes the proof of assertion (vii). ⌋

Appendix 1 (Lifting of semistable convolution semigroups)

Let G be a topological group and let $\tau \in \mathrm{Aut}(G)$. Moreover let G be complete and admit a countable basis for its topology. In view of Proposition 1.5 then the same holds true for $\tilde{C}(\tau)$. Thus by Kuratowski's theorem (observe Corollary 1.3 a)) the Borel field $\tilde{\mathfrak{B}}$ of $\tilde{C}(\tau)$ coincides with the trace of the Borel field $\mathfrak{B}$ of G in $C(\tau)$. Consequently, every probability measure μ on $\mathfrak{B}$ supported by $C(\tau)$, may also be considered as a probability measure $\tilde{\mu}$ on $\tilde{\mathfrak{B}}$.

Now let $(\mu_t)_{t>0}$ be a τ-semistable continuous convolution semigroup of probability measures on $\mathfrak{B}$ with coefficient $c \in]0,1[$ i.e. $\tau(\mu_t) = \mu_{ct}$ for all $t>0$ (cf. [3]). Then we have $\mu_t(C(\tau)) = 1$ for all $t>0$; and for every $U \in \mathfrak{U}_0(G)$ there exists some $\alpha = \alpha(U) \in \mathbb{R}_+^*$ such that

$$\mu_t(\complement U) \leq \alpha t \quad \text{and hence} \quad \sum_{k\in\mathbb{N}} \mu_t(\tau^{-k}(\complement U)) \leq \alpha t/(1-c)$$

for all $t>0$.

[This follows with the same arguments as in [8], Appendix: By assumption G is a Polish group. Hence $(\mu_t)_{t>0}$ admits a Lévy measure (cf. [7], Section 1).]

In view of $C(\tau) \setminus U_0 \subset \bigcup_{k\in\mathbb{N}} \complement\tau^{-k}(U)$ this yields

$$\tilde{\mu}_t(\complement U_0) \leq \alpha t/(1-c) \qquad \text{for all } t>0.$$

Hence $(\tilde{\mu}_t)_{t>0}$ is a $\tilde{\tau}$-semistable continuous convolution semigroup of probability measures on $\tilde{\mathfrak{B}}$ with coefficient c.

Appendix 2 (Continuous automorphism groups)

Let G be a topological group with Borel field $\mathfrak{B}$. Moreover, let $T = (\tau_t)_{t>0}$ be a one-parameter subgroup of Aut(G). Then a continuous convolution semigroup $(\mu_t)_{t>0}$ of probability measures on $\mathfrak{B}$ is said to be T-stable if $\tau_s(\mu_t) = \mu_{st}$ for all $s,t \in \mathbb{R}_+^*$ (cf. [3]).

Now let T be continuous in the sense that $t \longrightarrow \tau_t(x)$ is continuous for every $x \in G$. Moreover let G be metrizable. Then T is jointly continuous i.e. $(t,x) \longrightarrow \tau_t(x)$ is a continuous mapping of $\mathbb{R}_+^* \times G$ into G ([2], Theorem 1). Hence the contraction group $C(T) = \{x \in G: \lim_{t \downarrow 0} \tau_t(x) = e\}$ of T coincides with $C(\tau_s)$ for every $s \in]0,1[$ (cf. [8], Remark 5.8, 3.).

In analogy with the discrete case $\{\tau^n: n \in \mathbb{Z}\}$ one can endow C(T) with a topology $\mathcal{O}_T$ induced by T: If $U_t = C(T) \cap \bigcap_{t \leq r} \tau_r(U)$ for all $t \in \mathbb{R}_+^*$ and for all $U \in \mathcal{U}(G)$ then $\mathcal{O}_T$ is the unique topology turning C(T) into a topological Hausdorff group $\tilde{C}(T)$ such that $\mathcal{V}_T = \{U_t: t \in \mathbb{R}_+^*, U \in \mathcal{U}_o(G)\}$ is a basis for $\mathcal{U}_o(\tilde{C}(T))$.

But $\mathcal{O}_T$ coincides with $\mathcal{O}_{\tau_s}$ for every $s \in]0,1[$.

[Of course $\mathcal{O}_{\tau_s} \subset \mathcal{O}_T$. Conversely, let U_t be given as above. Choose some $n_o \in \mathbb{Z}$ such that $s^{n_o} \leq t$. Moreover, by the joint continuity of T, there exists some $V \in \mathcal{U}(G)$ such that $\tau_c(V) \subset U$ for all $c \in [s,1]$. Now let $x \in V_{n_o}$ and $r \geq t$. Then there exists some $n \leq n_o$ such that $s^n \leq r < s^{n-1}$. Hence $r = c^{-1}s^n$ for some $c \in [s,1]$. Thus

$$\tau_r^{-1}(x) = \tau_{1/r}(x) = \tau_c((\tau_s)^{-n}(x)) \in \tau_c(V) \subset U$$

i.e. $x \in U_t$ (observe $C(T) = C(\tau_s)$) and so $V_{n_o} \subset U_t$. Therefore we also have $\mathcal{O}_T \subset \mathcal{O}_{\tau_s}$.]

Moreover it easy to see that $\tilde{T} = (\tilde{\tau}_t)_{t>o}$ is a continuous automorphism group on $\tilde{C}(T)$.

By routine arguments one checks that all the results of Section 1 remain valid mutatis mutandis if τ is replaced by T. Furthermore Examples 2.1 and 2.5 can be adapted to this case (cf. [8], Example 5.2; and [6] respectively).

Finally, if G is complete and admits a countable basis for its topology, then every T-stable continuous convolution semigroup $(\mu_t)_{t>o}$ of probability measures on $\mathfrak{B}$ is supported by C(T), hence may be lifted to a $\tilde{T}$-stable continuous convolution semigroup $(\tilde{\mu}_t)_{t>o}$ on $\tilde{\mathfrak{B}}$.

References

1. Bourbaki,N.: Eléments de Mathématique X. Topologie Générale. Chap.10. Actual.Scient.Ind.1084. Paris: Hermann 1967
2. Chernoff,P., Marsden,J.: On continuity and smoothness of group actions. Bull.Amer.Math.Soc.76, 1044-1049 (1970)
3. Hazod,W.: Remarks on [semi-]stable probabilities. In: Probability Measures on Groups VII. Proceedings, Oberwolfach 1983, pp.182-203. Lecture Notes in Math. Vol.1064. Berlin-Heidelberg-New York-Tokyo: Springer 1984
4. Hazod,W., Siebert,E.: Continuous automorphism groups on a locally compact group contracting modulo a compact subgroup and applications to stable convolution semigroups. Semigroup Forum 33, 111-143 (1986)
5. Hewitt,E., Ross,K.A.: Abstract Harmonic Analysis I. Berlin-Göttingen-Heidelberg: Springer 1963
6. Jurek,Z.J.: Polar coordinates in Banach spaces. Bull.Polish Acad.Sci.32, 61-66 (1984)
7. Siebert,E.: Jumps of stochastic processes with values in a topological group. Probab.Math.Statist.5, 197-209 (1985)
8. Siebert,E.: Contractive automorphisms on locally compact groups. Math.Z.191, 73-90 (1986)

Eberhard Siebert
Mathematisches Institut
der Universität Tübingen
Auf der Morgenstelle 10

D-7400 Tübingen 1

Bundesrepublik Deutschland

------- THE CENTRAL LIMIT THEOREM FOR THE SYMMETRIC SPACE OF GL(3) -------

Audrey Terras
Math. Dept. C-012, U.C.S.D.
La Jolla, CA 92093 U.S.A.

1. INTRODUCTION.

We present a naive harmonic analyst's approach to central limit theorems for rotation-invariant independent identically distributed random variables on the symmetric space of the general linear group. The discussion makes use of properties of the Fourier transform on the symmetric space. One must also know the first few terms in the Taylor expansion of spherical functions for the symmetric space. These are similar to expansions obtained by A.T. James [11] for matrix argument analogues of the hypergeometric functions ${}_0F_0$. One surprise is that, in contrast to the case of SL(2,$\mathbb{R}$), the Fourier transform of the limiting density for our normalized product of random variables is somewhat different from the Fourier transform of the fundamental solution of the heat equation on the symmetric space of GL(3).

Our methods are special to the case of limit theorems for groups on which one can do harmonic analysis. There many other methods that can be used. Our results should be connected with limit theorems for products of random variables. However, it is rather difficult to compare the results obtained since many quantities have divergent formulations in the non-commutative group setting. We have attempted to follow the methods given by Karpelevich, Tutubalin and Shur [12] for the case of the Poincaré upper half plane. Some discussions of other methods as well as references can be found in Bougerol and Lacroix [1] and Cohen, Kesten and Newman [2].

In order to understand our discussion, we need to recall the Fourier analyst's proof of the classical central limit theorem. This requires the following <u>formulas from classical Fourier analysis</u> on the real line $\mathbb{R}$:

$$(f * g)(x) = \int_{\mathbb{R}} f(x-y)\, g(y)\, dy \quad ,$$

$$\hat{f}(x) = \int_{\mathbb{R}} f(y) \exp(-2\pi ixy)\, dy ,$$

$$\hat{\hat{f}}(x) = -f(x) , \quad \widehat{(f')}(x) = (2\pi ix)\hat{f}(x) ,$$

$$\widehat{f * g} = \hat{f} \cdot \hat{g} .$$

Here the functions f and g on $\mathbb{R}$ must be assumed suitably nice. The details can be found, for example, in Dym and McKean [5] or Terras [20, Vol. I].

Now recall the classical

CENTRAL LIMIT THEOREM.

Suppose that X_n, n=1,2,3, ... , are random variables in $\mathbb{R}$, each with density f(x). And suppose that the variables are independent, with mean 0 and standard deviation 1; i.e., that

$$\int_{\mathbb{R}} f(x)\, dx = 1, \quad \int_{\mathbb{R}} xf(x)\, dx = 0 , \quad \int_{\mathbb{R}} x^2 f(x)\, dx = 1 .$$

Then $n^{-1/2} (X_1 + \cdots X_n)$ is nearly Gaussian or normal with mean 0 and standard deviation 1, as n approaches infinity; i.e.,

$$\int_{a\sqrt{n}}^{b\sqrt{n}} (f * \cdots * f)(x)\, dx \underset{n \to \infty}{\sim} (2\pi)^{-1/2} \int_a^b \exp(-x^2/2)\, dx.$$

DISCUSSION.

By Lévy continuity as in Feller [6, p. 508] (or see Dym and McKean [5, p. 119]), it suffices to prove the Fourier transform of the central limit theorem which is:

$$\hat{f}(s\, n^{-1/2})^n \underset{n \to \infty}{\sim} \exp(-2\pi^2 s^2) .$$

To prove this, one needs the properties of the Fourier transform, the hypotheses on the density f(x), plus the Taylor expansion of the exponential:

$$\hat{f}(s\, n^{-1/2})^n = \left\{ \int_{\mathbb{R}} \exp(-2\pi i\, s n^{-1/2} x)\ f(x)\, dx \right\}^n$$

$$\sim \left\{ \int_{\mathbb{R}} \left\{ 1 - 2\pi i s n^{-1/2} x - 2\pi^2 s^2 n^{-1} x^2 \right\} f(x)\, dx \right\}^n$$

$$\sim \left\{ \int_{\mathbb{R}} f(x)\, dx - 2\pi i s n^{-1/2} \int_{\mathbb{R}} x\, f(x)\, dx - 2\pi^2 s^2 n^{-1} \int x^2 f(x) dx \right\}^n$$

$$= \left\{ 1 - 2\,\pi^2\, s^2\, n^{-1} \right\}^n \sim \exp(-2\pi^2 s^2), \quad \text{as } n \to \infty .$$

$\blacksquare$

2. NON-EUCLIDEAN HARMONIC ANALYSIS AND THE CENTRAL LIMIT THEOREM ON THE POINCARE UPPER HALF PLANE.

In this section we merely sketch the results of non-Euclidean harmonic analysis on the symmetric space of the special linear group SL(2,$\mathbb{R}$) of 2x2 real matrices of determinant one. More details can be found in Helgason [8 a,b] and Terras [20, Vol. I], for example.

Let $H = \{ x+iy \mid x,y \in \mathbb{R},\ y > 0 \}$ be the Poincaré upper half plane with the non-Euclidean arc length:

$$ds^2 = y^{-2}(dx^2 + dy^2), \qquad \text{for } z=x+iy.$$

It is not hard to see that the geodesics (curves minimizing distance) for this arc length are straight lines and circles orthogonal to the real axis. The geometry is certainly non-Euclidean, since an infinite number of geodesics through a fixed point z fail to meet a given geodesic L. See the figure on page 126 of [20, Vol. I].

There is a <u>group action</u> of the special linear group $G = SL(2,\mathbb{R})$ on $z\epsilon H$ via

$$z \to (az+b)/(cz+d).$$

This action sends H to H and preserves the arc length ds above.

The <u>G-invariant area element</u> on H is:

$$d\mu = y^{-2}\, dxdy,$$

and the <u>Laplacian</u> is:

$$\Delta = y^2 \left(\frac{\partial^2}{\partial x^2} + \frac{\partial^2}{\partial y^2} \right).$$

If you prefer, you can think of H as a <u>quotient space</u>:

$$H \cong G/K,$$

where K is the special orthogonal group SO(2) of 2x2 rotation matrices of determinant one. That is, an element of K looks like:

$$k_u = \begin{pmatrix} \cos u & \sin u \\ -\sin u & \cos u \end{pmatrix}.$$

The group K is the subgroup of G fixing the point $i = \sqrt{-1}$ in the upper half plane. The identification map of H with G/K is the map sending the point gi in H to the coset gK in G/K.

Next we need to consider an analogue of the Fourier transform for H. We define the <u>Helgason transform</u> of a function $f:H \to \mathbb{C}$ by:

$$\mathcal{H}f(s,k) = \int_H f(z)\, \overline{\mathrm{Im}(kz)^s}\, d\mu \qquad , \text{ for } s\epsilon\mathbb{C},\ k\epsilon K.$$

This transform has the following properties analogous to those of the Fourier transform on $\mathbb{R}$ (assuming that the functions involved are sufficiently nice).

PROPERTIES OF THE HELGASON TRANSFORM ON H.

1) <u>Inversion</u> (Harish-Chandra and Helgason).

Using the definition of $k_u \epsilon K$ given above, we have

$$f(z) = (4\pi)^{-1} \int_{t\epsilon\mathbb{R}} \frac{1}{2\pi} \int_{u=0}^{2\pi} \mathcal{H}f(1/2+it,k_u)\, \mathrm{Im}(k_u z)^{1/2+it}\, t \tanh \pi t\, du\, dt.$$

2) Convolution.

Suppose that $f,g:H \to \mathbb{C}$ with one function, say f, K-invariant; i.e., $f(kz)=f(z)$ for all $k\epsilon K$, $z\epsilon H$. Consider f and g as functions on $G = SL(2,\mathbb{R})$ via $f(a)=f(ai)$, for $a\epsilon G$, and define the convolution of f and g by that on the group G; i.e.,

$$(f*g)(a) = \int_G f(b)\, g(b^{-1}a)\, db \quad ,$$

where db is a Haar measure on G (i.e., a measure invariant under multiplication by elements of G). Then

$$\mathcal{H}(f * g) = \mathcal{H}f \cdot \mathcal{H}g \quad .$$

3) Differentiation.

$$\mathcal{H}(\Delta f)(s) = s(s-1)\, \mathcal{H}f(s) \quad .$$

DISCUSSION.

Parts 2) and 3) are easy. For example, part 3) follows from the fact that $\Delta y^s = s(s-1)\, y^s$; i.e., the fact that y^s is an eigenfunction of the non-Euclidean Laplacian. Thus part 1) says that any sufficiently nice function can be expanded in eigenfunctions of the Laplacian. So it can be viewed as the spectral theorem for Δ. And then $t \tanh \pi t\, dt$ is the spectral measure. To sketch the proof that this is so, we follow Helgason's discussion in his Battelle Rencontres lectures [8a]. First one reduces by G-invariance to the case of K-invariant functions $f(kz)=f(z)$ for all $k\epsilon K$, $z\epsilon H$. For such functions the Helgason transform is:

$$\mathcal{H}f(s) = \hat{f}(s) = \int_H f(z)\, \overline{y^s}\, d\mu = 2\pi \int_{r=0}^{\infty} f(e^{-r}i)\, \overline{P_{s-1}(\cosh r)}\, \sinh r\, dr,$$

using geodesic polar coordinates $z=ke^{-r}i$, $k\epsilon K$, $r>0$. Here the Legendre or spherical function is:

$$P_{s-1}(\cosh r) = h_s(z) = (2\pi)^{-1} \int_{u=0}^{2\pi} \operatorname{Im}(k_u z)^s \, du \ , \text{ for } z=e^{-r}i.$$

The Fourier inversion formula for K-invariant functions on H is then the <u>Mehler-Fock inversion formula</u>:

$$f(z) = (4\pi)^{-1} \int_{t\in\mathbb{R}} \hat{f}(1/2+it)\, P_{-1/2+it}(\cosh r)\ t \tanh \pi t \ \ dt.$$

The Kodaira-Titchmarsh formula relates the spectral measure for a differential operator like the radial Laplacian with the Green's function or resolvent kernel (see Dunford and Schwartz, Vol. II [4]) and can thus be used to prove the Mehler-Fock inversion formula. This does not seem to be a possible method of proof in the case of GL(n), since, as far as I know, no one has managed to find a formula for the Green's function for the radial Laplacian in that case. Harish-Chandra found a shortcut using only <u>two properties of the spherical function</u> P_s:

i) <u>FUNCTIONAL EQUATION</u>. $\quad P_{s-1} = P_{-s}$.

ii) <u>ASYMPTOTICS</u>. $P_s(x) \sim \pi^{-1/2} \dfrac{\Gamma(s+1/2)}{\Gamma(s+1)} \ (2x)^s$, as $x \to \infty$, if Re $s > 1/2$.

Then one can look at the inversion kernel asymptotically as $x,y \to \infty$:

$$V_R(x,y) = \int_0^R t \tanh \pi t \ \ P_{-1/2+it}(x)\, P_{-1/2+it}(y)\, dt$$

$$\sim \pi^{-1} \int_0^R y^{-1/2-it} \ \ x^{-1/2+it}\, dt \ ,$$

which is the kernel for Mellin inversion. Note that the spectral measure is there just to cancel out the gamma functions in the asymptotics of the spherical functions.

This concludes our discussion of the properties of the Helgason on H. More details can be found in Terras [20, Vol. I, §3.2].

■

THE FUNDAMENTAL SOLUTION OF THE HEAT EQUATION ON H.

Now we can use the Helgason transform to solve the <u>heat equation</u> on H and thus obtain an analogue of the <u>normal distribution</u>. By solving the heat equation on H, we mean that, if we are given an initial heat distribution $f(z)$ on H, which we assume is K-invariant, we will find a function $u(z,t)$ such that

$$u_t = \Delta_z u$$
$$u(z,0) = f(z).$$

We proceed formally, applying the Helgason transform to both sides of the partial differential equation, obtaining:

$$\frac{\partial}{\partial t}\hat{u}(s,t) = s(s-1)\hat{u},$$

which implies that

$$\hat{u}(s,t) = \hat{f}(s)\, e^{s(s-1)t}.$$

It follows that

$$u(z,t) = f * G_t,$$

where the <u>fundamental solution of the heat equation</u> is:

$$G_t(k\, e^{-r} i) = (4\pi)^{-1} \int_{v\in\mathbb{R}} \exp\{-(v^2+1/4)t\}\, P_{-1/2+iv}(\cosh r)\, v \tanh \pi v\, dv$$

$$= (4\pi t)^{-3/2} \sqrt{2}\, e^{-t/4} \int_r^\infty \frac{b \exp(-b^2/4t)\, db}{\sqrt{\cosh b - \cosh r}} .$$

It follows that G_t is positive and that G_t approaches the Dirac delta function δ_i, as t approaches 0 from above. Unfortunately there does not appear to be a simpler formula for G_t than the integrals given above (although the fundamental solution of the heat equation for $SL(2,\mathbb{C})$ is an elementary function).

Next we seek to discuss the central limit theorem for rotation invariant, independent, identically distributed random variables on H. The result has been discussed by a very large number of authors. Our treatment is based on the work of Karpelevich, Tutubalin and Shur [12]. A few other references are Heyer [10] and Papanicolaou [17], as well as those mentioned at the beginning of the introduction.

Motivation for studying this problem can be obtained by considering an engineering problem studied by Gertsenshtein and Vasil'ev [7]. A long lossless transmission line has random inhomogeneities giving rise to reflected waves. How much power is reflected? Now, it is seen in courses on microwave engineering that a reflection coefficient corresponds to a random variable Z in H. If the non-Euclidean distance of Z to i is large, this means that almost all power is reflected. Moreover, 2 inhomogeneities combine by multiplying the matrices in $SL(2,\mathbb{R})$. See any microwave engineering text for the details; e.g. Collin [3]. Other references are Helton [9] and Terras [20, Vol. I, pp. 127-134]. We should note that engineers usually work on the unit disc, $|z|<1$, rather than the upper half plane H. Of course, the Cayley transform maps one to the other. The Smith Chart is the graph paper that microwave engineers use to solve their problems. The lines are the images of a rectangular grid on H under the Cayley transform. See the figure in Terras [20, Vol. I, p. 129].

So we want a <u>central limit theorem</u> for some normalized

$$S_n = Z_1 \circ \cdots \circ Z_n ,$$

with Z_j = independent SO(2)-invariant random variables in H, each with non-Euclidean density $f(z)=f(kz)$, for all $k\in K$, $z\in H$. That is, for any measurable subset A of H, we have:

$$P(Z\in A) = \int_{z\in A} f(z)\, d\mu(z) .$$

Here we define the composition $Z_i \circ Z_j$ by multiplying the corresponding matrices in G. To an analyst the composition of these independent K-invariant random variables means the convolution of the corresponding densities, thought of as functions on G in the manner described above.

As in the classical case, it suffices to prove the Fourier transform of the central limit theorem. And as in the classical case, this reduces to knowing the asymptotics of the special function appearing in the Fourier transform. In our case, we need the asymptotic expansion:

$$P_{-1/2+ip}(\cosh r) \sim 1 - \frac{1}{4}\left(\frac{1}{4} + p^2\right) r^2 , \text{ as } r \to 0.$$

Note that the eigenvalue of the non-Euclidean Laplacian corresponding to this spherical function is $-(\frac{1}{4} + p^2)$; that is,

$$\Delta P_{-1/2+ip}(\cosh r) = -\left(\frac{1}{4} + p^2\right) P_{-1/2+ip}(\cosh r).$$

So let $S_n^\#$ = the normalized S_n obtained by dividing the geodesic radial variable by $\sqrt{n}$. So the Helgason transform of the associated density is:

$$\phi_{S_n^\#}(p) = \left\{ 2\pi \int_{r>0} f(e^{-r}i)\, P_{-1/2+ip}(\cosh r\, n^{-1/2}) \sinh r\, dr \right\}^n$$

$$\underset{n\to\infty}{\sim} \left\{ 2\pi \int_{r>0} f(e^{-r}i) \left\{1 - \frac{1}{4}\left(\frac{1}{4} + p^2\right) \frac{r^2}{n} \right\} \sinh r\, dr \right\}^n$$

$$\underset{n\to\infty}{\sim} \left\{ 1 - \frac{1}{4n}\left(\frac{1}{4} + p^2\right) d \right\}^n ,$$

where d is <u>an analogue of σ^2</u> defined by:

$$d = 2\pi \int_{r>0} r^2 f(e^{-r}i) \sinh r\, dr .$$

It follows that

$$\phi_{S_n^\#}(p) \sim \exp\left\{-\frac{d}{4}\left(\frac{1}{4} + p^2\right)\right\} , \quad \text{as } n \to \infty.$$

So, by a Lévy continuity type argument (see Terras [20, Vol. I, p. 161]), we find that

$$S_n^\# \to \text{Gaussian with density } G_{d/4} , \quad \text{as } n \to \infty.$$

This gives

THE NON-EUCLIDEAN CENTRAL LIMIT THEOREM FOR ROTATION INVARIANT RANDOM VARIABLES ON H.

Suppose that Z_n, $n \geq 1$, are independent, SO(2)-invariant random variables in H, each having the same density function f(z). Let $S_n = Z_1 \circ \cdots \circ Z_n$ be normalized as above. The normalized random variable $S_n^{\#}$ has density function $f_n^{\#}$ given by:

$$f_n^{\#}(e^{-r}i) = \sqrt{n}\ (f* \cdots * f)(e^{-r\sqrt{n}}\ i)\ \sinh(r\sqrt{n})/\sinh r\ ,$$

where the convolution of f's is n-fold and $k \in SO(2)$, $r>0$. Then for measurable sets $A \subset H$, we have:

$$\int_A f_n^{\#}(z)\ d\mu \sim \int_A G_{d/4}(z)\ d\mu \quad , \quad \text{as} \quad n \rightarrow \infty.$$

Here G_t is the fundamental solution of the non-Euclidean heat equation on H.

This leaves us with the

PROBLEM. Find the mean reflection coefficient (non-Euclidean distance to i) which is the integral:

$$2\pi \int_{r>0} r\ G_c(e^{-r}i)\ \sinh r\ \ dr.$$

Gertsenshtein and Vasil'ev [7] find that the integral is $\geq e^{2c}$. Thus we can conclude that almost all of the power is reflected as the length of the transmission line increases, since c approaches infinity with the length of the line.

REMARKS.

In our first discussion of this theorem (Terras [20, Vol. I, p. 160]) we erred in attempting to use an asymptotic relation between the Legendre function and the J-Bessel function to prove the central limit theorem. For one needs to study second order terms in the expansions and not just first order terms. The central limit theorem is, after all, a story about 2nd order terms.

Many people have questioned our normalization $S_n^{\#}$ of the random variable S_n. Our justification for the normalization is that we seek a non-Euclidean analogue of the classical normalization. Thus we do everything with respect to the non-Euclidean measure on the symmetric space. So our normalized density

$f_n^{\#}$ has the property:

$$\int_H f_n^{\#}(z)\, h(z)\, d\mu = 2\pi \int_{r>0} (f * \cdots * f)(e^{-r}i)\, h(e^{-r/\sqrt{n}}i) \sinh r\, dr,$$

for any radial integrable function h (where the convolution of f's is taken n times). If h is the indicator function of some set such as

$A = \{ ke^{-r}i \mid k\varepsilon K,\ r\varepsilon[a,b] \}$,

then this integral is:

$$2\pi \int_{a\sqrt{n}}^{b\sqrt{n}} (f * \cdots * f)(e^{-r}i) \sinh r\, dr,$$

which is an integral that can be viewed an an analogue of that which occurred in the classical central limit theorem on $\mathbb{R}$. Our central limit theorem then says that this integral must approach:

$$2\pi \int_a^b G_{d/4}(e^{-r}i) \sinh r \ \ dr$$

as n approaches infinity, where d is the integral:

$$d \ = \ 2\pi \int_{r>0} f(e^{-r}i)\, r^2 \sinh r\, dr.$$

Our results are reminiscent of those in Kingman [13]. Our differences with some other treatments come from the fact that the Helgason-Fourier transform does not transform nicely under dilation.

3. THE GL(N)-ANALOGUE OF ALL THIS.

Some references for this section are: Helgason [8a,b], Maass [15], Muirhead [16], Selberg [19], and Terras [20, Vol. II].

Our symmetric space is the space of positive matrices:

$$\mathcal{P}_n = \left\{ Y \in \mathbb{R}^{n\times n} \;\middle|\; {}^tY = Y > 0 \right\}.$$

Here Y positive means that all its principal minors (or equivalently all its eigenvalues) are positive.

The group action is by elements of the general linear group $GL(n,\mathbb{R})$ of nonsingular nxn real matrices via

$$Y[g] = {}^tgYg, \quad \text{for} \quad Y\epsilon\mathcal{P}_n \text{ and } g\epsilon GL(n,\mathbb{R}).$$

And we can identify our symmetric space with a quotient space as follows:

$$\mathcal{P}_n \to K\backslash G,$$

$$Y=I[g] \to Kg,$$

where $G=GL(n,\mathbb{R})$ and K is the orthogonal group $O(n)$ of all nxn real matrices g such that $I[g]=I$. The data which describe the geometry of $\mathcal{P}_n$ are:

the arc length,

$$ds^2 = \operatorname{Tr}((Y^{-1}dY)^2), \quad \text{where} \quad dY = (dy_{ij}),\ Y=(y_{ij}),$$

the G-invariant measure,

$$d\mu = |Y|^{-(n+1)/2} \prod_{1\le i\le j\le n} dy_{ij}, \quad |Y|=\text{determinant of } Y,$$

the Laplacian,

$$\Delta = \operatorname{Tr}\left(\left(Y\frac{\partial}{\partial Y}\right)^2\right), \quad \text{with} \quad \frac{\partial}{\partial Y} = \left(\frac{1}{2}(1+\delta_{ij})\frac{\partial}{\partial y_{ij}}\right)_{1\le i,j\le n}.$$

One finds, for example, that the geodesics through I are of the form given by the matrix exponential $\exp(tH)$, $t\epsilon\mathbb{R}$, where H is some nxn symmetric real matrix.

We are interested in harmonic analysis on $\mathcal{P}_n$ and thus in eigenfunctions of the Laplacian. The basic eigenfunctions are analogues of the function $\operatorname{Im}(z)^s$, for $z\epsilon H$. We call them power functions and they are defined for $s\epsilon\mathbb{C}^n$ and $Y\epsilon\mathcal{P}_n$ with

$$Y = \begin{pmatrix} Y_j & * \\ * & * \end{pmatrix}, \qquad Y_j \in \mathcal{P}_j ,$$

by products of powers of principal minors $|Y_j|=\det(Y_j)$:

$$p_s(Y) = \prod_{j=1}^{n} |Y_j|^{s_j} .$$

The Helgason-Fourier transform of a K-invariant function $f:\mathcal{P}_n \to \mathbb{C}$ is defined to be:

$$\hat{f}(s) = \int_{Y\in\mathcal{P}_n} f(Y)\, \overline{p_s(Y)}\, d\mu(y) \quad , \quad \text{for } s\in\mathbb{C}^n.$$

We can rewrite this in geodesic polar coordinates as a transform involving the spherical function:

$$h_s(Y) = \int_{k\in K} p_s(Y[k])\ dk.$$

Now geodesic polar coordinates for $Y\in\mathcal{P}_n$ are given by $Y=a[k]$, where a is a positive diagonal matrix and $k\in K=O(n)$. Y has such a decomposition by the spectral theorem. And the Jacobian of geodesic polar coordinates is equal to

$$J(a) = \prod_{j=1}^{n} a_j^{-(n-1)/2} \prod_{1\le i<j\le n} |a_i-a_j| \quad ,$$

where a is positive diagonal with jth entry a_j. Thus, one finds that the Helgason transform of a K-invariant function on $\mathcal{P}_n$ is:

$$\hat{f}(s) = \kappa_n \int_{H\in\mathfrak{a}} f(e^H)\, \overline{h_s(e^H)}\, J(e^H)\, dH, \qquad \text{where}$$

$$\mathfrak{a} = \left\{ H = \begin{pmatrix} h_1 & & 0 \\ & \ddots & \\ 0 & & h_n \end{pmatrix} \middle| \ h_j\in\mathbb{R} \right\} ,$$

and κ_n is a positive constant involving a power of π and gamma functions (see Terras [20, Vol.II, p. 35]).

The inversion formula for the Helgason transform of (sufficiently nice) K-invariant functions $f(Y)$, $Y \in \mathcal{P}_n$ is:

$$f(Y) = \omega_n \int_{\mathrm{Re}\, s = -\rho} \hat{f}(s)\, h_s(Y)\, |c_n(s)|^{-2}\, ds \quad ,$$

where $c_n(s)$ is the Harish-Chandra c-function, which has an expression involving products of quotients of beta functions and ω_n is a positive constant whose formula contains a power of π and various gamma functions (see Terras [20, Vol. II, p. 88]).

The proof of the inversion formula for the Helgason transform of K-invariant functions on $\mathcal{P}_n$ works as for the case of rotation-invariant functions on the Poincare upper half plane. One need only know the functional equations of the spherical functions h_s, plus the asymptotics of $h_s(a)$, as the quotients of adjacent diagonal entries of a approach 0 from above.

There is also a convolution property of the Helgason transform, assuming that one of the functions is K-invariant. And the Helgason transform changes G-invariant differential operators to multiplication by a polynomial.

It follows that one can find the fundamental solution of the heat equation on $\mathcal{P}_n$ (at least as an inverse Helgason transform). The heat equation is:

$$\frac{\partial u}{\partial t}(Y,t) = \Delta_Y u(Y,t), \qquad u(Y,0) = f(Y), \text{ K-invariant.}$$

Noting that if $p_s(Y)$, the power function, is normalized by writing

$$p_s(I[t]) = \prod_{j=1}^{n} t_j^{2r_j + j - (n+1)/2} \quad ,$$

where

$$t = \begin{pmatrix} t_1 & & * \\ & \ddots & \\ 0 & & t_n \end{pmatrix}, \text{ upper triangular with } t_j > 0,$$

then we have the following formula for the eigenvalue of the Laplacian:

$$\Delta p_s = \lambda_\Delta(s)\, p_s, \quad \lambda_\Delta = r_1^2 + \cdots + r_n^2 + (n-n^3)/48.$$

See Terras [20, Vol. II, p. 49]. One finds also that the inversion formula for the Helgason transform involves integrals over Re $r_j = 0$. So, now, taking the Helgason Fourier transform of the heat equation on $\mathcal{P}_n$, we obtain, as before:

$$\frac{\partial}{\partial t} \hat{u}(s,t) = \lambda_\Delta(s)\, \hat{u}(s,t)$$

and thus

$$\hat{u}(s,t) = \hat{f}(s)\exp(t\,\lambda_\Delta(s))$$

which implies

$$u = f * G_t\ .$$

Here G_t is the <u>fundamental solution of the heat equation</u>; i.e., G_t is a K-invariant function on $\mathcal{P}_n$ with Helgason - Fourier transform

$$\hat{G}_t(s) = \exp(t\,\lambda_\Delta(s)).$$

Of course, Fourier inversion gives an integral formula for G_t itself. One can see that G_t approaches the Dirac delta distribution at the identity as t approaches 0 from above.

To obtain a <u>central limit theorem</u> for K-invariant independent random variables Y_n in $\mathcal{P}_n$, each with density f, we need to know the Taylor expansion of the spherical function $h_s(Y)$ at Y=I. I only managed this for n=3, but recently Richards [18] has obtained the expansion for general n using a result of Kushner [14]. The expansion in question is almost the same as James' expansions of the matrix argument hypergeometric functions ${}_0F_0$ which can be found in Muirhead [16]. Here we consider only the case of $\mathcal{P}_3$. Suppose that H is a 3x3 real diagonal matrix with ith diagonal entry h_i. And let

$$s_j + \cdots + s_n = r_j + (2j-(n+1))/4.$$

Then the spherical function has the following <u>Taylor expansion near the identity</u>:

$$\begin{aligned} h_s(e^H) \sim\ & 1 + \frac{1}{3}\sum_{i=1}^{3} r_i \sum_{j=1}^{3} h_j \\ & + \frac{1}{30}\left\{3\sum_{i=1}^{3} r_i^2 + 2\sum_{i<j} r_i r_j - 1\right\}\sum_{k=1}^{3} h_k^2 \\ & + \frac{1}{30}\left\{2\sum_{i=1}^{3} r_i^2 + 8\sum_{i<j} r_i r_j + 1\right\}\sum_{k<q} h_k h_q \\ & + \text{higher order terms.} \end{aligned}$$

Some of the details of the proof are in Terras [20, Vol. II, pp. 72-82].

Now we can discuss the central limit theorem for K-invariant independent identically distributed random variables in $\mathcal{P}_3$. Suppose that each random variable Y_n has density the K-invariant function f. Then if S is a measurable subset of $\mathcal{P}_3$, the probability that the random variable Y_n is in S is:

$$\int_{Y \in S} f(Y)\, d\mu(Y) = \int_{I[g] \in S} f(I[g])\, dg ,$$

where dg denotes Haar measure on $GL(3,\mathbb{R})$.

The <u>composition</u> $Y_i \circ Y_j$ of two of these independent identically distributed random variables is defined to be that coming from multiplication of the corresponding group elements. The usual proof shows that the composition $Y_i \circ Y_j$ has density given by the convolution $f * f$; where

$$(f*g)(x) = \int_{y \in G} f(y)\, g(y^{-1}x)\, dy.$$

We can use the K-invariance of f to see that:

$$\int_{\mathcal{P}_3} f(Y)\, d\mu = \kappa_3 \int_{H \in \mathfrak{a}} f(e^H)\, J(e^H)\, dH = 1 ,$$

$$\kappa_3 \int_{H \in \mathfrak{a}} h_j\, f(e^H)\, J(e^H)\, dH = 0 , \tag{*}$$

$$\kappa_3 \int_{H \in \mathfrak{a}} h_i h_j\, f(e^H)\, J(e^H)\, dH = \omega , \quad \text{for } i \neq j,$$

$$\kappa_3 \int_{H \in \mathfrak{a}} h_j^2\, f(e^H)\, J(e^H)\, dH = \nu ,$$

where the four integrals are over real 3x3 diagonal matrices H with ith entry h_i. Here $J(e^H)$ is the Jacobian and κ_3 is the constant of the integral formula for polar coordinates. The numbers ν and ω are constants depending on f.

Consider the random variable $S_n = Y_1 \circ \cdots \circ Y_n$, which has the corresponding density $(f * \cdots * f)$ on G. Normalize S_n by replacing each h_j by $n^{-1/2}h_j$ and call the result $S_n^{\#}$. Then the Helgason-Fourier transform of the density corresponding to $S_n^{\#}$ (which we may call the characteristic function of $S_n^{\#}$) is:

$$\phi_{S_n^{\#}}(s) = \left(\kappa_3 \int_{\mathcal{P}_3} f(e^H)\, h_s(e^{n^{-1/2}H})\, J(e^H)\, dH \right)^n .$$

We justify this normalization as we did the corresponding normalization for SL(2,ℝ).

Now plug the Taylor expansion for h_s into this as well as formulas (*) above. The result is:

$$\phi_{S_n^{\#}}(s) \sim \left\{ 1 + \frac{\nu}{10n}\left(3 \sum_{j=1}^{3} r_j^2 + 2 \sum_{i<j} r_i r_j - 1\right) + \frac{\omega}{10n}\left(2 \sum_{j=1}^{3} r_j^2 + 8 \sum_{i<j} r_i r_j + 1 \right) \right\}^n,$$

as n approaches infinity. It follows that the result is asymptotic to:

$$\exp\left\{ \frac{1}{10}\left((3\nu+2\omega) \sum_{j=1}^{3} r_j^2 + (2\nu+8\omega) \sum_{i<j} r_i r_j + (\omega-\nu) \right)\right\} .$$

Unless $\nu+4\omega = 0$, this is <u>not</u> the Fourier transform of the fundamental solution of the heat equation, although it doesn't really differ all that much. All the same, this is a rather unsettling way to end our discussion. The basic reason that this result differs slightly from the expected result is the difference between various analogues of the moments. That is, the integrals in (*) which we are using as analogues of moments, differ from derivatives of Fourier transforms.

So finally we have the

CENTRAL LIMIT THEOREM ON $\mathcal{P}_3$.

Suppose that Y_n, $n \geq 1$, are independent O(3)-invariant random variables in $\mathcal{P}_3$, each having the same density f(Y) satisfying the equations (*) above. Let $S_n = Y_1 \circ \cdots \circ Y_n$ be normalized as above. The normalized variable $S_n^{\#}$ has density $f_n^{\#}$. Then for any measurable set S in $\mathcal{P}_3$ we have:

$$\int_S f_n^{\#}(Y)\, d\mu(Y) \sim e^{\omega-\nu+a/2} \int_S (G_a * F_b)(Y)\, d\mu(Y) \quad , \quad \text{as} \quad n \to \infty,$$

where $a=3\nu+2\omega$, $b=2\nu+8\omega$. Here G_a is the fundamental solution of the heat equation on $\mathcal{P}_3$ and F_b is defined to have Helgason transform

$$\hat{F}_b(s(r)) = \exp\left\{b \sum_{i<j} r_i r_j\right\}.$$

Here the Helgason transform of G_a is:

$$\hat{G}_a(s(r)) = \exp\left\{a \sum_{i=1}^{3} r_i^2 - \frac{1}{2}\right\} .$$

Note that we have reparametrized the Helgason-Fourier transform using the change of variables:

$$s_1 = r_1 - r_2 - 1/2 ,$$
$$s_2 = r_2 - r_3 - 1/2 ,$$
$$s_3 = r_3 + 1/2 .$$

Of course, one can give explicit formulas for F_b and G_a as inverse Helgason transforms which are integrals over purely imaginary values of r.

BIBLIOGRAPHY

1. P. Bougerol and J. Lacroix, Products of Random Matrices with Applications to Schrödinger Operators, Birkhäuser, Boston, 1985.

2. J.E. Cohen, H. Kesten, and C.M. Newman (Eds.), Random Matrices and Their Applications, Contemporary Math., Vol. 50, A.M.S., Providence, 1986.

3. R. E. Collin, Foundations for Microwave Engineering, McGraw-Hill, N.Y., 1966.

4. N. Dunford and J.T. Schwartz, Linear Operators, Vol. II, Wiley-Interscience, N.Y., 1963.

5. H. Dym and H.P. McKean, Fourier Series and Integrals, Academic, N.Y., 1972.

6. W. Feller, Introduction to Probability Theory and its Applications, Vol. II, Wiley, N.Y., 1966.

7. M.E. Gertsenshtein and V.B. Vasil'ev, Waveguides with random inhomogeneities and Brownian motion in the Lobachevsky plane, Theory of Probability and its Applications, 4 (1959), 391-398.

8a. S. Helgason, Lie groups and symmetric spaces, in C.M. DeWitt and J.A. Wheeler (Eds.), Battelle Rencontres, Benjamin, N.Y., 1968, pp. 1-71.

8b. S. Helgason, Groups and Geometric Analysis, Academic, N.Y., 1984.

9. J.W. Helton, Non-Euclidean functional analysis and electronics, Bull. A.M.S., 7 (1982), 1-64.

10. H. Heyer, An application of the method of moments to the central limit theorem, Lecture Notes in Math., Vol. 861, Springer-Verlag, N.Y., 1981, pp. 65-73.

11. A.J. James, Special functions of matrix and single argument in statistics, in R. Askey (Ed.), Theory and Applications of Special Functions, Academic, N.Y., 1975, pp. 497-520.

12. F.I. Karpelevich, V.N. Tutubalin, and M.G. Shur, Limit theorems for the compositions of distributions in the Lobachevsky plane and space, Theory of Probability and its Applications, 4 (1959), 399-402.

13. J.F.C. Kingman, Random walks with spherical symmetry, Acta Math., 109 (1963), 11-53.

14. H.B. Kushner, The linearization of the product of two zonal polynomials, SIAM J. Math. Anal., 19 (1988), 687-717.

15. H. Maass, Siegel's Modular Forms and Dirichlet Series, Lecture Notes in Math., Vol. 216, Springer-Verlag, N.Y., 1971.

16. R.J. Muirhead, Aspects of Multivariate Statistical Theory, Wiley, N.Y., 1978.

17. G.C. Papanicolaou, Wave propagation in a one-dimensional random medium, SIAM J. Appl. Math., 21 (1971), 13-18.

18. D. St. P. Richards, The central limit theorem on spaces of positive definite matrices, to appear in J. Multivariate Analysis.

19. A. Selberg, Harmonic analysis and discontinuous groups in weakly symmetric Riemannian spaces with applications to Dirichlet series, J. Indian Math. Soc., 20 (1956), 47-87.

20. A. Terras, Harmonic Analysis on Symmetric Spaces and Applications, Vols. I, II, Springer-Verlag, N.Y., 1985, 1988.

ROOTS OF HAAR MEASURE AND TOPOLOGICAL HAMILTONIAN GROUPS

Gerhard Turnwald

Introduction. Let G be a compact group with normalized Haar measure λ and let Q denote the group of quaternions $\{\pm 1, \pm i, \pm j, \pm k\}$ (with $i^2 = j^2 = k^2 = -1$, $ij = k$, $jk = i$, $ki = j$). In the sequel, all measures are to be interpreted as bounded linear functionals on the vector space of complex-valued continuous functions on G (i.e., we restrict ourselves to $M(G)$ in the terminology of [5]).

Theorem 1. The following conditions are equivalent:

(a) There is a probability measure $\mu \neq \lambda$ such that $\mu * \mu = \lambda$.

(b) G is neither abelian nor isomorphic with $Q \times E$ where E is a product of 2-element groups.

In order to prove that (b) implies (a), we shall use the first part of the next result which, in a slightly weaker form, is due to Strunkov.

Theorem 2. Every nonabelian locally compact Hamiltonian group G is topologically isomorphic with $Q \times E \times F$ where E is a locally compact (abelian) group of exponent two and F is a totally disconnected periodic locally compact abelian group which has no 2-element other than unity; the converse also holds.

(A topological group G is called Hamiltonian if every closed subgroup is normal; an element $x \in G$ is called p-element if the powers x^{p^n} converge to the unit element e (but by a 2-element group we understand a group with two elements); G is called periodic if all elements are compact; G is said to have exponent n if $x^n = e$ for all $x \in G$. In general we employ the terminology of [5].)

Section 1 is devoted to a proof of Theorem 1 which was enunciated in [2] for separable G (cf. Remark 1.5). In section 2 we prove Theorem 2 by making use of a simplified version of Strunkov's proof (cf. Remark 2.20). Lemma 2.14 (with $n = 1$) provides alternative descriptions of the group F occuring in Theorem 2. In Remark 2.22 it is shown that a nonabelian topological Hamiltonian group which is not locally compact need not have a subgroup isomorphic with Q; thus a reasonable analogue of Theorem 2 (indicated by Lemma 2.13) does not hold in general.

§1. Proof of Theorem 1

1.1 Lemma. Condition (a) is equivalent to:

(c) For some irreducible unitary representation U of G, the real algebra A_U, generated by all matrices $\bar{U}(g)$, has nilpotent elements.

Proof. Recall that every bounded measure μ is uniquely determined by its Fourier-Stieltjes transform $\hat{\mu} : U \longmapsto \int \bar{U}\, d\mu$ (with componentwise integration) where U is taken from a system of representatives of equivalence classes of (finite-dimensional) irreducible unitary representations; $\hat{\lambda}(U) = 1$ if U is equivalent to the trivial representation and $\hat{\lambda}(U) = O$ otherwise (O means the matrix with all entries equal to zero); $\widehat{\mu * \nu}(U) = \hat{\mu}(U)\,\hat{\nu}(U)$. (Cf. [5], 28.36)

Assume that (a) holds. Then $\hat{\mu}(U) \neq \hat{\lambda}(U)$ for some U not equivalent to the trivial representation. Then (c) follows from $\hat{\mu}(U)^2 = \hat{\lambda}(U) = O$ and $\hat{\mu}(U) \in A_U$. (Note that A_U may be regarded as a subspace of a real 8-dimensional vector space. If all matrices $\bar{U}(g)$ belong to a hyperplane of this vector space then so does $\hat{\mu}(U)$ for every real measure μ; hence $\hat{\mu}(U) \in A_U$.)

If (c) holds then we may choose a matrix $M \in A_U$ with $M \neq O$ and $M^2 = O$. (If $M^{r-1} \neq O$ and $M^r = O$ with $r \geq 2$ then M^{r-1} has the required properties.) Put $f(x) = tr(U(x)'M + \bar{U}(x)'\bar{M})$ (tr and $'$ meaning trace and transpose, respectively). Clearly, f is real-valued and continuous. Let ν denote the (bounded) measure with density f. A simple calculation (using the orthogonality relations for the coordinate-functions of U and taking into account that $M = A^{-1}\bar{M}A$ if $\bar{U}(g) = A^{-1}U(g)A$ for all $g \in G$) yields $\hat{\nu}(U) = d_U^{-1}M$ (and $\hat{\nu}(\bar{U}) = d_U^{-1}\bar{M}$) if U and $\bar{U}$ are not equivalent and $\hat{\nu}(U) = 2d_U^{-1}M$ otherwise (d_U meaning the degree of the representation U); for every representation V not equivalent to U or $\bar{U}$ we have $\hat{\nu}(V) = O$. Choose $\varepsilon > 0$ such that $1 + \varepsilon f \geq 0$ and put $\mu = \lambda + \varepsilon\nu$. Then μ is a probability measure and $\mu * \mu = \lambda$ (since the Fourier-Stieltjes transforms are the same); obviously, $\mu \neq \lambda$.

1.2 Lemma. If (c) fails, then G is Hamiltonian.

Proof. Let H be a closed subgroup with normalized Haar measure μ. If $\varepsilon_{g^{-1}} * \mu * \varepsilon_g = \mu$ then $g^{-1}Hg = H$, since the left-hand side is Haar measure of $g^{-1}Hg$. Hence it is sufficient to prove that, for every irreducible unitary representation U, $X = \hat{\mu}(U)$ commutes with $Y = \hat{\varepsilon_g}(U)$ (for all $g \in G$). Note that $X^2 = X$ (since $\mu * \mu = \mu$) implies $(XY - XYX)^2 = O$ and $(YX - XYX)^2 = O$. From this we conclude $XY = XYX = YX$, since A_U has no nilpotent elements (and $X, Y \in A_U$).

1.3 Proposition. (a) implies (b).

Proof. Suppose that (a) holds. By Lemma 1.1 there is an irreducible representation U such that A_U has nilpotent elements. Note that $d_U > 1$ since otherwise A_U is a subalgebra of $\mathbf{C}$; hence G is not abelian. Assume $G = Q \times E$ where E has exponent two. Then U is equivalent to $(x, y) \longmapsto V(x)\chi(y)$ where V is some irreducible unitary representation of Q and χ is a continuous character of E (cf. [5], 27.43). From $d_U > 1$ we obtain $d_V > 1$. Thus V is equivalent to the representation

$$\pm 1 \longmapsto \pm\begin{pmatrix} 1 & 0 \\ 0 & 1 \end{pmatrix}, \quad \pm i \longmapsto \pm\begin{pmatrix} i & 0 \\ 0 & -i \end{pmatrix}$$

$$\pm j \longmapsto \pm\begin{pmatrix} 0 & -1 \\ 1 & 0 \end{pmatrix}, \quad \pm k \longmapsto \pm\begin{pmatrix} 0 & -i \\ -i & 0 \end{pmatrix}$$

(cf. [5], 27.61e; take $m = 2$, $a = i$, $b = j$). Hence $V(1), V(i), V(j), V(k)$ are linearly independent over the reals and their linear combinations form an algebra without nilpotent elements, since it is isomorphic with the (division-)algebra of quaternions. Observe that all matrices $V(x)\chi(y)$ belong to this algebra, since $y^2 = e$ implies $\chi(y) = \pm 1$. From this we conclude that A_U has no nilpotent elements, a contradiction.

1.4 Proposition. (b) implies (a).

Proof. Assume that (a) fails. By Lemma 1.1 and Lemma 1.2, G is Hamiltonian. Assume that G is nonabelian. Then, by Corollary 2.18, G is topologically isomorphic with $Q \times E \times F$ where E is a product of 2-element groups and every continuous character χ of F has odd order. Note that $\chi(x) = 1$ if $\chi(x)$ is real (since then $\chi(x)^2 = 1$ and $\chi(x)^n = \chi^n(x) = 1$ for some odd n). If F is nontrivial then we may choose a nontrivial character χ; hence $\chi(x) = a + ib$ with $b \neq 0$ for some $x \in F$. Let V be the irreducible unitary representation of Q mentioned in the proof of Proposition 1.3 . Then $b^{-1}V(j)\chi(x) - ab^{-1}V(j) - V(k) = \begin{pmatrix} 0 & 0 \\ 2i & 0 \end{pmatrix}$ is nilpotent. Hence A_U has nilpotent elements for the irreducible unitary representation $U = V \otimes 1 \otimes \chi$ of $Q \times E \times F$ (cf. [5], 27.43). Lemma 1.1 thus implies that (a) holds, a contradiction. Hence F is trivial and (b) fails.

1.5 Remark. The preceding proof of Theorem 1 essentially reproduces section 2 of [2], except for the proof that (a) implies (b) which is missing in that paper. Instead of appealing to Strunkov's theorem (apparently unknown to the authors) Diaconis and Shahshahani attempted to establish an analogue for separable compact Hamiltonian groups. Their proof is marred by some errors and omissions.

They assert (p.344) that a finite group is Hamiltonian if and only if it is isomorphic with $Q \times F$ where F is a finite abelian group with no element of order 4; obviously, this can only hold for nonabelian groups. Thus the proof of Lemma 5 is insufficient, since there seems to be no reason why one could assume that G/T (to which the above assertion is applied) is nonabelian.

Next (p.345) it is claimed that for every separable compact group G there is a sequence of (finite-dimensional) representations ρ_n such that the intersection of their kernels is trivial. Note that, under this assumption, G is topologically isomorphic with a closed subgroup of the direct product of the compact groups $\rho_n(G)$. Since every (matrix-)group $\rho_n(G)$ has countable base, this also holds for G. Hence every separable compact group without countable base (e.g., the dual of an uncountable subgroup of the discrete torus T_d; cf. [5], 24.15 and 24.32) furnishes a counterexample. (The claim is correct, however, for a group with countable base.)

The decomposition in the last-but-two line of Lemma 7 does not follow from the preceding arguments and needs some additional justification. The description of the projective limit of the groups $\mathbf{H} \times F_n''$ on p.347 which is said to hold by definition seems to require a repetition of a part of the preceding arguments applied to certain modifications of the given representations.

Finally, Theorem 2 of [2] is obviously incorrect for abelian G and the proof of the converse (which is not needed for the proof of Theorem 1) is missing.

Using the first part of the proof of Lemma 5 (relying on some results about Lie-groups) together with some new ideas (e.g., in a compact Hamiltonian group G every closed subgroup isomorphic with the torus T is contained in the centre $Z(G)$) one may prove that $\rho(G)$ is abelian or finite if G is a compact Hamiltonian group and ρ is a (finite-dimensional) representation. This result can then be used instead of Lemma 5. The rest of the proof can be completed along similar lines as in [2]. In this way it is possible to give a proof of Theorem 2 for compact G with countable base that is substantially shorter (but not more elementary) than the following proof in section 2.

1.6 Remark. If $\mu * \mu = \lambda$ then, for every $r \geq 2$, the r-fold convolution power μ^r is equal to λ, too. Conversely, if $r \geq 2$ is the least positive integer with $\mu^r = \lambda$ then $\nu \neq \lambda$ and $\nu * \nu = \lambda$ for $\nu = \mu^k$ if k is the least integer with $2k \geq r$. Hence (a) is equivalent to the existence of a probability measure $\mu \neq \lambda$ such that $\mu^r = \lambda$.

It would be interesting to classify the compact groups G such that for fixed $r \geq 2$ there is a probability measure μ with $\mu^{r-1} \neq \lambda$ and $\mu^r = \lambda$.

§2. Locally compact Hamiltonian groups

In the sequel we freely use the elements of the theory of topological groups and duality theory of locally compact abelian groups. For the convenience of the reader references to [5] are provided for some details which perhaps are not so well known.

It is easy to verify that $[x, y] = \pm 1$ for $x, y \in Q$ ($[x, y] = x^{-1}y^{-1}xy$); this property will be used implicitly at several places.

2.1 Lemma. Every nonabelian discrete Hamiltonian group is isomorphic with the direct product of Q with an abelian group of exponent two and an abelian group in which every element is of (finite) odd order; the converse also holds.

Proof. This is a classical result due to Dedekind and Baer; cf. [4], Th. 12.5.4. (Frequently, Hamiltonian groups are defined to be nonabelian as in [4]. Some authors also say "Dedekind group" instead of "Hamiltonian group".)

2.2 Lemma. Every subgroup and every quotient group (with respect to a closed normal subgroup) of a topological Hamiltonian group is a topological Hamiltonian group.

Proof. Let H be a (not necessarily closed) subgroup of the topological Hamiltonian group G. Let C be a subgroup of H which is closed in H. Then C is a normal subgroup of H since $C = \overline{C} \cap H$ and $\overline{C}$ is normal. The second part is equally simple.

2.3 Lemma. Let G be a topological Hamiltonian group. Then, for all $a, b \in G$, $[a, b] \in \overline{\langle a \rangle} \cap \overline{\langle b \rangle}$ (hence $[a, b]$ commutes with a, b) and $[a^m, b^n] = [a, b]^{mn}$ for all integers m, n.

Proof. We have $b^{-1}ab \in \overline{\langle a \rangle}$, since $\overline{\langle a \rangle}$ is a normal subgroup. Hence $[a, b] \in \overline{\langle a \rangle}$; similarly, $[a, b] \in \overline{\langle b \rangle}$.

Assume that $[a^m, b] = [a, b]^m$. Then $[a^{m+1}, b] = [a, b]b^{-1}a^{-1}b[a, b]^m b^{-1}ab = [a, b]^{m+1}$, since $[a, b]$ commutes with a and b. Then inductively we obtain $[a^m, b] = [a, b]^m$ for all nonnegative integers m (the case $m = 0$ being trivial) and $[a^{-m}, b] = [a^{-1}, b]^m = [a, b]^{-m}$ since $[a^{-1}, b][a, b] = a[a, b]b^{-1}a^{-1}b = e$, i.e. $[a^m, b] = [a, b]^m$ holds for all integers m. Hence $[a^m, b^n] = [a, b^n]^m = [b^n, a]^{-m} = [a, b]^{mn}$.

2.4 Lemma. Every subgroup generated by a noncentral element of a topological Hamiltonian group is finite or non-discrete.

Proof. Let a be a noncentral element. Assume that $\langle a \rangle$ is discrete, hence closed. Choose b with $[a, b] \neq e$. Then $b^{-1}ab = a^n$ for some $n \neq 1$. Hence $a^{n(n-1)} = (b^{-1}ab)^{n-1} = b^{-1}a^{n-1}b = a^{n-1}$, i.e. a has finite order (note that, by Lemma 2.3, b commutes with $a^{n-1} = [a, b]$).

2.5 Lemma. Let G be a compact Hamiltonian group. If $c = [a, b] \neq e$ then $c \notin \overline{\langle c^r \rangle}$ for some $r \neq 0$.

Proof. Assume that c has infinite order, the assertion being trivial otherwise. Then, by Lemma 2.4, $\langle a \rangle$ is non-discrete since $c \in \overline{\langle a \rangle}$.

Choose a neighbourhood U of e such that $c \notin \overline{U}$. For every $x \in G$ there are neighbourhoods V_x and W_x of e and x, respectively, with $V_x^{-1}W_x^{-1}V_x W_x \subset U$. Choose a finite subcovering $W_{x_1}, \ldots, W_{x_s}$. Then $V = V_{x_1} \cap \ldots \cap V_{x_s}$ is a neighbourhood of e with $V^{-1}x^{-1}Vx \subset U$ for all $x \in G$. Choose $r \neq 0$ with $a^r \in V$. Then $c \notin \overline{\langle c^r \rangle}$ since (cf. Lemma 2.3) $c^{rk} = [a^r, b^k] \in V^{-1}b^{-k}Vb^k \subset U$ for every integer k.

2.6 Lemma. Let H be a closed subgroup of a locally compact abelian group G. If H is topologically isomorphic with T (the one-dimensional torus) then $G = HK$ and $\{e\} = H \cap K$ for a suitable closed subgroup K.

Proof. Every topological isomorphism $\varphi : H \longrightarrow T$ can be extended to a continuous character χ of G ([5], 24.12). Let K be the kernel of χ. Then $H \cap K = \{e\}$ and $x \in HK$ for arbitrary $x \in G$ since $\chi(x^{-1}\varphi^{-1}(\chi(x))) = \chi(x^{-1})\chi(x) = 1$.

2.7 Lemma. Let G be a locally compact Hamiltonian group and $c = [a, b] \neq e$. Then $H = \overline{\langle a, b \rangle}$ is compact; H is totally disconnected if c has finite order.

Proof. By Lemma 2.4, $\langle a \rangle$ is not topologically isomorphic with the (discrete) group of integers; hence $\overline{\langle a \rangle}$ is compact ([5], 9.1). Similarly, $\overline{\langle b \rangle}$ is compact. Hence the product of the normal subgroups $\overline{\langle a \rangle}, \overline{\langle b \rangle}$ is compact and $H = \overline{\langle a \rangle}\,\overline{\langle b \rangle}$.

Assume that c has finite order. Every commutator in H belongs to $\langle c \rangle$ since, obviously, $H/\langle c \rangle$ is abelian. The mapping $x \longmapsto [x, y]$ defined on H with values in the discrete group $\langle c \rangle$ is continuous for every $y \in H$. Hence the component H_0 of the identity e has image $\{e\}$, i.e. H_0 is contained in the centre $Z(H)$.

If H_0 is nontrivial then there is a continuous character χ of H_0 whose kernel N does nor contain c. N is a normal subgroup of H (since N is closed) and H_0/N is the component of the identity in H/N ([5], 7.12). Note that H_0/N is topologically isomorphic with T, since $\chi(H_0)$ is a compact connected nontrivial subgroup of T (i.e. equal to T). The elements aN, bN, cN of the Hamiltonian group H/N satisfy the same hypotheses as the elements a, b, c of H; hence, in order to arrive at a contradiction, we may suppose $H_0 \cong T$.

By Lemma 2.6, there is a closed subgroup K of $Z(H)$ such that $Z(H) = H_0K$ and $\{e\} = H_0 \cap K$. K is totally disconnected, since it is topologically isomorphic with the subgroup $Z(H)/H_0$ of the totally disconnected group H/H_0. Hence we may choose an open subgroup L of K that does not contain c. Let n be the order of c. Then $a^n \in Z(H)$ since, by Lemma 2.3, $[a^n, b] = c^n = e$. Thus $a^n \in h_0^n K$ for suitable $h_0 \in H_0$ (recall that $H_0 \cong T$). Since K/L is finite (being discrete and compact), we conclude that $ah_0^{-1}L$ has finite order in H/L; note that L is closed, hence normal in H.

Let r be the smallest positive integer with $(ah_0^{-1})^r \in H_0L$. Choose $a_1 \in ah_0^{-1}H_0 = aH_0$ such that $a_1^r \in L$ and let $\tilde{h}$ be an element of $\widetilde{H_0} = H_0L/L$ of order r; this is possible since $\widetilde{H_0} \cong H_0/H_0 \cap L \cong H_0$. Writing $\tilde{x}$ for the coset xL, we have $\tilde{b}^{-1}\tilde{a_1}\tilde{b} = \tilde{a_1}^j$ for some j with $0 \leq j < r$, since $\tilde{a_1}$ has order r and $\tilde{H}$ is Hamiltonian; note that $[\tilde{a_1}, \tilde{b}] = [\tilde{a}, \tilde{b}] = \tilde{c} \neq \tilde{e}$ implies $j > 1$ (we have used the fact that $\widetilde{H_0}$ lies in the centre of $\tilde{H}$). Similarly, $\tilde{b}^{-1}\tilde{a_1}\tilde{h}\tilde{b} = (\tilde{a_1}\tilde{h})^k$ for some k with $0 \leq k < r$ (since $\tilde{a_1}\tilde{h}$ has order $\leq r$). Thus $\tilde{a_1}^j\tilde{h} = (\tilde{a_1}\tilde{h})^k$ and, consequently, $a_1^{j-k} \in H_0L$ which (by definition of r) entails $j = k$. Hence $\tilde{h} = \tilde{h}^j$, which is impossible since $\tilde{h}$ has order r.

2.8 Lemma. If c is a commutator in a locally compact Hamiltonian group G then $c^2 = e$ and $c \in Z(G)$.

Proof. If $c \neq e$ has finite order then $H = \overline{\langle a, b\rangle}$ is compact and totally disconnected by Lemma 2.7. Hence for every neighbourhood U of unity we may choose an open normal subgroup $N \subset U$. Then $c^2 \in N$ since the commutator cN of the finite Hamiltonian group H/N has order ≤ 2 by Lemma 2.1. Hence $c^2 = e$.

Assume that $c = [a, b]$ has infinite order. Then $[a^2, b] = c^2 \neq e$ (cf. Lemma 2.3) and from Lemma 2.7 and Lemma 2.5 we obtain $c^2 \notin \langle c^{2r}\rangle$ for some $r \neq 0$. Hence the image of c in $H/\langle c^{2r}\rangle$ is a commutator of finite order greater than two, contradicting the first part of the proof.

Assume $c \neq e$. Then, for every $x \in G$, $x^{-1}cx \neq e$; since $\langle c\rangle = \{e, c\}$ is normal, we conclude $x^{-1}cx = c$.

2.9 Lemma. Let G be a nilpotent compact totally disconnected group. If, for p prime, G_p denotes the set of all p-elements of G and G_p' denotes the closure of the subgroup generated by all q-elements with $q \neq p$ (q prime), then G_p and G_p' are closed normal subgroups with $G = G_pG_p'$ and $\{e\} = G_p \cap G_p'$.

Proof. Recall that a group is nilpotent if, for some positive integer c, all commutators $[x_1, \ldots, x_{c+1}]$ $(=[[x_1, \ldots, x_c], x_{c+1}])$ are equal to unity. A finite group is nilpotent if and only if it is the direct product of its Sylow subgroups. (Cf. [4], 10.3)

Let H denote the closure of the subgroup generated by G_p. Every neighbourhood of unity in H contains an open normal subgroup N, since H is compact and totally disconnected. H/N is a finite nilpotent group generated by p-elements, hence is a p-group. Thus every element of H is a p-element, i.e. G_p is a closed subgroup. Obviously, G_p is normal.

For arbitrary $c \in G$, the character group $\hat{C}$ of $C = \overline{\langle c \rangle}$ is periodic (since C is totally disconnected) and thus is isomorphic with the weak direct product of its primary parts $\hat{C}_q$ (for all primes q; cf. [5], A.3). Hence C is topologically isomorphic with the direct product of the character groups of these groups. Observe that every element of a compact totally disconnected group with q-primary character group is a q-element, since the (finite) quotient group with respect to any open normal subgroup is the character group of a finite q-group (hence is a finite q-group itself). Hence $c = ab$ with suitable $a \in C_p$ and $b \in C'_p$. Since $C_p \subset G_p$ and $C'_p \subset G'_p$, we conclude $G = G_p G'_p$.

For every $x \in G_p \cap G'_p$ and every open normal subgroup N of G, xN is a p-element of G/N and also is a product of elements whose order is not divisible by p, i.e. $x \in N$ (since G/N is nilpotent). Thus $G_p \cap G'_p = \{e\}$, which completes the proof since, obviously, G'_p is normal.

2.10 Lemma. Every nonabelian locally compact Hamiltonian group has a subgroup isomorphic with Q.

Proof. Let a and b be elements with $[a,b] \neq e$. By Lemma 2.7 and Lemma 2.8, $H = \overline{\langle a,b \rangle}$ satisfies the hypotheses of Lemma 2.9. Hence H is the direct product of H_2 and H'_2. Putting $a = a_1 a_2, b = b_1 b_2$ with $a_1, b_1 \in H'_2$ and $a_2, b_2 \in H_2$, we obtain $[a,b] = [a_1,b_1][a_2,b_2]$. From $[a,b],[a_2,b_2] \in H_2$ (recall that $[a,b]^2 = e$) and $[a_1,b_1] \in H'_2$ we conclude $[a_1,b_1] = e$ and $[a_2,b_2] = [a,b]$.

If $a_2^4 \neq e$ then we may choose an open normal subgroup of H_2 (since H_2 is compact and totally disconnected) that does not contain a_2^4 or $[a_2,b_2]$. The quotient group is a finite nonabelian Hamiltonian group with a 2-element of order greater than 4, which is impossible by Lemma 2.1. Thus $a_2^4 = e$ and, similarly, $b_2^4 = e$. Hence the product of the (normal) subgroups $\langle a_2 \rangle$ and $\langle b_2 \rangle$ is finite and thus (by Lemma 2.1) contains a subgroup isomorphic with Q.

2.11 Lemma. Let G be a discrete group with subgroups Q_1, Q_2 isomorphic with Q. If G is the direct product of a subgroup A with Q_1, then G is also the direct product of A with Q_2 provided that A has no subgroup isomorphic with Q.

Proof. If $Q_1 \cap Q_2 = \{e\}$ then $Q_1 Q_2 / Q_1 \cong Q_2$ is a subgroup of $G/Q_1 \cong A$, a contradiction. Hence $Q_1 \cap Q_2$ is a nontrivial subgroup of Q_2 and thus contains the unique element of order 2. If the subgroup $A \cap Q_2$ of Q_2 is non-trivial then it also contains the unique element of order 2, which is impossible since $Q_1 \cap A = \{e\}$. Hence $A \cap Q_2 = \{e\}$ and $AQ_2/A \cong Q_2 \cong Q_1 \cong G/A$, i.e. $AQ_2 = G$.

2.12 Lemma. Let G be a topological Hamiltonian group that contains Q. If there exists an open subgroup N such that $N \cap Q = \{e\}$ then G is topologically isomorphic with $Q \times A$ for some topological group A.

Proof. N is a normal subgroup and $\tilde{G} = G/N$ is a nonabelian discrete Hamiltonian group which (by Lemma 2.1 and Lemma 2.11) is the direct product of $\tilde{Q} = QN/N$ with an abelian group $\tilde{A}$. Let A be the pre-image of $\tilde{A}$ (with respect to the canonical homomorphism) . From $Q \cap A \subset N$ we conclude $Q \cap A = \{e\}$; $\tilde{G} = \tilde{Q}\tilde{A}$ yields $G = QA$. Hence G is topologically isomorphic with $Q \times A$ (since Q and A are normal subgroups and A is open).

2.13 Lemma. The direct product of Q with an abelian topological group A is Hamiltonian if and only if $a^2 \in \overline{\langle a^4 \rangle}$ for all $a \in A$.

Proof. Obviously, $Q \times A$ is Hamiltonian if $y^{-1}(xa)y \in \overline{\langle xa \rangle}$ for all $x, y \in Q \times \{e\}$ and $a \in \{1\} \times A$. Since $\langle x \rangle$ is a normal subgroup whose order divides 4, we have $y^{-1}xy = x$ or $y^{-1}xy = x^{-1}$. The assertion is trivial in the first case; in the second case it follows from $y^{-1}(xa)y = x^{-1}a = (xa)^{-1}a^2$ and $a^2 \in \overline{\langle a^4 \rangle} \subset \overline{\langle xa \rangle}$ (since $a^4 = (xa)^4$).

Conversely, if $Q \times A$ is Hamiltonian then $i^3 a = j^{-1}(ia)j \in \overline{\langle ia \rangle}$ for all $a \in A$ (we identify A and Q with $\{1\} \times A$ and $Q \times \{e\}$, respectively). Let U be a neighbourhood of e in A. Then $i^n a^n = (ia)^n \in i^3 a U$ for suitable n. From this we get $i^n = i^3$ and $a^n \in aU$. Hence $n = 3 + 4r$ (with integral r) and $a^{3+4r} \in aU$, i.e. $(a^4)^{r+1} \in a^2 U$.

2.14 Lemma. For every locally compact abelian group A and every positive integer n the following assertions are equivalent:

(i) $a^n \in \overline{\langle a^{2n} \rangle}$ for all $a \in A$.
(ii) $\chi^n \in \overline{\langle \chi^{2n} \rangle}$ for all $\chi \in \hat{A}$.
(iii) A is a totally disconnected periodic group and $a^n = e$ for every 2-element $a \in A$.

Proof. Assume that (i) holds. If $\chi^{2n}(a) = 1$ then from the continuity of χ we obtain $\chi^n(a) = \chi(a^n) \in \overline{\langle \chi(a^{2n}) \rangle} = \{1\}$, i.e. $\chi^n(a) = 1$. Since every character of $\hat{A}$ has the form $\chi \longmapsto \chi(a)$ for some $a \in A$, (ii) follows. Also, taking into account that A is topologically isomorphic with the character group of $\hat{A}$, (ii) implies (i).

Let A satisfy (i). If the component A_0 of the identity is not trivial then we may choose a nontrivial continuous character χ of A_0. Then $\chi(A_0)$ is a nontrivial connected subgroup of T, hence equal to T. Thus there is an element $a \in A_0$ with $\chi(a)^n = -1$ and from $a^n \in \overline{\langle a^{2n} \rangle}$ we obtain $-1 = \chi(a^n) \in \overline{\langle \chi(a)^{2n} \rangle} = \{1\}$, a contradiction. Hence A is totally disconnected. A is periodic since, obviously, no subgroup $\langle a \rangle$ is infinite and discrete (cf. [5], 9.1).

Let $a \in A$ be a 2-element. For every open subgroup N, the image $\tilde{a}$ of a in A/N is an element whose order is a power of two. On the other hand, from $a^n \in \overline{\langle a^{2n} \rangle}$ we conclude $\tilde{a}^n = \tilde{a}^{2nr}$ for some r. Hence $\tilde{a}^n = \tilde{e}$, i.e. $a^n \in N$. This implies $a^n = e$, since A is totally disconnected. Hence (iii) holds.

Now suppose that (iii) holds and choose $a \in A$. By Lemma 2.9 we may write $a = a_1 a_2$ with $a_1 \in C_2'$ and $a_2 \in C_2$, where $C = \overline{\langle a \rangle}$. By assumption we have $a_2^n = e$; hence $a^n \in C_2'$. The image of a^n in $C_2' / \overline{\langle a^{2n} \rangle}$ has order ≤ 2. Since every finite image of this (totally disconnected) group has odd order, we conclude $a^n \in \overline{\langle a^{2n} \rangle}$.

2.15 Corollary. Let A be a locally compact abelian group. Then $Q \times A$ is Hamiltonian if and only if $Q \times \hat{A}$ is Hamiltonian.

Proof. This follows immediately from Lemma 2.13 and Lemma 2.14 (with $n = 2$).

2.16 Lemma. The following assertions are equivalent:

(i) A is a locally compact abelian group such that $a^2 \in \overline{\langle a^4 \rangle}$ for all $a \in A$.
(ii) A is topologically isomorphic with the direct product of a locally compact group E of exponent two and a locally compact abelian group F such that $x \in \overline{\langle x^2 \rangle}$ for all $x \in F$.

Proof. Assume that (i) holds and define $E = \{a \in A : a^2 = e\}$, $F = \{a \in A : a \in \overline{\langle a^2 \rangle}\}$. By Lemma 2.14, A is a totally disconnected periodic group in which every 2-element has order ≤ 2. Applying Lemma 2.9 to the compact group $C = \overline{\langle a \rangle}$ (for some fixed $a \in A$) yields $a \in C = C_2' C_2$. Clearly, E is a closed subgroup and $C_2 \subset E$.

In order to show that $C_2' \subset F$, it is sufficient to prove that F contains every closed subgroup H whose discrete quotient groups are generated by elements of odd order. As H is a totally disconnected locally compact abelian group, it is sufficient to prove that for every $h \in H$ and every open subgroup N of H we have $h^{2r} \in hN$ for suitable r, which is clear since every element of H/N has odd order. This argument also shows that the closed subgroup generated by F is contained in F. Thus F is a closed subgroup and $G = EF$.

Since it is obvious that $E \cap F = \{e\}$, it remains to prove that $(x, y) \longmapsto xy$ is an open mapping from $E \times F$ into G. As G is totally disconnected we only have to show that $(E \cap N)(F \cap N)$ is open if N is an open subgroup in which case the first part of the proof (applied to N) yields $N = (E \cap N)(F \cap N)$, thus completing the proof of (ii).

The converse is trivial (note that every group of exponent two is abelian).

2.17 Proof of Theorem 2. Let G be a nonabelian locally compact Hamiltonian group. By Lemma 2.10 we may assume that Q is a subgroup of G (and we identify 1 and e).

Let C be the centralizer of Q in G. For $g \notin C$ without loss of generality we may assume that g does not commute with i; then $gig^{-1} = i^{-1}$ and $jgi = ji^{-1}g = ijg$. Thus $jg \in C$ if jg also commutes with j. Otherwise (similarly as above) ijg commutes with j (and, of course, with i), i.e. $ijg \in C$. Hence $G = QC$.

Assume $c^4 = e$ for some $c \in C$. Then from $[i, jc] = [i, j] \neq e$ and $(jc)^4 = e$ we conclude $i^{-1}jci = (jc)^{-1}$ which yields $c^2 = e$. Hence C has no element of order 4. Thus C (which is closed, hence a locally compact Hamiltonian group) is abelian by Lemma 2.10.

Next we show that G is totally disconnected. The component G_0 of unity is abelian and has no element of order 4, since it is contained in the open subgroup C (note that C has finite index). If $G_0 \neq \{e\}$ then there is a nontrivial continuous character χ whose kernel N does not contain $[i, j]$. Then $\chi(G_0)$ is equal to T (being a nontrivial connected subgroup) and thus G_0/N is (topologically) isomorphic with T. Hence the component of unity in G/N (which clearly contains G_0/N) has elements of order 4, which contradicts our observation made above (with G instead of G/N; note that G/N is nonabelian). Thus G_0 is trivial.

Since every neighbourhood of e contains an open subgroup, we may apply Lemma 2.12 and conclude that G is topologically isomorphic with $Q \times A$ for some topological group A. Clearly, $\{e\} \times A$ belongs to the centralizer of $Q \times \{e\}$; hence, as we have seen above, A is abelian. The assertion thus follows from Lemma 2.13, Lemma 2.16, and Lemma 2.14 (with $n = 1$).

If, conversely, $G = Q \times E \times F$ where E and F have the stated properties, then G is Hamiltonian by Lemma 2.13, since $a^2 \in \overline{\langle a^4 \rangle}$ for every $a \in A = E \times F$ by Lemma 2.14 (with n=1) and (the trivial part of) Lemma 2.16.

2.18 Corollary. Every nonabelian compact Hamiltonian group is topologically isomorphic with $Q \times E \times F$ where E is a product of 2-element groups and every continuous character of the compact abelian group F has odd order; the converse also holds.

Proof. It remains to prove that (i) a compact group E has exponent two if and only if it is a product of 2-element groups and (ii) a compact abelian group F is totally disconnected and has no 2-element except e if and only if every continuous character of F has odd order.

Let E have exponent two. Then $\hat{E}$ has exponent two and is thus a weak direct product of 2-element groups (since $\hat{E}$ may be regarded as a vector space over $\mathbf{Z}/2\mathbf{Z}$). Hence E is topologically isomorphic with the direct product of the corresponding 2-element character groups. (This also follows from [5], 25.9.) The converse is obvious, thus proving (i).

Taking into account that $\hat{F}$ is a torsion group if F is totally disconnected, (ii) follows easily from Lemma 2.14 (with $n = 1$).

2.19 Corollary. Every nonabelian locally compact Hamiltonian group is periodic and totally disconnected; every 2-element has order ≤ 4, every central 2-element has order ≤ 2, every element of finite order not divisible by 4 is central, and the commutator subgroup has precisely two elements.

Proof. This follows immediately from Theorem 2; note that E is totally disconnected since $\hat{E}$ is periodic (since it has exponent two).

2.20 Remark. The proofs of Lemmas 2.2 - 2.5 more or less reproduce Strunkov's proofs of Lemmas 1 - 4 in [7]. Lemma 2.6 (for which in the proof of Lemma 5 Strunkov refers to [8]) is a very special case of a theorem in [1] (cf. also [5], 6.22(a); the related result 25.31 is not sufficient for our purposes). Lemma 2.7 corresponds to Lemma 5 in [7]; the last part of the proof (starting with $a^n \in Z(H)$) is simpler than Strunkov's argument (which moreover requires some modification since the assertion that the order of $a'x'^{-1}z'$ is divisible by 8 is incorrect). Also, instead of appealing to a result in [3], we give a very simple direct proof that H_0 is contained in $Z(H)$. Lemma 2.8 is almost identical with Lemma 6 in [7].

Lemma 2.9 (which is a special case of Lemma 12.7 in [3]) is all what is actually needed from [3] for the proof of Lemma 2.10 (which corresponds to Lemma 7). Lemma 2.12 and a part of 2.17 are essentially taken from the proof of the Theorem in [7]. Strunkov erroneously asserts that (in the notation of Lemma 2.12) $\tilde{Q}$ is equal to $\tilde{Q}_1$ if $\tilde{G}$ is the direct product of the subgroups $\tilde{Q}_1 \cong Q$ and $\tilde{A}$ ($\tilde{A}$ abelian); Lemma 2.11 is needed to rectify the argument (cf. Remark 2.21).

Apart from simplifying Strunkov's arguments (and correcting some minor errors), a major aim of §2 is to provide a reasonably self-contained proof of Strunkov's result that only refers to the (western) standard monograph [5] (and at two instances to [4]); the contents of [7] and [3] seem to be available in Russian only (which might be the reason why Strunkov's theorem is apparently little known).

Strunkov showed that every nonabelian locally compact Hamiltonian group is of the form $Q \times A$ where A satisfies condition (iii) of Lemma 2.14 with $n = 2$. The stronger result in Theorem 2 has also been stated by Mukhin ([6], Corollary 7). As a corollary of his investigations of the group of topological automorphisms (of a locally compact group) that leave invariant every closed subgroup, he proved that every nonabelian locally compact Hamiltonian group is totally disconnected. His claim that (using Lemma 2.1) this easily implies the assertion should be contrasted with the fact that this conclusion (for compact G with countable base) occupies 2 pages in [2] (and is still simplified by several unjustified assumptions; cf. Remark 1.5). It seems that Lemma 12.7 in [3] has to be used for completing the argument. The example in Remark 2.22 shows that the conclusion can fail for a group which is not locally compact (even if it has arbitrarily small open normal subgroups).

2.21 Remark. Let G be the (inner) direct product of subgroups Q, E, F with the properties specified in Theorem 2. Then $QE = \{x \in G : x^4 = e\}$ and $F = \{x \in G : x \in \overline{\langle x^2 \rangle}\}$ are uniquely determined. The following example shows that neither $Q, E,$ nor EF need be uniquely determined. Note, however, that these subgroups are determined up to topological isomorphisms since $E \cong \{x \in G : x^2 = e\}/G'$ and $EF \cong Z(G)/G'$.

Let H be the subgroup of $G = Q \times (\mathbf{Z}/2\mathbf{Z})$ generated by $(i, \overline{1})$ and $(j, \overline{1})$ ($\overline{n}$ denoting the residue class of n mod $2\mathbf{Z}$). It is easy to see that $H = \{(\pm i^m j^n, \overline{m}+\overline{n}) : m, n \in \mathbf{Z}\}$. Hence $(1, \overline{1}) \notin H$ and $H \neq G$. Since H is a nonabelian Hamiltonian group, it is isomorphic with a product $Q \times A$. Hence $|A| = 1$, $H \cong Q$ (but $H \neq Q$), and G is the direct product of H with $\{e, x\}$ where $x \notin H$ is an element of order two, i.e. $x = (1, \overline{1})$ or $x = (-1, \overline{1})$.

2.22 Remark. Theorem 2 and Lemma 2.13 naturally raise the question whether every topological Hamiltonian group G is isomorphic with $Q \times A$ where A is an abelian topological group such that $a^2 \in \overline{\langle a^4 \rangle}$ for all $a \in A$. The following example demonstrates that this is not the case.

Let p be an odd prime. The group $\mathbf{Z}_p$ of p-adic integers is a torsion-free compact abelian group with uncountably many elements. Hence we may choose $a \in \mathbf{Z}_p$ such that $m + na = 0$ (with integers m, n) implies $m = n = 0$. Note that $G = Q \times \mathbf{Z}_p$ is a compact Hamiltonian group (by Lemma 2.13, for example). The subgroup H generated by $(i, 1)$ and (j, a) is a topological Hamiltonian group by Lemma 2.2. It is easy to see that $H = \{(\pm i^m j^n, m + na) : m, n \in \mathbf{Z}\}$. Hence H has precisely two elements of finite order: $(1, 0)$ and $(-1, 0)$. Thus H contains no subgroup isomorphic with Q although it is nonabelian and has arbitrarily small open normal subgroups (since this holds for G); note that the corresponding discrete quotient groups have the form $Q \times A$.

References

[1] P. R. Ahern and R. I. Jewett: Factorization of locally compact abelian groups, Illinois J. Math. 9 (1965), 230-235.

[2] P. Diaconis and M. Shahshahani: On square roots of the uniform distribution on compact groups, Proc. Amer. Math. Soc. 98 (1986), 341-348.

[3] V. M. Gluškov: Locally nilpotent locally bicompact groups (Russian), Trudy Moskov. Mat. Obšč. 4 (1955), 291-332.

[4] M. Hall, Jr.: The Theory of Groups, Macmillan, New York, 1959.

[5] E. Hewitt and K. A. Ross: Abstract Harmonic Analysis I, II, Springer-Verlag, 1963 (2nd ed. 1979), 1970.

[6] Yu. N. Mukhin: Automorphisms which leave fixed the closed subgroups of a topological group (Russian), Sibirsk. Mat. Ž. 16 (1975), 1231-1239; Engl. transl.: J. Sov. Math. 16 (1975), 944-950.

[8] S. P. Strunkov: Topological Hamiltonian groups (Russian), Uspehi Mat. Nauk 20 (1965), no. 6(126), 157-161.

[9] N. Ya. Vilenkin: The theory of topological groups II (Russian), Uspehi Mat. Nauk 5 (1950), no. 4(38), 19-74.

Mathematisches Institut der Universität, Auf der Morgenstelle 10,
D-7400 Tübingen, Federal Republic of Germany.

NEGATIVE DEFINITE FUNCTIONS ON COMMUTATIVE HYPERGROUPS

Michael Voit
Institut für Mathematik
Technische Universität München
Arcisstr. 21
D-8000 München 2

Bloom, Heyer and Lasser ([BH], [H], [L3]) studied convolution semigroups on strong commutative hypergroups K. Lasser proved that for every convolution semigroup there exists an associated negative definite function on $\hat{K}$. The definition of such a function on $\hat{K}$ is a natural generalization of the notation used in [BF] and [BCR] for commutative groups and semigroups respectively. Unfortunately, they could not prove the converse statement, i.e. that any negative definite function f on $\hat{K}$ (where f satisfies some additional conditions, see Ch. 2) implies the existence of an associated convolution semigroup on K. In this paper we give a particular solution of this problem.

1. PRELIMINARIES

Let K be a locally compact Hausdorff space, $M_b(K)$ the space of all bounded Radon measures on K, $M_b^+(K)$ the subspace of all nonnegative measures contained in $M_b(K)$, and $M^1(K)$ the subset of all probability measures on K. The point measure at $x \in K$ will be abbreviated by δ_x, and *supp* μ $\mu \in M_b(K)$, denotes the support of the measure μ. Moreover, $C(K)$ denotes the space of all continous functions $f: K \to \mathbb{C}$, and $C_b(K)$ the space of all bounded functions in $C(K)$.

Following e.g. [H], [J] and [L1], a nonvoid locally compact Hausdorff space K is called a commutative hypergroup, if $M_b(K)$ is equipped with a convolution $*$ such that $(M_b(K), *)$ denotes a commutative Banach algebra with properties similar to the Banach algebra $(M(G), *)$ which is associated with a locally compact abelian group G. To that, there exists a neutral element $e \in K$ which satisfies $\delta_e * \mu = \mu \ \forall \ \mu \in M_b(K)$. Furthermore, there exists an involutive homeomorphism $.^- : K \to K$ such that $x = \bar{y} \Leftrightarrow e \in supp(\delta_x * \delta_y)$ for all $x, y \in K$ holds. Further details on the hypergroup-axioms can be found in [H], [J] or [L1].

Througout this paper let K be a commutative hypergroup. For any $x, y \in K$ and any measurable function $f : K \to \mathbb{C}$, we define

$$f(x * y) := \int_K f(z)\, d(\delta_x * \delta_y)(z) \ .$$

The dual space $\hat{K}$ is given by

$$\hat{K} = \{\alpha \in C_b(K) : \ \alpha \not\equiv 0, \ \alpha(x)\alpha(y) = \alpha(x * y), \ \alpha(\bar{x}) = \overline{\alpha(x)} \ \forall \ x, y \in K\} \ .$$

$\hat{K}$ endowed with the topology of uniform convergence on compact subsets of K is a locally compact Hausdorff space. We say that $\hat{K}$ is a hypergroup with respect to pointwise multiplication, if for any $\alpha, \beta \in \hat{K}$ there exists a measure $\delta_\alpha * \delta_\beta \in M^1(K)$ such that $\alpha(x)\beta(x) = \int_{\hat{K}} \gamma(x)\, d(\delta_\alpha * \delta_\beta)(\gamma) \ \forall \ x \in K$ holds, and if $\hat{K}$ is a hypergroup with this convolution and with the complex conjugation as involution. If $\hat{K}$ is a hypergroup with respect to pointwise multiplication, then $K \subset \hat{\hat{K}}$ in a natural way, see [J], Theorem 12.4B. If in addition $\hat{\hat{K}} = K$ holds, we call K a strong hypergroup.

The Fourier-Stieltjes transform $.^\wedge : M_b(K) \to C_b(\hat{K})$ and the inverse transform $.^\vee : M(\hat{K}) \to C_b(K)$ are defined by

$$\hat{\mu}(\alpha) = \int_K \overline{\alpha(x)}\, d\mu(x) \ \ (\alpha \in \hat{K}) \ \text{ and } \ \check{\mu}(x) = \int_{\hat{K}} \alpha(x)\, d\mu(x) \ \ (x \in K).$$

In a commutative hypergroup there always exists a Haar measure m which is, up to a multiplicative constant, uniquely determined (see [S] and [J]). For a fixed Haar measure and any $p \in [1, \infty]$ we use the abbreviation $L^p(K) = L^p(K, m)$. For further details see [J], Ch. 12.

2. CONVOLUTION SEMIGROUPS AND THEIR ASSOCIATED NEGATIVE DEFINITE FUNCTIONS

2.1. Definition. A family $(\mu_t)_{t \geq 0}$ of measures in $M_b^+(K)$ is called a convolution semigroup on K , iff

(a) $\mu_s * \mu_t = \mu_{s+t} \ \forall \ s, t \geq 0$, (b) $\|\mu_t\| \leq 1 \ \forall \ t \geq 0$ and

(c) $\lim_{t \to 0} \mu_t = \mu_0 = \delta_e$ with respect to the vaguous topology on $M_b(K)$.

2.2. Definition.

(a) A function $f \in C(K)$ is said to be positive definite iff

$$\sum_{i,j=1}^{n} c_i \bar{c}_j f(x_i * \bar{x}_j) \geq 0 \quad \forall \, n \in \mathbb{N}, \ x_1, \ldots, x_n \in K, \ c_1, \ldots, c_n \in \mathbb{C} \ .$$

By $P(K)$ we denote the set of all bounded positive definite functions on K .

(b) A function $\phi \in C(K)$ is called negative definite , iff $\phi(\bar{x}) = \overline{\phi(x)} \ \forall \ x \in K$, $\phi(e) \geq 0$

and

$$\sum_{i,j=1}^{n} c_i\bar{c}_j\phi(x_i * \bar{x}_j) \leq 0 \quad \forall\, n \in \mathbb{N},\ x_1,\ldots,x_n \in K,\ c_1,\ldots,c_n \in \mathbb{C} \text{ with} \sum_{i=1}^{n} c_1 = 0.$$

By $N(K)$ we denote the set of all negative definite functions on K satisfying $\mathrm{Re}\phi(x) \geq \phi(e)\ \forall\, x \in K$.

2.3. Remarks.

(a) For a positive definite function f the following properties are satisfied:

$$f(e) \geq 0,\ \ f(x * \bar{x}) \geq 0\ \forall\, x \in K,\ \ f(\bar{x}) = \overline{f(x)}\ \forall\, x \in K,$$

and, if f is bounded, additionally, $f(e) = \|f\|_\infty$. Moreover, for a function $f \in P(K)$, there exists exactly one positive measure $\mu \in M_b^+(\hat{K})$ satisfying $\check{\mu} = f$. For the proof of this Bochner theorem and related results see Jewett [J].

(b) In [L3], 3.1, negative definite functions on K are defined in another way. But, by [L3], 3.1.1, this definition is equivalent to our definition.

(c) For a locally compact abelian group G, every positive definite function is bounded, and every negative definite function ϕ satisfies $\mathrm{Re}\phi(x) \geq \phi(e)\ \forall\, x \in K$ (see [BF]). These facts are not valid for every commutative hypergroup (for examples see Section 4.8(a)).

The connection between convolution semigroups on K and positive and negative definite functions on $\hat{K}$ is given by the following theorems:

2.4. Theorem (Lasser, [L3], 2.2). Assume that $\hat{K}$ is a hypergroup with respect to pointwise multiplication. If $(\mu_t)_{t\geq 0}$ denotes a convolution semigroup on K, then there exists exactly one function $f \in N(\hat{K})$ such that

$$\hat{\mu}_t = \exp(-tf) \quad \forall\, t \geq 0$$

Using Lasser's notation, we say that f is associated with $(\mu_t)_{t\geq 0}$. Unfortunately, Lasser could only prove a rather weak converse implication:

2.5. Theorem (Lasser, [L3], 2.3). Let K be a strong hypergroup. Let $f \in N(\hat{K})$ be a negative definite function such that $\exp(-tf)$ is positive definite for any $t > 0$. Then there exists a unique convolution semigroup $(\mu_t)_{t\geq 0}$ on $\hat{K}$ such that f is associated with $(\mu_t)_{t\geq 0}$.

2.6. Remarks.

(a) In the proof of 2.5, the fact, that K denotes a strong hypergroup, is necessary to show that there are measures $\mu_t \in M_b^+(K)$ satisfying $\hat{\mu}_t = \exp(-tf)$ $(t > 0)$. If $\hat{K}$ is a hypergroup with respect to pointwise multiplication, [L2], Theorem 1 and Proposition 4, imply that K is a strong hypergroup if and only if for every bounded positive definite function $g \in P(\hat{K})$ there is a $\mu \in M_b^+(K)$ with $\hat{\mu} = g$. Hence, the assumption in

Theorem 2.5 that K is a strong hypergroup seems to be natural.

(b) Let K be a strong hypergroup. In order to prove a stronger version of Theorem 2.5 we will consider the problem which assumptions on K imply the following statement:

$$\phi \in N(\hat{K}) \Longrightarrow \exp(-t\phi) \in P(\hat{K}) \ \forall\, t \geq 0. \qquad (P)$$

(c) If $K = G$ is a locally compact abelian group, it is a well known fact that any negative definite function ϕ on K satisfies

$$\exp(-t\phi) \in P(\hat{K}) \ \forall\, t \geq 0 \quad \text{(see e.g. [BF], Theorem 7.8).}$$

The proof of this statement is elementary, but not available for hypergroups. The crucial problem is that, for any $\phi \in N(\hat{K})$ and $t \geq 0$, we obtain

$$\sum_{i,j=1}^{n} c_i \bar{c}_j \exp(-t\phi(x_i * \bar{x}_j)) = \sum_{i,j=1}^{n} c_i \bar{c}_j \exp\left(-t \int_{\hat{K}} \phi(z)\, d(\delta_{x_i} * \delta_{\bar{x}_j})(z)\right) \geq 0. \qquad (1)$$

But, in order to prove $\exp(-t\phi) \in P(\hat{K})$, we need the fact

$$\sum_{i,j=1}^{n} c_i \bar{c}_j (\exp(-t\phi))(x_i * \bar{x}_j) = \sum_{i,j=1}^{n} c_i \bar{c}_j \int_{\hat{K}} \exp(-t\phi(z))\, d(\delta_{x_i} * \delta_{\bar{x}_j})(z) \geq 0. \qquad (2)$$

Only for groups (1) and (2) are trivially equivalent. For hypergroups, it seems to be impossible to prove the equivalence of (1) and (2) in a direct way. Thus, in order to solve (P) for some special cases, we have to apply other methods.

3. BOUNDED NEGATIVE DEFINITE FUNCTIONS

Let K be always a commutative hypergroup. The functions $f \in L^\infty(K)$ can be identified with the continuous linear functionals on the Banach-$*$-algebra $L^1(K)$ by the canonical relation $\phi_f(g) := \int_K fg\, dm$. The functionals associated with bounded, positive or negative definite functions have some remarkable properties.

3.1. Definition. Let A be a commutative Banach-$*$-algebra (over $\mathbb{C}$) and A^* the space of all continuous functionals on A.

(a) $f \in A^*$ is called positive, iff $f(x * x^*) \geq 0 \ \forall\, x \in A$ holds.

(b) $f \in A^*$ is called positive on a maximal ideal I of A, iff $f(x^*) = \overline{f(x)} \ \forall\, x \in A$ and $f(x * x^*) \geq 0 \ \forall\, x \in I$ holds.

3.2. Lemma. For $f \in C_b(K)$ the following assertions are valid:

(a) f is positive definite, iff $\phi_f \in L^1(K)^*$ denotes a positive functional.

(b) If f is negative definite, the functional $(-\phi_f) \in L^1(K)^*$ is positive on the maximal ideal $I^1 := \{g \in L^1(K) : \int_K g(x)\, dm(x) = 0\}$.

Proof. (a) is proved e.g. in [J], 11.1B and 11.5B.

(b) Since the measures with finite support are dense in $M_b(K)$ with respect to the weak topology (see e.g. [J], 2.2A, 2.2D), the weak continuity of the convolution implies that, for any $g \in I^1$ and any $\epsilon > 0$, there is a measure $\nu = \sum_{i=1}^{n} c_i \delta_{x_i} \in M_b(K)$ ($c_1, \ldots, c_n \in \mathbb{C}$, $x_1, \ldots, x_n \in K$) satisfying

$$\left| \int_K f\, d\nu - \int_K fg\, dm \right| < \epsilon, \quad \left| \int_K f\, d\nu * \nu^* - \int_K f(g * g^*) dm \right| < \epsilon$$

and $|\nu(K)| < \epsilon$. Then, $\nu_0 := \nu - |\nu(K)|\delta_e$ satisfies $\int_K f\, d\nu_0 * \nu_0^* \leq 0$ and $|\int_K f\, d\nu_0 * \nu_0^* - \int_K f(g * g^*)\, dm| \leq \epsilon + 2\epsilon(\epsilon + \int_K fg\, dm) + \epsilon^2 f(e)$. Hence, for $\epsilon \to 0$, we obtain $\int_K f(g*g^*) dm \leq 0$. Analogously, for $g \in L^1(K)$, we have $\int_K fg^* dm = \overline{\int_K fg\, dm}$.

3.3. Proposition. Let K be a commutative hypergroup. Then the maximal ideal I^1 (see 3.2(b)) contains a bounded approximative unit $(w_\alpha)_{\alpha \in \mathcal{A}}$ (i.e. $(w_\alpha)_{\alpha \in \mathcal{A}} \subset I^1$ with $\sup_\alpha \|w_\alpha\| < \infty$ and $\lim_\alpha g * w_\alpha = g \ \forall\, g \in I^1$).

If $\hat{K}$ is a hypergroup with respect to pointwise multiplication, this proposition is proved in [CR], Theorem 3.3. In general, Proposition 3.3 follows by [W], Theorem 5.1.7 and 5.1.8.

3.4. Corollary. Let I^1 and $(w_\alpha)_{\alpha \in \mathcal{A}}$ be defined as in 3.3. For any positive functional ϕ on I^1 there exists a constant $M \geq 0$ such that

$$|\phi(g)|^2 \leq M \cdot \phi(g * g^*) \quad \forall\, g \in I^1.$$

Proof. The mapping $I^1 \times I^1 \to \mathbb{C}$, $(g,h) \mapsto \phi(g * h^*)$ is a positive definite sesquilinear form, for which the Cauchy-Schwartz-inequality is valid. Hence,

$$|\phi(g * h^*)|^2 \leq \phi(g * g^*)\, \phi(h * h^*) \quad \forall\, g, h \in I^1(K), \quad \text{and}$$

$$|\phi(g)|^2 = \lim_\alpha |\phi(g * w_\alpha)|^2 \leq \phi(g * g^*) \cdot M \quad \text{with} \quad M = \limsup_\alpha \phi(w_\alpha * w_\alpha^*) < \infty.$$

3.5. Corollary. Let $\alpha_1 \in \Delta_s(L^1(K))$ be the homomorphism defined by $\alpha_1(g) = \int_K g\, dm$ ($g \in L^1(K)$) and $I^1 = kern\ \alpha_1$ the maximal ideal as in 3.3. If $\phi \in L^1(K)^*$ is positive on I^1, then there is a constant $L > 0$ such that $(\phi + L\alpha_1) \in L^1(K)^*$ is positive.

Proof. If M denotes the constant from 3.4, we define $L := M + 3\|\phi\|$. We choose $g \in L^1(K)$ and $\epsilon > 0$. Then there exists an $u \in L^1(K)$ satisfying $\|u * g^* - g^*\| < \epsilon$ and $\alpha_1(u) = \|u\|_1 = 1$ (If V denotes a suitable neighbourhood of $e \in K$, take $u \in C(K)$ with $supp\ u \subset V$, $u \geq 0$

and $\int_K u\,dm = 1$). For $h := g - \alpha_1(g)u \in kern\ \alpha_1 = I_1$, we obtain $g = h + \alpha_1(g)u$ and

$$\begin{aligned}(\phi + L\alpha_1)(g * g^*) &= \phi(h * h^*) + 2\mathrm{Re}\,(\alpha_1(g)\phi(h^* * u)) + |\alpha_1(g)|^2\phi(u * u^*) + L|\alpha_1(g)|^2 = \\ &= \left[\phi(h * h^*) + 2\mathrm{Re}\,\alpha_1(g)\phi(h^*) + M|\alpha_1(g)|^2\right] + 2\mathrm{Re}\left[\alpha_1(g)(\phi(h^* * u) - \phi(h^*))\right] + \\ &\quad + |\alpha_1(g)|^2\phi(u^* * u) + |\alpha_1(g)|^2 \cdot 3\|\phi\| \geq \\ &\geq \left[\phi(h * h^*) - 2|\alpha_1(g)||\phi(h * h^*)^{1/2}M^{1/2} + M|\alpha_1(g)|^2\right] + 2\mathrm{Re}[\alpha_1(g)\phi(g^* * u - g^*)] \\ &\quad - 2\mathrm{Re}[|\alpha_1(g)|^2(\phi(u * u^*) - \phi(u))] + |\alpha_1(g)|^2\phi(u * u^*) + 3|\alpha_1(g)|^2\|\phi\| \geq \\ &\geq -2|\alpha_1(g)|\|\phi\|\epsilon + |\alpha_1(g)|^2[-2\mathrm{Re}\,\phi(u) - \phi(u * u^*) + 3\|\phi\|] \geq -2|\alpha_1(g)|\|\phi\|\epsilon.\end{aligned}$$

For $\epsilon \to 0$, $(\phi + L\alpha_1)(g * g^*) \geq 0$ follows.

3.6. Corollary. Let K be a commutative hypergroup.

(a) If $f \in N(K)$ is bounded, there exists a constant $L \geq 0$ such that $(L - f)$ is positive definite (and bounded).

(b) Let $f \in N(K)$ be bounded. Then there exists a measure $\mu \in M_b^+(\hat{K})$ satisfying

$$f(x) = f(e) + \int_{\hat{K}} (1 - \alpha(x))\,d\mu(\alpha) \quad \forall\, x \in K.$$

(c) Let $f \in L^\infty(K)$ be a function such that

$$\int_K fg^*\,dm = \overline{\int_K fg\,dm}\ \forall\, g \in L^1(K) \quad \text{and} \quad \int_K f(g * g^*)\,dm \leq 0\ \forall\, g \in I^1.$$

Then there is a bounded function $\tilde{f} \in N(K)$ satisfying $f = \tilde{f}$ locally almost everywhere.

Proof. (a) follows from 3.2 and 3.5. (c) follows from 3.2, 3.5 and [J], 11.5B.

(b) By (a), f can be written as $f = L - g$ where L is a positive constant and $g \in P(K)$. By [J], 12.3b, there is a measure $\mu \in M^+(\hat{K})$ with $g(x) = \int_{\hat{K}} \rho(x)d\mu(\rho)\ \forall\, x \in K$. Hence, $f(x) = L - \mu(\hat{K}) + \int_{\hat{K}}(1 - \rho(x))d\mu(\rho)$. For $x = e$, we obtain $f(e) = L - \mu(\hat{K})$.

Using 3.6, our problem from Section 2 is solved for bounded negative definite functions:

3.7. Theorem. Let K be a strong hypergroup. For any bounded negative definite function $\phi \in N(\hat{K})$ there exists a convolution semigroup $(\mu_t)_{t\geq 0}$ on K such that ϕ is associated with this semigroup.

Proof. By 3.6(a), there exists a constant L satisfying $L - f \in P(\hat{K})$. The set $P(\hat{K})$ is a (with respect to the topology of the uniform convergence on compact subsets of $\hat{K}$) closed cone contained in $C_b(K)$. Moreover, since K is a strong hypergroup, the product of functions $f, g \in P(\hat{K})$ satisfies $fg \in P(\hat{K})$. Hence, the power series of the exponential function implies $\exp(-tf) = \exp(t(L - f)) \cdot \exp(-tL) \in P(\hat{K})\ \forall\, t > 0$. Now we can apply Theorem 2.5.

3.8. Remark. Let K be a strong hypergroup. If $\tilde{N}(\hat{K}) \subset N(\hat{K})$ denotes the space of all functions which can be described as uniform limits on compact subsets of $\hat{K}$ of bounded negative definite functions, Theorem 3.7 and the proof of 2.4 imply a one-to-one correspondence between convolution semigroups on K and functions contained in $\tilde{N}(\hat{K})$. Unfortunately, we cannot prove $\tilde{N}(\hat{K}) = N(\hat{K})$ in general.

3.9. Examples.

(a) Let K be a discrete commutative hypergroup such that $\hat{K}$ is a hypergroup with respect to pointwise multiplication. Then, by [L1], 1.1.2, K is a strong hypergroup, and $\hat{K}$ is compact. Hence, by 3.7, we see that convolution semigroups on K and negative definite functions on $\hat{K}$ correspond in a one-to-one way. Moreover, 3.6 implies that all convolution semigroups on these hypergroups are Poisson-semigroups (for similar results see [GG] and [H]). We give some concrete examples:

(b) For fixed $\alpha, \beta \in \mathbb{R}$ with $\alpha = \beta \geq -1/2$ or $\alpha \geq \beta \geq -1/2$, $\alpha + \beta \geq 0$, the Jacobi polynomials $P_n^{(\alpha,\beta)}$ (which are normalized by $P_n^{(\alpha,\beta)}(1) = 1 \; \forall \, n \in \mathbb{N}_0$) generate strong hypergroup structures on $\mathbb{N}_0$ (see [L1], 3(a) and 4(a)). These hypergroups are studied in [H] and [GG], too.

(c) If G denotes a compact group, the dual space $\hat{G}$ which consists of all equivalence classes of the irreducible unitary representations of G can be endowed with a natural hypergroup structure. These examples yield strong hypergroups.

(d) Using the p-adic numbers (p prime), Dunkl and Ramirez constructed in [DR] commutative hypergroup structures on $\mathbb{N}^*$, the one-point-compactification of $\mathbb{N}$. They obtained these hypergroup structures by the operating of the compact group Δ_p of the units on the additive group of the p-adic integers ([DR], Ch. 2 and 3). By [DR], Theorem 3.6, it can be easily seen that the dual space $\hat{\mathbb{N}}^*$, which can be identified with $\mathbb{N}_0$ in a natural way, is a hypergroup with respect to pointwise multiplication. Moreover, one obtains that $\mathbb{N}^*$ is a strong hypergroup. Hence, $\hat{\mathbb{N}}^* = \mathbb{N}_0$ is a strong hypergroup for which the above results are available.

(e) If $K = \mathbb{N}^*$ denotes a hypergroup associated with the p-adic integers according to (d), for any negative definite function f on $\hat{K} = \mathbb{N}_0$ the statement of Theorem 3.7 is true. To prove this, we have $\exp(-tf) \in P(\mathbb{N}_0) \; \forall \, t > 0$ to show. Choose $n \in \mathbb{N}$ and $x_1, \ldots, x_n \in \mathbb{N}_0 = \hat{K}$. Since for $m := \max\{x_1, \ldots, x_n\}$ the set $K_m := \{0, 1, \ldots, m\}$ is a finite subhypergroup of $\hat{K}$ (this follows e.g. by [DR], Theorem 3.6), $f|_{K_m}$ is bounded and negative definite on K_m. Applying 3.7, we obtain $\exp(-tf)|_{K_m} \in P(K_m) \; \forall \, t > 0$. Moreover, 2.3(a) implies $\|\exp(-tf)|_{K_m}\|_\infty \leq \exp(-tf(0)) \; \forall \, t > 0$. Hence, Definition 2.2(a) yields $\exp(-tf) \in P(\hat{K}) \; \forall \, t > 0$.

4. NEGATIVE DEFINITE FUNCTIONS ON POLYNOMIAL HYPERGROUPS

4.1. Let $(P_n)_{n\in\mathbb{N}_0}$ be a sequence of orthogonal polynomials which are defined by the recursion formula

$$P_0 = 1,\ P_1(x) = ax + (1-a),\ P_1 P_n = a_n P_{n+1} + b_n P_n + c_n P_{n-1} \tag{1}$$

where $a > 0$, $a_n, c_n > 0$, $b_n \geq 0$, $a_n + b_n + c_n = 1$ and $n \in \mathbb{N}$. Then the linearization coefficients $g_{m,n,k}$ $(m, n, k \in \mathbb{N}_0,\ |m-n| \leq k \leq m+n)$ are uniquely determined by $P_m P_n = \sum_{k=|m-n|}^{m+n} g_{m,n,k} P_k$. If all these linearization coefficients are nonnegative, we can define a commutative hypergroup structure on $\mathbb{N}_0$. For this we define the convolution of point measures δ_m, δ_n $(m, n \in \mathbb{N}_0)$ by $\delta_m * \delta_n = \sum_{k=|m-n|}^{m+n} g_{m,n,k}\delta_k$. Then the neutral element and the involution on $\mathbb{N}_0$ are given by 0 and by the identical mapping respectively. For $D_s := \{x \in \mathbb{R} : (P_n)_{n\in\mathbb{N}_0} \textit{ is bounded}\}$, the mapping $D_s \to \hat{\mathbb{N}}_0$, $x \mapsto \alpha_x$, $\alpha_x(n) = P_n(x)$, describes a homeomorphism between D_s (equipped with the natural topology) and the dual $\hat{\mathbb{N}}_0$ of our polynomial hypergroup. For details see [L1].

Now we study negative definite functions on polynomial hypergroups $\mathbb{N}_0$. Since the involution on $\mathbb{N}_0$ is the identical mapping, all positive and negative definite functions are real valued (see 2.2 and 2.3), and, without lost of generality, we can always assume that the coefficients c_k used in Definition 2.2 are real valued. This fact will be applied later.

Besides the polynomial hypergroup structure $(\mathbb{N}_0, *)$, we use the semigroup structure $(\mathbb{N}_0, +)$ on $\mathbb{N}_0$ which shall be endowed with the identical mapping as involution. For this semigroup with involution, the (bounded) characters are given by $\mathbb{N}_0 \to \mathbb{R}$, $n \mapsto Q_n(x) = x^n$, $x \in [-1, 1]$. Positive and negative definite functions on this semigroup are defined analogously as in Definition 2.2 (see [BCR], Ch.4, too). Next, for a given positive or negative definite function $f : \mathbb{N}_0 \to \mathbb{R}$ on the polynomial hypergroup structure $(\mathbb{N}_0, *)$ or on the semigroup $(\mathbb{N}_0, +)$, we can define an associated linear functional T_f on $\mathbb{R}[x]$, the $\mathbb{R}$-vector space of all polynomials, by

$$T_f : \mathbb{R}[x] \longrightarrow \mathbb{R},\quad T_f(P_n) := f(n)\ \ (T_f(Q_n) := f(n) \text{ respectively}),\ n \in \mathbb{N}_0.$$

Then T_f has some properties which are independent of $(\mathbb{N}_0, *)$ and $(\mathbb{N}_0, +)$ respectively.

4.2. Proposition. Let either $(P_n)_{n\in\mathbb{N}_0}$ be a sequence of orthogonal polynomials which generates a polynomial hypergroup $(\mathbb{N}_0, *)$ according to 4.1, or let $P_n(x) = x^n$, $n \in \mathbb{N}_0$, be the sequence of polynomials which generates the semigroup $(\mathbb{N}_0, +)$. Then we obtain:

(a) A function $f : \mathbb{N}_0 \to \mathbb{R}$ with $f(0) \geq 0$ is negative definite iff the associated functional T_f on $\mathbb{R}[x]$ satisfies

$$P \in \mathbb{R}[x],\ P(1) = 0,\ P(x) \geq 0\ \ \forall\, x \in \mathbb{R} \quad \Rightarrow \quad T_f(P) \leq 0. \tag{N}$$

(b) $f : \mathbb{N}_0 \to \mathbb{R}$ is positive definite iff T_f satisfies

$$P \in \mathbb{R}[x],\ P(x) \geq 0\ \ \forall\, x \in \mathbb{R} \quad \Rightarrow \quad T_f(P) \geq 0.$$

(c) $f : \mathbb{N}_0 \to \mathbb{R}$ is positive definite and bounded, iff T_f satisfies

$$P \in \mathbb{R}[x],\ P(x) \geq 0\ \ \forall\, x \in D_s = \{x \in \mathbb{R} : (P_n)_{n\in\mathbb{N}_0} \text{ is bounded}\} \quad \Rightarrow \quad T_f(P) \geq 0.$$

Proof.

(a) Let $f : \mathbb{N}_0 \to \mathbb{R}$ be negative definite . Without lost of generality, we suppose $f(0) = 0$ (otherwise take $f - f(0)$ instead of f). If $P \in \mathbb{R}[x]$ is defined by $P = \sum_{k=0}^{n} a_k P_k$ ($a_0, \ldots, a_n \in \mathbb{R}$) and if P satisfies $P(1) = 0$ and $P(x) \geq 0\ \forall x \in \mathbb{R}$, [BCR], 6.2.1, implies that there are polynomials $Q, R \in \mathbb{R}[x]$ with $P = Q^2 + R^2$, $Q = \sum_{k=0}^{n} b_k P_k$ and $R = \sum_{k=0}^{n} c_k P_k$ (with $b_0, \ldots, b_n, c_0, \ldots, c_n \in \mathbb{R}$). Hence, $Q(1) = R(1) = 0$ and $P_k(1) = 1\ \forall\, k \in \mathbb{N}_0$ yields $\sum_{k=0}^{n} b_k = \sum_{k=0}^{n} c_k = 0$. Moreover, by

$$\sum_{k,l=0}^{n} (b_k b_l + c_k c_l) P_k P_l = \sum_{k=0}^{n} a_k P_k = P \quad \text{we get} \quad \sum_{k,l=0}^{n} (b_k b_l + c_k c_l)\delta_k * \delta_l = \sum_{k=0}^{n} a_k \delta_k.$$

Using (1) and the negative definiteness of f, we obtain

$$T_f(P) = \sum_{k=0}^{n} a_k f(k) = \sum_{k,l=0}^{n} (b_k b_l + c_k c_l) f(k * l) \leq 0.$$

Converse, if T_f has property (N), then we choose $n \in \mathbb{N}_0$ and $a_0, \ldots, a_n \in \mathbb{R}$ satisfying $\sum_{k=0}^{n} a_k = 0$. Then, for $P := \sum_{k,l=0}^{n} a_k a_l P_k P_l \in \mathbb{R}[x]$, we obtain $P(1) = 0$ and $P(x) \geq 0\ \forall\, x \in \mathbb{R}$. Thus, $0 \geq T_f(P) = \sum_{k,l=0}^{n} a_k a_l f(k * l)$ follows.

(b) can be proved similar to (a).

(c) follows for hypergroups by [J], 12.3B (note that $D_s \simeq \hat{\mathbb{N}}_0$ is compact, and T_f defines a positive measure on $\hat{\mathbb{N}}_0$). For $(\mathbb{N}_0, +)$, (c) follows e.g. by [BCR], Ch.6, 2.4.

4.3. Proposition. Let $(\mathbb{N}_0, *)$ be a polynomial hypergroup. If $f : \mathbb{N}_0 \to \mathbb{R}$ denotes a negative definite function, then there exists a sequence $f_k : \mathbb{N}_0 \to \mathbb{R}$ of positive definite functions satisfying $f(n) = \lim_{k\to\infty}(k - f_k(n))\ \ \forall\, n \in \mathbb{N}_0$.

Proof. Let $(P_n)_{n\in\mathbb{N}_0}$ be the sequence of orthogonal polynomials associated with $(\mathbb{N}_0, *)$. Moreover, define $(Q_n)_{n\in\mathbb{N}_0}$ by $Q_n(x) = x^n$, $n \in \mathbb{N}_0$. Then (see 4.1) $Q_n(1) = P_n(1) = 1$ $\forall\, n \in \mathbb{N}_0$. We define the connection coeffizients $b_{n,k}, c_{n,k} \in \mathbb{R}$ ($n, k \in \mathbb{N}_0$, $0 \leq k \leq n$) by

$$Q_n = \sum_{k=0}^{n} b_{n,k} P_k \qquad \text{and} \qquad P_n = \sum_{k=0}^{n} c_{n,k} Q_k.$$

If f denotes a negative definite function on $(\mathbb{N}_0, *)$, by 4.2(a) we see that the associated functional T_f has the property (N). If we define $g : \mathbb{N}_0 \to \mathbb{R}$ by $g(n) := T_f(Q_n)$, 4.2(a) implies that g is a negative definite function on $(\mathbb{N}_0, +)$. Moreover, we have

$g(n) = \sum_{k=0}^{n} b_{n,k} f(k)\ \forall\, n \in \mathbb{N}_0$. Now, e.g. [BCR], p. 99, shows that the functions $g_k : \mathbb{N}_0 \to \mathbb{R}$, $g_k(n) := k \cdot \exp(-g(n)/k)$, $k \in \mathbb{N}$, are positive definite. Furthermore, for $n \in \mathbb{N}_0$, we have $g(n) = \lim_{k\to\infty}(k - k \cdot \exp(-g(n)/k)) = \lim_{k\to\infty}(k - g_k(n))$. If we define the functions $f_k : \mathbb{N}_0 \to \mathbb{R}$ by $f_k(n) = \sum_{l=0}^{n} c_{n,l} g_k(l)$, a twofold application of Proposition 4.2(b) implies that these functions are positive definite on $(\mathbb{N}_0, *)$. Moreover, $Q_n(1) = P_n(1) = 1\ \forall\, n \in \mathbb{N}_0$ yields

$$\sum_{k=0}^{n} b_{n,k} = \sum_{k=0}^{n} c_{n,k} = 1 \quad \text{and} \quad \lim_{k\to\infty}(k - f_k(n)) = \lim_{k\to\infty}\left(\sum_{l=0}^{n} c_{n,l}(k - g_k(n))\right) = f(n).$$

Using certain assumptions on the polynomial hypergroup $(\mathbb{N}_0, *)$, Proposition 4.3 can be strengthened. We say that the associated sequence $(P_n)_{n\in\mathbb{N}_0}$ of orthogonal polynomials has property (L), iff $[-1,1] \subset D_s$ and $b_{n,k} \geq 0$ $(n, k \in \mathbb{N}_0,\ k \leq n)$ holds where the coefficients are defined as in the proof of 4.3.

If $(\mathbb{N}_0, *)$ has this property (L), and if $f : \mathbb{N}_0 \to \mathbb{R}$ denotes a negative definite function on $(\mathbb{N}_0, *)$ satisfying $f(n) \geq f(0) \geq 0\ \forall\, n \in \mathbb{N}_0$, then the function g defined as in the proof of 4.3 is a negative definite function on $(\mathbb{N}_0, +)$ with $g(n) \geq g(0) \geq 0\ \forall\, n \in \mathbb{N}_0$. Then the functions g_k, $k \in \mathbb{N}$, are bounded positive definite functions on $(\mathbb{N}_0, +)$. Hence, 4.2(c) implies that the functions f_k, $k \in \mathbb{N}$, are bounded positive definite functions on $(\mathbb{N}_0, *)$. Thus, we have proved

4.4. Theorem. Let $(\mathbb{N}_0, *)$ be a polynomial hypergroup such that the associated sequence $(P_n)_{n\in\mathbb{N}_0}$ of orthogonal polynomials has property (L). If $f : \mathbb{N}_0 \to \mathbb{R}$ is a negative definite function satisfying $f(n) \geq f(0) \geq 0\ \forall\, n \in \mathbb{N}_0$, then there exists a sequence $(f_k)_{k\in\mathbb{N}}$ of bounded positive definite functions on $(\mathbb{N}_0, *)$ with $f(n) = \lim_{k\to\infty}(k - f_k(n))\ \forall\, n \in \mathbb{N}_0$.

4.5. Corollary. Let $(\mathbb{N}_0, *)$ and $(P_n)_{n\in\mathbb{N}_0}$ be given as in 4.4. For $f : \mathbb{N}_0 \to \mathbb{R}$ with $f(0) \geq 0$, the following assertions are equivalent:

(a) f is negative definite on $(\mathbb{N}_0, *)$ and satisfies $f(n) \geq f(0) \geq 0\ \forall\, n \in \mathbb{N}_0$.

(b) For $n \in \mathbb{N}$, $c_0, \ldots, c_n \in \mathbb{R}$ with $\sum_{k=0}^{n} c_k = 0$, $\sum_{k=0}^{n} c_k P_k(x) \geq 0\ \forall\, x \in D_s$ implies

$$\sum_{k=0}^{n} c_k f(k) \leq 0.$$

(c) The functional T_f satisfies

$$P \in \mathbb{R}[x],\ P(1) = 0,\ P(x) \geq 0\ \forall\, x \in D_s \quad \Rightarrow \quad T_f(P) \leq 0.$$

Proof.

(a) $\Rightarrow$ (b) Using 4.4, there are bounded positive definite functions $f_k : \mathbb{N}_0 \to \mathbb{R}$ $(k \in \mathbb{N})$ with $f(n) = \lim_{k\to\infty}(k - f_k(n))\ \forall\, n \in \mathbb{N}_0$. Thus, for $m, k \in \mathbb{N}$ and $c_0, \ldots, c_m \in \mathbb{R}$,

4.2(c) yields

$$\sum_{l=0}^{m} c_l P_l(x) \geq 0 \;\forall\, x \in D_s \Longrightarrow \sum_{l=0}^{m} c_l f_k(l) \geq 0.$$

Hence, for $k, m \in \mathbb{N}$ and $c_0, \ldots, c_m \in \mathbb{R}$ satisfying $\sum_{l=0}^{m} c_l = 0$, we obtain

$$\sum_{l=0}^{m} c_l P_l(x) \geq 0 \;\;\forall\, x \in D_s \Longrightarrow \sum_{l=0}^{m} c_l(k - f_k(l)) \leq 0.$$

(b) $\Leftrightarrow$ (c) is obvious.

(c) $\Rightarrow$ (a) If we apply 4.2(a), we have to show $f(n) \geq f(0) \geq 0 \;\forall\, n \in \mathbb{N}_0$. For this reason, for any $n \in \mathbb{N}_0$, we define $c_0 = 1$, $c_1 = c_2 = \ldots = c_{n-1} = 0$ and $c_n = -1$. For $x \in D_s$ we obtain $P_0(x) - P_n(x) = 1 - P_n(x) \geq 0$. Hence, (c) implies $f(n) - f(0) \geq 0 \;\forall\, n \in \mathbb{N}_0$.

4.6. Corollary. Let $(\mathbb{N}_0, *)$ and $(P_n)_{n\in\mathbb{N}_0}$ be chosen as in 4.4. Moreover, we assume that any product of bounded positive definite functions g_1, g_2 on $(\mathbb{N}_0, *)$ is a (bounded) positive definite function. Then, for $f : \mathbb{N}_0 \to \mathbb{R}$ with $f(0) \geq 0$, the following statements are equivalent:

(a) f is negative definite and satisfies $f(n) \geq f(0) \geq 0 \;\forall\, n \in \mathbb{N}_0$.

(b) $\exp(-tf)$ is a bounded positive definite function on $(\mathbb{N}_0, *)$ for any $t > 0$.

Proof. (b) $\Rightarrow$ (a) is proved in [L3], 1.5.
(a) $\Rightarrow$ (b) follows by 4.5, (a) $\Rightarrow$ (b), [B1], Theorem 5, and by 4.2(c).

4.7. Examples.

(a) If the coefficients b_n, $n \in \mathbb{N}$, used in the recursion formula (1), satisfy $b_n = 0 \;\forall\, n \in \mathbb{N}$, then we can define P_1 by $P_1(x) = x$ (the parameter $a > 0$ in (1) has no influence on the hypergroup structure $(\mathbb{N}_0, *)$!). Then, by induction on n, (1) implies the nonnegativity of the connection coefficients $b_{n,k}$ ($n, k \in \mathbb{N}_0$, $k \leq n$) which are defined by $x^n = \sum_{k=0}^{n} b_{n,k} P_k(x)$. Moreover, for all known examples of such hypergroups, we obtain $D_s = [-1, 1]$ and thus property (L) (for an extensive list of examples see [L1]).

(b) For the following examples the suppositions of Corollary 4.7 are satisfied:

(b1) For fixed $\alpha \geq -1/2$, the ultraspherical polynomials $(P_n^{(\alpha)})_{n\in\mathbb{N}_0}$ define a strong hypergroup structure on $\mathbb{N}_0$ (see [L1], 3(a) and 4(a)). For $\alpha = \frac{n-3}{2}$, $n \geq 3$, $n \in \mathbb{N}$, the dual hypergroup structure which is defined on $D_s = [-1, 1]$ is isomorphic with the Gelfand pair $(SO(n), SO(n-1))$. Hence, using 2.4, 2.5 and 4.6, $\hat{\mu}_t = \exp(-tf)$ describes an one-to-one correspondence between convolution semigroups $(\mu_t)_{t\geq 0}$ on these hypergroup structures on $[-1, 1]$ and negative definite functions f on $\mathbb{N}_0$ with $f \geq f(0) \geq 0$.

(b2) Generalized Tchebichef polynomials $(T_n^{(\alpha,\beta)})_{n\in\mathbb{N}_0}$, $\alpha - 1 \geq \beta \geq -1/2$ (see [L1], 3(f) and 4(f)) and Cartier polynomials $(P_n^a)_{n\in\mathbb{N}_0}$, $a \in \mathbb{N}$, $a \geq 2$ (see [Le] and [L1], 3(d)) generate hypergroup structures on $\mathbb{N}_0$ which satisfy the assumptions in 4.4 - 4.6.

4.8. Remarks.

(a) Let $(\mathbb{N}_0,*)$ and $(P_n)_{n\in\mathbb{N}_0}$ be a polynomial hypergroup and the associated orthogonal polynomials according to 4.1 respectively. Then, for $x \in \mathbb{R} \setminus D_s$, the functions $\alpha_x : \mathbb{N}_0 \to \mathbb{R}$, $n \mapsto P_n(x)$, define positive definite functions on $(\mathbb{N}_0,*)$ which are unbounded. Moreover, $f : \mathbb{N}_0 \to \mathbb{R}$ defined by $f(n) := P_n'(x)$ is a negative definite function on $(\mathbb{N}_0,*)$ satisfying $f(n) \geq f(0) = 0 \;\forall\, n \in \mathbb{N}_0$. Especially, by [L3], 1.11, f is a so-called quadratic form on $(\mathbb{N}_0,*)$. Thus, it is easy to see that $(-f)$ is a negative definite function, too, which satisfies $f(n) < 0 \;\forall\, n \in \mathbb{N}$.

(b) The Theorems 4.4-4.6 are not applicable to the polynomial hypergroups which are associated with the Jacobi polynomials $(P_n^{(\alpha,\beta)})_{n\in\mathbb{N}_0}$ where $\alpha > \beta \geq -1/2$ (these polynomials generate strong polynomial hypergroups, see [L1], 3(a) and 4(a)). The correctness of the statements in 4.4 - 4.6 is, for these examples, still unknown.

(c) The results presented in this paper show that negative definite functions on commutative hypergroups are more difficult to handle than thpose on locally compact abelian groups. On the other hand, we can use our definition of negative definite functions on the dual $\hat{K}$ if and only if $\hat{K}$ is a hypergroup with respect to pointwise multiplication. Hence, if we will consider convolution semigroups and their associated negative definite functions on $\hat{K}$ for any commutative hypergroup, we have to apply another definitions of positive and negative definite functions on $\hat{K}$ (see e.g. [B1], [B2] and [L2]).

References

[B1] Berg, C.: Studies definies negatives et espaces Dirichlet sur la sphere, Sem. Brelot-Choquet-Deny, Theorie du Potential, 13e annee, 1969/1970.

[B2] Berg, C.: Dirichlet Forms on Symmetric Spaces, Ann. Inst. Fourier, 23.1, 135-156 (1973).

[BCR] Berg, C., Christensen, J.P.R., Ressel, P.: Harmonic Analysis on Semigroups. New York-Berlin-Heidelberg-Tokyo: Springer, 1984.

[BF] Berg, C., Forst, G.: Potential Theory on Locally Compact Abelian Groups, Berlin-Heidelberg-New York. Springer, 1975.

[BH] Bloom, W., Heyer, H.: Convolution semigroups and resolvent families of measures on hypergroups. Math. Z. 188, 449-474 (1985).

[CR] Chilana, A.K., Ross, K.A.: Spectral Synthesis in Hypergroups, Pac. J. Math.76, 313-328(1978).

[DR] Dunkl, C.F., Ramirez, D.E.: A family of countable compact P^*-Hypergroups, Trans. Am. Math. Soc. 202, 339-356(1975).

[GG] Gallardo, L., Gebuhrer, O.: Lois de probabilite infinement divisibles sur les hypergroupes commutatifs, discrets, denomerables. Probability Measures on Groups, Oberwolfach, 1983, 116-130. Lecture Notes in Math., Vol. 1064, Berlin-Heidelberg-New York. Springer, 1984.

[H] Heyer, H.: Probability theory on hypergroups: A survey. Probability Measures on Groups, Oberwolfach, 1983, 481-550. Lecture Notes in Math., Vol. 1064, Berlin-Heidelberg-New York. Springer, 1984.

[J] Jewett, R.I.: Spaces with an abstract convolution of measures. Adv. Math. 18, 1-101 (1975).

[L1] Lasser, R.: Orthogonal polynomials and hypergroups. Rend. Math. Appl. 3, 185-209 (1983).

[L2] Lasser, R.: Bochner theorems for hypergroups and their application to orthogonal polynomial expansions, J. Approx. Th. 37, 311-327 (1983).

[L3] Lasser, R.: Convolution semigroups on hypergroups. Pacific J. of Math. 127, 353-371 (1987).

[Le] Letac, G.: Dual random walks and special functions on homogenous trees. Publications de l'Institut Elie Cartan, Nancy, Bd. 7 (1983).

[S] Spector, R.: Mesures invariantes sur les hypergroupes. Trans. Am. Math. Soc. 239, 147-165 (1978).

[W] Wolfenstetter, S.: Jacobi-Polynome und Bessel-Funktionen unter dem Gesichtspunkt der harmonischen Analyse, Dissertation, Technische Universität München, 1984.

L^p-IMPROVING MEASURES ON HYPERGROUPS

R.C. Vrem
Humboldt State University
Arcata, CA 95521 / USA

Introduction

Stein [15] notes that if μ is any regular Borel measure on the circle whose Fourier transform is $O(n^{-\alpha})$ for $\alpha>0$ then $\mu * L^p \subseteq L^r$ for some $1<p<r<\infty$. In fact, he describes a nonnegative measure which is singular with respect to Lebesgue measure which has this property of "improving" L^p. Stein poses the problem of characterizing such measures in terms of their "size." Measures of this type have been studied for compact abelian groups by Bonami [1]; Christ [2]; Ritter [13],[14]; Hare [6]; Graham, Hare and Ritter [5]. Hare [7] also discusses such measures for compact (nonabelian) groups.

In this paper, we extend some of these results to compact abelian hypergroups as well as provide some interesting examples which do not appear in the group case. A class of continuous measure which improve L^p is also constructed for the compact group $SU(2)$ and many of these measures are shown to be singular with respect the Haar measure of $SU(2)$.

For K a compact abelian hypergroup, we call a measure μ in $M(K)$ L^p-improving if there exist $1<p<r<\infty$ such that $\mu * L^p(K) \subseteq L^r(K)$. In §1, several of the basic results in [5] are extended to compact abelian hypergroups. In particular, we show that if μ is L^p-improving for some $p>1$ then μ is L^p-improving for all $p>1$. An analogue of Young's inequality is established for hypergroups and is used to show that any $L^q(K)$ function $(q>1)$ will be L^p-improving. An example is provided to show that, in contrast to the group case, point masses can be L^p-improving for hypergroups. This example is then used to provide a counter-example to a results in [5] in the hypergroup case. In §2 a discussion of the relationship of L^p-improving to lacunarity is provided. We first show that any measure whose Fourier-Stieltjes transform is supported on a $\Lambda(q)$ set $(1<q<\infty)$ must be an $L^q(K)$ function and hence L^p-improving. We then provide an example of a hypergroup where every point mass (except the point mass at the identity) is L^p-improving. Then results of Hare [6] are extended to hypergroups showing the relationship between L^p-improving measures and $\Lambda(p)$ sets.

§3 initiates a discussion of central L^p-improving measures on compact (nonabelian) groups. These measures are exploited to show that all continuous central measures on SU(2) are L^p-improving. In fact, we extend a result of Ragozin [11, 2.2] who showed that the convolution of any two continuous central measures on SU(2) is absolutely continuous with respect to Haar measure on SU(2) to show that the convolution of any two continuous central measures on SU(2) is in fact in $L^2(K)$. Finally, a class of continuous central singular measures is shown to be L^p-improving on SU(2).

Throughout this paper, K will denote a compact abelian hypergroup. The basic properties of hypergroups are found in Jewett [9]. We will also use Jewett's notation except the involution on K will be denoted by $x \to \check{x}$ and δ_x will denote the point mass at x. $\hat{K}$ will be the space of all continuous hypergroup characters defined on K and m will denote normalized Haar measure on K. L^p will denote $L^p(K)$ with respect to Haar measure on K.

The basic theory of lacunarity on compact hypergroups was established in [17]. We call $P \subseteq \hat{K}$ a <u>$\Lambda(p)$set</u> $(p>2)$ if there is a constant A such that $\|f\|_p \le A\|f\|_2$ whenever $f \in L^2$ and $\hat{f}(\psi) = 0$ for ψ not in P. The least such constant A is called the $\Lambda(p)$ constant for P and is denoted by $\Lambda(p,P)$. We call P a Λ set if there exists a constant A such that $\|f\|_p \le A\sqrt{p}\|f\|_2$ for all trigonometric polynomials f such that $\hat{f}(\psi) = 0$ for all ψ not in P.

1. <u>Basic Theory of L^p-Improving Measures</u>

The first result appears in [5, 0.2] and extends immediately to hypergroups.

<u>Theorem 1.1</u> Suppose that for some $1 < q_1 < q_2 < \infty$ and some $C > 0$, $\|\mu * f\|_{q_2} \le C\|f\|_{q_1}$ whenever $f \in L^{q_1}$. Let $0 < \alpha < 1$. Suppose that either $p_i = q_i\alpha^{-1}$ for $i = 1,2$ or $p_i = q_i[\alpha+(1-\alpha)q_i]^{-1}$ for $i = 1,2$. Then $\|\mu * f\|_{p_2} \le C^{\alpha}\|\mu\|^{1-\alpha}\|f\|_{p_1}$ for all $f \in L^{p_1}$.

Proof. Follows immediately from the Riesz-Thoren Interpolation Theorem (see e.g. [8, E.18]) and the observation $\mu * L^1 \subseteq L^1$ and $\mu * L^{\infty} \subseteq L^{\infty}$ with norm $\|\mu\|$ [9, 6.2B].

<u>Corollary 1.2</u> If μ is L^q-improving for some $1 < q < \infty$ then μ is L^p-improving for all $1 < p < \infty$.

In light of the preceding results, in order to show that a measure μ is L^p-improving it suffices to show that there are constants $p > 2$

and A so that $\|\mu*f\|_p \le A\|f\|_2$ for all $f\in L^2$ or equivalently, there exists $1<p'<2$ and A' such that $\|\mu*f\|_2 \le A'\|f\|_{p'}$ for all $f\in L^{p'}$.

In the case of compact groups, no point mass can be L^p-improving. However, for hypergroups point masses can be L^p-improving as we see in the next example.

<u>Example 1.3</u> Let H be any compact abelian hypergroup and let $K = H\vee\mathbb{Z}_2$, i.e., K is the join of H and $\mathbb{Z}_2 = \{0,1\}$ (see [9, 10.5] for information on joins). We will write 0 as the identity of K and we will show that δ_1 is L^p-improving. Let $f\in L^p$ with $1 < p < q < \infty$ and compute

$$(\delta_1*f)(x) = \begin{cases} f(1) \text{ if } x\in H \\ \int_H f\, dm_K \text{ if } x = 1 \end{cases}$$

and hence

$$\|\delta_1*f\|_q^q = |f(1)|^q m_K(H) + m_K(1)|\int_H f dm_K|^q$$

so that

$$\|\delta_1*f\|_q^q \le |f(1)| + \|f\|_1$$

and hence $\delta_1*f\in L^q$ for all $f\in L^p$ so δ_1 is L^p-improving.

The next proposition is modeled after a similar result of Ritter [14, Lemma 2].

<u>Proposition 1.4</u> Let K be a compact abelian hypergroup with $\mu\in M(K)$. If there is a positive integer N such that μ^N is L^p-improving then μ is also L^p-improving.

Proof. We assume there exists $A' > 0$, $1 < p' < 2$ such that

$$\|\mu^N*f\|_2 \le A'\|f\|_{p'}$$

for all $f\in L^{p'}$ and show there exists $A > 0$ and $1 < p < 2$ such that

$$\|\mu*f\|_2 \le A\|f\|_{p'}$$

for all $f\in L^p$. The technique is to use Stein's Analytic Interpolation Theorem [16, 4.1]. For $\mathrm{Re}(z)\ge 0$ define T^z on the simple integrable functions of K via

$$(T^z f)^\wedge(\gamma) = \hat{f}(\gamma)|\hat{\mu}(\gamma)|^{Nz}.$$

for each $\gamma\in\hat{K}$. Clearly, $\sum_{\gamma\in\hat{K}} k_\gamma|\hat{f}(\gamma)|^2(|\hat{\mu}(\gamma)|^{Nz})^2 < \infty$ since each $|\hat{\mu}(\gamma)|$ is bounded [18, 3.2] so that $T^z f\in L^2$. Using [18, 3.4] we have for simple functions f and g

$$\int_K (T^z f)g\, dm = \sum_{\gamma\in\hat{K}} k_\gamma \hat{f}(\gamma)\hat{g}(\gamma)|\hat{\mu}(\gamma)|^{Nz}$$

so by the Cauchy-Schwartz inequality the map $z \to \int_K (T^z f) g \, dm$ is analytic for $\mathrm{Re}(z) > 0$. If $\mathrm{Re}(z) = 1$, [18, 3.4] implies

$$\|T^z f\|_2^{\,2} \le \|\mu^N * f\|_2^{\,2} \le (A')^2 \|f\|_{p'}^{\,2} .$$

Thus if $\mathrm{Re}(z) = 1$, T^z is bounded from $L^{p'}$ to L^2 with constant A' independent of $\mathrm{Im}(z)$. If $\mathrm{Re}(z) = 0$ then

$$\|T^z f\|_2^{\,2} \le \sum_{\gamma \in \hat{K}} k_\gamma |\hat{f}(\gamma)|^2 = \|f\|_2$$

so the result now follows from Stein's Theorem by imitating the remaining argument in Ritter [14, Lemma 2].

The following theorem generalizes [5, 1.1] but the proof in the hypergroup case is slightly more delicate because the convolution of point masses is only a probability measure and not necessarily another point mass.

<u>Theorem 1.5</u> Let K be a compact abelian hypergroup. Let $\mu \in M^+(K)$ be L^p-improving and let $\alpha = \sup\{qp^{-1} : \mu * L^p \subseteq L^r,\ p>1\}$. Let β be the index conjugate to α. If $r > \beta$ and $g \in L^r(\mu)$ then $\rho = g\mu$ is L^p-improving.

Proof. The proof here closely follows the proof in [5]. Select s so that $s^{-1} + r^{-1} = 1$ which implies $s < \alpha$. There exist $1 < p, q$ and $A > 0$ such that $ps < q$ and $\|\mu * f\|_q \le A\|f\|_p$ for all $f \in L^p$. For $f \in C(K)$ we apply Hölder's inequality to obtain

$$\|\rho * f\|_q^{\,q} \le \int_K \{[\int_K |f(\check{y} * x)|^s d\mu(y)]^{1/s} [\int_K |g(y)|^r d\mu(y)]^{1/r}\}^q dm(x)$$
$$\le \|g\|^q_{L^2(\mu)} \int_K [\int_K (|f|(\check{y} * x))^s d\mu(y)]^{q/s} dm(x).$$

Since K is compact and $\delta_{\check{y}} * \delta_x$ is a probability measure another application of Hölder's inequality implies $(|f|(\check{y} * x))^s \le |f|^s(\check{y} * x)$ so that

$$\|\rho * f\|_q^{\,q} \le \|g\|^q_{L^r(\mu)} \int_K [\int_K |f|^s(\check{y} * x) d\mu(y)]^{q/s} dm(x)$$
$$= \|g\|^q_{L^r(\mu)} \|\,|f|^s * \mu\|^{q/s}_{q/s} \le \|g\|^q_{L^r(\mu)} \|\,|f|^s * \mu\|_q^{q/s}$$
$$\le \|g\|^q_{L^r(\mu)} A^{q/s} \|f\|^q_{ps} .$$

Thus,

$$\|\rho * f\|_q \le \|g\|_{L^r(\mu)} A^{1/s} \|f\|_{ps}$$

and hence $\rho * f$ is L^p-improving since $ps < q$.

<u>Corollary 1.6</u> Let K be a compact abelian hypergroup. Let $1<p<r<\infty$ and suppose $\mu * L^p \subseteq L^r$. If f is any bounded Borel function on K then $f\mu * L^p \subseteq L^r$.

Proof. Let $r = \infty$.

The next corollary is Young's Inequality for compact abelian hypergroups.

Corollary 1.7 Let K be a compact abelian hypergroup with $1 \leqslant p < \infty$, $1 \leqslant q < \infty$, $r^{-1} = p^{-1} + q^{-1} - 1 \geqslant 0$. Then $f*g \in L^r$ and $\|f*g\|_r \leq \|f\|_p \|g\|_q$ whenever $f \in L^p$ and $g \in L^q$.

Proof. Let $\mu = m$ which convolves L^1 to L^∞ and gives $\alpha = \infty$.

Corollary 1.8 If K is a compact abelian hypergroup then each L^q function $(q > 1)$ is L^p-improving.

Graham, Hare and Ritter go on to prove that for every nonzero L^p-improving measure μ on a compact abelian group there exists a measure $\nu, \nu \ll \mu$ such that ν is not L^p-improving. It is easily seen that this result is false in general for hypergroups, simply let μ be any L^p-improving point mass.

2. L^p-Improving Measures and Lacunarity

In this section we begin a discussion of L^p-improving measures and their relationship to $\Lambda(p)$ sets. We first have a lemma.

Lemma 2.1 Let K be a compact hypergroup with $P \subseteq \hat{K}$ and $1 < q < \infty$. Then P is a $\Lambda(q)$ set if and only if $\mu \in M_p(K)$ (i.e., $\mu \in M(K)$ and $\hat{\mu}(\psi) = 0$ for all ψ not in P) implies $\mu = fm$ for some $f \in L^q$.

Proof. This follows from a standard argument (see e.g. [10, 5.3]) since L^1 has an approximate unit $\{h_\alpha\}$ consisting of trigonometric polynomials satisfying $\|h_\alpha\|_1 = 1$ ([18, 2.12]) for all α.

Theorem 2.2 If K is a compact abelian hypergroup and $P \subseteq \hat{K}$ is a $\Lambda(q)$ set for some $1 < q < \infty$ then each $\mu \in M_P(K)$ is L^p-improving.

Proof. Follows immediately from Lemma 2.1 and Corollary 1.8.

Example 2.3 Let $K_a = \{0,1,2...,\infty\}$ be the countable, compact abelian hypergroups introduced by Dunkle and Ramirez [3] where $0 < a \leqslant 1/2$. In this case δ_∞ is the point mass at the identity. If $k \in \{0,1,2,...\}$ then we will show that δ_k is L^p-improving. We have $\hat{K}_a = \{\psi_n : n=0,1,2...\}$ and it easily follows that

$$\hat{\delta}_k(\psi_n) = \begin{cases} 0 & \text{if } n > k+1 \\ \frac{a}{a-1} & \text{if } n = k+1 \\ 1 & \text{if } n \leqslant k \end{cases}$$

Thus $\operatorname{supp}(\hat{\delta}_k) = \{\psi_0, ..., \psi_{k+1}\}\}$ is finite and hence a $\Lambda(q)$ set for all $1 < q < \infty$. Hence Theorem 2.2 shows that δ_k is L^p-improving for all

$1 < q < \infty$. Hence Theorem 2.2 shows that δ_k is L^p-improving for all $k \in \{0,1,2,...\}$.

Given a measure μ in $M(K)$ define the sets

$$E(\varepsilon) = \{\gamma \in \hat{K} : |\hat{\mu}(\gamma)| \geq \varepsilon\}$$

for $\varepsilon > 0$. These sets were defined by Hare [6] and used to provide a nice characterization of L^p-improving measures. The next theorem is simply a restatement of Hare's main result [6] in the context of compact abelian hypergroups.

Theorem 2.4 Let K be a compact abelian hypergroup with $\mu \in M(K)$ with $\|\mu\| \leq 1$. The following are equivalent:

(i) μ is L^p-improving

(ii) There exists $p > 2$, $\alpha \geqslant 1$ and a constant c such that for all $\varepsilon > 0$, $E(\varepsilon)$ is a $\Lambda(p)$ set with $\Lambda(p, E(\varepsilon)) \leqslant c\varepsilon^{-\alpha}$.

(iii) There exist constants c_1 and c_2 such that for all $\varepsilon > 0$ and for all $2 < q < \infty$, $E(\varepsilon)$ is a $\Lambda(q)$ set with $\Lambda(q, E(\varepsilon)) \leqslant \frac{c_1}{\varepsilon} q^{-c_2 \log \varepsilon}$.

Proof. Imitate the proof in [6] using basic facts about harmonic analysis on compact hypergroups found in [9] and [18] as well as Proposition 1.4.

The next corollary provides sufficient conditions for a measure to be L^p-improving and its analogue for compact abelian groups can be found in [6].

Corollary 2.5 If $\mu \in M(K)$ with $\sum_{\gamma \in \hat{K}} |\hat{\mu}(\gamma)|^r < \infty$ for some $r < \infty$ then μ is L^p-improving.

We complete this section with a second corollary which provides for the existence of infinite Λ sets if certain L^p-improving measures can be found.

Corollary 2.6 If $\mu \in M(K)$ is L^p-improving and $\hat{\mu} \notin c_0(\hat{K})$ then $\hat{K}$ contains an infinite Λ set.

Proof. Since μ is not in $c_0(\hat{K})$ select η so that $|E(\eta)|$ is infinite. Simply choose $\varepsilon = \min\{\eta, \exp(-1/2C_2)\}$ and appeal to (iii) of Theorem 2.4.

There are examples of compact abelian hypergroups with no infinite Λ sets so that any L^p-improving measure on such a hypergroup must vanish at infinity. We will discuss these ideas further in §3.

3. L^p-Improving Measures on Compact Groups

Hare [7,1.1] extends her main theorem in [6] to compact (nonabelian) groups. In this section we also discuss L^p-improving measures in the general context of compact groups. Let G be a compact (not necessarily abelian) group with corresponding compact abelian conjugary class hypergroup G_I (see [9, 8.4]). Let μ be a central measure on G, written $\mu \in ZM(G)$. Then μ is called <u>central L^p-improving</u> if there exist $1 < p < r < \infty$ such that $\mu * ZL^p \subseteq ZL^r$ (where ZL^q denotes all the central functions in L^q).

<u>Proposition 3.1</u> Let $\mu \in ZM(G)$. Then μ is central L^p-improving if and only if μ is L^p-improving on G_I.

Proof. Follows easily from [17, 3.1].

The next example will illustrate some of these ideas in the special case where $G = SU(2)$.

<u>Example 3.2</u> Let $G = SU(2)$ and note that $G_I = [0,\pi]$ and that $\hat{G}_I = \{\psi_n : n = 1,2,\ldots\}$ where

$$\psi_n(\theta) = \begin{cases} 1 & \text{if } \theta = 0 \\ (-1)^n & \text{if } \theta = \pi \\ \dfrac{\sin(n\theta)}{n \sin \theta} & \text{if } \theta\ (0,\pi) \end{cases}$$

This can be found in [4, 7.3]. Furthermore, $\hat{G}_I$ is a discrete hypergroup under pointwise multiplication of characters with a Haar measure $\hat{m}$ given by $\hat{m}(\psi_n) = n^2$. Note that if $\theta \in (0,\pi)$ then

$$|\psi_n(\theta)| \leq \frac{1}{n|\sin\theta|} .$$

In fact, if μ is any measure with supp $\mu \subseteq G_I \setminus Z$ (where Z is $ZG_I = \{0,\pi\}$), say supp $(\mu) \subseteq [a,b]$ where $0 < a < b < \pi$, then

$$|\hat{\mu}(\psi_n)| = |\int_{G_I} \psi_n(\check{\theta}) d\mu(\theta)| \leq \int_{G_I} \frac{1}{n|\sin\theta|} d|\mu|(\theta) .$$

Letting $t_\mu = \min\{a, \pi - b\}$, we have $|\hat{\mu}(\psi_n)| \leq \frac{1}{nt_\mu} |\mu|(G_I)$. In particular, if μ,ν are supported off $\{0,\pi\}$ then

$$|(\mu*\nu)\hat{}(\psi_n)| = |\hat{\mu}(\psi_n)||\hat{\nu}(\psi_n)| \leq \frac{1}{n^2 t_\mu t_\nu} |\mu|(G_I)|\nu|(G_I) .$$

Thus

$$\sum_{n=1}^{\infty} |(\mu*\nu)\hat{}(\psi_n)|^2 \hat{m}(\psi_n) \leq \sum_{n=1}^{\infty} \frac{[|\mu|(G_I)|\nu|(G_I)]}{n^4 t_\mu t_\nu} n^2 < \infty.$$

We now have the following theorem.

<u>Theorem 3.3</u> If $G = SU(2)$ then for all $\mu,\nu \in M(G_I)$ with supp(μ), supp(ν) contained in $G_I \setminus ZG_I$ there exists $f \in L^2(G_I)$ such that $\mu*\nu = f\sigma$ where σ is normalized Haar measure on G_I.

Proof. We have shown that $(\mu*\nu)\hat{} \in l^2(\hat{G}_I)$ so the result follows immediately from [18,3.4].

Corollary 3.4 If μ,ν are continuous central measures on $G = SU(2)$ then there exists $f \in ZL^2(G)$ such that $\mu * \nu = f d_{m_G}$.

Proof. We need only note that a measure is continuous and central on SU(2) if and only if it is central and supported off the center of SU(2) [11,2.1].

Corollary 3.5 Every continuous central measure on SU(2) is central L^p-improving and in fact L^p-improving

Proof. If μ is any continuous central measure on $G = SU(2)$ then $\mu^2 = \mu * \mu$ is L^p-improving on G_I by Corollary 1.8 and hence μ is L^p-improving on G_I by Proposition 1.4. Thus μ is central L^p-improving on SU(2) by Proposition 3.1. However, $\mu^* * \mu$ can also be viewed as an L^2 function on SU(2) so by Young's inequality for compact groups $\mu^* * \mu$ is L^p-improving on SU(2) . Using [7, Lemma 4] this implies μ is L^p-improving on G .

Note that if μ is any continuous central measure on SU(2) , using the fact that $L^2 * L^2 = A(SU(2)) \subseteq L^p(SU(2))$ for all $p > 1$, we have μ^4 is in L^p for all $p \geqslant 1$. In particular, if $a \in SU(2) \setminus Z$ then the continuous central measure μ^a defined by the equation

$$\int_{SU(2)} f d\mu^a = \int_{SU(2)} f(gag^{-1}) dm(g) \ , \ f \in C(SU(2))$$

is a singular measure [11, 2.3] which is L^p-improving. This measure corresponds to the point mass at the conjugacy class containing a on the hypergroup $SU(2)_I$. This provides another example of a hypergroup where all the point masses off the maximal subgroup are L^p-improving.

Generalizations of this example seem to hold for SU(n) and most likely for any compact connected simple Lie group. Details will appear in a subsequent paper. We note here that if G is any compact connected semisimple Lie group then G has no infinite central Sidon sets [12, Theorem 9] so that G has no infinite Sidon sets which implies G_I has no infinite Λ sets [19, 5.1]. This implies any central L^p-improving measure on such groups G must have a Fourier-Stieltjes transform which vanishes at infinity by Corollary 2.6.

References

1. Bonami, A.: Etude des coefficients de Fourier des fonctions de $L^p(G)$. Ann. Inst. Fourier, Grenoble 20,335-402 (1970).

2. Christ, M.: A convolution inequality concerning Cantor-Lebesgue measures. Revista Matemática Iberoamericana 1,79-83 (1985).

3. Dunkl, C.F., Ramirez, D.E.: A family of countable compact P_*-hypergroups. Trans. Amer. Math.Soc. 202, 339-356(1975).

4. Fournier, J.J.F., Ross, K.A.: Random Fourier series on compact abelian hypergroups. J. Austral. Math Soc. (Series A) 37, 45-81 (1984).

5. Graham, C.C., Hare, K.E., Ritter, D.L.: The size of L^p-improving measures. J. Func. Anal., to appear.

6. Hare, K.: A characterization of L^p-improving measures. Proc. Amer. Math. Soc. **131**, 143-155 (1988).

7. Hare, K.: L^p-improving measures on compact non-abelian groups. Preprint.

8. Hewitt, E., Ross, K.: Abstract Harmonic Analysis II. Berlin-Heidelberg-New York: Springer 1970.

9. Jewett, R.I.: Spaces with an abstract convolution of measures. Advances in Math **18**, 1-101 (1975).

10. Lopez, J., Ross, K.: Sidon Sets. Lecture Notes in Pure and Applied Mathematics **13**. New York: Marcel Dekker 1975.

11. Ragozin, D.L.: Central measures on compact simple Lie groups. J. Func. Anal. **10**, 212-229 (1972).

12. Rider, D.: Central lacunary sets. Monatsh. Math. **76**, 328-338 (1972).

13. Ritter, D.: Some singular measures on the circle which improve L^p spaces. Colloquium Math. **52**, 133-144 (1987).

14. Ritter, D.: Most Riesz product measures are L^p-improving. Proc. Amer. Math Soc. **97**, 291-295 (1986).

15. Stein, E.M.: Harmonic analysis on $\mathbb{R}^n$. In: Studies in Harmonic Analysis. Studies in Mathematics 13, pp 97-135. Washington, D.C.: Mathematical Association of America 1976.

16. Stein, E.M., Weiss, G.: Introduction to Fourier Analysis on Euclidean Spaces. Princeton: Princeton University Press 1971.

17. Vrem. R.C.: Lacunarity on compact hypergroups. Math. Z. **164**, 93-104 (1978).

18. Vrem, R.C.: Harmonic analysis on compact hypergroups. Pac. J. Math. **85**, 239-251 (1979).

19. Vrem, R.C.: Independent sets and lacunarity for hypergroups, Preprint.

Statistics of Rotations

Geoffrey S. Watson ,
Princeton University .

1 Introduction

Because of the Gaussian distribution , statisticians have always been interested in orthogonal transformations . In this connection James (1964) introduced us to the orthogonal , Stiefel and Grassmann manifolds . However non-uniform distributions on these manifolds came later . In this paper , distributions on O(q) and S(q) arise explicitly only in the last section on Bayesian methods . Our "data" will be a set of points on the unit sphere Ω_q in R^q which will be orthogonally transformed and observed (with error) again . The estimated transformation will be a random variable on O(q) . We will concentrate mainly on the rotation case .

MacKenzie (1957) first raised this problem in R^3 motivated by a problem in crystallography . Each of n known unit vectors $\mathbf{u}_1 , \ldots , \mathbf{u}_n$ in R^3 is given the same rotation **A** yielding vectors $\mathbf{v}_1 , \ldots , \mathbf{v}_n$. The rotation **A** must be estimated when the $\mathbf{v}_i=\mathbf{A}\mathbf{u}_i$'s are measured by $\mathbf{y}_i$'s where the (independent) $\mathbf{y}_i$'s have Fisher distributions with modal vectors $\mathbf{v}_i$ and common concentration parameter κ . This means that the probability density of $\mathbf{y}_i$ on the 3-sphere Ω_3 is given by the formula

$$(\kappa/4\pi \sinh \kappa) \exp \kappa \, \mathbf{v}_i^* \mathbf{y}_i , \qquad (1.1)$$

which we write briefly as $\mathbf{y}_i \sim F_3 (\kappa ,\mathbf{v}_i)$. Here * denotes the transpose . MacKenzie used the singular value decomposition (S.V.D.) of the 3x3 matrix $\mathbf{M}_n=n^{-1}\Sigma \mathbf{u}_i \mathbf{y}_i^* = \mathbf{L}\Lambda\mathbf{R}$ and gave the estimator $\hat{\mathbf{A}} = \mathbf{R}\mathbf{L}^*$ which is appropriate when det $\mathbf{M}_n$ is postive . We remark that the polar decomposition of $\mathbf{M}_n$ could have been used . MacKenzie actually approached this problem as a least squares problem because he did not think Fisher errors were a particularly obvious assumption . MacKenzie(1962) formulated another crystallographic problem involving rotations and used least squares again but this time he used the exponential form for a rotation to perform his calculations in much the same way that Chang (1986) used it theoretically . His 1957 paper was taken further , with the same methods by Stephens (1979) , when trying to formulate the idea of correlation between unit vectors ; he considered the case where det $\mathbf{M}_n$ might be negative . Moran (1976) re-solved the estimation problem using the quaternion representation of the rotation group and was motivated by the rotation of tectonic plates . This approach was recently completed (e.g. provided with confidence sets for the unknown rotation) and applied by Thompson &

Prentice (1988) to the same application . Chang (1986, 1987a , 1987b) with the same motivation completed Mackenzie's paper in greater generality and with many computations with real data (e.g. he used a general density $g(\boldsymbol{\mu}^*\mathbf{y})$ in any number q of dimensions , looked the asymptotics of n large and/or κ large ,gave tests of various hypotheses about **A** , found various confidence sets etc.). He also considered the case where one knows $\mathbf{x}_i$ not $\mathbf{u}_i$ where $\mathbf{x}_i$ is $\mathbf{u}_i$ measured subject to error- though unfortunately only when the error distributions of the $\mathbf{x}_i$'s and $\mathbf{y}_i$'s are the same . Rivest (1988) has recently been completing a paper on large κ asymptotics . Finally Chang & Bingham (1988) are completing a manuscript which uses the quaternion representation and introduces a Bingham distribution on Ω_4 as a prior on **A** . The present writer corresponded with his classmate MacKenzie in the early 60's and pointed out how one could take his work further but , having no specific applications in mind , dropped the topic until not long before the invitation to give this lecture came . Having worked out much of what follows , Prof Chang sent him all the circulating manuscripts ! Thus what was to have been research project has become largely , though not entirely , a review . We hope that , in our account , the main points of this new literature are derived more simply for statistical readers and seen in better historical perspective .

The discussion is self-contained except the reader is assumed to have some acquaintance with statistics on Ω_q - see e.g. Watson , (1983)- which we now sketch .

Thus we will need the generalization of (1.1) , variously called the Von-Mises -Fisher or Langevin distribution ,

$$a_q(\kappa)^{-1} \exp \kappa\boldsymbol{\mu}^*\mathbf{x} , \quad \kappa>0 , \quad \mathbf{x} , \boldsymbol{\mu} \in \Omega_q , \tag{1.2}$$

where the normalizing function is given by

$$a_q(\kappa) = (2\pi)^{q/2}\, I_{(q/2)-1}(\kappa)\, \kappa^{-(q/2)+1} . \tag{1.3}$$

We will write $\mathbf{x} \sim F_q(\kappa,\boldsymbol{\mu})$ and define the function

$$A_q(\kappa) = (d/d\kappa) \log a_q(\kappa) \approx 1-(q-1)/2\kappa , \text{ as } \kappa\to\infty . \tag{1.4}$$

Then

$$\left.\begin{aligned} E\mathbf{x} &= A_q(\kappa)\, \boldsymbol{\mu} , \\ \boldsymbol{\Sigma} &= \text{covariance matrix of } \mathbf{x} = E\,(\mathbf{x} - E\mathbf{x})(\mathbf{x}- E\mathbf{x})^* \\ &= A'_q \boldsymbol{\mu}\boldsymbol{\mu}^* + A_q(I - \boldsymbol{\mu}\boldsymbol{\mu}^*)/\kappa . \end{aligned}\right\} \tag{1.5}$$

Thus if the "data " $\mathbf{x}_1,\mathbf{x}_2, \ldots, \mathbf{x}_n$ are I.I.D. $F_q(\kappa,\boldsymbol{\mu})$, $\Sigma\mathbf{x}_i$ is the sufficient statistic , and the Central Limit Theorem tells us that $n^{-1/2} (\Sigma\mathbf{x}_i - nA_q(\kappa)\boldsymbol{\mu})$ is the Gaussian distribution with mean vector **0** and covariance matrix $\boldsymbol{\Sigma}$, or briefly $G_q(\mathbf{0},\boldsymbol{\Sigma})$. The likelihood of the data (i.e. the probability density evaluated at the observations) is , writing $X = \Sigma x_i$ and taking logarithms , $-n \log a_q(\kappa) + \kappa\mu^*X$. Thus the maximum likelihood (m.l.) estimates of κ and μ are

$$\hat{\mu} \parallel X \ , \ \hat{\kappa} \text{ the solution of } A_q(\hat{\kappa}) = \|X\|/n \ . \qquad (1.6)$$

However if μ is known , the m.l. κ' estimator of κ satisfies

$$A_q(\kappa') = \mu^*X /n \ . \qquad (1.7)$$

When κ is large , there are useful some approximations ; what is meant by large depends on q - see Watson (1987a) . If we write $x = x_{\parallel\mu} + x_{\perp\mu} = t\,\mu + (1-t^2)^{-1/2}\,\xi$, where $\xi \perp \mu$, $\|\xi\| = 1$, then as $\kappa \to \infty$,

$$2\kappa(1-t) \to U \equiv \chi^2_{q-1} \ ,$$
$$\kappa^{1/2}x_{\perp\mu} \to U^{1/2}\,\xi \ .$$

Thus

$$2\kappa(n-\mu^*X) \to \chi^2_{n(q-1)} \ , \ \kappa^{1/2}X_{\perp\mu}/n \to Z \sim G_q(0 \ , I-\mu\mu^*) \ . \qquad (1.8)$$

Some trivial consequences are : as $\kappa \to \infty$,

$$\mu^*X \to n \ , \ \|X_{\perp\mu}\| \to 0 \ , \ \|X\| \to n \ . \qquad (1.9)$$

These results lie behind some of the theory used below .

Many variants of MacKenzie's problem could arise . Firstly one could have $y_i \sim F(\kappa_i,\mu)$ i.e. variable accuracies . Unless there are repeated observations , this means we must assume that the κ_i are known . Secondly the u_i could be observed with error as mentioned earlier . Thirdly it might happen that u_i and y_i are axial or bivectors and that the probability density of y_i is

$$b_q(\kappa)^{-1} \exp \kappa\,(y_i^*Au_i)^2 \ , \qquad (1.10)$$

a density considered in Watson (1983) . In this case there is not an explicit formula for the estimate of A . Finally there are generalizations of the problem e.g. take , instead of unit vectors y_i , points Y_i on a Stiefel manifold where Y is qxm , $m \leq q$, $Y^*Y = I_m$. We will consider this elsewhere . Also we don't mention other applications which require the estimation of rotations e.g. in structural Geology . In Section 2 we will re-examine Mackenzie's problem . We will discuss the cases where n , the sample size is large and where n is not large but κ is i.e. the observational errors on the vectors are small. In Section 3 we will comment on the problem . In Section 4 we will briefly consider distributions on O(q) and the estimation of A by Bayesian methods .

2 Re-examination of the problem

Generalizing the MacKenzie problem to Ω_q , the log-likelihood of the data is

$$-n \log a_q(\kappa) + \kappa\,\Sigma\, y_i^*Au_i \ , \qquad (2.1)$$

so that the maximum likelihood (m.l.) estimator of κ , $\hat{\kappa}$, satisfies

$$A_q(\hat{\kappa}) = \Sigma\, \mathbf{y}_i{}^*\mathbf{A}\mathbf{u}_i\, /n \ . \qquad (2.2)$$

If κ is very large , $A_q(\kappa) \approx 1-(q-1)/2\kappa$ so

$$\hat{\kappa} = (q-1)\, n \,/\, 2(n- \Sigma \mathbf{y}_i{}^*\mathbf{A}\mathbf{u}_i\ ) \ . \qquad (2.3)$$

To find the orthogonal $\hat{\mathbf{A}}$ which maximizes $n^{-1}\Sigma \mathbf{y}_i{}^*\mathbf{A}\mathbf{u}_i = \mathrm{Tr}\ \mathbf{A}\mathbf{M}_n$, where $\mathbf{M}_n = n^{-1}\Sigma \mathbf{u}_i\mathbf{y}_i{}^*$, we use the S.V.D. of $\mathbf{M}_n$, namely

$$\mathbf{M}_n = \mathbf{L}\Lambda\mathbf{R}^* = \Sigma\, \lambda_j \mathbf{l}_j \mathbf{r}_j{}^* \ , \qquad (2.4)$$

the λ_j being non-negative and ordered with λ_q the least .We note that $\mathbf{M}_n$ is a sufficient statistic i.e. it wraps up all the information in the data about both κ and $\mathbf{A}$.

If det $\mathbf{M}_n$ is positive , it follows that det $\mathbf{R}$ det $\mathbf{L} = +1$ since Λ has positive elements . Hence $\mathbf{R}\mathbf{L}^*$ belongs to SO(q) (i.e. has determinant +1) and Tr $\mathbf{A}\mathbf{M}_n = \Sigma\lambda_j$. However

$$\mathrm{Tr}\mathbf{A}\mathbf{M}_n = \Sigma\, \lambda_j\, \mathbf{r}_j{}^*\mathbf{A}\mathbf{l}_j \leq \Sigma\lambda_j \ ,$$

since the multipliers are cosines of angles . Thus the choice

$$\hat{\mathbf{A}} = \mathbf{R}\mathbf{L}^* \qquad (2.5)$$

maximizes Tr $\mathbf{A}\mathbf{M}_n$ over all $\mathbf{A}$ such that det $\mathbf{A} = +1$. Now when κ is large we will expect det $\mathbf{M}_n$ to have the same sign as the unknown $\mathbf{A}$ since $\mathbf{M}_n = n^{-1}\Sigma\, \mathbf{u}_i\mathbf{y}_i{}^* \to n^{-1}\Sigma \mathbf{u}_i(\mathbf{A}\mathbf{u}_i)^* = n^{-1}\Sigma \mathbf{u}_i\mathbf{u}_i{}^*\mathbf{A}^*$ where $n^{-1}\ \Sigma \mathbf{u}_i\mathbf{u}_i{}^*$, which we (following Chang) will call Σ_n , is non-negative . For fixed κ but increasing n , suppose $\mathbf{\Sigma}_n \to \mathbf{\Sigma}$. Then $\mathbf{M}_n \to A_q(\kappa)\ \mathbf{\Sigma}\mathbf{A}^*$ so we may again expect that det $\mathbf{A}$ and det $\mathbf{M}_n$ will have the same sign . If neither κ or n is large , sampling fluctuations can lead to det$\mathbf{M}_n$ negative so (2.5) will not be an admissible solution if one is restricted to rotations . Following Stephen's idea , we for this case may construct the s.v.d. of $\mathbf{M}_n$ so that $\Lambda = \mathrm{diag}\,(\lambda_1,\ldots,\lambda_{q-1},-\lambda_q)$. Then (2.5) is correct and the maximum of Tr$\mathbf{A}\mathbf{M}_n = \lambda_1+\ldots+\lambda_{q-1}-\lambda_q$. Of course is we are not confined to SO(q) , the first solution should always be used .

Had we used the polar decomposition $\mathbf{M}_n = \mathbf{U}\mathbf{P}$, where $\mathbf{U}$ is orthogonal and $\mathbf{P}$ is symmetric (the unique positive square root of $\mathbf{M}_n\mathbf{M}_n{}^*$) , we would have found that $\hat{\mathbf{A}} = \mathbf{U}^*$.

Had we begun by seeking the least squares estimator of $\mathbf{A}$, we would have found it to minimize

$$\Sigma\ \| \mathbf{y}_i - \mathbf{A}\mathbf{u}_i \|^2 \ .$$

So if the Euclidean norm is used , this leads us back to the above m.l. estimator .There is clearly no limit to the number of estimators that can be suggested . M.l. estimation of $\mathbf{A}$ only has obvious merits if one has assumed the right error distribution , and usually only when the sample size is large since the theorems showing that m.l. estimation has merits assume $n \to \infty$. M.l estimators can be non-robust - i.e. sensitive to gross errors in the data . Clearly one would like an estimator that is robust , unbiased and with the smallest possible sampling error . In Section 3 we will discuss this issue .

When q=3 it is natural , when we are told that **A** is a rotation , to estimate the axis of this rotation $\hat{\mathbf{a}}$ using $\hat{\mathbf{A}}\hat{\mathbf{a}} = \hat{\mathbf{a}}$ and $\hat{\omega}$, the estimated angle of rotation about $\hat{\mathbf{a}}$ induced by $\hat{\mathbf{A}}$, from $\mathrm{Tr}\, \hat{\mathbf{A}} = 1 + 2\cos\hat{\omega}$. These equations show how the estimation error in $\hat{\mathbf{A}}$ is propagated into $\hat{\mathbf{a}}$ and $\hat{\omega}$. As $\kappa \to \infty$, $\hat{\mathbf{A}} \to \mathbf{A}$ so $\hat{\mathbf{a}} \to \mathbf{a}$ and $\hat{\omega} \to \omega$. However we can see that , if ω is very small , there will be problems . If $\hat{\omega}$ is close to zero then $\hat{\mathbf{A}}$ is close to the identity and $\hat{\mathbf{a}}$ will be poorly determined . The direct solution of the equation for $\hat{\omega}$ will give trouble i.e. lead to a bias in the angle . This is an inherent difficulty in the problem and appears whatever estimate of **A** is used .

We will for ignore these facts and continue to discuss the problem as first formulated .

For large samples , we have the usual m.l. estimation and likelihood ratio test (which also is only shown to have merit as $n \to \infty$) technologies . The former provides an automatic method for finding the asymptotic distributions of estimators , the latter a method for deriving tests and the asymptotic distributions of test statistics . (Chang derives methods from first principles because he considers a more general distribution of errors with density g ($\boldsymbol{\mu}^*\mathbf{x}$) , as is done , e.g. in Watson (1983) , for other problems , when n is large .) For example , we have the consistency of $\hat{\mathbf{A}}$ & $\hat{\kappa}$ (i.e. as $n \to \infty$, $\hat{\mathbf{A}} \to \mathbf{A}$, $\hat{\kappa} \to \kappa$), and further , Wilks' Theorem tell us , after discarding some terms which are asymptotically small , that

$$2\hat{\kappa} n\, (\mathrm{Tr}\hat{\mathbf{A}}\mathbf{M}_n - \mathrm{Tr}\mathbf{A}\mathbf{M}_n) = 2\hat{\kappa}\, (\Sigma \mathbf{y}_i{}^*\hat{\mathbf{A}}\mathbf{u}_i - \Sigma\, \mathbf{y}_i{}^*\mathbf{A}\mathbf{u}_i) \sim \chi_Q{}^2 \quad , (2.6)$$

$$Q = q(q-1)/2 ,$$

Q being the number of parameters required to specify **A** . We are here testing that each of these Q parameters has a prescribed value . (2.6) can be used to test the hypothesis that **A** is the true orthogonal matrix and to provide a confidence set for an unknown **A** . For the latter purpose it needs modification to be useful -a topic we return to below .

If q=3 and one wishes to test that **A** has a prescribed axis $\mathbf{a}_0$,say , one has merely to re-estimate **A** subject to this restriction . In the general case Chang (1986) suggested testing whether **A** belongs to a subgroup of O(q) , G say . If membership of G puts r restrictions on the Q parameters in **A** , and one returns to (2.1) and re-estimates **A** subject to belonging to G , getting the estimate $\hat{\mathbf{A}}_G$, say , then Wilks' Theorem says that

$$2\hat{\kappa}\,(\,\mathrm{Tr}\hat{A}M_n - \mathrm{Tr}\hat{A}_G M_n\,) \sim \chi^2{}_r\,, \qquad (2.7)$$

since we are really testing that these r restrictions are obeyed. It seems that it is the dimensionality of the manifold rather than the fact that G is a sub-group that matters here .

The results (2.6), (2.7) hold for large n but results are needed for fixed n and large κ . In a series of papers we have discussed the Fisher distribution for large κ . In particular in Watson (1956,1960) and Watson & Williams (1956) we evolved an "analysis of dispersion " , by analogy with analysis of variance , which provided , very simply , procedures for the large κ situation . It seems worthwhile to repeat the original argument before applying it here .

Suppose , temporally for this paragraph only , that the y_i's have independently the density (1.2) .Then $\hat{\mu} = \Sigma y_i \,/\|\Sigma y_i\|$ and since κ is large

$$\text{m.l est. of } \kappa \text{ if } \mu \text{ known} =(q-1)\, n\,/2\, (n-\mu^*\Sigma y_i)\,,$$

$$\text{m.l. est. of } \kappa \text{ if } \mu \text{ is est'd} =(q-1)\, n\,/2\, (n-\|\Sigma y_i\|\,. \qquad (2.8)$$

Since κ is a concentration parameter , its reciprocal is a measure of dispersion . It is therefore natural to write (possibly with a multiplier $2(q-1)^{-1}$) the identity

$$n - \mu^*\Sigma y_i \;=\; (\,n - \|\Sigma\, y_i\|\,) + (\,\|\Sigma y_i\| - \mu^*\Sigma y_i) \qquad (2.9)$$

and to interpret it as

sample dispersion about μ = sample dispersion about $\hat{\mu}$ + dispersion of $\hat{\mu}$ about μ

It follows from (1.8) that

$$2\kappa(n-\mu^*\Sigma y_i) = 2\kappa(n-\|\Sigma y_i\|) \;+\; 2\kappa(\|\Sigma y_i\| -\mu^*\Sigma y_i) \qquad (2.10)$$

is distributionally , as $\kappa\to\infty$,

$$\chi^2{}_{(q-1)n} \;=\; \chi^2{}_{(q-1)(n-1)} \;+\; \chi^2{}_{(q-1)}\,. \qquad (2.11)$$

Intuitively (q-1)n comes from n observations each with q-1 degrees of freedom .The q-1 d.f. comes from the q-1 d.f. in μ and the (q-1)(n-1) from fitting q-1 parameters to a dispersion with (q-1)n d.f. so ending up with (q-1)n -(q-1) d.f. . Thus

$$\frac{(n-1)\;(\;\|\Sigma y_i\|-\Sigma y_i)}{(\;n\;-\;\|\Sigma y_i\;\|} \;=\; F_{(q-1),(q-1)(n-1)}\;, \qquad (2.12)$$

as in Watson & Williams (1956) , where the r.h.s of (2.12) is a random variable with the well-known F-distribution with these particular d.f. .

Applying the same method here (reverting of course to our original notation) , we get from (2.3) and its version with A known , in place of (2.10) ,

$$2\kappa(\,n- \Sigma y_i{}^*Au_i) = 2\kappa(n-\Sigma y_i\hat{A}u_i) + 2\kappa(\Sigma y_i\hat{A}u_i \;-\Sigma y_i{}^*Au_i)\,, \quad (2.13)$$

with its distributional companion , obtained by the same logic ,

$$\chi^2_{(q-1)n} = \chi^2_{(q-1)n-Q} + \chi^2_Q \, . \qquad (2.14)$$

Hence

$$\frac{(q-1)n-Q}{Q} \; \frac{\Sigma y_i^* \hat{A} u_i - \Sigma y_i^* A u_i}{n - \Sigma y_i^* \hat{A} \hat{u}_i} = F_{Q,(q-1)n-Q} \, . \qquad (2.15)$$

This is a result proved by Chang (1987b) & Rivest (1988) ; the latter paper derives the asymptotic power function . I have not proved a "master theorem" that says my analysis always gives the right answer because I don't know how to frame it to cover every case where it can apparently be used ! We apply it again for q=3 and rotations to get a test for a given axis **a** , a_0. One gets immediately

$$\frac{(2n-3)}{2} \; \frac{\Sigma y_i^* \hat{A} u_i - \Sigma y_i^* \hat{A}_0 u_i}{n - \Sigma y_i^* \hat{A} u_i} = F_{2,2n-3} \, , \qquad (2.16)$$

where $\hat{A}_0$ is the m.l. estimate of **A** , given it has axis **a** . Again this has been proved by Chang (1987b) so the reader should share my confidence .

Suppose a sample of n(i) is summarized by the sufficient statistic $M_{n(i)}$ where i = 1, ... , k and n = Σn(i) . From each $M_{n(i)}$ find $\hat{A}_i$. Suppose the k populations have the same κ . To test the hypothesis

$$A_1 = \ldots = A_k \, , \qquad (2.17)$$

we need $\hat{A}$, the pooled estimator (i.e. the m.l. estimator of **A** , obtained from all the k samples ,when (2.17) is assumed) . It is clear that this is obtained from

$$M_n = n^{-1} \{ n(1)M_{n(1)} + \ldots + n(k)M_{n(k)} \} \, , \qquad (2.18)$$

by finding the s.v.d. of M_n . The analysis of dispersion immediately gives

$n\,(1 - Tr\hat{A}M_n)$ = dispersion of all samples about a common estimated **A** , ($\hat{A}$)

$= \Sigma\, n(i)(1 - Tr\hat{A}_i M_{n(i)}) + (\Sigma n(i) Tr\hat{A}_i M_{n(i)} - nTr\hat{A}M_n)$,

= dispersion within samples +
dispersion between sample estimates .

The matching equation for the degrees of freedom is

$$(q-1)n-Q = \Sigma\, ((q-1)n(i) - Q\,) + (k-1)Q \, . \qquad (2.19)$$

Thus

$$\frac{(q-1)n-kQ}{Q(k-1)} \; \frac{\Sigma n(i) Tr\hat{A}_i M_{n(i)} - n\, Tr\hat{A}M_n}{\Sigma\, n(i)\,(1 - Tr\hat{A}_i M_{n(i)}\,)} = F_{Q(k-1),(q-1)n-Qk} \qquad (2.20)$$

is the test statistic for hypothesis (2.17)

To test the hypothesis that the k orthogonal matrices belong to the prescribed subset G with r restrictions on the parameters, i.e.

$$A_1, A_2, \ldots, A_k \subset G \quad . \qquad (2.21)$$

Let $\hat{A}_G$ be the orthogonal matrix estimated from the matrix M_n defined in (2.18), using the hypothesis (2.21), by the method explained earlier Then the dispersion identity is

$$n(1-\mathrm{Tr}\hat{A}_G M_n) = \Sigma n(i)(1-\mathrm{Tr}\hat{A}_i M_{n(i)}) + (\Sigma n(i)\mathrm{Tr}\, \hat{A}_i M_{n(i)} - n\mathrm{Tr}\, \hat{A}_G M_n)$$

with the d.f. equation

$$(q-1)n-(Q-r) \quad = \quad (q-1)n-Qk \quad + \quad (k-1)Q+r \; .$$

Thus the F-test is

$$\frac{(q-1)n-Qk}{(k-1)Q+r} \; \frac{\Sigma n(i)\mathrm{Tr}\, \hat{A}_i M_{n(i)} - n\,\mathrm{Tr}\, \hat{A}_G M_n}{\Sigma\, n(i)\,(1-\mathrm{Tr}\, \hat{A}_i M_{n(i)})} = F_{(k-1)Q+r\,,\,(q-1)n-Qk} \, .$$

(2.22)

We now turn to the problem of getting, from the above results, some satisfactory forms for confidence sets. While it is true that the set of A's such that the l.h.s. of (2.15) is less than the r.h.s when F is taken as the upper 5% point in its distribution, this is not helpful even in 3 dimensions. This is serious in the case q=3 where we have some natural parametrizations and real problems. In these cases we will want confidence sets for these parameters and these have been found. Since I know of no actual problems for q>3 requiring a specific parametrization, we only do this in terms of the exponential representation of **A**, used first by MacKenzie (1962) and later by Chang(1986). It is well known that if **A** is orthogonal, we may write **A**= exp **S** where **S** is skew-symmetric, i.e. $\mathbf{S}^* = -\mathbf{S}$ with * meaning now the conjugate transpose. Let S be real. To show that det**A**=+1, we observe that $T = -i\mathbf{S}$ is hermitian and so has real eigen values and so can be written $T = \Sigma\tau_j \mathbf{v}_j \mathbf{v}_j$ with $\Sigma\tau_j = 0$ while $\mathbf{A} = \Sigma \exp(i\tau_j)\mathbf{v}_j\mathbf{v}_j^*$ Thus the product of the roots of **A** is unity which is the value of det**A**

Now $\hat{\mathbf{A}}$ will be close to **A** when either $n \to \infty$ or $\kappa \to \infty$ so we may then write

$$\mathbf{A} = \hat{\mathbf{A}} \exp \mathbf{S} \; , \qquad (2.23)$$

where **S** is a skew-symmetric qxq matrix with small elements. The Q elements above the diagonal, when arranged in some prescribed order, may be thought of as a vector **s** in R^Q so we could write S = **S**(**s**). We have

$$\mathbf{A} \approx \hat{\mathbf{A}}\,(\,I + \mathbf{S} + \mathbf{S}^2/2\,)\, .$$

so that the l.h.s. of the large sample result, (2.6), may be written

$$2\hat{\kappa}n\,\mathrm{Tr}\, M_n\,(\hat{\mathbf{A}} - \mathbf{A}) \approx 2\hat{\kappa}n\,(\,-\mathrm{Tr}\, M_n\hat{\mathbf{A}}\mathbf{S} \; - 2^{-1}\,\mathrm{Tr}\, M_n\hat{\mathbf{A}}\mathbf{S}^2\,)\, .$$

Now $M_n\hat{A} = L\Lambda L^*$, so

Tr $M_n\hat{A}S$ = Tr ΛL^*SL ,

and

Tr $M_n\hat{A}S^2$ = Tr$L\Lambda L^*S^2$ = Tr $SL\Lambda L^*S$ = Tr $L^*SL\Lambda L^*SL$.

The matrix $T = L^*SL$ is skew-symmetric since

$$(L^*SL)^* = L^*S^*L = - L^*SL .$$

Hence Tr $\Lambda L^*SL = 0$. Further if the elements of $T = L^*SL$ above its diagonal are thought of as a vector **t** in R^Q , $T = L^*SL$ represents an orthogonal transformation of **s** to **t** for $-2 \|s\|^2$ = TR S^2 and Tr T^2 = Tr S^2 . We write it as

$$t = L_Q s \;; \qquad (2.24)$$

it is easy to give an explicit formula for L_Q . When q=3 , $L_Q = L^*$.

Thus we have

$$2\hat{\kappa}n \,\mathrm{Tr}\, M_n(\hat{A} - A) \approx - \hat{\kappa}n \,\mathrm{Tr}\, T^2\Lambda ,$$
$$= \hat{\kappa}n \,\mathrm{Tr}\, T^*\Lambda T$$
$$\sim \chi^2_Q , \quad \text{for } n\to\infty \qquad (2.25)$$

or , with the sum over $i < j$,

$$\hat{\kappa}n \,\Sigma\, t_{ij}^2(\lambda_i+\lambda_j) \sim \chi^2_Q . \qquad (2.25')$$

Since n is large , κ and $\hat{\kappa}$ may be used interchangeably . The l.h.s. of (2.25') can be written as $\kappa n\, \mathbf{t}^*D(\lambda)\mathbf{t}$ where $D(\lambda)$ is a diagonal Q×Q matrix . Thus we can rewrite so (2.25') becomes

$$\kappa n\, \mathbf{s}^*L^*_Q D(\lambda) L_Q \mathbf{s} \approx \chi^2_Q . \qquad (2.26)$$

The large κ results (2.13) and (2.14) lead to the same results with κ instead of $\hat{\kappa}$. (2.15) may now be written

$$\{(q-1)n-Q\} \frac{n \,\mathrm{Tr}\, M_n(\hat{A} - A)}{n(1- \Sigma\lambda_i)} = \{(q-1)n-Q\} \frac{\mathrm{Tr}\, T^*\Lambda T}{2\, (1-\Sigma\lambda_i)} \qquad (2.27)$$

The r.h.s. of (2.27) will , when n is not too small , be approximately χ^2_Q and it can be expressed in tems of **s** . (2.27) may be a slightly better approximation than (2.25) .

Since $Ey = A_q(\kappa)\,\mu$, we have , as $n \to \infty$, $M_n \to A_q(\kappa)\,\Sigma$ if we , following Chang , set $n^{-1}\Sigma u_i u_i^* = \Sigma_n$ and assume that $\Sigma_n \to \Sigma$. Denote the eigen values of Σ by σ_i . Since Tr $\Sigma_n = 1$, $\Sigma\sigma_i = 1$. Thus

$$\lambda_i \approx A_q(\kappa)\, \sigma_i , \; 1-\Sigma\lambda_i \approx 1-A_q(\kappa) , \; D(\lambda) \approx A_q(\kappa)\, D(\sigma) \qquad (2.28)$$

and (2.26) can be written as

$$[\kappa n A_q(\kappa)]\, \mathbf{s}^*L_Q D(\sigma) L_Q^* \mathbf{s} \approx \chi_Q^2 , \qquad (2.29)$$

and (2.27) as

$$\{[(q-1)n-Q]A_q(\kappa)/2[1-A_q(\kappa)]\}\mathbf{s}^*L_QD(\sigma)L_Q{}^*\mathbf{s} \approx QF_{Q,(q-1)n-Q} \quad (2.30)$$

When n is appreciable , the r.h.s. of (2.30) is approximately $\chi_Q{}^2$. If also κ is large , the coefficient of $\mathbf{s}^*L_QD(\sigma)L_Q{}^*\mathbf{s}$ in (2.30) is approximately $2\kappa n$ so that (2.29) and (2.30) become the same - and the same as (2.25) .

We can summarize the last two paragraphs by saying that if one or both of κ and n is large (i.e. is κn is large) (2.25) is true . This is the same as saying that

$$\mathbf{s} \;\tilde{\sim}\; G_Q\,(\mathbf{0},\, \{n\kappa\, L_QD(\lambda)L_Q{}^*\}^{-1}\,)\ ,\ \text{as } n\kappa \to \infty\ . \quad (2.31)$$

Thus , since $\mathbf{s}$ measures the departure of $\mathbf{A}$ from $\hat{\mathbf{A}}$, we can find from (2.31) a confidence set for A . Chang (1986) illustrates this result when q=3 and $\mathbf{A}$ is a rotation when the set is an ellipsoid - of course he transforms it into the natural parameters , the latitude and longitude of the axis of rotation and the angle of rotation and the ellipsoid is distorted .

To complete our survey , we note that in Chang's (1986) paper it is only assumed there that the $\mathbf{y}_i$ have rotationally symmetric densities $g(\mathbf{y}_i{}^*\mathbf{A}\mathbf{u}_i)$. This generalization from Fisher to g is well-known - see e.g. Watson(1983) - and stems from a suggestion of Jon Wellner's . It is only useful for large n asymptotics . One begins with a chosen estimate and test statistic , say $\hat{\mathbf{A}}$ and $2\kappa n\mathrm{Tr}M_n(\hat{\mathbf{A}}-\mathbf{A})$, and deduces asymptotic results .

3 Discussion

(i) The above tests have been derived originally from Wilks' Theorem , after making some simplifications . Rao (1972,section6e) gives details of his efficient scores test and of Wald's test . All three test statistics have the same asymptotic distribution and applied workers use whichever is most convenient for the problem in hand . The Rao and Wald statistics are explicitly given as quadratic forms . Suppose we write

$$\mathbf{A} = \exp\, \mathbf{W}(\mathbf{w})\ ,\hat{\mathbf{A}} = \exp\, \hat{\mathbf{W}}(\hat{\mathbf{w}}) = \mathbf{RL}^*,$$

where $\mathbf{W}$ is skew symmetric and $\hat{\mathbf{w}}$ and $\mathbf{w}$ are as above vectors in R^Q. Then these forms will be constructed from the vector $\hat{\mathbf{w}}-\mathbf{w}$. But from (2.23) ,

$$\exp\, \mathbf{W} = \exp\hat{\mathbf{W}}\, \exp\, \mathbf{S}\ ,$$

so that

$$-\mathbf{s} = \hat{\mathbf{w}} - \mathbf{w}\ ,$$

which gives a direct interpretation of $\mathbf{s}$ as used above . Thus (2.26) must be asymptotically equivalent to the Rao and Wald tests which are not easy to construct here .

(ii) We first summarize the results of section2 for the case q=3 when $A_q(\kappa) = \coth\kappa - \kappa^{-1}$ .Then with $\mathbf{A} = \hat{\mathbf{A}}\exp\, \mathbf{S}(\mathbf{s})$, and κn large ,

$(n\kappa)^{1/2}$ is approximately $G_3(\mathbf{0}\ ,\ (LD(\lambda)L^*)^{-1}$. (3.1)

If n is large , κ may be replaced by $\hat{\kappa}$. $D(\lambda)$ is the diagonal matrix wit elements $\lambda_2+\lambda_3$, $\lambda_3+\lambda_1$, $\lambda_1+\lambda_2$, the λ_j's being the singular values of $M_n = n^{-1}\Sigma \mathbf{u}_i \mathbf{y}_i^*$. $L = [\ l_1, l_2, l_3]$ is the rotation matrix in the s.v.d. $M=L\Lambda R^*$. If the λ_j's are equal , the distribution of $\mathbf{s}$ is spherically symmetrical . Approximately the λ_i are equal to $(\coth\kappa-\kappa^{-1})\ \sigma_i$, wher the σ_i are the eigen values of $\boldsymbol{\Sigma}_n = n^{-1}\Sigma\ \mathbf{u}_i\mathbf{u}_i$. Tr $\boldsymbol{\Sigma}_n = \sigma_1+\sigma_2+\sigma_3 = 1$ so $\Sigma\lambda_i \cong \coth\kappa-\kappa^{-1}$. Equal λ_i correspond to σ_i's =1/3 . These ar approximately the eigen values one would get if the $\mathbf{u}_i$'s were a sizeable random sample of uniforms on the sphere , since $E\mathbf{u}\mathbf{u}^* = I_3/3$. If the $\mathbf{u}_i$ were uniformly distributed on a great circle , one eigen value would be zero and the other two about equal . If the $\mathbf{u}_i$'s formed a single cluster , one σ_i would be almost unity and the other two almost zero . In the two latter cases the covariance matrix of $\mathbf{s}$ will be ill-conditioned . Clear;ly the more evenly the $\mathbf{u}_i$'s are distributed , the better $\mathbf{A}$ will be estimated .

Now $\mathbf{s}$ is a vector describing the rotational error in $\hat{\mathbf{A}}$. $S(\mathbf{s})\mathbf{s}$ in three dimensions is $\mathbf{s}\times\mathbf{s}=\mathbf{0}$, so $\exp\{S(\mathbf{s})\}\mathbf{s}=\mathbf{0}$. Set $\mathbf{s}=\rho\mathbf{p}$ with $\|\mathbf{p}\|=1$. The $\mathbf{p}$ is the axis and ρ the angle of this error . Writing $\mathbf{V}=\mathbf{LDL}^*$, the marginal density of $\mathbf{p}$ on Ω_3 is (see e.g. Watson(1983), p110)

$$f(\mathbf{p})=[4\pi\det\mathbf{V}\{\mathbf{p}^*\mathbf{V}^{-1}\mathbf{p}\}^3]^{-1/2} \quad . \tag{3.2}$$

The marginal density of ρ is given by

$$g(\rho) = \{(2\pi)^{-3/2}(\det\mathbf{V})^{-1/2}\rho^2(n\kappa)^{3/2}\int_{\Omega_3} \exp[-\rho^2 n\kappa\mathbf{p}^*\mathbf{V}^{-1}\mathbf{p}/2]\omega_3(d\mathbf{p}) \tag{3.3}$$

where $\omega_3(.)$ is the usual areal element os the sphere . The integral in (3.3) is seen to be the normalization constant in a Bingham distribution (and so an awkward function) since the density of a Bingham is proportional to

$$\exp\ \mathbf{x}^*\mathbf{K}\mathbf{x}\ ,\ \mathbf{K}\ \text{symmetric}\ ,\ \|\mathbf{x}\| = 1\ . \tag{3.4}$$

If the λ_i are equal , to λ say , $\mathbf{V} = 2\lambda I_3$, $f(\mathbf{p})$ is the uniform density (meaning that the rotational errror axis has no preferred direction) and

$$g(\rho)= \frac{\rho^2 4\pi(n\kappa)^{3/2}}{(2\pi)^{3/2}\ (2^3\lambda^3)^{1/2}}\exp\{-n\kappa\rho^2/2\lambda\} \quad . \tag{3.5}$$

This density should be supported by $[0,\pi)$. The maximum of $g(\rho)$ occurs at $\rho = (2\lambda/n\kappa)^{1/2}$, a small value . If all the $\mathbf{u}_i$'s lie near and around a great circle , the elements $\lambda_2+\lambda_3$ and $\lambda_3+\lambda_1$ of $D(\lambda)$ are roughly equal , and $\lambda_1+\lambda_2$ twice as large . Thus f(p) is roughly rotationally symmetric about l_3 where it has a high peak . Thus the rotational error is largely a rotation about the normal to the plane of the great circle . This is exactly what one would predict from making a few drawings ! This also the way to guess what happens when the data is clustered . Formally we see then that f(p) has a ridge-like maximum .

A Princeton undergraduate , David Hull , has done extensive simulations with n=10,20 , κ=5,20,50 and with the initial points (a) randomly chosen from the uniform distribution , (b) clustered (axis of rotation 30 and 60 degrees away from it), and on a great circle ((c) axis of rotation on , and (d) perpendicular, to the plane of the circle) . The results will be discussed elsewhere but were mostly as expected except that I was (i) surprized at the frequency of negative det$\mathbf{M}_n$ (ii) disappointed in the accuracy of the $\chi^2{}_3$ approximation to $2\kappa n \mathrm{Tr}\, \mathbf{M}_n(\hat{\mathbf{A}}-\mathbf{A})$, (iii) convinced that one needs a bias correction for the angle .

(iii) For q=3 , the quaternion representation of Moran(1976) is

$$\mathbf{A} \quad \begin{matrix} u_0{}^2+u_1{}^2-u_2{}^2-u_3{}^2 , & 2(u_2u_2-u_0u_3) & , 2(u_0u_2+u_1u_2) \\ 2(u_0u_3+u_1u_2) \quad , & u_0{}^2+u_2{}^3-u_1{}^2-u_3{}^2, & 2(u_2u_3-u_0u_1) \\ 2(u_1u_3-u_0u_2) \quad , & 2(u_0u_1+u_2u_3) & , u_0{}^2+u_3-u_1{}^2-u_2{}^2 \end{matrix} \tag{3.7}$$

where $\mathbf{u}^*$ =$[u_0,u_1,u_2,u_3]$ and $-\mathbf{u}^*$ lead to the same $\mathbf{A}$. Substituting in the loglikelihood (2.1) , one finds that it may be written as

$$n\{-\log a_3(\kappa) +\kappa\, \mathbf{u}^*\mathbf{S}\mathbf{u}\}, \tag{3.8}$$

where $\mathbf{S}$ is the symmetric matrix

$$\begin{matrix} m_{11}+m_{22}+m_{33} , & m_{32}-m_{23}, & m_{13}-m_{31}, & m_{21}-m_{12} \\ & m_{11}-m_{22}-m_{33} , & m_{12}+m_{21} , & m_{31}+m_{13} \\ & & m_{22}-m_{11}-m_{33}, & m_{23}+m_{32} \\ & & & m_{33}-m_{11}-m_{22} \end{matrix}$$

where m_{ij} = (i,j)-element in $\mathbf{M}_n$. Thus $\hat{\mathbf{u}}$ is the eigen vector $\mathbf{v}_0$ os $\mathbf{S}$ corresponding to its largest eigen value ε_0 say , and $\hat{\mathbf{A}}$ is obtained by substitution in (3.6) .Thompson & Prentice (1987) indicate how to complete the solution . The estimated angle of rotation ψ , colatitude and longitude θ ,φ are found from $\hat{u}_0$ =$\cos\psi/2$,$\hat{u}_1/\hat{u}_2$ =$\tan\hat{\varphi}$,$\hat{u}_3$=$\sin\psi/2\cos\hat{\theta}$. They give an asymptotic($n\to\infty$) covariance matrix for these estimators which depends upon the reciprocals of $\varepsilon_0-\varepsilon_1,\varepsilon_0-\varepsilon_2,\varepsilon_0-\varepsilon_3$ and the eigen vectors $\mathbf{v}_1,\mathbf{v}_2,\mathbf{v}_3$.

Mathematically the two methods are equivalent . It would be interesting to see whether they differ in numerical stability . The covariance matrix blows up when the one or more of the other eigen values approaches ε_0 as it should since then v_0 is not well determined. The dependence of the ε's on the distribution of the $\mathbf{u}_i$'s is not so easy to determine .

(iv) These methods , being equivalent to least squares ,are probably senitive to outliers . More robust methods might seek $\mathbf{A}$ to minimize $\Sigma\ g(\|\mathbf{y}_i-\mathbf{A}\mathbf{u}_i\|)$ with g other than the square function .Any other choice - even g(z)=z- leads to iterative numerical methods and more difficult asymptotics . Finding $\mathbf{A}$ to maximize $\Sigma(\mathbf{Y}^*\mathbf{A}\mathbf{u}_i)^2$ should lead (I beleive) to a more robust method at little loss of efficiency if the Langevin assumption is correct .This assumption is often gratuitous

Hull and I tried a variety of methods based on getting angle and axis estimates from all pairs $(\mathbf{u}_i,\mathbf{y}_i),(\mathbf{u}_j,\mathbf{y}_j)$. It would be of interest to find the Influence function ,Hampel et al. (1987), for the m.l. estimator .

(v) More important seems to be the need for more reliable confidence regions . The natural methods to try are the Bootstrap method ,Efron(1982) , and empirical likelihood method of Owen (1987a,1987b) . These would probably require the use of non-parametric density estimation on the sphere - Diggle&Fisher (1985) , Cabrera, Hall &Watson(1988) .

(v) Chang investigated when the $\mathbf{u}_i$'s are known with error-call them $\mathbf{x}_i$- but unfortunately used the same error distribution as for the $\mathbf{y}_i$. This was reasonable in the problem he was dealing with . It is rather better to assume e.g. that the $\mathbf{x}_i$are $F_3(\kappa_1,\mathbf{u}_i)$ and the $\mathbf{y}_i$ are $F(\kappa_2,\mathbf{A}\mathbf{u}_i)$. Then the loglikelihood of the data involves the $\mathbf{u}_i$ in the terms

$$\kappa_1\Sigma\mathbf{x}_i^*\mathbf{u}_i+\kappa_2\Sigma\mathbf{y}_i^*\mathbf{A}\mathbf{u}_i\ . \tag{3.8}$$

The m.l. estimator of $\mathbf{u}_i$ is then parallel to $\kappa_1\mathbf{x}_i+\kappa_2\mathbf{A}^*\mathbf{y}_i$. Even if th κ's were known we would , to find $\mathbf{A}$ have to maximize

$$\Sigma\ \|\kappa_1\mathbf{x}_i+\kappa_2\mathbf{A}^*\mathbf{y}_i\|\ , \tag{3.9}$$

an awkward problem . If one did not have to normalize $\hat{\mathbf{u}}_i$, we would be lead to maximizing $\Sigma\ \mathbf{Y}_i^*\mathbf{A}\mathbf{x}_i$, an intuitively sensible method . This is the estimator Chang studies . The simplest result one could hope to be approximately true is that

$$2\kappa n\mathrm{Tr}\mathbf{M}_n(\hat{\mathbf{A}}-\mathbf{A}) \cong \chi^2{}_3\ , \tag{3.10}$$

holds with κ some function of κ_1 and κ_2 . Theoretical approximatic and arithmetic done with Hull suggest that

$$1/\kappa = 1/\kappa_1+1/\kappa_2\ . \qquad (3.1$$

This agrees with Chang result for equal κ's .

4 Bayesian estimation of A

We will assume that κ is known . The Bayesian approach to MacKenzie's problem then requires only a prior distribution over O(q) or SO(q) . To get one with a mode at A_0 , start with the intuitive notion that the density with respect to invariant measure (see e.g. James (1954) or Muirhead (1982)) might be proportional to

$\exp\lambda' \| A - A_0 \|^2$ or

$$c(\lambda)^{-1} \exp\lambda \mathrm{Tr} A^* A_0 \ , \ \lambda > 0 \ . \tag{4.1}$$

This gets more concentrated as λ increases . Combining this with the likelihood of the data , the posterior density for A is proportional to

$$\exp \mathrm{Tr}\{n\kappa M^*_n + \lambda A_0\} \ , \tag{4.2}$$

the normalization constant being the inverse of the integral of (4.2) over O(q) which is given later . To find the modal value of A for this distribution , we must find the S.V.D. of $n\kappa M_n + \lambda A_0{}^*$.

Mardia & Khatri (1977) , following Downs (1972) , (see also Prentice (1982)) consider the generalization of the Fisher distribution to Stiefel manifolds . Specializing to O(q) , one is lead to consider

$$c(K)^{-1} \exp \mathrm{Tr} K^* A \ , \tag{4.3}$$

where K is a qxq matrix with S.V.D.

$$\left. \begin{array}{l} K = FGH^* \ , \ G \text{ diagonal with positive elements} \\ F^*F = H^*H = I \ . \end{array} \right\} \tag{4.4}$$

Then

$$\begin{aligned} 2c(K) &= {}_0F_1\,(q/2 \, ; KK^*/4) \ , \\ &= {}_0F_1\,(q/2 \, ; G^2/4) \ , \end{aligned} \tag{4.5}$$

where ${}_0F_1$ is one of the hypergeometric functions with matrix argument - see e.g. Muirhead (1982) . Further the modal value of A is $A_0 = FH^*$. As the elements of G increase , the distribution of A concentrates on A_0 . As G tends to zero , the distribution of A becomes uniform on O(q) . Using (4.3) as a prior for A , the above argument leads to a posterior density of A given by

$$c^{-1} \exp \mathrm{Tr} \ (n\kappa M_n{}^* + K)^* A \ , \tag{4.6}$$

where $$2c = {}_0F_1 \, \{q/2 \, ; (n\kappa M^*_n + K)(n\kappa M + K^*)/4 \}. \tag{4.7}$$

This clearly generalizes the formulae in the first paragraph .

To proceed further one would use a parametrization of A . If , for example , q=3 and we are concerned only with rotations , we use the axis **a** and the angle ω of rotation and specify the unit vector **a** by its polar coordinates (θ,φ) , then the invariant measure is

$$\sin^2 \omega/2 \ d\omega \ \sin\theta \ d\theta \ d\varphi \ . \tag{4.8}$$

Thus if **A** in (4.6) is expressed in terms of ω , θ , φ , we will obtain their posterior distribution .

If $n\kappa$ is large enough , the data will dominate the prior and the posterior density is appreciable only near $\hat{\mathbf{A}} = \mathbf{RL^*}$ and is proportional to $\exp \mathrm{Tr}\, n\kappa M_n \mathbf{A}$. Thus , as in the derivation of (2.25) , we may set $\mathbf{A} \approx \hat{\mathbf{A}}(\,I + \mathbf{S} + \mathbf{S}^2/2\,)$, $\mathbf{S^*} = -\mathbf{S}$, with small elements which may as before be thought of as a vector **s** in R^Q . This leads to a posterior density is proportional to

$$\exp(-n\kappa)\{\mathbf{s^*L^*}_Q D(\lambda) \mathbf{L}_Q \mathbf{s}\} \ , \qquad (4.9)$$

or that

$$\mathbf{s} \sim G_Q(\mathbf{0}, [2\, n\kappa \mathbf{L}_Q D(\lambda) \mathbf{L}_Q{}^*]^{-1}) \ , \qquad (4.10)$$

the same results as we obtained in Section 2 !

Chang & Bingham (1988) observe that , using the quaternion version of **A** , the joint density of the data is the exponential of (3.7) . This is a Bingham distribution on Ω_4 so they choose a Bingham prior on Ω_4 so that the posterior distribution is also Bingham . This may be equivalent to my discussion .

5 Acknowledgements

The writing of this paper was partially supported by Grant MCS 84211301 from the National Science Foundation . The writer is grateful to Prof. Chang for preprints of his papers and of that by Rivest.

References

J. Cabrera, P. Hall & G. S. Watson, (1988). "Kernel density estimation with spherical data," *Biometrika*

T. Chang, (1986). "Spherical regression," *Ann. Stat.* **14**, pp. 907-24.

T. Chang, (1987). "On the statistical properties of estimated rotations," *J. Geophys. Res.* **92**, pp. 6319-29.

T. Chang, (1988) "Spherical regression with errors in variables,. (MS .)

T.Chang, C.Bingham, (1988) "On approximations to the Bingham distribution and their use in spherical regression" , (MS)

P. J. Diggle & N. I. Fisher, (1985). "SPHERE: A contouring program for spherical data," *Computers and Geosciences* **11**, pp. 725-66.

Downs, T.D. (1972) . Orientation statistics , *Biometrika* , **59** , 665-676

B.Efron(1982) *The Jacknife , the Bootstrap and other resampling methods* , Cof. Series in Appl. Math. **38** , Philadelphia , SIAM , pp61.

N. I. Fisher, B. J. J. Embleton & T. L. Lewis, (1987). *Statistical Analysis of Spherical Data* , Cambridge Univ. Press, Cambridge, p. 329.

I. M. Gel'fand, R. A. Minlos & Z. Ya. Shapiro, (1963). *Representations Of the Rotation and Lorentz Groups and Their Applications* , Macmillan, New York, p. 366.

F. R. Hampel, E. M. Ronchetti, P. J. Rousseeuw & W. A. Stahel, (1987) . *Robust Statistics: An Approach Based on Influence Functions,* John Wiley & Sons, New York,

A.T.James , (1954) "Normal multivariate analysis and the orthogonal group " *Ann.Math.Stat.* , 25,40-75

P. E. Jupp & K. V. Mardia, (1980). "A general correlation coefficient for directional data and related regression problems," *Biometrika* 67, pp. 163-73.

J. K. Mackenzie, (1957). "The estimation of an orientation relationship," *Acta. Cryst.* 10, pp. 61-2.

J.K.Mackenzie, (1962) . "The estimation of an orientation relationship from traces of known planes", *Acta.Cryst.* 15 ,pp979-82.

Mardia, K.V., Khatri, C.G.(1977) Uniform distribution on the Stiefel manifold , *J.Mult. Analysis* , 7 ,468-473

P. A. P. Moran, (1976). "Quaternions, Haar measure and the estimation of a palaeomagnetic rotation," *Perspectives in Prob. and Stat.* , J. Gani, Ed., Sheffield, Applied Probability Trust, pp. 295-301.

R.Muirhead, (1982) . *Aspects of Multivariate statistic al analysis* , John Wiley & Sons , New York , pp673

A.B. Owen (1987a) Empirical Likelihood Ratio confidence intervals for a single functional , *Tech.Rep.* # 271 , Stanford University.

A.B.Owen (1987b) Empirical Likelihood ratio confidence regions , *Tech. Rep.* ,#283 , Stanford University .

M.J.Prentice, (1982). Antipodally symmetric distributions for orientation statistics , *J.Stat. Planning & Inference , 205-214*

L.-P. Rivest, (1986). "Small sample inference for spherical regression, (To appear.)

Thompson,R.,Prentice,M.J.(1987)"Alternative method of calculating finite plate rotations", *Phys. Earth&Planetary Interiors* ,48 ,pp79-83

M. A. Stephens, (1979),"Vector correlation," *Biometrika* 66, 41-48.

G. S. Watson, (1956). "Analysis of dispersion on a sphere," *Monthly Notices Roy. Astro. Soc. Geophys. Suppl.* 7, pp. 153-59.

G. S. Watson & E. J. Williams, (1956). "On the construction of significance tests on the circle and the sphere, *Biometrika* 43, pp. 2344-52.

G. S. Watson, (1960). "More significance tests on the sphere," *Biometrika* 47, pp. 87-91.

G. S. Watson, (1983).*Statistics On Spheres* , John Wiley & Sons, New York, p. 237.

G.S.Watson (1987a) The Langevin distribution on high dimensional spheres (to appear)

G. S. Watson, (1987b) . "Permutation tests for the independence of a scalar and a unit vector," (to appear)

COMPLETELY BOUNDED AND RELATED RANDOM FIELDS ON LOCALLY COMPACT GROUPS

Kari Ylinen
Department of Mathematics, University of Turku
SF-20500 Turku, Finland

1. Introduction

Let G be a locally compact group and H a (complex) Hilbert space with inner product $(\cdot|\cdot)$. Any function $\phi\colon G \to H$ will be called a random field; we refer to [13] for background and motivation. Various classes of weakly continuous random fields (right, left, and two-way homogeneous, hemihomogeneous, and V-bounded or weakly harmonizable) were discussed in [13]. All these can be characterized in terms of the covariance function $R\colon G\times G \to \mathbb{C}$ defined by $R(s,t) = (\phi(s)|\phi(t))$ for $\phi\colon G \to H$. In this note we briefly consider two more classes of random fields which where defined in [15] also in terms of the covariance function: the strongly harmonizable and the completely bounded ones. We mention and complement results obtained in [15]; the new material is in Section 3.

2. Completely bounded bilinear and multilinear forms on products of group C*-algebras

We give a motivating discussion which goes somewhat beyond the needs of the applications to random fields. For the locally compact group G, let $C^*(G)$ denote its group C^*-algebra (containing $L^1(G)$ for a fixed left Haar measure as a dense subspace). Let $B(G)$ be the Fourier-Stieltjes algebra, i.e., the linear span of the set of continuous positive-definite functions on G. When we identify $B(G)$ with $C^*(G)^*$ via the bijection $T\colon B(G) \to C^*(G)^*$ satisfying $\langle\mu,Tf\rangle = \int f d\mu$ for all $\mu \in L^1(G)$, $C^*(G)^*$ inherits from the pointwise multiplication of $B(G)$ a Banach algebra product which can naturally be called convolution, see [3].

If G is commutative, $C^*(G)$ identifies with $C_o(\Gamma)$, the C^*-algebra of continuous functions vanishing at infinity, defined on the dual group Γ of G, and we get the usual convolution of measures in $C_o(\Gamma)^* = M(\Gamma)$ by the method described above. In the case of two locally compact abelian groups G_1 and G_2, Graham and Schreiber [4] showed that for bounded bilinear forms B on $C_o(\Gamma_1)\times C_o(\Gamma_2)$ convolution may be defined by means of pointwise multiplication of their natural Fourier transforms $\hat{B}\colon G_1\times G_2 \to \mathbb{C}$. The key is the representation of $\hat{B}$ in the form

(1) $\hat{B}(s,t) = (\pi_1(s)\pi_2(t)\xi|\eta)$

where π_1 and π_2 are continuous unitary representations of G_1 and G_2, respectively, on the same Hilbert space K, and $\xi, \eta \in K$. (Our formulation differs slightly from [4] but is equivalent.) Alternatively (see [15]), we may write B itself as

(2) $B(x,y) = (\pi_1(x)\pi_2(y)\xi|\eta)$

where π_i is a nondegenerate *-representation of $C_o(\Gamma_i)$ on K.

In the case of any finite number of not necessarily abelian locally compact groups $G_1,\dots,G_n$, (2) leads to a consideration of mappings $M: C^*(G_1)\times\dots\times C^*(G_n) \to \mathbb{C}$ with the representation

(3) $\quad M(x_1,\dots,x_n) = (\pi_1(x_1)\cdots\pi_n(x_n)\xi|\eta)$

where each π_i is a nondegenerate *-representation of $C^*(G_i)$ on a Hilbert space K, and $\xi, \eta \in K$. Such mappings are the <u>completely</u> <u>bounded</u> n-linear functionals on the product $C^*(G_1)\times\dots\times C^*(G_n)$ (introduced with a different but equivalent definition in a more general situation in [2]). Any bounded linear functional is completely bounded; if $f \in C^*(G)^*$, there is a *-representation π of $C^*(G)$ on some Hilbert space K with vectors $\xi, \eta \in K$ such that

(4) $\quad f(x) = (\pi(x)\xi|\eta), \quad x \in C^*(G).$

Moreover, $\|f\|$ is the infimum of $\|\xi\|\|\eta\|$ over all representations (4) (see [3]). Analogously, for completely bounded multilinear forms $M: C^*(G_1)\times\dots\times C^*(G_n) \to \mathbb{C}$ the completely bounded norm $\|M\|_{cb}$ is the infimum of $\|\xi\|\|\eta\|$ over all representations (3). For references and details, see [15].)

The key to the representation (1) in [4] (or (2)) is the Grothendieck inequality, "the fundamental theorem of the metric theory of tensor products" of [5]. In fact, one way of expressing the content of the Grothendieck inequality (apart from a consideration of the Grothendieck constant) is to say that any bounded bilinear form on the product of two commutative C*-algebras is completely bounded. At least from the point of view of harmonic analysis it may be argued that even here the completely bounded norm (which in this case is equivalent to the usual supremum norm) is the natural one to use, and the notion of complete boundedness (rather than just boundedness) is the one to be used in the general case. Indeed, the completely bounded norm behaves properly with respect to the natural convolution in the commutative two-variable case, and even in the general situation the convolution of completely bounded multilinear forms makes sense. For details we refer to [15].

3. <u>Random fields</u>

Throughout the sequel G is a locally compact group, $C^*(G)$ its group C*-algebra, and $\omega: C^*(G) \to L(H_\omega)$ the universal representation. The corresponding continuous unitary representation of G (on H_ω) is also denoted by ω, and so is its natural extension to the convolution *-algebra $M(G) = C_o(G)^*$ of measures. The enveloping von Neumann algebra of $C^*(G)$ is denoted by $W^*(G)$ $(\subset L(H_\omega))$ and identified with the bidual $C^*(G)^{**}$ of $C^*(G)$. The Fourier transform $\hat{B}: G\times G \to \mathbb{C}$ of a bounded bilinear form B on $C^*(G)\times C^*(G)$ was defined in [11] by the formula $\hat{B}(s,t) = \tilde{B}(\omega(s),\omega(t))$ where $\tilde{B}$ is the unique separately weak* continuous extension of B to $W^*(G)\times W^*(G)$.

As before, H will be a Hilbert space and $\phi: G \to H$ a random field with the covariance function $R: G\times G \to \mathbb{C}$, $R(s,t) = (\phi(s)|\phi(t))$. The following definition is a list of properties expressible by means of the covariance function. It refines the commonly accepted terminology in the case of $G = \mathbb{R}$ (see e.g. [1, 8]). Note that the definition of weak harmonizability used here is equivalent to the one given in [13], see [13, Theorem 2.3]. For a "concrete" treatment of weak harmonizability in the case of a compact group we refer to [14].

3.1. DEFINITION. Define $R': G \times G \to \mathbb{C}$ by the formula $R'(s,t) = R(s,t^{-1})$.

(a) ϕ is right (resp. left) homogeneous if $R(s,t) = R(su,tu)$ (resp. $R(s,t) = R(us,ut)$) for all $s, t, u \in G$. If both conditions hold, ϕ is two-way homogeneous.

(b) ϕ is hemihomogeneous if there are two continuous positive-definite functions $\rho_1, \rho_2: G \to \mathbb{C}$ such that $(\phi(s)|\phi(t)) = \rho_1(t^{-1}s) + \rho_2(st^{-1})$ for all $s, t \in G$.

(c) ϕ is weakly harmonizable if R' is the Fourier transform of some bounded bilinear form on $C^*(G) \times C^*(G)$.

(d) ϕ is strongly harmonizable if R' belongs to the Fourier-Stieltjes algebra $B(G \times G)$.

(e) ϕ is completely bounded if R' is the Fourier transform of a completely bounded bilinear form on $C^*(G) \times C^*(G)$.

It is easily seen that in the abelian setting weak harmonizability means that R' is the Fourier transform of a (bounded) bimeasure on the product of the dual group of G with itself, whereas in the strongly harmonizable case this bimeasure is actually a (complex) measure. (For a study and references on these notions, see e.g. [1].) In general, strong harmonizability implies complete boundedness which in turn implies weak harmonizability [15, Theorem 7.2]. Moreover, weak harmonizability follows from hemihomogeneity, and it is immediate that a weakly continuous random field which is left or right homogeneous is also hemihomogeneous. The importance of hemihomogeneity stems in part from the fact (generalizing the main result of [7]) that any weakly harmonizable random field has a hemihomogeneous dilation [12]. A similar relation between complete boundedness and right homogeneity was established in [15, Theorem 7.4 (b)]: A random field $\phi: G \to H$ is completely bounded if, and only if, there exists a Hilbert space K containing H as a subspace and a continuous right homogeneous random field $\psi: G \to K$ such that $\phi = P_H \circ \psi$ where P_H is the projection onto H.

We prepare the proof of a characterization of completely bounded random fields with a lemma. If $f \in C^*(G)^*$, we let f also denote its canonical extension to $W^*(G)$.

3.2. LEMMA. *Suppose A is a subspace of $W^*(G)$ which separates $C^*(G)^*$. Let $\Phi: C^*(G) \to H$ be a bounded linear map and $f: C^*(G) \to \mathbb{C}$ a positive linear form. If for the bitranspose $\Phi^{**}: W^*(G) \to H$*

$$(5) \qquad \| \Phi^{**}(x) \|^2 \leq f(xx^*)$$

whenever $x \in A$, then (5) also holds for all $x \in W^*(G)$.

Proof. Being a separating subspace, A is $\sigma(W^*(G), C^*(G)^*)$-dense [9, p. 125], i.e., σ-weakly dense in $W^*(G)$ [10, p. 122]. But on $W^*(G)$ the σ-weakly continuous linear functionals are the same as the σ-strongly* continuous ones [10, p. 70], and so A is also σ-strongly* dense in $W^*(G)$ (see [9, p.65]). Now suppose that $x \in W^*(G)$ and $\xi \in H$, $\|\xi\| \leq 1$. Choose a net (x_i) in A converging σ-strongly* to x. The mapping $(u,v) \to f(uv)$ is separately σ-weakly continuous on $L(H_\omega)$, hence jointly σ-strongly* continuous (see

[6, p. 95]), and so $\lim_i f(x_i x_i^*) = f(xx^*)$. Since Φ^{**} is $\sigma(W^*(G),C^*(G)^*)$-$\sigma(H,H^*)$-continuous, $\lim_i (\Phi^{**}(x_i)|\xi) = (\Phi^{**}(x)|\xi)$, and so $|(\Phi^{**}(x)|\xi)|^2 \leq f(xx^*)$, since (5) holds for each x_i. Taking the supremum over ξ yields (5) for x.□

3.3. THEOREM. The following conditions are equivalent for a weakly continuous random field ϕ: $G \to H$:

(i) ϕ is completely bounded;

(ii) there is a continuous positive-definite function ρ: $G \to C$ such that

$$(6) \qquad \| \sum_{i,j=1}^{n} c_i \phi(s_i) \|^2 \leq \sum_{i=1}^{n} \sum_{j=1}^{n} c_i \bar{c}_j \rho(s_i s_j^{-1})$$

for all finite sequences $s_1,\dots,s_n \in G$, $c_1,\dots,c_n \in \mathbb{C}$.

Proof. Denote by $M_{dd}(G)$ the subspace of the finitely supported measures in $M(G)$. The condition (6) is equivalent to

$$(7) \qquad \| \int \phi d\mu \|^2 \leq \int \rho d(\mu * \mu^*)$$

for all $\mu \in M_{dd}(G)$. Now suppose (i) holds. There is a bounded linear map Φ: $C^*(G) \to H$ such that $\phi(s) = \Phi^{**}(\omega(s))$ for all $s \in G$ [15, Theorem 7.2 (b)]. From [15, Theorem 7.4] it follows that there is a positive linear form f: $C^*(G) \to \mathbb{C}$ such that $\| \Phi^{**} x \|^2 = \| \Phi x \|^2 \leq f(xx^*)$ for all $x \in C^*(G)$, and so Lemma 3.2 shows that $\| \Phi^{**} x \|^2 \leq f(xx^*)$ for all $x \in W^*(G)$, in particular for all $x = \omega(\mu)$, $\mu \in M_{dd}(G)$. Now $\rho = T^{-1} f$ (see Section 2) is a continuous positive-definite function on G, and we have

$$(8) \qquad \| \int \phi d\mu \|^2 = \| \Phi^{**}(\omega(\mu)) \|^2 \leq f(\omega(\mu)\omega(\mu^*)) = \int \rho d(\mu * \mu^*)$$

for all $\mu \in M_{dd}(G)$ (see [11, p.362]), i.e., (7) and thus (6) holds for ρ. Next assume (ii). For all $\mu \in M_{dd}(G)$ $\| \int \phi d\mu \|^2 \leq \int \rho d(\mu * \mu^*) = \langle T\rho, \omega(\mu)\omega(\mu^*) \rangle \leq \| T\rho \| \, \| \omega(\mu) \|^2$ (see Section 2), and since ϕ is weakly continuous, there is a bounded linear operator Φ: $C^*(G) \to H$ such that $\Phi^{**}(\omega(s)) = \phi(s)$ for all $s \in G$ (see [11, p. 379]). Denoting $f = T\rho$ we get $\| \Phi^{**}(\omega(\mu)) \|^2 \leq f(\omega(\mu)\omega(\mu^*))$ for all $\mu \in M_{dd}(G)$ (see (7) and [11, Lemma 3.4]), and so by Lemma 3.2 $\| \Phi(x) \|^2 = \| \Phi^{**}(x) \|^2 \leq f(xx^*)$ for all $x \in C^*(G)$. Thus (i) holds by [15, Theorem 7.4 (a)].□

REFERENCES

[1] Chang, D. K., Rao, M. M.: Bimeasures and nonstationary processes. In: Real and Stochastic Analysis, edited by M. M. Rao, New York-Chichester-Brisbane-Toronto-Singapore: John Wiley & Sons 1986, pp. 7-118.

[2] Christensen, E., Sinclair, A. M.: Representations of completely bounded multilinear operators. J. Funct. Anal. 72, 151-181 (1987).

[3] Eymard, P.: L'algèbre de Fourier d'un groupe localement compact. Bull. Soc. Math. Fr. 92, 181-236 (1964).

[4] Graham, C. C., Schreiber, B. M.: Bimeasure algebras on LCA groups. Pac. J. Math. 115, 91-127 (1984).

[5] Grothendieck, A.: Résumé de la théorie métrique des produits tensoriels topologiques. Bol. Soc. Mat. São Paulo 8, 1-79 (1956).

[6] Haagerup, U.: The Grothendieck inequality for bilinear forms on C*-algebras. Adv. Math., 56, 93-116 (1985).

[7] Niemi, H.: On stationary dilations and the linear prediction of certain stochastic processes. Soc. Sci. Fenn. Comment. Phys.-Math. 45, 111-130 (1975).

[8] Rao, M. M.: Harmonizable processes: structure theory. Enseign. Math., II. Ser. 28, 295-351 (1982).

[9] Schaefer, H. H.: Topological vector spaces. New York: The Macmillan Company, London: Collier-Macmillan Limited 1966.

[10] Takesaki, M.: Theory of operator algebras I. New York-Heidelberg-Berlin: Springer Verlag 1979.

[11] Ylinen, K.: Fourier transforms of noncommutative analogues of vector measures and bimeasures with applications to stochastic processes. Ann. Acad. Sci. Fenn., Ser. A I 1, 355-385 (1975).

[12] Ylinen, K.: Dilations of V-bounded stochastic processes indexed by a locally compact group. Proc. Am. Math. Soc. 90, 378-380 (1984).

[13] Ylinen, K.: Random fields on noncommutative locally compact groups. In: Probability Measures on Groups VIII, Proceedings of a Conference held in Oberwolfach, November 10-16, 1985, pp. 365-386. Lecture Notes in Math. Vol. 1210, Berlin-Heidelberg-New York-London-Paris-Tokyo: Springer-Verlag 1986.

[14] Ylinen, K.: Random fields on compact groups. In: The Very Knowledge of Coding, Studies in honor of Aimo Tietäväinen on the occasion of his fiftieth birthday, July 6, 1987, edited by H. Laakso and A. Salomaa, pp. 143-156. Turku: Turun yliopiston offsetpaino 1987.

[15] Ylinen, K.: Noncommutative Fourier transforms of bounded bilinear forms and completely bounded multilinear operators. J. Funct. Anal. 79, 144-165 (1988).

ON THE MAX-DIVISIBILITY OF TWO DIMENSIONAL NORMAL RANDOM VARIABLES

András Zempléni

Summary It is known /see the book by Resnick [5] / that the two dimensional normal distribution with nonnegative correlation is max-infinitely divisible. Here we show that in the same structure the two dimensional normal distribution with negative correlation is irreducible.

§1. Introduction

In this paper we discuss max-factorization of two dimensional normal distributions. If X,Y are independent $\mathbb{R}^2$ -valued random variables with probability distribution functions F and G then the probability distribution function of max(X,Y) is the pointwise product F·G. The set of n-dimensional distribution functions with this operation form a semigroup which we shall denote by $\mathcal{M}_n$. In this structure occurs a lot of arithmetical questions /which are in the convolution case answered by the famous theorem of Cramér/: is the normal distribution max-infinitely divisible, what can we say about its decompositions of the form

$$\Phi = F\,G.$$

It is obvious that in the one-dimensional case /n=1/ every probability distribution function is max-infinitely divisible. In higher dimensions the infinitely divisible elements of $\mathcal{M}_n$ were characterized by Balkema and Resnick /[1]/. Namely they proved that a probability distribution function F is max-infinitely divisible if and only if there is a σ-finite measure μ on $[-\infty,+\infty)^n$ such that with

$$H(\underline{x}) = \mu\left([-\infty,\infty)^n \setminus [-\infty,x_1] \times \ldots \times [-\infty,x_n]\right),$$
$$F(\underline{x}) = \exp\left\{-H(\underline{x})\right\}$$

holds for any $\underline{x} \in \mathbb{R}^n$ /together with some additional conditions which ensure the limit properties of $F(\underline{x})$/. For absolutely continuous probability distribution functions on $\mathbb{R}^2$ they gave an other necessary and sufficient condition which is easier to verify: F is max-infinitely divisible /inf. div./ if and only if

$$\frac{\partial F(x,y)}{\partial x}\cdot\frac{\partial F(x,y)}{\partial y} \leq \frac{\partial^2 F(x,y)}{\partial y\partial x}\cdot F(x,y) \qquad (1)$$

holds on $\mathbb{R}^2$. Using this fact Resnick proved in [5] the infinite divisibility of the two dimensional normal probability distribution with nonnegative correlation.

To formulate our main result we recall the notion of max-irreducible distributions: a distribution function $F\in\mathcal{M}_2$ is max-irreducible if it has no decomposition of the form $F=G\cdot H$. Our theorem is a sufficient condition for absolutely continuous distribution functions in $\mathcal{M}_2$ to be irreducible, which implies that a normal distribution with negative correlation is always irreducible. Other sufficient conditions for a distribution function to be max-irreducible were proved in [6].

The interest was focused on the class of irreducible distributions /in the convolution semigroup/ after publishing Hincin's decomposition theorem about the factorization of any probability distribution into the convolution of /at most countable/ irreducible distributions and an antiirreducible one. /A distribution is called antiirreducible if it has no irreducible divisor./ See for example [2]. In $\mathcal{M}_2$ we do not know anything about the existence of such a decomposition theorem. I.Z. Ruzsa and G.J. Székely proved some general results for decompositions in abstract semigroups /[3], [4]/ but most of them fail to be applicable to our semigroup $\mathcal{M}_2$ since here the set of divisors of elements may not be shift-compact /which is one of their essential assumptions/. This fact causes that in $\mathcal{M}_2$ the antiirreducible elements are not always infinitely divisible [7]. We show in §2. that the two dimensional normal distribution cannot be antiirreducible in $\mathcal{M}_2$.

§2. Irreducibility of the normal distribution with negative correlation

Remark. We consider such properties of the normal distribution that obviously do not depend on changes of the scala and shift parameters. Hence it is enough to consider the following density function:

$$f(x,y) = \frac{\sqrt{1-b^2}}{2\pi}\exp\left\{-\left(x^2-2bxy+y^2\right)/2\right\} \qquad (2)$$

where $|b|<1$ /b is the correlation between the components/.

First we formulate a sufficient condition for an absolutely continuous probability distribution function in $\mathcal{M}_2$ to be irreducible.

Theorem. Let $F\in\mathcal{M}_2$ be an absolutely continuous distribution function with density function f such that

$$F(x,y)\frac{\partial F(x,y)}{\partial x}\cdot\frac{\partial F(x,y)}{\partial y} > 0$$

holds for any $(x,y) \in \mathbb{R}^2$, and suppose that

$$\lim_{y\to+\infty} \frac{\partial F(x,y)}{\partial x} \cdot \frac{\partial F(x,y)}{\partial y} \Big/ \big(f(x,y)\cdot F(x,y)\big) = +\infty \qquad (3)$$

uniformly in every interval $x \in (-k,+\infty)$. Then F is max-irreducible. /This statement remains obviously true if the roles of x and y are changed./

Before proving this theorem we show the following corollary:

Corollary. For $b<0$ the normal distribution given by (2) is irreducible.

Proof of the corollary. We have to verify condition (3). In our case

$$\frac{f(x,y)\cdot F(x,y)}{\frac{\partial F(x,y)}{\partial x}\cdot\frac{\partial F(x,y)}{\partial y}}$$

$$= \frac{\int_{-\infty}^{y}\int_{-\infty}^{x} \exp\{-(u^2-2buv+v^2)/2\}dudv}{\int_{-\infty}^{y}\int_{-\infty}^{x} \exp\{-(u^2-2buv+v^2)/2-b(x-u)(y-v)\}dudv}. \qquad (4)$$

Let k be fixed. Then we majorate the right side of 4 by

$$1\Big/\int_{-\infty}^{0}\int_{-\infty}^{-k} \exp\{-(u^2-2buv+v^2)/2\}dudv\cdot\exp\{-by\},$$

which obviously tends to 0 if $y_o \to +\infty$ /$b<0$/.

Proof of the theorem. First we mention that an absolutely continuous distribution function for which $F(\underline{x})>0$ /for every $\underline{x}\in\mathbb{R}^2$/ can only have absolutely continuous divisors. Otherwise there would be a decomposition $F=G\cdot H$ and an $S\subset\mathbb{R}^2$ with $\lambda_2(S)=0$ for which $P_G(S)>0$. We may suppose the existence of an $\underline{x}_o\in\mathbb{R}^2$ for which $\underline{x}_o<\underline{s}$ holds for every $\underline{s}\in S$. Then

$$P_F(S) \geq H(\underline{x}_o)\cdot P_G(S) \geq F(\underline{x}_o)\cdot P_G(S) > 0$$

which contradicts the absolute continuity of F.

Now we turn to the proof of the irreducibility of F. We use again an indirect method. Let $F=G_1\cdot G_2$ be a supposed decomposition of F. Then

$$\frac{\partial(F/G_1)}{\partial x} = \Big(\frac{\partial F}{\partial x}\cdot G_1 - \frac{\partial G_1}{\partial x} F\Big)/G_1^2 \geq 0, \qquad (5)$$

$$\frac{\partial(F/G_1)}{\partial y} = \Big(\frac{\partial F}{\partial y}\cdot G_1 - \frac{\partial G_1}{\partial y}\cdot F\Big)/G_1^2 \geq 0. \qquad (6)$$

$$0 \leq \frac{\partial^2(F/G_1)}{\partial y\partial x} = \frac{G_1^2\big(f\cdot G_1 + \frac{\partial F}{\partial x}\cdot\frac{\partial G}{\partial y} - g_1 F - \frac{\partial G_1}{\partial x}\cdot\frac{\partial F}{\partial y}}{G_1^4}$$

$$- \frac{2G_1\frac{\partial G}{\partial y}\left(\frac{\partial F}{\partial x}G_1 - \frac{\partial G_1}{\partial x}\cdot F\right)}{G_1^4}$$

$$\leq \frac{fG_1^2 - \frac{\partial F}{\partial x}\cdot\frac{\partial G_1}{\partial y}G_1 - \frac{\partial G_1}{\partial x}\cdot\frac{\partial F}{\partial y}G_1 + 2\frac{\partial G_1}{\partial y}\cdot\frac{\partial G_1}{\partial x}\cdot F}{G_1^3} . \qquad (7)$$

Let us consider (7) where $y \to +\infty$. Here we may use the estimation $f\cdot F \leq \varepsilon\cdot\frac{\partial F}{\partial x}\cdot\frac{\partial F}{\partial y}$ by (3). Let us denote $\frac{\partial \log F}{\partial x}$, $\frac{\partial \log F}{\partial y}$, $\frac{\partial \log G}{\partial x}$, $\frac{\partial \log G}{\partial y}$ by A, B, a, b, respectively. Thus /multiplying (7) by G_1/F / we get

$$0 \leq A\cdot B\cdot \varepsilon - A\, B - B\cdot a + 2a\cdot b. \qquad (8)$$

$A \geq a$, $B \geq b$ are obvious. We have supposed $A\cdot B > 0$, so we may divide (8) by $A\cdot B$:

$$0 \leq \varepsilon - \frac{b}{B} - \frac{a}{A} + 2\frac{a}{A}\cdot\frac{b}{B} . \qquad (9)$$

We show that inequality (9) can be true only if $\frac{a}{A} \leq \varepsilon$ and $\frac{b}{B} \leq \varepsilon$ or if $\frac{a}{A} \geq 1-2\varepsilon$, $\frac{b}{B} \geq 1-2\varepsilon$. Let us suppose $\frac{b}{B} \leq \frac{a}{A}$. If $\frac{b}{B} < \frac{1}{2}$ then (9) implies $\frac{b}{B} \leq \varepsilon$, so we only have to consider the case $b/B \geq 1/2$. Then we get

$$0 \leq \varepsilon - \frac{b}{B} + \frac{a}{A}\left(2\frac{b}{B} - 1\right)$$

thus

$$0 \leq \varepsilon - 2\frac{b}{B} + 2\left(\frac{b}{B}\right)^2$$

which implies that $\frac{b}{B} \geq 1-\varepsilon$ /if ε is small enough/ and hence by (9) we get that

$$\frac{a}{A} \geq 1 - \varepsilon - \varepsilon = 1 - 2\varepsilon.$$

We can reformulate our result in the following inequalities: in our decomposition

$$\log F(x,y) = \log G_1(x,y) + \log G_2(x,y)$$

for any point (x,y) we can choose an i /$1 \leq i \leq 2$/ such that for this divisor G_i

$$\frac{\partial \log G_i(x,y)}{\partial x} \leq 2\varepsilon(x,y)\cdot\frac{\partial \log F(x,y)}{\partial x}$$

and /for the same i/

$$\frac{\partial \log G_i(x,y)}{\partial y} \leq 2\varepsilon(x,y)\cdot\frac{\partial \log F(x,y)}{\partial y} . \qquad (10)$$

The continuity of our functions and the positivity of $\frac{\partial \log F}{\partial\cdot}$ implies that for N large enough on the set $H = \{(x,y) : x \geq -k,\ y \geq N\}$ for any $(x,y) \in H$ the same G_i is small, because on H we can suppose $\varepsilon(x,y) < 1/10$ and so if $(x_1,y_1) \in H$ and

$$\frac{\partial \log G_1(x_1,y_1)}{\partial x} \Big/ \frac{\partial \log F(x_1,y_1)}{\partial x} \leq 1/10 ,$$

in another $(x_2,y_2) \in H$

$$\frac{\partial \log G_2(x_2,y_2)}{\partial x} \Big/ \frac{\partial \log F(x_2,y_2)}{\partial x} \leq 1/10$$

held then somewhere on H $\partial \log G_1(x,y)/\partial x = \frac{1}{2}\partial \log F(x,y)/\partial x$ would hold which contradicts (10). Let us suppose that G_1 is this small component.

$$\int_{-k}^{+\infty} \frac{\partial \log G_1(x,y_n)}{\partial x} dx \leq 2\varepsilon \int_{-k}^{+\infty} \frac{\partial \log F(x,y_n)}{\partial x} dx$$

$$= 2\varepsilon\left(\log F(+\infty,y_n) - \log F(-k,y_n)\right)$$

which tends to 0 if $y_n \to +\infty$ because $-\log F(-k,y_n)$ is a bounded sequence. Hence $\log G_1(+\infty,y_n) - \log G_1(-k,y_n) \to 0$ which implies $\log\left(G_1(-k,y_n)\right) \to 0$ for any $k \in \mathbb{N}$ and it contradicts our assumptions /i.e. G_1 cannot be a probability distribution function/.

<u>Remark</u>. A slight modificiation of this theorem enables us to prove the effective irreducibility of some bounded /from below/ random variables. /This means that they can only be decomposed into the form F=F G./ For example Gumbel's exponential distribution

$$F(x,y) = 1-e^{-x}-e^{-y}+e^{-x+y+\vartheta xy} \qquad /\vartheta > 0/$$

on the positive quadrant $x > 0$, $y > 0$ is effective irreducible.

<u>Statement</u>. The two dimensional normal distribution is not max-anti-irreducible.

<u>Proof</u>. The statement is a simple consequence of the following theorem (8): an absolutely continuous inf. div. distribution function F in $\mathcal{M}_2$ is antiirreducible if and only if there is a sequence of rectangles

$$(T_k)_{k=1}^{\infty} = [x_{k+1},x_k)\times[y_{k+1},y_k) \quad /x_k \searrow -\infty, y_k \searrow -\infty,\ k\to\infty/$$

where $x_1=+\infty$, $y_1=+\infty$ is not excluded, with the following properties:

/i/ $\operatorname{supp} P_F \subset \bigcup_{k=1}^{\infty} T_k$

/ii/ $\operatorname{supp} P_F \cap T_k \neq \emptyset$ for all $k \in \mathbb{N}$.

<u>Remark</u>. As one can guess from the form of the quoted description of infinitely divisible antiirreducible distribution functions - only a very "artificial" irreducible divisor of such distributions which do not fulfill the properties above is constructed in the proof. It would be interesting in the case of normal distributions to find a more nat-

ural irreducible divisor of it.

Finally I express my thanks to my colleagues T.F. Móri and G.J. Székely for inspiring this research and for their valuable remarks.

References

[1] Balkema, A.A. - Resnick, S.I. /1977/ Max-infinite divisibility, J. Appl. Prob. 14. 309-319.

[2] Linnik, Ju.V. - Ostrowski, I.V. Decomposition of Random Variables and Vectors, Trans. of Math. Monographs Vol. 48, Providence R. I., 1977 /Russian ed.: Nauka, Moscow, 1972/.

[3] Ruzsa, I.Z. - Székely, G.J. /1985/ Theory of decomposition in semigroups, Advances in Mathematics 56, 9-27.

[4] Ruzsa, I.Z. - Székely, G.J. Algebraic Probability Theory, Wiley /to appear/ Preprint of the Math. Inst. Hung. Acad. Sci. 1987/29-30.

[5] Resnick, S.I. Extreme Values, Regular Variation and Point Processes, Springer /1987/.

[6] Zempléni, A. /1987/ On the arithmetical properties of the multiplicative structure of probability distribution functions, in Proc. of the 5th Pannonian Symp. on Mathematical Statistics /W. Grossmann, J. Mogyoródi, I. Vincze, W. Wertz eds./ 221-233. Akadémiai Kiadó, Budapest.

[7] Zempléni, A. /1988/ On the multidimensional multiplicative structure of p.d.f.'s, Proc. of the 4th European Meeting of Young Statisticians /Varna, 1985 - to appear/.

[8] Zempléni, A. /1988/ The description of the class I_o in the max-structure of probability distribution functions, in Proc. of the 6th Pannonian Symp. on Mathematical Statistics /M.L. Puri, P. Révész, W. Wertz eds./ 291-305. Reidel Publ. Co.

Properties of the cosh hypergroup*

by

Hansmartin Zeuner
Mathematisches Institut
der Universität Tübingen
Auf der Morgenstelle 10
D–7400 Tübingen
Federal Republic of Germany

1. Introduction

In this article the structure of the hypergroup defined on $\mathbb{R}_+$ by the formula $\varepsilon_x * \varepsilon_y := \frac{\cosh(x+y)}{2\cosh x \cosh y}\varepsilon_{x+y} + \frac{\cosh(x-y)}{2\cosh x \cosh y}\varepsilon_{|x-y|}$ for $x, y \in \mathbb{R}_+$ will be studied. This is one of the most elementary examples of a hypergroup not related to a group. However, although some of its subhypergroups were studied by Gilewski, Urbanik [5] and Bloom, Selvanathan [1], it did not appear in the literature so far. Unlike many hypergroups on $\mathbb{R}_+$, where the characters and even the operation itself can only be defined indirectly as solutions of certain differential equations (see Chébli [2], [3], and Zeuner [11]), in the case of this hypergroup everything can be calculated in closed form. It is therefore an easily accessible example for some of the strange properties of hypergroups.

Two special problems of commutative hypergroups will be examplified in this article: First it will be shown in detail why there is no dual hypergroup structure $\star$ on the dual $\hat{K}$ of this hypergroup related to the original hypergroup by the formula $\xi(x)\chi(x) = \int_{\hat{K}} \tau(x)\, d(\varepsilon_\xi \star \varepsilon_\chi)$ for all $\xi, \chi \in \hat{K}$, $x \in \mathbb{R}_+$. It turns out that a probability measure $\varepsilon_\xi \star \varepsilon_\chi$ with this property does not exists for all ξ and χ, and when it exists it does not fulfill the requirement of a compact support. Secondly this hypergroup serves as an example to show the problems in defining a factor hypergroup with respect to a *non compact* normal subhypergroup.

* Die Arbeit wurde mit Unterstützung eines Stipendiums des Wissenschaftsrats der NATO durch den DAAD ermöglicht

In contrast to this situation we study (in completion of earlier results of Mizony [9]) the dual structure of Naimark's hypergroup. In this case a dual convolution product in the sense of the preceding paragraph exists for *all* elements of $\hat{K}$; however the support of this measure is not compact except in the trivial cases and there is no involution such that condition (HG 5) of Heyer [6] is fulfilled.

2. Hypergroups on the halfline

Let $(K, *)$ be a commutative hypergroup (see Heyer [6] and Jewett [7] for reference) and $\hat{K}$ its dual, that is the set of bounded continuous functions φ on K such that $\varphi(x) \cdot \varphi(y) = \int_K \varphi \, d\varepsilon_x * \varepsilon_y$ and $\varphi(x^\vee) = \overline{\varphi(x)}$ for $x, y \in K$. Then the Fourier transform of a function f which is integrable with respect to a Haar measure ω of K is the complex valued function $\mathcal{F}f : \varphi \to \int_K f \cdot \varphi \, d\omega$ on $\hat{K}$. The Plancherel measure is the unique measure $\hat{\omega}$ on $\hat{K}$ such that $\int_K |f|^2 \, d\omega = \int_{\hat{K}} |\mathcal{F}f|^2 \, d\hat{\omega}$ for all $f \in \mathcal{L}^1(\omega) \cap \mathcal{L}^2(\omega)$.

In the case that K is the halfline $\mathbb{R}_+ = [0, \infty[$, many different hypergroups structures are known. In all known examples the characters $\varphi \in \hat{K}$ are the solutions φ_λ of the differential equation $\varphi_\lambda'' + \frac{A'}{A}\varphi_\lambda' + (\rho^2 + \lambda^2)\varphi_\lambda = 0, \quad \varphi_\lambda(0) = 1, \quad \varphi_\lambda'(0) = 0$, where A is a C^∞-function on $\mathbb{R}_+$ satisfying certain conditions (see Chébli [2], [3]), $\rho := \frac{1}{2} \lim_{x \to \infty} \frac{A'(x)}{A(x)}$ and $\lambda \in \mathbb{R}_+ \cup i\,[0, \rho]$.

Although it is not known yet whether all hypergroups on the halfline are of this form, some general facts valid for all of these can easily be proved (see Zeuner [11] for details): every hypergroup on $\mathbb{R}_+$ is commutative, the inversion is given by $x^\vee = x$, 0 is the neutral element, and after a suitable normalization we obtain $\max supp(\varepsilon_x * \varepsilon_y) = x + y$ and $\min supp(\varepsilon_x * \varepsilon_y) = |x - y|$ for $x, y \in \mathbb{R}_+$. In all known examples the support of $\varepsilon_x * \varepsilon_y$ is the whole intervall $[|x - y|, x + y]$ except in two cases where the support is the two element set $\{|x - y|, x + y\}$ for $x, y > 0$ (see Zeuner [11], proposition 4.2): the convolution $\varepsilon_x * \varepsilon_y := \frac{1}{2}(\varepsilon_{x+y} + \varepsilon_{|x-y|})$ corresponding to the function $A(x) = 1$ and the hypergroup with $A(x) = (\cosh x)^2$ for all $x \in \mathbb{R}_+$ which will be studied in this article.

3. The cosh hypergroup and its characters

Let us define a convolution $*$ on $\mathbb{R}_+$ by the following formula:

$$\varepsilon_x * \varepsilon_y := \frac{\cosh(x + y)}{2 \cosh x \cosh y} \varepsilon_{x+y} + \frac{\cosh(x - y)}{2 \cosh x \cosh y} \varepsilon_{|x-y|} \quad \text{for } x, y \in \mathbb{R}_+.$$

In order to show that this operation defines the structure of a hypergroup on $\mathbb{R}_+$ we have to prove that $*$ is associative. To do this we introduce the functions

$$\varphi_\lambda(x) := \frac{\cos \lambda x}{\cosh x} \quad (x \in \mathbb{R}_+)$$

for every $\lambda \in \mathbb{C}$. It is clear that $\varphi_{i\lambda}(x) = \frac{\cosh \lambda x}{\cosh x}$ for $x \in \mathbb{R}_+$ and $\lambda \in \mathbb{R}$ and especially $\varphi_i = 1$.

3.1. Lemma: *For every $x, y \in \mathbb{R}_+$ and $\lambda \in \mathbb{C}$ we have*

$$\int \varphi_\lambda \, d\varepsilon_x * \varepsilon_y = \varphi_\lambda(x)\, \varphi_\lambda(y).$$

Proof: This follows from $\cos \lambda(x+y) + \cos \lambda|x-y| = 2 \cos \lambda x \cos \lambda y$. ■

In other words, every φ_λ ($\lambda \in \mathbb{C}$) is a multiplicative functional — and it follows from Zeuner [11], proposition 4.2 that these are all of them. Let $x, y, z \in \mathbb{R}_+$. If we define μ_0 to be the measure with density $\frac{1}{\cosh}$ with respect to $(\varepsilon_x * \varepsilon_y) * \varepsilon_z$ and μ_1 the measure with the same density with respect to $\varepsilon_x * (\varepsilon_y * \varepsilon_z)$, it is a direct consequence of the lemma that $\int \cos \lambda t \, \mu_0(dt) = \int \cos \lambda t \, \mu_1(dt)$ for all $\lambda \in \mathbb{R}_+$. It follows from the uniqueness theorem for the Fourier cosine transformation that $\mu_0 = \mu_1$ and therefore $(\varepsilon_x * \varepsilon_y) * \varepsilon_z = \varepsilon_x * (\varepsilon_y * \varepsilon_z)$. This establishes the associativity of $*$. Using *supp* $\varepsilon_x * \varepsilon_y = \{|x-y|, x+y\}$ it is easily proved that $(\mathbb{R}_+, *)$ satisfies the axioms of a hypergroup. Since this hypergroup is closely related to the hyperbolic cosine function we will denote it as the *cosh hypergroup*.

Next we will calculate the Haar measure and the Plancherel measure for this hypergroup.

3.2. Proposition: *The measure ω with density $x \mapsto (\cosh x)^2$ with respect to the Lebesgue measure λ_+ on the halfline is a Haar measure for the cosh hypergroup.*

Proof: This can be verified by straightforward calculation. It also follows from Zeuner [11] proposition 4.2. ■

In order to obtain the dual $\hat{K}$ (i.e. the set of bounded multiplicative functions φ with $\varphi(x^{-}) = \overline{\varphi(x)}$ for all $x \in \mathbb{R}_+$) of the cosh hypergroup we simply observe that the function $x \mapsto \varphi_\lambda(x) = \frac{\cos \lambda x}{\cosh x}$ is bounded if and only if $|\Im \lambda| \leq 1$ and that $\varphi_\lambda(x)$ is real for all $x \in \mathbb{R}_+$ if and only if $\lambda \in \mathbb{R}$ or $\lambda \in i\mathbb{R}$. Since $\varphi_\lambda = \varphi_{-\lambda}$ we therefore have $\hat{K} = \mathbb{R}_+ \cup i\,[0,1]$.

3.3. Proposition: $\frac{2}{\pi}\lambda_+$ *is the Plancherel measure on* $\hat{K}$.

Proof: Let $f \in \mathcal{L}^1(\omega) \cap \mathcal{L}^2(\omega)$ and $f_1(x) := f(|x|) \cdot \cosh x$. Then the Fourier transform of f is $\mathcal{F}f(\lambda) = \int_0^\infty f\varphi_\lambda \, d\omega = \frac{1}{2}\int_{-\infty}^\infty f_1(x)\cos(\lambda x)\,dx = \frac{1}{2}\hat{f}_1(\lambda)$ where $\hat{}$ is the usual Fourier transform on the real line. Therefore we obtain $\int_0^\infty |\mathcal{F}f(\lambda)|^2\,d\lambda = \frac{1}{4}\int_0^\infty |\hat{f}_1(\lambda)|^2\,d\lambda = \frac{\pi}{2}\int_0^\infty |f_1(x)|^2\,dx = \frac{\pi}{2}\int_0^\infty |f(x)|^2\,\omega(dx)$ by the Plancherel formula and $\frac{2}{\pi}\lambda_+$ is the Plancherel measure of the cosh hypergroup. ∎

4. The convolution structure of $\hat{K}$

A commutative hypergroup K is said to have a dual hypergroup with respect to pointwise multiplication if for every $\xi, \chi \in \hat{K}$ there exists a probability measure $\varepsilon_\xi \star \varepsilon_\chi$ on $\hat{K}$ such that $\xi(x)\chi(x) = (\varepsilon_\xi \star \varepsilon_\chi)^\vee(x) := \int \tau(x)\,\varepsilon_\xi \star \varepsilon_\chi(d\tau)$ and such that $(\hat{K}, \star)$ is a hypergroup. In this paragraph we will study in detail why the cosh hypergroup does not have a dual hypergroup with respect to pointwise multiplication.

4.1. Lemma: *Let K be a Chébli-Trimèche hypergroup with $\rho > 0$, μ be a probability measure on $\hat{K} = \mathbb{R}_+ \cup i[0,\rho]$ and $a \in]0,\rho]$.*

a) *If $\mu(i]a,\rho]) = 0$ then*

$$\lim_{x\to\infty} \frac{\mu^\vee(x)}{\varphi_{ia}(x)} = \mu(\{ia\}).$$

b) *If $\mu(i]a,\rho]) > 0$ then* $\lim_{x\to\infty} \frac{\mu^\vee(x)}{\varphi_{ia}(x)} = \infty.$

Proof: a) It follows from Voit [10], equation (17) and corollary 2.8 that $\lim_{x\to\infty} \frac{\varphi_\lambda(x)}{\varphi_{ia}(x)} = 0$ for all $\lambda \in i[0,a[\cup\mathbb{R}_+$. Furthermore $|\frac{\varphi_\lambda(x)}{\varphi_{ia}(x)}| \le 1$ by Zeuner [12], 4.9 b). Therefore the theorem of majorized convergence implies

$$\lim_{x\to\infty} \frac{\mu^\vee(x)}{\varphi_{ia}(x)} = \int \lim_{x\to\infty} \frac{\varphi_\lambda(x)}{\varphi_{ia}(x)}\,\mu(d\lambda) = \mu(\{ia\}).$$

The proof of b) is similar using the fact that $\lim_{x\to\infty} \frac{\varphi_\lambda(x)}{\varphi_{ia}(x)} = \infty$ for $\lambda \in i]a,\rho]$. ∎

4.2. Proposition: Let $(K, *)$ be the cosh hypergroup.

a) *If $\xi, \chi \in \mathbb{R}_+$ then the measure μ with Lebesgue-density*

$$\lambda \mapsto \frac{1}{4}\Big(\frac{1}{\cosh\frac{\pi}{2}(\lambda+\xi+\chi)} + \frac{1}{\cosh\frac{\pi}{2}(\lambda+\xi-\chi)} + \\ + \frac{1}{\cosh\frac{\pi}{2}(\lambda-\xi+\chi)} + \frac{1}{\cosh\frac{\pi}{2}(\lambda-\xi-\chi)}\Big)$$

on $\mathbb{R}_+$ *is the unique probability measure on* $\hat{K}$ *such that* $\varphi_\xi \cdot \varphi_\chi = \mu^\vee$.

b) *If* $\xi \in \mathbb{R}_+$ *and* $\chi \in [0,1[$ *then the measure* μ *with Lebesgue-density*

$$\lambda \mapsto \frac{\cos\frac{\pi}{2}\chi}{2}\Big(\frac{\cosh\frac{\pi}{2}(\lambda+\xi)}{\cosh^2\frac{\pi}{2}(\lambda+\xi)\cos^2\frac{\pi}{2}\chi + \sinh^2\frac{\pi}{2}(\lambda+\xi)\sin^2\frac{\pi}{2}\chi} + \frac{\cosh\frac{\pi}{2}(\lambda-\xi)}{\cosh^2\frac{\pi}{2}(\lambda-\xi)\cos^2\frac{\pi}{2}\chi + \sinh^2\frac{\pi}{2}(\lambda-\xi)\sin^2\frac{\pi}{2}\chi}\Big)$$

on $\mathbb{R}_+$ *is the unique probability measure on* $\hat{K}$ *such that* $\varphi_\xi \cdot \varphi_{i\chi} = \mu^\vee$.

c) *If* $\xi, \chi \in [0,1]$ *and* $\xi + \chi < 1$ *then the measure* μ *with Lebesgue-density*

$$\lambda \mapsto \frac{\cosh\frac{\pi}{2}\lambda}{2}\Big(\frac{\cos\frac{\pi}{2}(\xi+\chi)}{\cosh^2\frac{\pi}{2}\lambda\cos^2\frac{\pi}{2}(\xi+\chi) + \sinh^2\frac{\pi}{2}\lambda\sin^2\frac{\pi}{2}(\xi+\chi)} + \frac{\cos\frac{\pi}{2}(\xi-\chi)}{\cosh^2\frac{\pi}{2}\lambda\cos^2\frac{\pi}{2}(\xi-\chi) + \sinh^2\frac{\pi}{2}\lambda\sin^2\frac{\pi}{2}(\xi-\chi)}\Big)$$

on $\mathbb{R}_+$ *is the unique probability measure on* $\hat{K}$ *such that* $\varphi_{i\xi} \cdot \varphi_{i\chi} = \mu^\vee$.

d) *If* $\xi, \chi \in]0,1[$ *and* $\xi + \chi = 1$ *then the measure* $\mu := \frac{1}{2}(\varepsilon_0 + \nu)$ *where* ν *has the Lebesgue-density*

$$\lambda \mapsto \frac{\cosh\frac{\pi}{2}\lambda\sin\pi\xi}{\cosh^2\frac{\pi}{2}\lambda\sin^2\pi\xi + \sinh^2\frac{\pi}{2}\lambda\cos^2\pi\xi}$$

on $\mathbb{R}_+$ *is the unique probability measure on* $\hat{K}$ *such that* $\varphi_{i\xi} \cdot \varphi_{i\chi} = \mu^\vee$.

e) *If* $\xi, \chi \in]0,1[$ *and* $\xi + \chi > 1$ *then there exists no probability measure* μ *on* $\hat{K}$ *such that* $\varphi_{i\xi} \cdot \varphi_{i\chi} = \mu^\vee$.

Proof: In order to prove assertions a), b), c), and d) we use the following formula which can easily be obtained by the residual calculus:

$$\int_{-\infty}^{\infty} \frac{\cos\lambda x}{\cosh\frac{\pi}{2}(\lambda+\xi)}\,d\lambda = \frac{2\cos\xi x}{\cosh x} \text{ for } x \in \mathbb{R}_+ \text{ and } \xi \in \mathbb{C} \text{ with } |\Im\xi| < 1 \qquad (1)$$

(see Erdéli [4], p. 30, 1.9 (1)).

a) If μ is defined as in a) then $\mu^\vee(x) =$
$= \frac{1}{4\cosh x}\int_{-\infty}^{\infty} \frac{\cos\lambda x}{\cosh\frac{\pi}{2}(\lambda+\xi+\chi)} + \frac{\cos\lambda x}{\cosh\frac{\pi}{2}(\lambda+\xi-\chi)}\,d\lambda = \frac{\cos(\xi+\chi)x + \cos(\xi-\chi)x}{2\cosh^2 x} = \varphi_\xi(x)\cdot\varphi_\chi(x)$.
Since the density function of μ is positive and integrates to 1 (this can be obtained from the last formula by setting $x := 0$), μ is a probability measure. The uniqueness follows from Jewett [7], 12.2 A.

b) By replacing χ by $i\chi$ and combining the first and the third term resp. the second and forth term in the density in a) we obtain μ defined as in b). $\mu^\vee = \varphi_\xi \cdot \varphi_{i\chi}$ follows from (1) since $|\Im(\xi \pm i\chi)| < 1$. The same argument as in a) shows that μ is a probability measure.

c) is proved in the same way.

d) follows from c) by continuity.

e) Assume that μ is a probability measure on $\hat{K} = \mathbb{R}_+ \cup i\,[0,1]$ such that $\mu^\vee(x) = \varphi_{i\xi}(x) \cdot \varphi_{i\chi}(x)$. Since $\varphi_{i\lambda}(x) \sim e^{-(1-\lambda)x}$ the right hand side is $O(e^{-(2-\xi-\chi)x})$ as x tends to infinity and it follows from part b) of the lemma that $\mu(i]a,1]) = 0$ where $a := \xi + \chi - 1 > 0$. From part a) of the lemma we obtain

$$\mu(\{ia\}) = \lim_{x\to\infty} e^{(1-a)x}\mu^\vee(x) = \lim_{x\to\infty} e^{(2-\xi-\chi)x}\,\frac{\cosh\xi x \cdot \cosh\chi x}{\cosh^2 x} = 1.$$

But that would imply $\mu = \varepsilon_{ia}$ (since μ is a probability) and $\cosh\xi x \cdot \cosh\chi x = \cosh(1-\xi-\chi)x \cdot \cosh x$ for all $x \in \mathbb{R}_+$ which is a contradiction unless $\xi = 1$ or $\chi = 1$. ∎

4.3. Remark: Since by proposition 3.3 the support of the Plancherel measure is $\mathbb{R}_+ \neq \hat{K}(= \mathbb{R}_+ \cup i[0,1])$ the cosh hypergroup cannot have a dual with respect to pointwise multiplication. This also follows from proposition 4.2 by the following two reasons:

1. There are elements $\xi, \chi \in \hat{K}$ such that $\varphi_\xi \cdot \varphi_\chi = \mu^\vee$ for no probability measure μ on $\hat{K}$.

2. In cases a), b), c), and d) of the proposition $\varepsilon_\xi \star \varepsilon_\chi$ exists; however the support of this measure is not compact.

5. Subhypergroups and factors

5.1. Proposition: *The cosh hypergroup has exactly the following non trivial (closed) subhypergroups:* $H_a := a\mathbb{N} := \{0, a, 2a, 3a, \cdots\}$ *for* $a > 0$.

Proof: If H $(\neq \{0\}, K)$ is a subhypergroup then $x, y \in H$ implies $|x-y|$, $x+y \in H$ by the definition of the convolution. Since H is closed $a := \inf\{|x-y| > 0 \,:\, x, y \in H\}$ is an element of H, and $a > 0$ (since $a = 0$ would imply $H = K$). Therefore $\{0, a, 2a, 3a, \cdots\}$ is contained in H and by the choice of a both sets are equal. ∎

5.2. Remark: The hypergroups H_a have been studied in Gilewski, Urbanik [5] and more recently in Bloom, Selvanathan [1]. It is surprising that H_a and H_b are not isomorphic as hypergroups unless $a = b$.

5.3. Remark: Proposition 5.1 is also true for the hypergroup defined by $\varepsilon_x \bullet \varepsilon_y := \frac{1}{2}\varepsilon_{x+y} + \frac{1}{2}\varepsilon_{|x-y|}$. For all hypergroups on $\mathbb{R}_+$ with *supp* $\varepsilon_x * \varepsilon_y = [|x-y|, x+y]$ the only subhypergroups are $\mathbb{R}_+$ and $\{0\}$. This follows from $x \in H \Rightarrow [0, 2x] \subset H$.

5.4. Remark: If H is a *compact* subhypergroup of a hypergroup H a hypergroup operation on the double coset space $K/\!\!/H$ can be defined by $\varepsilon_{HxH} \star \varepsilon_{HyH} := \int_H \varepsilon_{HxhyH}\, \omega_H(dh)$ (where ω_H is the normalized Haar measure of H). Since H_a is not compact this method to define a hypergroup structure on the factor space is not possible.

There exist commutative hypergroups K where a factor hypergroup with respect to H can be determined by another approach: If there exists a hypergroup homomorphism h from K onto another hypergroup $(J, \circ)$ (i.e. $h(\varepsilon_x * \varepsilon_y) = \varepsilon_{h(x)} \circ \varepsilon_{h(y)}$ for $x, y \in K$) such that $H = \{x \in K : h(x) = e\}$ we might call J the factor hypergroup of K with respect to H even if H is not compact. For example for the hypergroup on $K := \mathbb{R}_+$ defined by $\varepsilon_x \bullet \varepsilon_y := \frac{1}{2}\varepsilon_{x+y} + \frac{1}{2}\varepsilon_{|x-y|}$ and the subhypergroup $H := \mathbb{N}$ the factor hypergroup is $J := [0,1]$ with the operation $\varepsilon_x \circ \varepsilon_y := \frac{1}{2}\varepsilon_{1-|1-x-y|} + \frac{1}{2}\varepsilon_{|x-y|}$. The homomorphism $h : K \to J$ is $h(x) = 2d(x, \mathbb{N})$ in this case.

However, in the situation of the cosh hypergroup K even this approach is not possible: assume that there is an homomorphism h from K onto J such that H_a is the kernel of h. Then

$$h(\tfrac{a}{4}) = h(\tfrac{3a}{4}) \qquad \text{but} \qquad h(\varepsilon_{\frac{a}{4}} * \varepsilon_{\frac{a}{4}}) = \frac{\cosh a/2}{2\cosh^2 a/4}\varepsilon_{h(a/2)} + \frac{1}{2\cosh^2 a/4}\varepsilon_{h(0)} \neq$$
$$\neq \frac{\cosh a/2}{2\cosh a/4 \cosh 3a/4}\varepsilon_{h(a/2)} + \frac{\cosh a}{2\cosh a/4 \cosh 3a/4}\varepsilon_{h(0)} = h(\varepsilon_{\frac{a}{4}} * \varepsilon_{\frac{3a}{4}}).$$

This example shows that the restriction is essential that a subhypergroup has to be compact in order to allow a factor hypergroup.

6. The dual of Naimark's hypergroup

A hypergroup on $\mathbb{R}_+$ which is closely related to the cosh hypergroup is defined by the convolution

$$\varepsilon_x * \varepsilon_y := \frac{1}{2\sinh x \sinh y} \int_{|x-y|}^{x+y} \varepsilon_t \sinh t \, dt.$$

The Haar measure has the Lebesgue density $\sinh^2$, the characters are the functions $\varphi_\lambda(x) := \frac{\sin \lambda x}{\lambda \sinh x}$ where $\lambda \in \hat{K} = \mathbb{R}_+ \cup i[0,1]$, and the Plancherel measure has the density $\lambda \mapsto \frac{2}{\pi}\lambda^2$ with respect to the Lebesgue measure on $\mathbb{R}_+$ (the different form of the Plancherel measure in Jewett [7], 9.5 comes from a different parametrization of $\hat{K}$).

In this situation there exists a dual convolution product $\varepsilon_\xi \star \varepsilon_\chi \in \mathcal{M}^1(\hat{K})$ such that $\varphi_\xi(x)\varphi_\chi(x) = \int_{\hat{K}} \varphi_\lambda(x)\,\varepsilon_\xi \star \varepsilon_\chi(d\lambda)$ for *all* $\xi,\chi \in \hat{K}$. In the case of $\xi,\chi \in \mathbb{R}_+$, the support of the Plancherel measure, this follows from Koornwinder [8], 8.3 and the explicit calculation of the density of $\varepsilon_\xi \star \varepsilon_\chi$ can be found in Mizony [9], p. 11.

6.1. Proposition: a) *If $\xi,\chi \in \mathbb{R}_+$ then the measure μ with Lebesgue-density*

$$\lambda \mapsto \frac{\sinh \pi\xi \sinh \pi\chi}{\xi\chi} \frac{\lambda \sinh \pi\lambda}{(\cosh \pi\lambda + \cosh \pi\xi \cosh \pi\chi)^2 - \sinh^2 \pi\xi \sinh^2 \pi\chi} =$$

$$= \frac{\lambda}{\xi\chi}\Big(\tanh \frac{\pi}{2}(\lambda - \xi + \chi) + \tanh \frac{\pi}{2}(\lambda + \xi - \chi) -$$

$$- \tanh \frac{\pi}{2}(\lambda + \xi + \chi) - \tanh \frac{\pi}{2}(\lambda - \xi - \chi)\Big)$$

on $\mathbb{R}_+$ is the unique probability measure on $\hat{K}$ such that $\varphi_\xi \cdot \varphi_\chi = \mu^\vee$.

b) *If $\xi \in \mathbb{R}_+$ and $\chi \in [0,1[$ then the measure μ with Lebesgue-density*

$$\lambda \mapsto \frac{\sinh \pi\xi \sin \pi\chi}{\xi\chi} \frac{\lambda \sinh \pi\lambda}{(\cosh \pi\lambda + \cosh \pi\xi \cos \pi\chi)^2 + \sinh^2 \pi\xi \sin^2 \pi\chi}$$

on $\mathbb{R}_+$ is the unique probability measure on $\hat{K}$ such that $\varphi_\xi \cdot \varphi_{i\chi} = \mu^\vee$.

c) *If $\xi,\chi \in [0,1]$ and $\xi + \chi < 1$ then the measure μ with Lebesgue-density*

$$\lambda \mapsto \frac{\sin \pi\xi \sin \pi\chi}{\xi\chi} \frac{\lambda \sinh \pi\lambda}{(\cosh \pi\lambda + \cos \pi(\xi - \chi))(\cosh \pi\lambda + \cos \pi(\xi + \chi))}$$

on $\mathbb{R}_+$ is the unique probability measure on $\hat{K}$ such that $\varphi_{i\xi} \cdot \varphi_{i\chi} = \mu^\vee$.

d) *If $\xi,\chi \in]0,1[$ and $\xi + \chi = 1$ then the measure μ with the Lebesgue-density*

$$\lambda \mapsto \frac{\sin \pi\xi \sin \pi\chi}{\xi\chi} \frac{\lambda \coth \frac{\pi}{2}\lambda}{\cosh \pi\lambda + \cos \pi(\xi - \chi)}$$

on $\mathbb{R}_+$ is the unique probability measure on $\hat{K}$ such that $\varphi_{i\xi} \cdot \varphi_{i\chi} = \mu^\vee$.

e) *If $\xi,\chi \in]0,1[$ and $\xi + \chi > 1$ then the measure $\mu := \frac{\xi+\chi-1}{\xi\chi} \cdot \varepsilon_{\xi+\chi-1} + \nu$ where ν has the Lebesgue-density*

$$\lambda \mapsto \frac{\sin \pi\xi \sin \pi\chi}{\xi\chi} \frac{\lambda \sinh \pi\lambda}{(\cosh \pi\lambda + \cos \pi(\xi - \chi))(\cosh \pi\lambda + \cos \pi(\xi + \chi))}$$

is the unique probability measure on $\hat{K}$ *such that* $\varphi_{i\xi} \cdot \varphi_{i\chi} = \mu^\vee$.

Proof: It follows from Erdéli [4], p. 88, 2.9. (7) that $\int_0^\infty \frac{\cosh ax - \cosh bx}{\sinh x} \sin \lambda x \, dx = \frac{\pi(\cos \pi b - \cos \pi a)}{2} \frac{\sinh \pi\lambda}{(\cosh \pi\lambda + \cos \pi a)(\cosh \pi\lambda + \cos \pi b)}$ if $\Re a$, $\Re b < 1$. By the inversion theorem for the Fourier sine transformation we obtain for $a := \xi - \chi$ and $b := \xi + \chi$

$$\varphi_{i\xi}(x)\varphi_{i\chi}(x) =$$

$$= \int_0^\infty \frac{\sin \pi\xi \sin \pi\chi}{\xi\chi} \frac{\lambda \sinh \lambda x}{(\cosh \pi\lambda + \cos \pi(\xi - \chi))(\cosh \pi\lambda + \cos \pi(\xi + \chi))} \varphi_\lambda(x) \, d\lambda. \quad (2)$$

From this we conclude a), b), c), and d) with the same arguments as in the proof of 4.2.

e) Since $\Re(2 - \xi - \chi)$, $\Re(\xi - \chi) < 1$ we may apply (2) in order to obtain

$$\varphi_{i\xi}(x)\varphi_{i\chi}(x) - \frac{\xi + \chi - 1}{\xi\chi} \varphi_{\xi+\chi-1}(x) = \frac{\cosh(2 - \xi - \chi)x - \cosh(\xi - \chi)x}{2\xi\chi \sinh^2 x} =$$

$$= \int_0^\infty \frac{\sin \pi\xi \sin \pi\chi}{\xi\chi} \frac{\lambda \sinh \lambda x}{(\cosh \pi\lambda + \cos \pi(\xi - \chi))(\cosh \pi\lambda + \cos \pi(2 - \xi - \chi))} \varphi_\lambda(x) \, d\lambda.$$

But from $\cos \pi(2 - \xi - \chi)x = \cos \pi(\xi + \chi)x$ the assertion follows.

■

Bibliography

[1] W.R. Bloom, S. Selvanathan, Hypergroup structures on the set of the natural numbers. *Bull. Austral. Math. Soc.*, **33** (1986), 89-102.

[2] H. Chébli: *Positivité des opérateurs de "translation généralisée" associées à un opérateur de Sturm-Liouville et quelques applications à l'analyse harmonique.* Thèse, Université Louis Pasteur, Strasbourg I (1974).

[3] H. Chébli: Opérateurs de translation généralisée et semi-groupes de convolution. In: *Théorie du Potentiel et Analyse Harmonique.* Edité par J. Faraut. Lecture Notes in Mathematics **404**, Springer Verlag Berlin-Heidelberg-New York, 1974.

[4] A. Erdélyi et al. *Tables of Integral Transforms*, Vol. I. McGraw-Hill, New York-Toronto-London (1954).

[5] J. Gilewski, K. Urbanik: Generalized convolutions and generating functions. *Bull. Acad. Polon. Sci. Sér. Sci. Math. Astronom. Phys.* **16** (1968), 481-487.

[6] H. Heyer: Probability theory on hypergroups: A survey. In: *Probability Measures on Groups VII*, edited by H. Heyer. Lecture Notes in Mathematics **1064**. Springer Verlag, Berlin-Heidelberg-New York-Tokyo (1984).

[7] R.I. Jewett: Spaces with an abstract convolution of measures. *Advances in Mathematics* **18** (1975), 1-101.

[8] T. Koornwinder. Jacobi functions and analysis on noncompact semisimple Lie groups. In: *Special Functions: Group Theoretical Aspects and Applications*, edited by R.A. Askey, T.H. Koornwinder, W. Schempp, 1-85.

[9] M. Mizony. Algèbres et noyaux de convolution sur le dual sphérique d'un groupe de Lie semi-simple, non compact et de rang 1. *Publications du Département de Mathématiques de Lyon* **13-1** (1976), 1-14.

[10] M. Voit. Positive characters on commutative hypergroups and some applications. *Math. Z.* **198** (1988), 405-421.

[11] Hm. Zeuner: One-dimensional hypergroups. To appear in: *Advances in Mathematics* (1989).

[12] Hm. Zeuner. Laws of large numbers for hypergroups on $\mathbb{R}_+$. To appear.

PAPERS PRESENTED AT THE CONFERENCE

BUT NOT INCLUDED IN THIS VOLUME

T. Byczkowski: Functional methods in the theory of stochastic processes with values in a locally compact group

E. Dettweiler: Representation of Banach space valued martingales as stochastic integrals

O. Gebuhrer: L^1 ergodic theorems on ${}^K A(G)^K$ algebras

J. Hilgert: Lie semigroups and their applications to Gaussian semigroups

F. Hirsch: On the flow of a stochastic differential equation

H. Rindler: Groups of measure preserving transformations

I.Z. Ruzsa, G.J. Székely: Decomposition of probability measures on groups

G.J. Székely: Generalized generating functions

K. Urbanik Limit sets of probability measures

W. Woess: A converse to the mean value property on homogeneous trees

LIST OF PARTICIPANTS

C. Berg	Kobenhavn, Denmark
M.S. Bingham	Hull, England
N.H. Bingham	Egham, England
Ph. Bougerol	Vandoeuvre les Nancy, France
T. Byczkowski	Wroclaw, Poland
H. Carnal	Bern, Switzerland
A. Derighetti	Lausanne, Switzerland
Y. Derriennic	Brest, France
E. Dettweiler	Ankara, Turkey
J.L. Dunau	Toulouse, France
L. Elie	Paris, France
Ph. Feinsilver	Carbondale, USA
A. Figà-Talamanca	Rome, Italy
L. Gallardo	Brest, France
O. Gebuhrer	Strasbourg, France
P. Gerl	Salzburg, Austria
W. Hazod	Dortmund, Germany
H. Heyer	Tübingen, Germany
T. Hida	Nagoya, Japan
J. Hilgert	Darmstadt, Germany
F. Hirsch	Cachan, France
G. Högnäs	Åbo, Finland
E. Kaniuth	Paderborn, Germany
F. Kinzl	Salzburg, Austria
J. Kisyński	Lublin, Poland
A. Kumar	Delhi, India
R. Lasser	Oberschleißheim, Germany
E. Le Page	Rennes, France
G. Letac	Toulouse, France
M. McCrudden	Manchester, England
A. Mukherjea	Tampa, USA
N. Obata	Nagoya, Japan
M.M. Rao	Riverside, USA

P. Ressel	Eichstätt, Germany
H. Rindler	Wien, Austria
I.Z. Ruzsa	Budapest, Hungary
L.K. Schmetterer	Wien, Austria
R. Schott	Vandoeuvre les Nancy, France
M. Schürmann	Heidelberg, Germany
H. Senateur	Toulouse, France
E. Siebert	Tübingen, Germany
G. Szekely	Budapest, Hungary
A. Terras	La Jolla, USA
G. Turnwald	Tübingen, Germany
K. Urbanik	Wrocław, Poland
M. Voit	München, Germany
R.C. Vrem	Arcata, USA
W. von Waldenfels	Heidelberg, Germany
G.S. Watson	Princeton, USA
W. Woess	Leoben, Austria
K. Ylinen	Turku, Finland
Hm. Zeuner	Tübingen, Germany

Vol. 1201: Curvature and Topology of Riemannian Manifolds. Proceedings, 1985. Edited by K. Shiohama, T. Sakai and T. Sunada. VII, 336 pages. 1986.

Vol. 1202: A. Dür, Möbius Functions, Incidence Algebras and Power Series Representations. XI, 134 pages. 1986.

Vol. 1203: Stochastic Processes and Their Applications. Proceedings, 1985. Edited by K. Itô and T. Hida. VI, 222 pages. 1986.

Vol. 1204: Séminaire de Probabilités XX, 1984/85. Proceedings. Edité par J. Azéma et M. Yor. V, 639 pages. 1986.

Vol. 1205: B.Z. Moroz, Analytic Arithmetic in Algebraic Number Fields. VII, 177 pages. 1986.

Vol. 1206: Probability and Analysis, Varenna (Como) 1985. Seminar. Edited by G. Letta and M. Pratelli. VIII, 280 pages. 1986.

Vol. 1207: P.H. Bérard, Spectral Geometry: Direct and Inverse Problems. With an Appendix by G. Besson. XIII, 272 pages. 1986.

Vol. 1208: S. Kaijser, J.W. Pelletier, Interpolation Functors and Duality. IV, 167 pages. 1986.

Vol. 1209: Differential Geometry, Peñíscola 1985. Proceedings. Edited by A.M. Naveira, A. Ferrández and F. Mascaró. VIII, 306 pages. 1986.

Vol. 1210: Probability Measures on Groups VIII. Proceedings, 1985. Edited by H. Heyer. X, 386 pages. 1986.

Vol. 1211: M.B. Sevryuk, Reversible Systems. V, 319 pages. 1986.

Vol. 1212: Stochastic Spatial Processes. Proceedings, 1984. Edited by P. Tautu. VIII, 311 pages. 1986.

Vol. 1213: L.G. Lewis, Jr., J.P. May, M. Steinberger, Equivariant Stable Homotopy Theory. IX, 538 pages. 1986.

Vol. 1214: Global Analysis – Studies and Applications II. Edited by Yu.G. Borisovich and Yu.E. Gliklikh. V, 275 pages. 1986.

Vol. 1215: Lectures in Probability and Statistics. Edited by G. del Pino and R. Rebolledo. V, 491 pages. 1986.

Vol. 1216: J. Kogan, Bifurcation of Extremals in Optimal Control. VIII, 106 pages. 1986.

Vol. 1217: Transformation Groups. Proceedings, 1985. Edited by S. Jackowski and K. Pawalowski. X, 396 pages. 1986.

Vol. 1218: Schrödinger Operators, Aarhus 1985. Seminar. Edited by E. Balslev. V, 222 pages. 1986.

Vol. 1219: R. Weissauer, Stabile Modulformen und Eisensteinreihen. III, 147 Seiten. 1986.

Vol. 1220: Séminaire d'Algèbre Paul Dubreil et Marie-Paule Malliavin. Proceedings, 1985. Edité par M.-P. Malliavin. IV, 200 pages. 1986.

Vol. 1221: Probability and Banach Spaces. Proceedings, 1985. Edited by J. Bastero and M. San Miguel. XI, 222 pages. 1986.

Vol. 1222: A. Katok, J.-M. Strelcyn, with the collaboration of F. Ledrappier and F. Przytycki, Invariant Manifolds, Entropy and Billiards; Smooth Maps with Singularities. VIII, 283 pages. 1986.

Vol. 1223: Differential Equations in Banach Spaces. Proceedings, 1985. Edited by A. Favini and E. Obrecht. VIII, 299 pages. 1986.

Vol. 1224: Nonlinear Diffusion Problems, Montecatini Terme 1985. Seminar. Edited by A. Fasano and M. Primicerio. VIII, 188 pages. 1986.

Vol. 1225: Inverse Problems, Montecatini Terme 1986. Seminar. Edited by G. Talenti. VIII, 204 pages. 1986.

Vol. 1226: A. Buium, Differential Function Fields and Moduli of Algebraic Varieties. IX, 146 pages. 1986.

Vol. 1227: H. Helson, The Spectral Theorem. VI, 104 pages. 1986.

Vol. 1228: Multigrid Methods II. Proceedings, 1985. Edited by W. Hackbusch and U. Trottenberg. VI, 336 pages. 1986.

Vol. 1229: O. Bratteli, Derivations, Dissipations and Group Actions on C*-algebras. IV, 277 pages. 1986.

Vol. 1230: Numerical Analysis. Proceedings, 1984. Edited by J.-P. Hennart. X, 234 pages. 1986.

Vol. 1231: E.-U. Gekeler, Drinfeld Modular Curves. XIV, 107 pages. 1986.

Vol. 1232: P.C. Schuur, Asymptotic Analysis of Soliton Problems. VIII, 180 pages. 1986.

Vol. 1233: Stability Problems for Stochastic Models. Proceedings, 1985. Edited by V.V. Kalashnikov, B. Penkov and V.M. Zolotarev. VI, 223 pages. 1986.

Vol. 1234: Combinatoire énumérative. Proceedings, 1985. Edité par G. Labelle et P. Leroux. XIV, 387 pages. 1986.

Vol. 1235: Séminaire de Théorie du Potentiel, Paris, No. 8. Directeurs: M. Brelot, G. Choquet et J. Deny. Rédacteurs: F. Hirsch et G. Mokobodzki. III, 209 pages. 1987.

Vol. 1236: Stochastic Partial Differential Equations and Applications. Proceedings, 1985. Edited by G. Da Prato and L. Tubaro. V, 257 pages. 1987.

Vol. 1237: Rational Approximation and its Applications in Mathematics and Physics. Proceedings, 1985. Edited by J. Gilewicz, M. Pindor and W. Siemaszko. XII, 350 pages. 1987.

Vol. 1238: M. Holz, K.-P. Podewski and K. Steffens, Injective Choice Functions. VI, 183 pages. 1987.

Vol. 1239: P. Vojta, Diophantine Approximations and Value Distribution Theory. X, 132 pages. 1987.

Vol. 1240: Number Theory, New York 1984–85. Seminar. Edited by D.V. Chudnovsky, G.V. Chudnovsky, H. Cohn and M.B. Nathanson. V, 324 pages. 1987.

Vol. 1241: L. Gårding, Singularities in Linear Wave Propagation. III, 125 pages. 1987.

Vol. 1242: Functional Analysis II, with Contributions by J. Hoffmann-Jørgensen et al. Edited by S. Kurepa, H. Kraljević and D. Butković. VII, 432 pages. 1987.

Vol. 1243: Non Commutative Harmonic Analysis and Lie Groups. Proceedings, 1985. Edited by J. Carmona, P. Delorme and M. Vergne. V, 309 pages. 1987.

Vol. 1244: W. Müller, Manifolds with Cusps of Rank One. XI, 158 pages. 1987.

Vol. 1245: S. Rallis, L-Functions and the Oscillator Representation. XVI, 239 pages. 1987.

Vol. 1246: Hodge Theory. Proceedings, 1985. Edited by E. Cattani, F. Guillén, A. Kaplan and F. Puerta. VII, 175 pages. 1987.

Vol. 1247: Séminaire de Probabilités XXI. Proceedings. Edité par J. Azéma, P.A. Meyer et M. Yor. IV, 579 pages. 1987.

Vol. 1248: Nonlinear Semigroups, Partial Differential Equations and Attractors. Proceedings, 1985. Edited by T.L. Gill and W.W. Zachary. IX, 185 pages. 1987.

Vol. 1249: I. van den Berg, Nonstandard Asymptotic Analysis. IX, 187 pages. 1987.

Vol. 1250: Stochastic Processes – Mathematics and Physics II. Proceedings 1985. Edited by S. Albeverio, Ph. Blanchard and L. Streit. VI, 359 pages. 1987.

Vol. 1251: Differential Geometric Methods in Mathematical Physics. Proceedings, 1985. Edited by P.L. García and A. Pérez-Rendón. VII, 300 pages. 1987.

Vol. 1252: T. Kaise, Représentations de Weil et GL_2 Algèbres de division et GL_n. VII, 203 pages. 1987.

Vol. 1253: J. Fischer, An Approach to the Selberg Trace Formula via the Selberg Zeta-Function. III, 184 pages. 1987.

Vol. 1254: S. Gelbart, I. Piatetski-Shapiro, S. Rallis. Explicit Constructions of Automorphic L-Functions. VI, 152 pages. 1987.

Vol. 1255: Differential Geometry and Differential Equations. Proceedings, 1985. Edited by C. Gu, M. Berger and R.L. Bryant. XII, 243 pages. 1987.

Vol. 1256: Pseudo-Differential Operators. Proceedings, 1986. Edited by H.O. Cordes, B. Gramsch and H. Widom. X, 479 pages. 1987.

Vol. 1257: X. Wang, On the C*-Algebras of Foliations in the Plane. V, 165 pages. 1987.

Vol. 1258: J. Weidmann, Spectral Theory of Ordinary Differential Operators. VI, 303 pages. 1987.

Vol. 1259: F. Cano Torres, Desingularization Strategies for Three-Dimensional Vector Fields. IX, 189 pages. 1987.

Vol. 1260: N.H. Pavel, Nonlinear Evolution Operators and Semigroups. VI, 285 pages. 1987.

Vol. 1261: H. Abels, Finite Presentability of S-Arithmetic Groups. Compact Presentability of Solvable Groups. VI, 178 pages. 1987.

Vol. 1262: E. Hlawka (Hrsg.), Zahlentheoretische Analysis II. Seminar, 1984–86. V, 158 Seiten. 1987.

Vol. 1263: V.L. Hansen (Ed.), Differential Geometry. Proceedings, 1985. XI, 288 pages. 1987.

Vol. 1264: Wu Wen-tsün, Rational Homotopy Type. VIII, 219 pages. 1987.

Vol. 1265: W. Van Assche, Asymptotics for Orthogonal Polynomials. VI, 201 pages. 1987.

Vol. 1266: F. Ghione, C. Peskine, E. Sernesi (Eds.), Space Curves. Proceedings, 1985. VI, 272 pages. 1987.

Vol. 1267: J. Lindenstrauss, V.D. Milman (Eds.), Geometrical Aspects of Functional Analysis. Seminar. VII, 212 pages. 1987.

Vol. 1268: S.G. Krantz (Ed.), Complex Analysis. Seminar, 1986. VII, 195 pages. 1987.

Vol. 1269: M. Shiota, Nash Manifolds. VI, 223 pages. 1987.

Vol. 1270: C. Carasso, P.-A. Raviart, D. Serre (Eds.), Nonlinear Hyperbolic Problems. Proceedings, 1986. XV, 341 pages. 1987.

Vol. 1271: A.M. Cohen, W.H. Hesselink, W.L.J. van der Kallen, J.R. Strooker (Eds.), Algebraic Groups Utrecht 1986. Proceedings. XII, 284 pages. 1987.

Vol. 1272: M.S. Livšic, L.L. Waksman, Commuting Nonselfadjoint Operators in Hilbert Space. III, 115 pages. 1987.

Vol. 1273: G.-M. Greuel, G. Trautmann (Eds.), Singularities, Representation of Algebras, and Vector Bundles. Proceedings, 1985. XIV, 383 pages. 1987.

Vol. 1274: N. C. Phillips, Equivariant K-Theory and Freeness of Group Actions on C*-Algebras. VIII, 371 pages. 1987.

Vol. 1275: C.A. Berenstein (Ed.), Complex Analysis I. Proceedings, 1985–86. XV, 331 pages. 1987.

Vol. 1276: C.A. Berenstein (Ed.), Complex Analysis II. Proceedings, 1985–86. IX, 320 pages. 1987.

Vol. 1277: C.A. Berenstein (Ed.), Complex Analysis III. Proceedings, 1985–86. X, 350 pages. 1987.

Vol. 1278: S.S. Koh (Ed.), Invariant Theory. Proceedings, 1985. V, 102 pages. 1987.

Vol. 1279: D. Ieşan, Saint-Venant's Problem. VIII, 162 Seiten. 1987.

Vol. 1280: E. Neher, Jordan Triple Systems by the Grid Approach. XII, 193 pages. 1987.

Vol. 1281: O.H. Kegel, F. Menegazzo, G. Zacher (Eds.), Group Theory. Proceedings, 1986. VII, 179 pages. 1987.

Vol. 1282: D.E. Handelman, Positive Polynomials, Convex Integral Polytopes, and a Random Walk Problem. XI, 136 pages. 1987.

Vol. 1283: S. Mardešić, J. Segal (Eds.), Geometric Topology and Shape Theory. Proceedings, 1986. V, 261 pages. 1987.

Vol. 1284: B.H. Matzat, Konstruktive Galoistheorie. X, 286 pages. 1987.

Vol. 1285: I.W. Knowles, Y. Saitō (Eds.), Differential Equations and Mathematical Physics. Proceedings, 1986. XVI, 499 pages. 1987.

Vol. 1286: H.R. Miller, D.C. Ravenel (Eds.), Algebraic Topology. Proceedings, 1986. VII, 341 pages. 1987.

Vol. 1287: E.B. Saff (Ed.), Approximation Theory, Tampa. Proceedings, 1985–1986. V, 228 pages. 1987.

Vol. 1288: Yu. L. Rodin, Generalized Analytic Functions on Riemann Surfaces. V, 128 pages, 1987.

Vol. 1289: Yu. I. Manin (Ed.), K-Theory, Arithmetic and Geometry. Seminar, 1984–1986. V, 399 pages. 1987.

Vol. 1290: G. Wüstholz (Ed.), Diophantine Approximation and Transcendence Theory. Seminar, 1985. V, 243 pages. 1987.

Vol. 1291: C. Mœglin, M.-F. Vignéras, J.-L. Waldspurger, Correspondances de Howe sur un Corps p-adique. VII, 163 pages. 1987

Vol. 1292: J.T. Baldwin (Ed.), Classification Theory. Proceedings, 1985. VI, 500 pages. 1987.

Vol. 1293: W. Ebeling, The Monodromy Groups of Isolated Singularities of Complete Intersections. XIV, 153 pages. 1987.

Vol. 1294: M. Queffélec, Substitution Dynamical Systems – Spectral Analysis. XIII, 240 pages. 1987.

Vol. 1295: P. Lelong, P. Dolbeault, H. Skoda (Réd.), Séminaire d'Analyse P. Lelong – P. Dolbeault – H. Skoda. Seminar, 1985/1986. VII, 283 pages. 1987.

Vol. 1296: M.-P. Malliavin (Ed.), Séminaire d'Algèbre Paul Dubreil et Marie-Paule Malliavin. Proceedings, 1986. IV, 324 pages. 1987.

Vol. 1297: Zhu Y.-l., Guo B.-y. (Eds.), Numerical Methods for Partial Differential Equations. Proceedings. XI, 244 pages. 1987.

Vol. 1298: J. Aguadé, R. Kane (Eds.), Algebraic Topology, Barcelona 1986. Proceedings. X, 255 pages. 1987.

Vol. 1299: S. Watanabe, Yu.V. Prokhorov (Eds.), Probability Theory and Mathematical Statistics. Proceedings, 1986. VIII, 589 pages. 1988.

Vol. 1300: G.B. Seligman, Constructions of Lie Algebras and their Modules. VI, 190 pages. 1988.

Vol. 1301: N. Schappacher, Periods of Hecke Characters. XV, 160 pages. 1988.

Vol. 1302: M. Cwikel, J. Peetre, Y. Sagher, H. Wallin (Eds.), Function Spaces and Applications. Proceedings, 1986. VI, 445 pages. 1988.

Vol. 1303: L. Accardi, W. von Waldenfels (Eds.), Quantum Probability and Applications III. Proceedings, 1987. VI, 373 pages. 1988.

Vol. 1304: F.Q. Gouvêa, Arithmetic of p-adic Modular Forms. VIII, 121 pages. 1988.

Vol. 1305: D.S. Lubinsky, E.B. Saff, Strong Asymptotics for Extremal Polynomials Associated with Weights on $\mathbb{R}$. VII, 153 pages. 1988.

Vol. 1306: S.S. Chern (Ed.), Partial Differential Equations. Proceedings, 1986. VI, 294 pages. 1988.

Vol. 1307: T. Murai, A Real Variable Method for the Cauchy Transform, and Analytic Capacity. VIII, 133 pages. 1988.

Vol. 1308: P. Imkeller, Two-Parameter Martingales and Their Quadratic Variation. IV, 177 pages. 1988.

Vol. 1309: B. Fiedler, Global Bifurcation of Periodic Solutions with Symmetry. VIII, 144 pages. 1988.

Vol. 1310: O.A. Laudal, G. Pfister, Local Moduli and Singularities. V, 117 pages. 1988.

Vol. 1311: A. Holme, R. Speiser (Eds.), Algebraic Geometry, Sundance 1986. Proceedings. VI, 320 pages. 1988.

Vol. 1312: N.A. Shirokov, Analytic Functions Smooth up to the Boundary. III, 213 pages. 1988.

Vol. 1313: F. Colonius, Optimal Periodic Control. VI, 177 pages. 1988.

Vol. 1314: A. Futaki, Kähler-Einstein Metrics and Integral Invariants. IV, 140 pages. 1988.

Vol. 1315: R.A. McCoy, I. Ntantu, Topological Properties of Spaces of Continuous Functions. IV, 124 pages. 1988.

Vol. 1316: H. Korezlioglu, A.S. Ustunel (Eds.), Stochastic Analysis and Related Topics. Proceedings, 1986. V, 371 pages. 1988.

Vol. 1317: J. Lindenstrauss, V.D. Milman (Eds.), Geometric Aspects of Functional Analysis. Seminar, 1986–87. VII, 289 pages. 1988.

Vol. 1318: Y. Felix (Ed.), Algebraic Topology – Rational Homotopy. Proceedings, 1986. VIII, 245 pages. 1988

Vol. 1319: M. Vuorinen, Conformal Geometry and Quasiregular Mappings. XIX, 209 pages. 1988.